Cell Engineering

Volume 11

Series Editor
Mohamed Al-Rubeai, University College Dublin, Belfield, Dublin 4, Ireland

Integrating advances in molecular biology into bioprocesses presents a continuous challenge to scientists and bioengineers. This series is conceived to help meet this challenge. It examines and assesses the feasibility of new approaches for the modification of cellular function such as gene expression, protein processing, secretion, glycosylation, immortalisation, proliferation, and apoptosis as well as the systematic study of the metabolic genotype-phenotype relationship. The series provides detailed coverage of the methodology for improving cellular properties of cells used in the production of biopharmaceuticals, gene and cell therapies and tissue engineering. It also seeks to explain the cellular mechanisms underlying *in vitro* physiological activity and productivity.

Ralf Pörtner
Editor

Biopharmaceutical Manufacturing

Progress, Trends and Challenges

Springer

Editor
Ralf Pörtner
Institute of Bioprocess and Biosystems Engineering
Hamburg University of Technology
Hamburg, Germany

ISSN 1389-6946 ISSN 2542-9515 (electronic)
Cell Engineering
ISBN 978-3-031-45668-8 ISBN 978-3-031-45669-5 (eBook)
https://doi.org/10.1007/978-3-031-45669-5

This Springer imprint is published by the registered company Springer Nature Switzerland AG
The registered company address is: Gewerbestrasse 11, 6330 Cham, Switzerland

Paper in this product is recyclable.

Preface

Biopharmaceutical manufacturing, in this volume specifically the production of recombinant therapeutic proteins by mammalian cells and to some extend by plant cell culture as well as applications for cell and gene therapy, is continuously demanding new and improved bioprocessing technologies to deliver a desired product quality profile at reduced costs. This volume is aimed to provide high-quality contributions reviewing cutting-edge progress in this area. It is the third volume in the book series Cell Engineering dedicated to the application of cell culture techniques.

Cell Engineering 9 – Animal Cell Culture released in 2015 reviewed the state of the art with in-depth assessments that emphasize the practical aspects of efficient operation of cell culture techniques, including selection and characterisation of cell lines, suitable types of bioreactor system, optimal physical and chemical environment and appropriate production mode, as well as the relevance of process analytical technology (PAT) in the biopharmaceutical industry.

Cell Engineering 10 – Cell Culture Engineering and Technology released in 2021 was dedicated to cell engineering, novel technologies and the use of omics, including products from animal cells, use of remote control for therapeutic designer cells, mRNA-techniques, omics' data-driven design of biopharmaceutical production, media optimisation, glycoengineering and cell engineering platforms.

This volume *Cell Engineering 11 – Biopharmaceutical Manufacturing: Progress, Trends and Challenges* is a source of the latest innovative research and technical development in biomanufacturing systems. It is organised into two parts:

- Part I: Manufacturing of recombinant therapeutic proteins (e.g. therapeutic antibodies, biosimilars/biogenerics)
- Part II: Manufacturing aspects of cell and gene therapy
- Each with selected chapters on the following topics for both up- and downstream
- Advanced process strategies, especially continuous manufacturing
- Advanced culture techniques, especially single-use systems
- Process transfer, scale-up/scale-down models

- Processing advances/manufacturing productivity/efficiency
- Model-assisted process understanding and development/Digital Twins
- Process controls and analytics
- Quality control, Quality by Design
- Facility design, full-scale commercial systems, manufacturing technology innovation

The volume comprises contributions of experts from academia and industry active in the field of cell culture development for the production of recombinant proteins, cell therapy and gene therapy, with consideration of Digital Twins and facility design. The knowledge and expertise of the authors cover disciplines like cell biology, engineering, biotechnology and biomedical sciences. Inevitably, some omissions will occur in the test, but the authors have sought to avoid duplications by extensive cross-referencing to chapters in other volumes of this series and elsewhere. We hope the volume provides a useful compendium of techniques for scientists in industrial and research laboratories active in this field.

I am grateful for the support of all the excellent contributors, the series editor Prof. Dr. Mohamed Al-Rubeai, University College Dublin (Ireland), and to the staff at Springer for their efficiency in preparing this volume, and to Dr. Gonzalo Cordova for his help and patience.

Hamburg, Germany
May 2023

Ralf Pörtner

Contents

About the Editor

Ralf Pörtner is a senior research associate and group leader at Technical University Hamburg, Institute of Bioprocess and Biosystems Engineering and Honorary Professor at TH Mittelhessen University of Applied Sciences, Giessen. His main research interests are development and optimization of bioreactor techniques and process strategies, in particular for cell cultures and microbial reactions, including model-based control concepts, and tissue engineering. He has in total published about 150 peer-reviewed papers.

Part I
Manufacturing of Recombinant Therapeutic Proteins

Chapter 1
Single-Use Systems in Biopharmaceutical Manufacture: State of the Art and Recent Trends

Valentin Jossen, Regine Eibl, Gilles Broccard, and Dieter Eibl

Abstract Single-use systems have become increasingly important in recent years. This applies in particular to their use in the manufacture of biopharmaceutical products based on mammalian cell cultures, such as antibodies for the therapy of cancer and autoimmune diseases. Today, developers and manufacturers of such biotherapeutics can choose from a large product portfolio of single-use systems from different suppliers. For example, there are single-use versions of storage bags, filters, mixers, bioreactors, connectors, harvesting and transfer systems, membrane adsorbers, freeze- and -thaw systems, to name just a few examples. Thus, the production of clinical samples in complete single-use production facilities (upstream processing to filling) is already a reality. Based on the definition of the term and the presentation of milestones in the development of single-use technology, the book chapter discusses the most important advantages and existing limitations of the single-use systems currently available on the market. Subsequently, the current state of development of single-use technology is presented on the basis of selected product examples. The focus is on the upstream area and single-use bioreactors. Finally, approaches for process intensification up to complete continuous production resulting from the increasing implementation of single-use systems are deduced.

Keywords Biopharmaceuticals · Downstreaming · Fill & finish · Process intensification · Single-use · Upstreaming

During the writing of the book chapter, Pall and Cytiva were two separate companies.

V. Jossen (✉) · G. Broccard
Ibex Mammalian Development Services, Lonza, Visp, Switzerland
e-mail: valentin.jossen1@lonza.com

R. Eibl · D. Eibl
Institute of Chemistry and Biotechnology, School of Life Sciences and Facility Management, Zurich University of Applied Sciences, Wädenswil, Switzerland

R. Pörtner (ed.), *Biopharmaceutical Manufacturing*, Cell Engineering 11,
https://doi.org/10.1007/978-3-031-45669-5_1

Abbreviations

1D	One-dimensional
2D	Two-dimensional
3D	Three-dimensional
ADC	Antibody Drug Conjugate
ATF	Alternating Tangential Flow Filtration
BHK	Baby Hamster Kidney cells
BPOG	BioPhorum Organisations Group
CAR	Chimeric Antigen Receptor
CFD	Computational Fluid Dynamics
CHO	Chinese Hamster Ovary
CR	Cell Retention
CSPR	Cell Specific Perfusion Rate
DECHEMA	Society for Chemical Engineering and Biotechnology
DNA	Desoxyribonucleic Acid
DO	Dissolved Oxygen
DP	Drug Product
DS	Drug Substance
DSP	Downstream Processing
E&L	Extractables and Leachables
E. coli	*Escherichia coli*
E-FL	Embryonic Feline Lung Fibroblasts
GMP	Good Manufacturing Practice
HCCF	Harvest Cell Culture Fluid
HCD	High Cell Density
HCP	Host Cell Protein
HEK	Human Embryonic Kidney
Hi-5	Trichoplusia ni
iPSC	Induced Pluripotent Stem Cells
k_La	Volumetric Oxygen Mass Transfer Coefficient
mAbs	Monoclonal Antibodies
MSC	Mesenchymal Stem Cells
NK	Natural Killer
NTU	Nephelometric Turbidity Units
P. pastoris	*Pichia pastoris*
QC	Quality Control
Re_{mod}	Modified Reynolds Number
rSEAP	Recombinant Secreted Alkaline Phosphatase
SBI	Segment Blade Impeller
Sf-9	*Spodoptera frugiperda*
SPTFF	Single-Pass Tangential Flow Filtration
SUBs	Single-Use Bioreactor
TFF	Tangential Flow Filtration

TIPS	Titerless Infected-Cell Preservation and Scale-Up
USP	Upstream processing
VCD	Viable Cell Density
VCV	Viable Cell Volume
V_w	Filling or Working Volume
WCB	Working Cell Bank

Introduction

Single-use (SU) systems and their constituent technologies are now integral parts of many biopharmaceutical production processes. They are used in every different stage of research and development, preclinical and clinical production, as well as commercial processes in which therapeutic proteins such as monoclonal antibodies (mAbs), hormones, vaccines, cell and gene therapeutics or antibody drug conjugates (ADC) are produced in small and medium volumes [1–3]. The host cells used in these systems can be insect cells or microorganisms, but the dominant ones are mammalian cells (e.g., Chinese Hamster Ovary {CHO} cells).

The term "single-use system" describes a utensil or device with a cell, medium or product-contact surface that is made of plastic and is disposed of (e.g., incinerated) after a single use. Consequently, the term "disposables" is also often used for the consumable components of SU systems, which are made of materials such as polyamide, polycarbonate, polyethylene, polystyrene, polypropylene, ethylene vinyl acetate, polyethersulfone, polysiloxane, and fluoropolymers such as polytetrafluoroethylene or polyvinylidene fluoride [4, 5]. The consumable components in SU systems are usually rigid shapes or flexible bags consisting of multilayer films and are often delivered pre-assembled and sterilized. Gamma radiation (25–50 kGy) is preferred for sterilization, but X-ray treatment is currently on the rise [6–9].

SU systems are not a new phenomenon; strictly speaking, they have been around for 50 years. The first SU cultivation systems (hollow fiber bioreactors and CellFactories) were developed in the 1970s [10–12] and were used in early mAb and vaccine production processes using adherently grown mammalian cells. Plastic filter capsules, sterile syringe filters and SU storage bags (including transport systems) have been commercially available since the 1990s [13]. An important milestone in the development of SU technology was the launch of the first wave-mixed SU bioreactor, the Wave™ Reactor (see section "Single-use bioreactors"), in 1998 [14]. The Wave™ Reactor not only paved the way for other scalable SU bioreactor types, but also led to the development of additional SU process equipment for biopharmaceutical production processes.

In early 2000, it became possible for the first time to couple sterile fluid pathways using aseptic connectors in a non-sterile environment [13], and in 2006 the first stirred SU bioreactors (HyClone™ S.U.B. and Xcellerex™ XDR-Disposable Stirred Tank Bioreactor) were introduced to the market [15]. This triggered a new wave of interest in the application and development of SU technology that was supported by various organizations (e.g., BioProcess Systems Alliance, BioPhorum Operations Group {BPOG}, International Society for Pharmaceutical Engineering, Society for Chemical Engineering and Biotechnology {DECHEMA} in the American and European regions). Developers and manufacturers of SU systems, together with users from industry and academia, began to address the weaknesses (see section "Advantages and disadvantages of single-use systems") of existing SU systems and further the development and design of facilities with SU systems.

Driven by the pressure to develop biopharmaceuticals and biosimilars ever faster and more cost-effectively, there has, since 2015, been an increasing focus on process intensification and continuous processes using SU systems, resulting in a revival in perfusion technology (see section "Trend intensified and continuous facilities"). In fact, the number of SU facilities for biopharmaceutical processes is increasing worldwide. There has been particular interest in SU technology in the Asian region over the last 5 years; developers and manufacturers of cell and gene therapeutics are keen to exploit the potential of SU technologies (see section "Advantages and disadvantages of single-use systems"); and the Covid 19 pandemic has driven a boom in SU technologies that continues to today [3].

Advantages and Disadvantages of Single-Use Systems

SU systems can be installed and put into operation quickly, are cheaper than their reusable counterparts, and are characterized by a high degree of modularity and flexibility [16]. As recent research has shown, they also support the implementation of intensified and continuous biopharmaceutical production processes (see section "Trend intensified and continuous facilities") [17–20]. Since there is no need to sterilize or clean SU systems on-site (no facilities for sterilization, no cleaning in place and no water for injection), the footprint of the facility can be reduced and the risk of microbial contamination, and in particular cross-contamination, is lower. The development time for new products can also be shortened, and the number of product batches per year can be increased by implementing SU systems. Various studies have shown that the use of SU systems also has a lower environmental impact than reusable systems [21–24]. However, the permanent renewal of the consumables in SU systems does generate a lot of waste. Therefore, in order to further improve the sustainability of SU systems, numerous manufacturers are currently looking into the feasibility of mechanical recycling of the SU system consumables that are used in the largest quantities, such as bioprocessing bags [9, 25].

Disadvantages of SU systems include the lack of standardization, which results in dependence on certain suppliers, limited scalability, possible leakages, the occurrence of extractables and leachables (E&L) and the high costs of consumables. Most of the disadvantages listed may be ascribed to material properties of the plastics used and their further processing/treatment. Interestingly though, the main reasons cited for rejecting SU systems in biopharmaceutical processes are E&L [26], which can occur at unacceptable concentrations.

While extractables are chemical substances formed under aggressive laboratory process conditions, leachables are formed under real process conditions. Nevertheless, leachables such as Bis(2,4-di-tert-butylphenyl)-phosphate [27, 28] or 3,5-dinitro-bisphenole A [29] are a subset of extractables. Safety information on known E&L is constantly updated by the Extractables and Leachables Safety Information Exchange Consortium and is available on the internet (www.elsiedata.org). Sources of E&L include plastic processing (the addition of lubricants, plasticizers, antioxidants, heat and UV stabilizers), irradiation, storage and transportation of consumables from SU systems. In fact, E&L can originate from in-process systems (e.g., bioprocess bags, filters), as well as primary (e.g., syringes, ampoules, vials) and secondary (e.g., wrappers, container labels) packaging components, potentially causing a negative impact on product quality, quantity, and stability [30]. In the worst case, E&L could be toxic in the concentrations present and endanger patient safety. Since patient safety is the top priority in the manufacturing of biopharmaceutical products, the closest process steps to the final drug product (DP) (see Fig. 1.1) that have the longest contact time are considered to be those with the greatest risk in terms of their E&L profile. This means that upstream processing (USP) poses the lowest risk in terms of critical E&L concentrations, while SU systems used in fill & finish processes pose the greatest risk, and downstream processing (DSP) steps fall in between.

USP includes cultivation of cells from the working cell bank (WCB), inoculum production, cell expansion, mAb expression in bioreactors, and harvesting (cell separation and clarification). In the majority of cases, DSP includes capture and polishing steps that deliver the final drug substance (DS). This is followed by formulation and, finally, fill & finish. The resulting DP is labelled, packaged and shipped after quality control (QC). The culture media preparation and production process steps in USP and the buffer production steps in DSP are not shown, but are discussed further in this chapter.

Interestingly, studies have shown that the concentration of leachables in USP and DSP can be drastically reduced by filtration steps or pre-flushing of the filtration system [31–33]. Therefore, it is advisable to determine the E&L risk profile for each step of the production process at an early stage and in collaboration with specialized toxicologists. Recommendations and standards published in the last 9 years, such as those from the BPOG [34–36], the American Society for Testing and Materials [37], and the U.S. Pharmacopeia [38] can be used for this purpose. It is worth mentioning that E&L data generated by the manufacturers of the SU equipment do not release the manufacturers of the DP from conducting their own studies relevant to their processes.

Nevertheless, the advantages of the current generation of SU systems available on the market far outweigh their weaknesses. An important prerequisite, however, is that the SU systems are selected and used correctly. Users can choose from a large portfolio of different products and even design platform processes with SU systems. Renowned manufacturers of SU systems such as Cytiva (ReadyToProcess series,

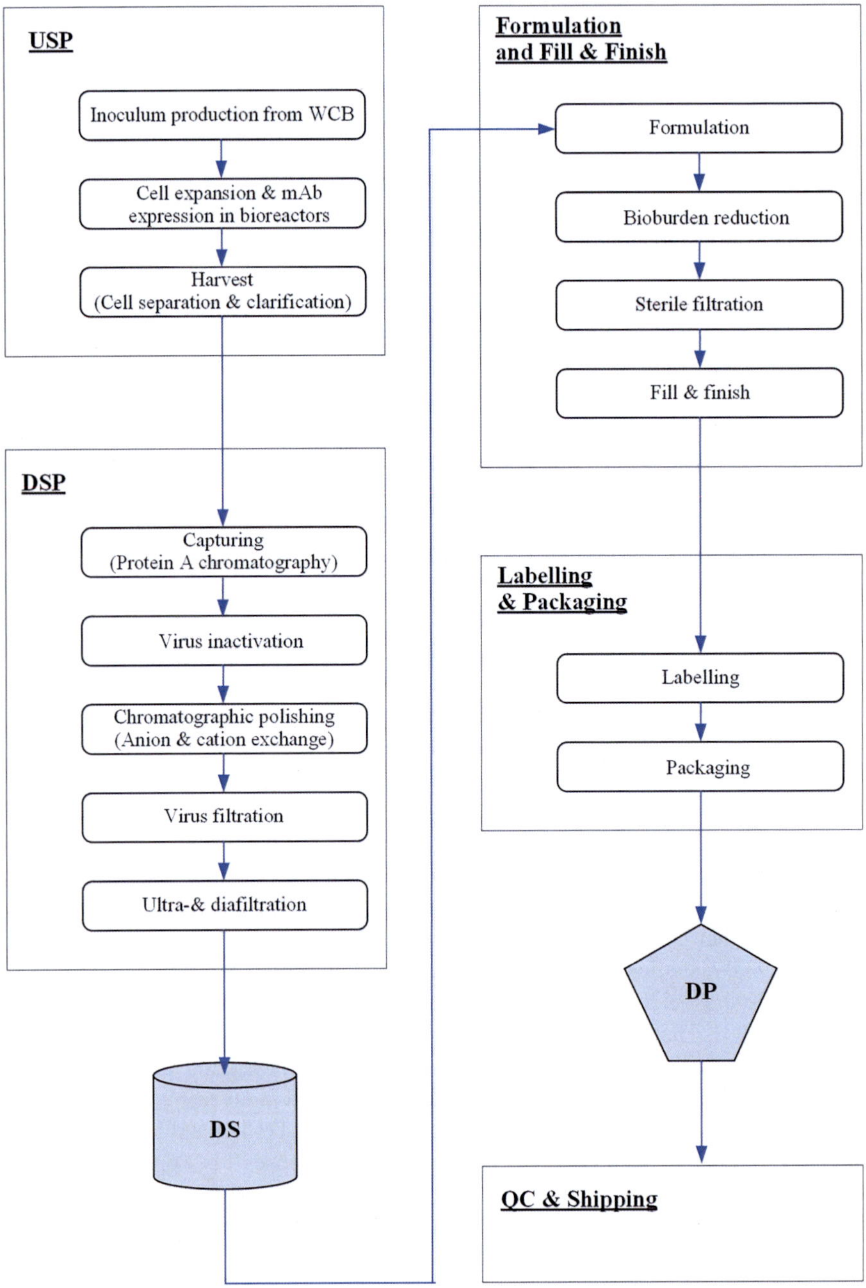

Fig. 1.1 Process steps and stages of a typical biopharmaceutical production process (example of mAb production)

FlexFactory™ series), Merck (Mobius® series), Pall (Allegro™ series), Sartorius (Flexact® series) and Thermo Fisher Scientific (HyPerforma™ series) offer platforms for USP, DSP, formulation and filling. State of the art processes based on these platforms have well-defined process sequences, which enable the development of efficient engineering workflows, improve supply security and offer potential for process optimization [39].

Single-Use Systems: Where We Are Today

Single-Use Upstream Processing

SU systems are omnipresent in USP of biopharmaceutical production processes. Complete upstream lines based on SU systems are already a reality in mAb production processes, where the following SU devices are used: tubing, aseptic connectors, sampling and transfer systems, bioprocess containers with storage bags, mixers, single-use bioreactors (SUBs), filtration systems, such as depth filters, micro-, ultra- and diafiltration systems, centrifuges and separators. In addition, typical laboratory utensils, such as vials, shake flasks and pipettes are also used. However, these types of utensils, as well as tubing, sampling systems, transfer systems and bioprocess containers with storage bags are not the subject of this chapter. Detailed information on these topics can be found in [40] (Chaps. 2 and 5).

Over the past 16 years mixers and bioreactors have been the most frequently implemented SU systems according to analysis by Langer [3]. For this reason, they and the most often used SU systems for primary recovery (cell separation / clarification) will be described in more detail. Because SUBs are at the heart of USP, they will be the focus of this section.

Single-Use Mixers

SU mixers are primarily used in USP for media production. Mechanically driven systems are used for this task and are commercially available up to the single-digit cubic meter range. They have either a cylindrical or cube-shaped plastic bag and are mostly stirred but can occasionally be oscillating systems (e.g., Cytiva ReadyToProcess™ Mixers, max. 50 L working volume) or disc-mixed systems (Thermo Fisher Scientific ImPULSE™ Mixers, max. 5000 L working volume). In stirred SU mixers, the rotating stirring element is often a magnetic stirrer. However, stirred SU mixers with a seal (top or bottom mounted) are also available, such as Pall's Allegro™ SU mixer (max. 1000 L working volume, see Fig. 1.2) or Thermo Fisher Scientific's HyPerforma™ SU mixer.

The HyPerforma™ SU mixer is available in 6 sizes (50–2000 L working volume) and has a top mounted stirrer. The system offers a 5:1 mixing volume turndown ratio and is available in heated and non-heated versions. Pall's Allegro™ SU mixer,

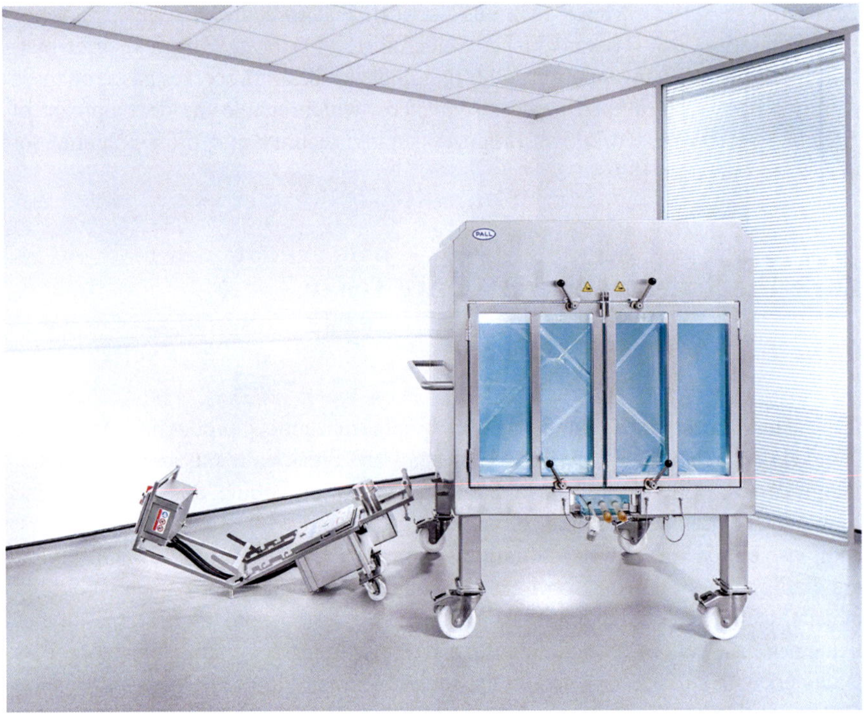

Fig. 1.2 Magnetic mixer system. (Image provided courtesy of Pall Corporation)

however, is available up to 1000 L working volume and is equipped with a bottom mounted stirrer.

Stirred SU mixers exist up to 7000 L working volume (ABEC's CSR® General Mixer). Currently, these are the largest SU mixers on the market. A more comprehensive overview of current SU mixing technology is provided by Werner and his co-authors [41] and by de Boulard and Kienle [42]. Both publications discuss other mechanically driven SU mixers (e.g., stirred systems based on levitation technology or tumbling stirrers) and hydraulically driven versions that are particularly suitable for DSP (see section "Single-use downstream processing") and product formulation (see section "Single-use formulation and fill and finish"). Unsurprisingly, most of the operating principles described for SU mixers are also found in SU bioreactors, and manufacturers of these devices also offer SU mixing technology.

Single-Use Bioreactors

SUBs, which are available from the mL scale to the cubic meter scale, are used for cell production (e.g., inoculum productions) and product expression (e.g., mAb) purposes. Based on their power input (see Fig. 1.3), they can be divided into four

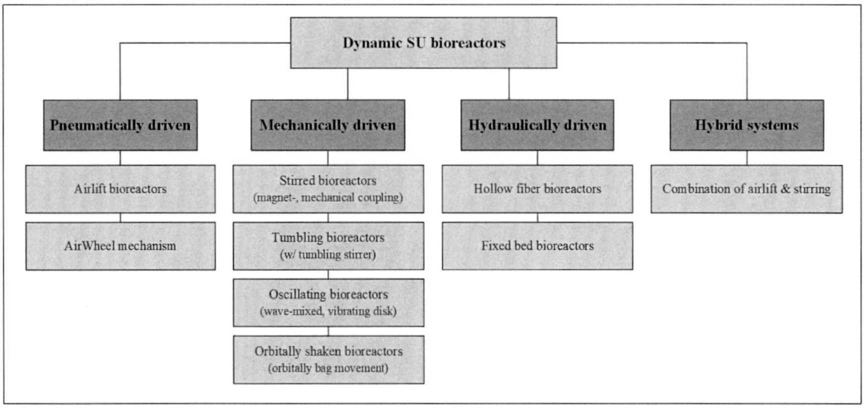

Fig. 1.3 Categorization of common SU bioreactor types depending on power input

categories: mechanically, pneumatically and hydraulically driven, and hybrid versions. Mechanically driven SUBs are the largest group and include stirred, wave-mixed and orbitally shaken SU versions.

The mechanical power input is achieved through stirring, an oscillating motion (wave-mixed motion) or orbital shaking. Direct aeration produces heat and mass transfer in pneumatic systems and a pump is used in hydraulic systems.

Stirred Single-Use Bioreactors

Decades of experience in the development of reusable stirred bioreactors for microbial and mammalian cells are reflected in the design of stirred SUBs. SUBs up to 6000 L working volume are now available from different suppliers (see Table 1.2). The systems consist of either rigid vessels (10 mL to 40 L working volume) or flexible bags (4.5–6000 L working volume) and have height to diameter ratios (H/D) of 1.5:1–2.2:1. The largest 8000 L SUBs (6000 L working volume) were launched in 2019 by ABEC. The rigid stirred SUB systems are typically free-standing, plastic vessels made of polycarbonate, which do not require a support container for shaping. Unlike bag-based systems, folding stresses, which can lead to layer breakup and therefore leakages, are normally not a problem when using rigid systems. When using culture bags, it is important that the bags are properly installed and fit perfectly into their holding device to achieve optimum performance, especially in terms of heat transfer. Cavities, pockets, and folds, which can be caused when unfolding the bag and during installation, should be prevented. Depending on the bioreactor scale, heating and cooling of the vessel is performed by heating blocks, block coolers, or double jackets. Gas vent filter heaters are normally used for stirred SUBs instead of exhaust gas condensers, even though their performance is lower. To reduce the risk

Table 1.1 Comparison of product titers for different mAbs at small (Ambr® 15), pilot (SUB 200 L) and production (SUB 1000 L) scale. Data from Lonza IBEX with GS Xceed® CHO platform

Cell line I Product (1–4 I A-D)	Ambr® 15 (Sartorius)	SUB 200 L (Sartorius)	SUB 1000 L (Sartorius)
	Normalized product titer ($c_{titer}/c_{titer,Ambr}$)		
1 I A	1.00	1.08	1.20
2 I B	1.00	1.23	1.19
3 I C	1.00	0.90	0.85
4 I D	1.00	1.01	1.03

of exhaust filter clogging due to filter fouling or foaming, several exhaust filters are connected in parallel in larger stirred SUBs.

A different principle is used for the BioBLU® 0.3c bioreactor (Eppendorf). A liquid-free Peltier element controls the temperature and condensation in the exhaust air. A similar approach is also used for the Ambr® 250 system (Sartorius) in order to reduce the loss of liquid dues to evaporation. The BioBLU® 0.3c and the Ambr® 250 are both stackable modular systems with up to 32 or more parallel cultivation vessels. The bioreactor vessels have working volumes of 100–250 mL and are suitable for screening experiments and process development. Both systems offer versions for mammalian or microbial cells, with different types of impellers (pitch blade impellers vs. Rushton impellers) used to meet the different requirements of the production organism. In addition to applications using mammalian (mainly CHO [58, 59]) and microbial (mainly *Escherichia coli* {*E. coli*} and *Pichia pastoris* {*P. pastoris}* [60–62]) cells, the systems are also used for the processing of human primary cells (i.e., mesenchymal stem cells {MSCs}, induced pluripotent stem cells {iPSCs} and CAR-T cells [43–45, 63–65]) due to their high degree of automation, their ability to run cultivations in parallel, and their low initial working volumes. Another popular stirred SUB system for screening studies, media optimization and clone selection is the Ambr® 15, which provides working volumes of between 10 and 15 mL, and automated liquid handling. Unlike other rigid stirred SUB systems, the Ambr® 15 has a cubical shape and no internals, apart from the optical sensor spots for pH and DO measurement, the impeller, and the impeller shaft. Nienow et al. [66] have shown that the Ambr® 15 is superior to conventional shake flasks, which are frequently used in early process development or for CHO clone selection. They demonstrated that the performance of an IgG4-expressing CHO cell line in the Ambr® 15 is more comparable to large-scale bioreactors than shake flasks are. Lonza data also show the comparability of product titers for different mAbs in the Ambr® 15 and larger SU bioreactors (see Table 1.1). Recently, the Ambr® 15 was also proofed for the process development of primary cells [67]. However, due to the higher technical complexity of the Ambr® 15 system, more effort is required to keep it running compared to classical shake flask or TubeSpin® bioreactors.

Benchtop-scale stirred SUBs, including the AppiFlex ST (Applikon), the BioBLU® 1c, 3c, 5c, and 10c (Eppendorf), the Mobius® CellReady 3 L (Merck Millipore) and the UniVessel® SU (Sartorius), are close to the classical design of reusable, stirred cell culture bioreactors. All mammalian versions are equipped with one or two top-mounted marine or segment blade impellers and can achieve specific power inputs in the range of 50–200 W/m^3, which are sufficiently high for common mammalian-based processes. The UniVessel® SU was the first commercially available rigid stirred SUB agitated by two-stage impellers. The three elements of the segment blade impeller (SBI) are similar in shape to the "Elephant ear" impeller of the BioBLU®, but they have a lower blade angle of 30° (instead of 45°), which results in lower power inputs. The UniVessel® SU has been extensively studied in the past in order to predict fluid flow, specific power inputs (0.4–435 W/m^3), mixing times (3–100 s), oxygen mass transfer (10–50 h^{-1}), and hydrodynamic stresses [46, 68, 69]. There are countless applications with mammalian cells [70, 71], insect cells [54, 72] and human primary cells [46, 47] reported in the scientific literature. Another well characterized stirred SU bioreactor is the Mobius® CellReady 3 L, which was introduced in summer 2009. The system is equipped with a marine impeller (top driven), a microsparger or open-pipe sparger, and standard sensors. Detailed information about the biochemical engineering parameters of the Mobius® CellReady 3 L, including fluid flow and hydrodynamic stresses, can be found in several research articles and text books [73–75]. Different authors have shown that similar cell densities and antibody titers can be achieved in the Mobius® CellReady 3 L and in stirred 3 L glass bioreactors [73, 76]. In addition, Cierpka et al. [77], Grein et al. [78] and Lawson et al. [79] have shown that the Mobius® CellReady 3 L and 50 L can be successfully used for processing stem cells. Due to the advances in 3D printing technology and the availability of biocompatible printing materials (USP class VI materials), Applikon launched the AppliFlex ST bioreactor, whose top plate and internal parts (including the impeller) are 3D printed. The customization of the bioreactor offers the user a new degree of flexibility allowing the size and position of the tip tubes/feed lines, as well as the type of impeller and its geometry to be adjusted. Thus, the bioreactor can be designed according to the process/application and not vice versa. However, due to technical limitations in 3D printing, the maximum bioreactor size is currently limited to benchtop scale (~ 10 L).

The SU stirred bag systems from Thermo Fisher Scientific (HyPerforma™ S.U.B) and Cytiva (Xcellerex™) are available for volumes up to 2000 L for mammalian cell cultures and provide real alternatives to stirred bioreactors made of stainless steel. Both bioreactor systems work with cylindrical bags in which low shear impellers and commonly used aeration devices such as microspargers and open-pipe spargers have been preinstalled. Relevant impeller (d/D of 1:3–1:2) and reactor geometries (H/D 1:1–2:1) of these bag bioreactors were replicated from their steel counterparts. Both systems possess an off-center, angular, axial flow impeller. The eccentrically located impeller eliminates the fluid vortex on the surface, removing the need for baffles [80]. The XDR bioreactor runs with a magnetically coupled, bottom-driven marine impeller, and the S.U.B is top driven and incorporates a seal. The XDR and S.U.B systems are currently used in many mAb and vaccine production processes using mammalian (e.g., CHO, hybridoma, Vero, baby hamster kidney {BHK}) and insect cells. Similar product quantities (middle to high cell densities and titers in g/L -range), and qualities have been achieved by Lonza. Furthermore, the interest in microbial versions of SUBs has resulted in the design of the S.U.F 300 and the XDR 200 MO.

The BIOSTAT® STR from Sartorius is available up to 2000 L and is as close as possible to the classical configuration of a modern reusable cell culture bioreactor. The bag can be equipped with a ring sparger or a microsparger and two axial flow SBIs, or a combination of one axial flow SBI and one radial flow disc impeller. Homogenous mixing in the bioreactor is achieved by the centered stirring system. Experimental and numerical studies performed by Löffelholz et al. [69] have shown the scalability of the BIOSTAT® STR bioreactors due to their comparable hydrodynamic properties resulting from their geometrical similarity to reusable systems. They reported specific power inputs in the range of 29–86 W/m^3 (u_{tip} ~ 1.8 m/s) and mixing times between 10 and 200 s for the BIOSTAT® STR 50 L, 200 L and 1000 L with 2 × SBI. BIOSTAT® STR bioreactors are used by Lonza in its IBEX single-use production facility in Visp (Switzerland). An example of process scalability in the BIOSTAT® STR bioreactors is shown in Fig. 1.4, when Lonza's GS Xceed® CHO platform technology is used.

Pall Life Sciences launched in 2018 with the Allegro™ STR system, a bioreactor which deviates from classical bioreactor designs. The cultivation container of the Allegro™ STR consists of a cube-shaped bag and is available at five different scales (Allegro™ 50, 200, 500, 1000, 2000 L). In contrast to classical, cylindrical stirred SUBs, the support container of the Allegro™ STR integrates notches on the sides of the bag which act as baffles. The bottom-mounted pitched-blade impeller delivers mixing times of <30 s and k_La-values of up to 40 h^{-1} for impeller speeds of up to

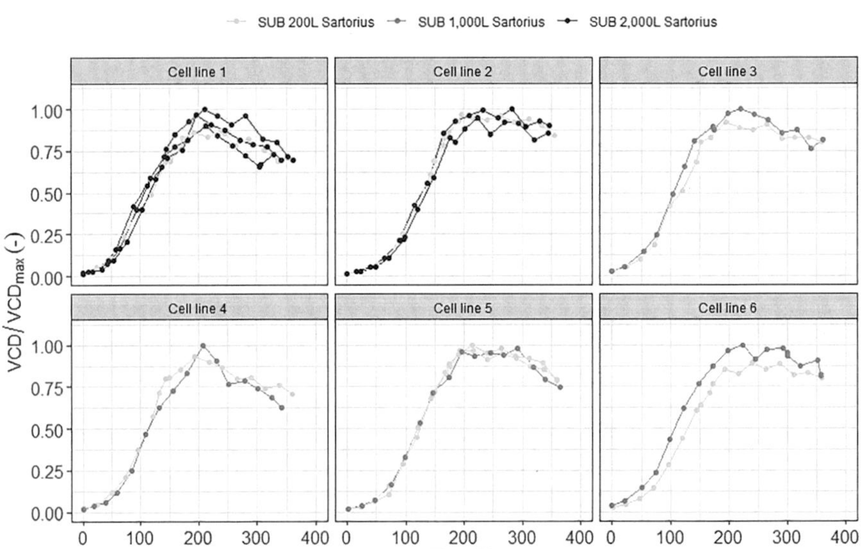

Fig. 1.4 Time-dependent profiles of the normalized viable cell density (VCD) in the BIOSTAT® STR bioreactors at different scales (200 L pilot scale, 1000 L and 2000 L manufacturing scale). Data represent Lonza fed-batch production process based on the GS Xceed® CHO platform technology

105 rpm [81]. Cultivation studies performed by Schirmer et al. [48] with an IgG-producing CHO cell line have shown that the Allegro™ STR 200 L bioreactor provides comparable cell densities (21.5 × 10^6 cells/mL) and mAb titers (1.9 g/L) to a similarly sized classical stirred SUB. A similar bioreactor concept based on a cuboidal vessel was recently introduced by Thermo Fisher Scientific: the HyPerforma™ DynaDrive™ S.U.B series (50, 500, 3000, 5000 L), which is specifically designed for intensified processes. Due to the cuboidal vessel shape and the resulting H/D ratios (2.5:1–2.9:1), high turn-down ratios of up to 10:1 and 20:1 are possible. In contrast to other stirred SUBs with rigid stirrer shafts, the impellers of the DynaDrive™ are mounted onto a flexible drive train. Depending on the size of the bioreactor, two to four marine impellers with diameters of 110–490 mm that strategically scale a flexible rope ladder are used with a sweep impeller at the base of the ladder. Detailed information about biochemical engineering parameters of the DynaDrive™ systems and the resulting fluid flow patterns can be found in [49, 82–84]. In brief, the DynaDrive™ offers a maximum P/V of up to 80 W/m^3, k_La-values of 40 to >50 h^{-1} and mixing times of <60 s. Cobia et al. [85] demonstrated the scalability of the DynaDrive™ systems (50, 500, 5000 L) for a 14-day long standard fed-batch run. They reported comparable VCDs (~20 × 10^6 cells/mL) and product titers (~ 3 g/L) over the different bioreactor scales (Fig. 1.5).

The Process Analytical Technology (PAT) initiative of the US Food and Drug Administration (FDA) requires various process parameters to be monitored to ensure the quality of biopharmaceutical products. Therefore, SU bioreactors must be equipped with appropriate sensor technology. Nowadays, most SU bioreactors only have basic instrumentation (i.e., pH, DO and temperature sensor, see Table 1.2). The systems are either directly equipped with SU sensors or can be

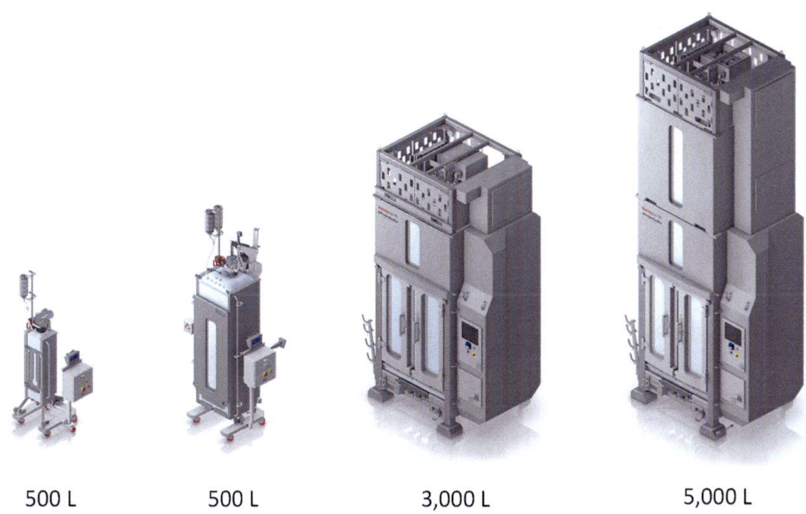

Fig. 1.5 Thermo Fisher Scientific's HyPerforma™ DynaDrive™ bioreactor. (Photo courtesy of Thermo Fisher Scientific)

Table 1.2 Overview of often used, instrumented SU bioreactors (from mL to m^3 scale) in biopharmaceutical productions [43–57]

Vendor	Bioreactor brand	V_w (L)	Basic instrumentation (pH/DO sensor)	Applications
Stirred with rigid container				
Applikon	AppliFlex ST	0.1–12	Vessels with classical pH and optical (LumiSens) DO sensors Biomass or optical density sensors	Mammalian and microbial cells for protein therapeutics (e.g., CHO, BHK) Human cells for cell therapeutics (e.g., MSCs, NK)
Eppendorf	BioBLU® c	0.65–40	Vessel with integrated optical DO and classical or optical pH sensors	Mammalian and microbial cells for protein therapeutics and vaccines (e.g., CHO, Vero, *E. coli*) Human cells for cell- and gene therapeutics (e.g., MSCs, iPSCs, HEK)
Merck Millipore	Mobius® CellReady 3 L	1–2.4	Vessel can be equipped with classical sensors (DO/pH)	Mammalian cells for protein therapeutics (e.g., CHO, BHK) Human cells for cell therapeutics (e.g., MSCs)
Sartorius	Ambr® (15 or 250)	0.01–0.25	Vessels equipped with optical sensor for DO and classical or optical sensor for pH	Mammalian and microbial cells for protein therapeutics (e.g., CHO, BHK, *E. coli, P. pastoris* Human cells for cell therapeutics (e.g., MSCs, T-, CAR-T)
	UniVessel® SU	0.6–2	Vessel equipped with optical sensors for DO and pH (optional use of classical sensors)	Mammalian and insect cells for protein therapeutics (e.g., CHO, BHK, *Sf*-9) Human cells for cell therapeutics (e.g., MSCs)
Stirred with flexible bag				
ABEC	CSR® series	50–6000	Bags can be configured with conventional or optical sensors	Mammalian cells for protein therapeutics (e.g., CHO)
Cytiva	Xcellerex™ XDR	4.5–2000	Bioreactors can be configured with conventional sensors	Mammalian cells for protein therapeutics and vaccines (e.g., CHO, Vero) Human cells for gene therapeutics (e.g., HEK)

(continued)

Table 1.2 (continued)

Vendor	Bioreactor brand	V_w (L)	Basic instrumentation (pH/DO sensor)	Applications
Merck Millipore	Mobius® CellReady	10–2000	SensorReady technology for inserting different sensors	Mammalian cells for protein therapeutics (e.g., CHO) Human cells for cell therapeutics (e.g., MSCs)
Pall Life Sciences	Allegro™ STR	10–2000	Ports for aseptic insertion of conventional and/or SU sensors	Mammalian cells for cell therapeutics (e.g., CHO)
Sartorius	BIOSTAT® STR	12.5–2000	Bags are equipped with SU sensors for DO and pH (optional integration of conventional sensors for DO, pH or pCO_2) Capacitance and conductivity measurement (ViaMass sensor)	Mammalian and microbial cells for protein therapeutics and vaccines (e.g., CHO, BHK, *E. coli*) Human cells for cell therapeutics (e.g., MSCs)
Stirred with flexible bag				
Thermo Fisher Scientific	HyPerforma™ SUB	10–2000	Ports for integration of conventional or SU sensors for DO and pH	Mammalian cells for protein therapeutics (e.g., CHO)
	HyPerforma™ DynaDrive™ SUB	5–5000	Ports for integration of conventional or SU sensors for DO and pH	Mammalian cells for protein therapeutics (e.g., CHO) Insect cells for gene therapeutics (e.g., *Sf*-9)
Wave-mixed (1D motion)				
Cytiva	ReadyToProcess WAVE™ 20/50	0.05–25	Bags are equipped with optical sensors for pH and DO	Mammalian cells for protein therapeutics (including inoculum production) (e.g., CHO)
	WAVE bioreactor™ 200	5–100	Bags are equipped with optical sensors for pH and DO	Mammalian cells for protein therapeutics and vaccines (including inoculum production) (e.g., CHO, Hi-5)
	Xuri™ cell expansion system	0.3–25	Bags are equipped with optical sensors for pH and DO	Human cells for cell therapeutics (e.g., T, NK)

(continued)

Table 1.2 (continued)

Vendor	Bioreactor brand	V_w (L)	Basic instrumentation (pH/DO sensor)	Applications
Sartorius	BIOSTAT® RM	0.1–100	Bags are equipped with optical SU sensors for pH and DO	Mammalian cells for protein therapeutics and vaccines (including inoculum production) (e.g., CHO, BHK, E-FL, *Sf*-9, Hi-5) Human cells for cell therapeutics (e.g., T, NK)
Thermo Fisher Scientific	HyPerforma™ rocker bioreactor	5–25	TruFlor sensors for pH and DO	Mammalian cells for protein therapeutics (including inoculum production) (e.g., CHO)
Wave-mixed (2D motion)				
BIOTECH BV	CELL-tainer®	0.25–200	Bags are equipped with optical sensors for DO and pH Possible to integrate glucose and lactate sensors	Mammalian and microbial for protein therapeutics (including inoculum production) (e.g., CHO, *E. coli*)
Wave-mixed (3D motion)				
Pall Life Sciences	Allegro™ XRS	3–20	Bags are equipped with optical sensors for pH and DO	Mammalian cells for protein therapeutics (e.g., CHO)
Orbitally shaken with flexible bag				
Kühner AG	OrbShake™	3–2500	Bags are equipped with optical sensors for pH and DO	Mammalian cells for protein therapeutics (e.g., CHO)

equipped with classical sensors using aseptic connections. Some SU bioreactors also include built-in capacitance sensors, which allow non-invasive online measurement of viable cell volume (VCV). Further information on the current status of SU sensors and their measurement principles can be found in [86]. Current trends are focused on the introduction of non-invasive spectroscopic measurement techniques (i.e., UV/VIS, infrared, fluorescence, Raman spectroscopy), the most interesting of which is Raman spectroscopy. Several biologically relevant molecules, including glucose, lactate, acetate, formate, glutamine, glutamate and ammonia, are Raman active, and the technology has the potential to measure viable cell density, total cell density and product concentration if good models can be built based on appropriate process training sets. For an example of how Raman spectroscopy can be used to monitor a process and control feed addition during a fed-batch process using Lonza's CHOK1SV GS-KO® cells, please see [87, 88].

Wave-Mixed Single-Use Bioreactors

All wave-mixed bioreactors operate with 2D- or 3D bags made of a multilayer film [89]. The bag is often made from polyethylene, with the contact layer usually made from ethylene vinyl acetate (EVA). Mixing in the bag is achieved through a wave-induced mixing process resulting from the oscillating movement of a rocking platform. The wave characteristics inside the bag depend on the bag geometry, the rocking angle, the rocking rate, the bag translation, the filling volume, and the fluid properties (i.e., fluid density and viscosity). Various studies in the past have shown that wave-induced mixing is particularly beneficial for shear sensitive suspension cells [90–92]. Using Computational Fluid Dynamics (CFD) simulations, Werner et al. [93] found that energy dissipation and shear patterns were more homogenous than in stirred cell culture bioreactors equipped with Rushton turbines and paddle impellers. Due to the wave movement, oxygen is introduced into the fluid from the headspace within the bag through bubble-free surface aeration, with the medium surface continuously being renewed. The oxygen transfer rate (which is described by the volumetric oxygen transfer coefficient k_La) and its influence on cultivation results have been investigated for most available wave-mixed bioreactor systems [94, 95]. Wave-mixed systems differ in terms of their bag designs (i.e., material, geometry), their control mechanisms, the types of sensors installed, and the platform motion, which can be either 1D, 2D or 3D. The intensity of the mass and energy transfer and thus cell growth and product expression can be directly controlled by wave generation and propagation. However, a good understanding of the system is required to adequately control the various influencing parameters. Therefore, a technical characterization of the systems using biochemical engineering methods is useful before the systems are deployed. Detailed information about the technical characterization of wave-mixed bioreactors, including guidelines (e.g., DECHEMA, BPOG) for the methods used, can be found in the scientific literature [74, 96–98].

The 1D oscillatory mixing concept is most often used in the biopharmaceutical industry. The fluid flow inside 1D motion bags can be characterized by means of a modified Reynolds number (Re_{mod}), which was introduced by Eibl et al. [97] in 2009. They found that turbulent conditions in 1D motion bags occur for $Re_{mod} > 1000$, with Re_{mod} depending on the working volume, the length of the culture bag, the liquid level, the rocking rate, the kinematic viscosity of the fluid, and an empirical bag-dependent constant. In contrast to the Re_{mod} concept, Oncül et al. [91] introduced the non-dimensional Womersley number in combination with the parameter β to quantify the unsteady nature of the flow. They found that turbulent conditions occur in oscillating wave-mixed bioreactors for $\beta > 700$ and Womersley numbers >8.5. Reported mixing times (4–800 s) and k_La values (5–50 h^{-1}) make wave-mixed bioreactors with a 1D motion suitable for most cell cultures (e.g., CHO and BHK cells), especially for those with low and medium oxygen demands and Newtonian fluid flow properties [89, 97, 99, 100]. For 1D motion bags, it was demonstrated that increasing the rocking angle and rocking rate increases the k_La more effectively than increasing the aeration rate. Limiting oxygen transfer conditions have been reported for 1D motion bags, when aerobic microorganisms are grown [101, 102]. The oxygen mass transfer for aerobic microorganism cultivations in wave-mixed bioreactors can be increased by increasing the film layer formation, as shown by Hitchcock [103] for bacterial human papilloma virus vaccine production involving high optical cell densities. For this purpose, Hitchcock used a BIOSTAT® RM 50 L bag with 5 L working volume. In contrast, Dreher et al. [104] demonstrated that by using a linear feeding strategy based on a model for oxygen consumption, an OD_{600} of 130 can be achieved for *E. coli* BL21 in a BIOSTAT® RM 10 L (w/5 L working volume, 10° and 42 rpm).

2D and 3D motion bag systems (e.g., CELL-tainer® or Allegro™ XRS, see Table 1.2) provide significantly higher k_La values of up to 73 and 450 h^{-1}. The adapted bag motion increases turbulence and surface renewal. Therefore, these systems are suitable for cultures with high oxygen demands or with special mixing requirements, such as non-Newtonian culture broths. The CELL-tainer® combines a vertical rocking motion with a horizontal translation, whereas the Allegro™ XRS has a 3D bag and uses a simultaneous biaxial rocking motion that gives the flow a tumbling characteristic. Detailed information on the movement of the CELL-tainer® bioreactor, including biochemical engineering and CFD simulation data, can be found in [95]. Westbrook et al. [105] successfully used the CELL-tainer® bioreactor system to cultivate *E. coli* BL21 and produce the recombinant glutathione S-transferase-hCD83ext fusion protein, achieving comparable results to stirred bioreactors. Junne et al. [101] demonstrated the linear scalability of the CELL-tainer® system by maintaining the principal geometrical features (ratio between angle and plate width) and were able to produce comparable fluid characteristics and shear stresses at various different scales (12–120 L).

It is undisputed that wave-mixed bioreactors with 1D motion have been more thoroughly investigated and are more often used in the biopharmaceutical industry than 2D and 3D motion systems. In fact, CFD simulations have revealed the presence of laminar flow conditions and low shear stress levels far below those reported for stirred cell culture bioreactors [93, 106]. This fact has also encouraged different users to employ wave-mixed 1D motion bags for cell therapeutical manufacturing, such as the expansion of human mesenchymal stem cells [53, 107] and human immune cells (e.g. CAR-T or NK cells) [108]. In light of this, Sartorius and Cytiva launched wave-mixed bioreactor systems (BIOSTAT® RM TX and Xuri™ Cell Expansion System) with special bag designs for primary cell processing. In addition to these applications, 1D motion bags have also proven themselves in animal cell-based production processes of viruses for gene therapies, as well as for vaccines and for virus-like particle vaccines [55, 109, 110].

Orbitally Shaken Single-Use Bioreactors

While engineering parameters and scale-up criteria can generally be established without any significant problem for stationary, stirred, directly aerated SU bioreactors, the free fluid surface makes this more difficult for orbitally shaken bioreactors.

Differing flow pattern engineering values that are available for shake flasks (e.g., Erlenmeyer flasks) are not transferable to orbitally shaken tubes (e.g., TubeSpin® bioreactor 15, 50, 450 and 600) and bag-based systems (e.g., OrbShake™ bioreactor). The best investigated single-use, orbitally shaken bioreactor system is the OrbShake™ bioreactor (Kühner AG), which consists of a 3D cylindrical SU bag, a supporting stainless steel container, and a peripheral measurement and control unit. The OrbShake™ system is available at benchtop (V_w 12 L) and production (V_w 2500 L) scales, with two additional system scales in between (SB50-X and SB200-X). The mixing of the fluid is achieved by the orbital motion of the shaker platform. Fluid dynamics are highly affected by the selected operating conditions, since different combinations of reactor diameter, orbital diameter, fluid height, rotational shaker speed, and fluid kinematic viscosity of the fluid create different flow/vortical structures that ultimately control shear stress, turbulence, aeration, convection, mixing, and the suspension process. Detailed information about biochemical engineering data and the fluid flow dynamics in the OrbShake™ bioreactor can be found in [111–115].

Dittler et al. [116] reported the successful use of the SB10-X system for establishing a titerless infected-cells preservation and scale-up (TIPS) process using *Sf-9* cells and the recombinant baculovirus *Autographa californica multicapside nucleopolyhedrovirus* that produces secreted alkaline phosphatase (rSEAP). By using the SB10-X system and the TIPS method, they achieved a rSEAP activity of 144–158 U/mL at 4 days post-infection, saving 2–3 days compared to classical two-phase processes. Recently, Coronel et al. [116] equipped the SB10-X with tangential flow filtration (TFF) and alternating tangential flow filtration (ATF) devices to run the system in perfusion mode (see also section "Trend intensified and continuous facilities") and produce human influence A virus vaccines. The authors demonstrated that high viable cell densities of up to 50×10^6 cells/mL can be achieved in orbitally shaken bioreactors. Anderlei et al. [117] reported k_La values of up to 45 h^{-1} for the SB200-X with 100 L working volume, and mixing times between 25 and 70 s depending on the shaking frequency (50–70 rpm). Lower k_La values ranging from ~2 to 25 h^{-1}, depending on the working volume (500–2500 L) and shaking frequency (35–55 rpm), were reported for the SB2500-X.

Systems for Primary Recovery/Clarification

Most biopharmaceutical products (especially mAbs) are secreted into the cell culture medium. Therefore, a primary clarification step is required to separate the cells and large contaminants from the product. For this purpose, the most common separation strategies are either based on filtration or centrifugation. After primary clarification, secondary and/or tertiary clarification is performed to remove finer particles and bioburden. These steps can consist of one or multiple grade depth filters, followed by a sterilizing-grade membrane filter. Suitable SU filters for different clarification ranges are available from Merck Millipore (Millistak+® Pod system), 3M™ (Zeta Plus™), Sartorius (Sartoclear® L-Drum technology) and Pall (Stax™ technology).

The filters consist of diatomaceous earth, cellulose or fully synthetic fibers (e.g., polyacrylic fibers), inorganic filter aids and, in some cases, charged moieties. The fully synthetic filters (e.g., Millistak+® HC Pro) offer advantages in terms of process consistency and require significantly less flushing before use [118]. The filter surface of some filters is positively charged to reduce negatively charged contaminants by electrokinetic adsorption. The adsorptive properties allow the filtration to be tailored to a wide range of cell culture broths. For example, the 3M™ Emphaze™ and Harvest RC are AEX based adsorptive filters that greatly reduce rDNA in HCCF [119–121]. In general, SU depth filters offer advantages due to their direct scale-up, modular design and the availability of fully closed filter sheets/capsules for easy and safe handling.

SU centrifuges represent an alternative to SU filters for entire primary clarification and the secondary clarification step, especially when large amounts of biomass need to be separated (e.g., intensified fed-batch or perfusion processes). In recent years, new SU centrifuges (GEA kytero®, Alfa Laval CultureOne™) have come onto the market, complementing those (CARR UniFuge and U2k, Sartorius Ksep®) that have been available for some time. The Ksep® system from Sartorius uses counter flow centrifugation with fluidized bed technology to separate the cells from the supernatant [122]. The separation process consists of repeating cycles of loading, washing and discharging of buffer and cells. The system is available in two different sizes that allow maximum flow rates of 114 L/h (Ksep® 400; processing volumes of 10-200 L) and 720 L/h (Ksep®6000 s; processing volumes of 50–2000 L). Saballus et al. [123] recently demonstrated the applicability of the Ksep® system for clarifying high cell density CHO processes (100×10^6 cells/mL). By using a post-centrifugal filtration step, they were able to reduce turbidity from 8000 to <5 NTU, while maintain a mAb recovery of >90%. Other SU centrifuges, such as the CultureOne™ or the kytero® (see Fig. 1.6), are based on classical disc stack technology [124–126]. In these systems, the suspension with the higher cell density is concentrated at the periphery of the separator bowl and the purified supernatant is removed from the central upper part. Thus, a continuous discharge of highly concentrated cells and clear supernatant is achieved. Both SU centrifuge systems are available in two sizes; the CultureOne™ Primo and kytero® 500 are recommended for cell harvesting from bioreactors up to 500 L in size (processing capacity up to 200 L/h) and are thus used at pilot scale [124, 126]. In contrast, the CultureOne™ Maxi and kytero® 2000 are designed for production scale and can process cell suspension volumes of up to 2000 L (processing capacity up to 600 L/h). The kytero® systems are driven entirely without shafts, mechanical bearings and seals. The rotating separator bowl is held in suspension by magnetic technology (breezeDrive®). In the CultureOne™ systems the process chamber is hermetically separated from the environment by mechanical seals. Ott et al. [127] recently demonstrated the applicability of the kytero® 500 SU centrifuge for clarifying a CHO cell culture suspension (fed-batch process with $VCD_{max} = 16.8 \times 10^6$ cells/mL). They were able to completely separate the cell suspension from a 200 L SUB in less than 2 h, achieving mAB recoveries of 98–99% and turbidities of 10–15 NTU.

Fig. 1.6 The kytero® SU centrifuge. (Copyright GEA AG)

Further options for cell separation include acoustic separation technologies that utilize SU acoustic chambers to efficiently reduce turbidity, particles and cells for bioreactor sizes up to 2000 L [128]. Macroscale acoustic separators work with multidimensional standing waves to remove cells and particles from the suspension. For this purpose, the cell suspension moves through an acoustic field in which physical forces retain the cells and cluster them into tightly packed groups at the intersection or nodes of the acoustic wave. As soon as a critical cell cluster size is reached the cell clusters start to sediment due to gravity. An example of such a system is the Cadence™ Acoustic Separator from Pall [129, 130].

Single-Use Downstream Processing

SU systems, which were initially used mainly in USP (see section "Single-use upstream processing"), have also become increasingly popular in DSP in recent years. Today, complex steps such as chromatography, centrifugation and TFF can be performed with SU systems. However, due to gradually increasing cell densities and product titers over the past two decades, process intensification has become an important aspect in DSP (see section "Trend intensified and continuous facilities"). The use of SU technology in DSP offers greater design flexibility and faster adaptation when process changes become necessary. In addition, the lead times for qualifying and commissioning a process are shorter, resulting in a reduction in costs and effort for change-control documentation. This can, therefore, have a major impact on time-to-market for new biopharmaceutical products. However, SU technology in DSP is still limited in terms of scalability and therefore does not support production scales above 4000 L [128]. Hence, there is a trend towards larger disposable modules that support larger production volumes or further process intensification.

Chromatography is the key technology used in the purification of biopharmaceuticals. For example, three chromatographic steps are usually performed to purify a mAb. The first step, called capture chromatography, is used to isolate, concentrate and stabilize the product from the clarified cell supernatant. This is followed by two polishing steps to further deplete host cell protein {HCP}, deoxyribonucleic acid {DNA} and high molecular weight material (e.g., aggregated product). Using disposable columns and SU chromatography skids enables turn-around times to be faster and increases facility productivity by eliminating cleaning processes and validation. SU chromatography systems are available from Cytiva (ÄKTA™ ready SU system) and Pall (Allegro™ SU chromatography system). Both systems work with fully disposable flow paths, pre-packed traditional or SU columns and membrane-based chromatography products. This provides considerable flexibility, especially for multi-product facilities, and reduces processing costs in the long term. The use of pre-packed disposable columns allows the labor-intensive and highly technical packing process to be outsourced. This eliminates the need for a column packing system and buffers for column packing and cleaning. However, the size of the pre-packed columns is limited. GMP-compliant columns are available in different resins and formats from Cytiva and Repligen. SU membrane adsorbers (e.g., Sartobind® from Sartorius, Mustang® from Pall) represent an alternative to column-based chromatography and are of special interest for DSP process intensification (see section "Trend intensified and continuous facilities").

Virus removal is based on at least two orthogonal steps, as required by the International Conference on Harmonization of Technical Requirements for Registration of Pharmaceuticals for Human Use (ICH Q5A guidelines). Typically, this is achieved using a combination of a low pH virus inactivation step and a filtration and chromatography virus removal step. Specially engineered SU mixing systems are available for the virus inactivation step. The virus filtration steps often employ filtration capsules (e.g., Planova™, Viresolve®, Virosart®) with membrane cut-offs of 15–20 nm and are run on dedicated SU processing platforms (e.g., SU virus filtration systems from Agilitech, Sartorius, Cytiva, Merck Millipore).

Single-Use Formulation and Fill and Finish

The aim of the formulation step is to concentrate and to exchange the buffer in the product solution to achieve appropriate volume reduction and buffer conditions. TFF is typically used for this purpose. Using TFF, the solution can be pre-concentrated to the ideal concentration for diafiltration. Buffer exchange can be accomplished by diafiltration, and the product can be concentrated to the final concentration for further processing. Generally, diafiltration volumes of 5–7 are used to achieve a buffer exchange of up to 99.9%. Nowadays, different TFF systems (e.g., Pall, Repligen, Merck Millipore, Parker) are available on the market which provide SU flow paths and SU cassettes with high membrane areas (up to 10 m^2). Once the final formulation is homogenous, it should be sterilized and filled without affecting the stability or functionality of the DP and with minimal product loss. To meet all these requirements, the following equipment is needed: thawing unit, pooling or conjugation unit (e.g., vaccines) mixing unit, automated formulation control and dosing unit, filtration unit and final filling machine. SU freezing (e.g., Cellsius® Pak, RoSS-pFTU) and filling systems (e.g., SciLog® FD, Linkit® AX, Allegro™ SU filling, RoSS.FILL) are nowadays available from different manufacturers (e.g., Meissner, Parker, Pall, Sartorius, Single-Use Support). A case study of SU filling line setup is available in the following reference [131].

Trend Intensified and Continuous Facilities

Konstantinov and Cooney [132] defined a continuous unit operation as a subunit of a process, in which a continuous flow input for a prolonged period of time and minimal internal hold volume is present. The output can either be continuous or discretized into small batches, which are then produced in a cyclic manner. Based on this definition, two different types of continuous processes can be defined: (1) the hybrid continuous process, in which batch and continuous unit operations are combined, and (2) the fully integrated unit operation or end-to-end process. Nowadays, the hybrid continuous process is already a reality in biopharmaceutical production, where three different types can be distinguished: (1) continuous upstream followed by batch downstream, (2) continuous upstream and capture with batch downstream and (3) batch upstream followed by continuous downstream. The first type is not new and has been used for decades for the production of therapeutic enzymes, blood factors and certain mAbs [133]. The development of this approach to a continuous capture step represents an important improvement that addresses the high costs associated with DSP, where the capture step alone accounts for approximately 25% of the costs [134]. The continuous capturing approach means that large holding tanks are no longer required after the clarification step and the size of the Protein A column can be theoretically reduced by two orders of magnitude, which also leads to a significant reduction in buffer utilization. In the third approach cell cultivation is normally performed in traditional fed-batch mode, while one or two DSP operations (e.g., capture, polishing) are operated continuously, which results in smaller continuous chromatography columns (= cost reduction).

It is undisputed that the availability of a wide range of SU systems has driven the development of more intensified and continuous mAb production processes in recent years [3, 135, 136]. This fact is not surprising, since existing SU technology facilitates the execution of intensified unit operations. However, a distinction must be made between USP and DSP intensification. Upstream intensification usually requires the use of perfusion technology, in which cells are retained in the bioreactor while part of the cell culture medium is continuously exchanged. The highest VCDs reported in the scientific literature for perfusion processes are up to 200×10^6 cells/mL [137, 138]. To achieve such high VCDs with minimal medium consumption, Ozturk proposed using the cell-specific perfusion rate (CSPR) to control the perfusion process. The CSPR describes the amount of cell culture medium supplied to the bioreactor per cell per day and is a good indicator of the nutrient supply per cell. Since the cells are retained in the bioreactor during perfusion, the bioreactor must be equipped with either an internal or external perfusion device for cell retention (see Table 1.3). The selection of a suitable cell retention device is an important consideration in the development of perfusion processes, and any SU cell retention device should meet the following characteristics: (1) high robustness, (2) high cell retention efficiency w/o cell damage, (3) high product yield, (4) scalability, (5) low running costs, and (6) continuous harvesting. Cell retention devices include internal perfusion membrane filters, TFF systems, ATF systems, acoustic wave separators and centrifuges. Apart from wave-mixed bioreactor systems with internal perfusion membrane filters, all other systems are available up to manufacturing scale. Wave-mixed systems are easy to use and are therefore often applied for N-1 (i.e., seed stage) perfusion processes. The availability of SU perfusion systems has contributed over the last years to the development of high-seed fed-batch production and concentrated fed-batch processes. For more information on these types of intensified processes, please see [136]. In addition to the previously mentioned intensified processes, a fully continuous perfusion process is of great interest. The goal of such a process, in which spent cell culture medium and product are continuously discharged, is to achieve a steady state in which productivity and product quality can be maintained over a long period of time with only minimal deviation. Unlike N-1 perfusion processes, cell bleeds have to be performed regularly in order to maintain the necessary constant VCD and high cell viability. Currently, target run times for such processes range from 30 to 90 days; the longer the process runs under stable conditions, the higher the space-time yield of the bioreactor. By using continuous USP processes, specific productivities of 1 g/L/d have been reported [141]. The long run times clearly show the high demands on the SU bioreactor and its internal and external peripherals in terms of resistance and sterility. An example of a stirred SU bioreactor, specifically designed for continuous perfusion processes and intensified fed-batch operations is the HyPerforma™ DynaDrive™ series from Thermo Fisher Scientific, which was shown in section "Single-use bioreactors".

Because intensified USP can result in higher concentrations of solid and cellular contaminants (e.g., HCP, DNA), and protein aggregates, harvesting and subsequent processing is more difficult than in conventional fed-batch processes. However, there are two main approaches to DSP intensification: (1) increasing the binding capacity of chromatography columns or using them more effectively, and/or (2) increasing productivity by running each process step continuously. For the capture step in mAb production, Protein A is commonly used, which has already been optimized several times (e.g., increasing binding capacity, improving stability)

Table 1.3 Overview of currently used SU cell retention devices for process intensification. [18, 20, 127, 139, 140]. The advantages and disadvantages listed in the table refer to the technologies in general and do not represent an evaluation of individual suppliers

	Internal cell retention	External cell retention			
Working principle	Membrane filtration	TFF	ATF	Centrifugation	Acoustic wave separation
Cell recirculation	No	Yes	No		
CR efficiency [%]	100	100	100	99	85–99
VCD_{max} [[a]10^6 cells/mL]	200	200	200	20	42
Scalable to manufacturing scale	No	Yes	Yes	Yes	Yes (parallelization)
Ease of use	Very easy	Very easy	Very easy	Difficult	Difficult[b]
Operational costs	Low	Very high	Very high	High	Low
Advantages of technology	Easy to operate (direct integration of cell retention membrane)	Low residence times	Low residence times Reduced membrane fouling (alternating flow directions) Reduced shear stress	Separation of living and dead cells (only living cells remain in the system)	Ready for long-term operations
Disadvantages of technology	Possible filter clogging Dead cells remain in the system Only available for wave-mixed bioreactors	Possible membrane fouling and blockage Dead cell remain in the system	Problems with HCD (increased viscosity) Dead cells remain in the system	Rotor speed-related cell damage Continuous long-term operation may be challenging	Generated heat may cause cell damage above pilot scale
Suppliers (e.g., ...)	Sartorius (Flexsafe® RM series) Cytiva (Cellbag™ BioClear series)	Cytiva (Xcellerec™ APS) Hollow fibers from other vendors (e.g., Sartorius, Repligen)	Repligen (XCell ATF® system)[a]	Sartorius (Ksep®)	Pall (cadence™ acoustic separator)

[a]both reusable and single-use variants available
[b]ease of use strongly cell line dependent

since its discovery. A detailed overview of modifications to Protein A can be found in [142]. Alternatives to intensified capture steps are based on performing several cheaper chromatography steps instead of using affinity chromatography or performing the capture step exclusively using cation exchange chromatography [143]. Other purification options include precipitation [144–146], crystallization of the target product [147] and two-phase extraction [148]. Capture and polishing steps can also be performed using membrane chromatography instead of classical packed bed columns. Membrane chromatography allows higher flow rates (up to 50-fold) at lower operating pressures than packed bed columns. SU membrane capsules are available from various manufacturers (e.g., Natrix Separations, Pall, Sartorius), which can also be used in bind-elute or flow-through mode. According to investigations by Lim et al. [149], membrane technology is much cheaper than conventional packed bed columns and can result in cost savings of up to 65%. However, the disadvantage of membrane chromatography is its lower binding capacity compared to packed bed columns. For this reason, membrane technology is currently not widely used for the capture step in DSP. Casey et al. [150] proposed the addition of fivefold single-pass TFF (=SPTFF) to the capture step in order to reduce the size of the capture column. Such an approach reduces resin and buffer usage by up to 80%, which results in a significant cost reduction. Another approach to intensifying DSP is to use mixed-mode chromatography for the polishing step. Instead of performing several chromatography steps one after the other, several components with different affinities are used as the stationary phase of the chromatography column.

As mentioned above, the ultimate goal of process intensification is to create an end-to-end process. This includes a continuous capture step using multi-column or counter-current chromatography. In multi-column technology, several columns are connected in parallel and operated cyclically. Therefore, smaller column sizes are sufficient. Moreover, such an approach eliminates the hold-up step after the production bioreactor and reduces the amount of resin and buffer used. Good manufacturing practice {GMP} compliant multi-column chromatography systems are currently available from Merck Millipore (Mobius® Multi Column Capture System) and Sartorius (Resolute® BIOSMB platform) with full SU flow paths. Both systems can be used in bind-elute mode for the capture step, and in flow-through mode for polishing (e.g., ion-exchange, hydrophobic or mixed-mode chromatography). By binding the impurities instead of the target product, continuous operation is made possible. Once the column is fully loaded with impurities, impurity breakthrough occurs and the column must be replaced. However, this does not represent a true continuous mode of operation and is therefore referred to as pseudo-continuous. As mentioned previously, it is necessary to perform a virus inactivation and a virus filtration step during DSP. Two approaches exist for continuous virus inactivation: (1) adaptation of the classical approach with two SU mixers that are alternately operated (e.g., Cadence™ Virus Inactivation system from Pall) and (2) the implementation of a plug-flow reactor, where acid is added to the product flow at the reactor inlet. In the latter case, the low pH value is maintained for the product residence time in the plug-flow reactor and afterwards neutralized at the outlet by adding base. Continuous virus filtration is carried out using a multi-filter concept in which two individual filters are connected in parallel. As soon as one filter reaches its capacity (e.g., due to pressure limit) the system automatically switches to the other filter, so the first filter can be replaced aseptically. The formulation of mAb usually requires several TFF steps, in which the product is first concentrated by ultrafiltration, transferred to the appropriate buffer system via diafiltration, and brought to its final concentration by a second ultrafiltration step. Therefore, these processes are generally not continuous steps. SPTFF can be used for continuous processing for protein concentration and diafiltration. For this purpose, the filtration time can be increased by higher membrane areas and lower flow rates to such an extent, that one continuous filtration step is sufficient.

Summary and Conclusions

SU systems have become indispensable in biopharmaceutical production processes. The majority (approx. 80%) are used in research and development as well as pre- and clinical sample production [26]. They are characterized by their high modularity, flexibility and safety, as well as their potential to save costs. SU systems have become important, especially for contract manufacturing organizations producing therapeutic proteins such as mAbs and their biosimilars, as well as for vaccine manufacturers active in the Covid 19 pandemic. They can be found along the entire production process from USP to fill & finish. Despite the pandemic-related supply shortages of SU system consumables, the growth of SU systems is expected to continue. In fact, the global SU process market is expected to grow at a compound annual growth rate of 16.6% by 2030 [151]. Cell and gene therapy products that have already been approved and those under development will contribute to this growth. The current trend is towards intensified and continuous processes based on SU technology, in which perfusions are implemented and the targeted product titers and quantities are achieved in shorter times or in smaller facilities. The improvements (e.g., new film materials, easier handling of the systems) and innovations (e.g., mixers and bioreactors for ultra-high cell density processes, special perfusion systems) developed by the manufacturers of SU systems in recent years have contributed decisively to this trend. Nevertheless, many of the systems available on the market are currently not optimally suited for this purpose. Improved polymers that are more homogeneous and allow designs to be more innovative are required; SU sensors need to provide better long-term stability; further innovations are needed in tubing and connector technology; and ultimately SU products need to become more cost-effective. However, the latter could also be an opportunity for new producers of SU systems that could contribute to the increasing implementation of these systems beyond the biopharmaceutical sector (e.g., cosmetics and food industry).

Acknowledgments The authors would like to thank Emilie Martin and Loic Pasquier (Lonza Mammalian Development Services) for providing the bioreactor data.

Conflict of Interest The authors declare that they have no conflict of interest.

References

1. Shukla AA, Gottschalk U (2013) Single-use disposable technologies for biopharmaceutical manufacturing. Trends Biotechnol 31:147–154. https://doi.org/10.1016/j.tibtech.2012.10.004
2. Schmidhalter DR, Elzner S, Schmid R (2019) Progress in the development of single-use solutions in antibody–drug conjugate (ADC) manufacturing. In: Eibl R, Eibl D (eds) Single-use technology in biopharmaceutical manufacture. Wiley, pp 303–310
3. Langer ES (2022) Single-use technology in biopharmaceutical manufacture and beyond. Chemie Ingenieur Technik 94:1892–1901. https://doi.org/10.1002/cite.202200095
4. Eibl D, Eibl R (2019) Single-use equipment in biopharmaceutical manufacture. In: Single-use technology in biopharmaceutical manufacture. Wiley, pp 1–11
5. Samaras JJ, Micheletti M, Ding W (2022) Transformation of biopharmaceutical manufacturing through single-use technologies: current state, remaining challenges, and future development. Annu Rev Chem Biomol Eng 13:73–97. https://doi.org/10.1146/annurev-chembioeng-092220-030223
6. Bio-Process Systems Alliance (2007) Guide to irradiation and sterilization validation of single-use bioprocesses, part 1. BioPorcess Int 8:32–39
7. Bio-Process Systems Alliance (2007) Guide to irradiation and sterilization validation of single-use bioprocesses, part 2. Bioprocess Int 9:60–70
8. Blanck E, Parma H, Ganguli P, Lindskog E, Carter J (2015) Implementation of raw material control strategies in the manufacture of single-use bioprocessing containers. Biopharm Int 28: 22–29
9. Luu D-N, Barbaroux M, Dorez G, Mignot K, Doger E, Laurent A, Brossard J-M, Maier C-J (2022) Recycling of post-use bioprocessing plastic containers – mechanical recycling technical feasibility. Sustainability 14:15557. https://doi.org/10.3390/su142315557
10. Schwander E, Rasmusen H (2005) Scalable, controlled growth of adherent cells in a disposable multilayer format. Gen 25:29
11. Hopkinson J (1985) Hollow fiber cell culture systems for economical cell-product manufacturing. Nat Biotechnol 3:225–230. https://doi.org/10.1038/nbt0385-225
12. Knazek RA, Gullino PM, Kohler PO, Dedrick RL (1972) Cell culture on artificial capillaries: an approach to tissue growth in vitro. Science 178:65–67. https://doi.org/10.1126/science.178.4056.65
13. Martin JM (2011) A brief history of single-use manufacturing. Biopharm Int 24:5–7
14. Singh V (1999) Disposable bioreactor for cell culture using wave-induced agitation. Cytotechnology 30:149–158. https://doi.org/10.1023/A:1008025016272
15. Cameau E, Kornmann H, Charbaut E, de Abreu G, Desgeorges A (2010) Evaluation of a single-use bioreactor for fed-batch production of a monoclonal antibody. Biopharm Int 2010
16. Lütke-Eversloh T, Rogge P (2018) Biopharmaceutical manufacturing in single-use bioreactors. Pharm Ind 80:281–284
17. Zijlstra G, Touw K, Koch M, Monge M (2019) Design considerations towards an intensified single-use facility. In: Single-use technology in biopharmaceutical manufacture. Wiley, pp 181–192
18. Müller D, Klein L, Lemke J, Schulze M, Kruse T, Saballus M, Matuszczyk J, Kampmann M, Zijlstra G (2022) Process intensification in the biopharma industry: improving efficiency of protein manufacturing processes from development to production scale using synergistic approaches. Chem Eng Process Process Intensif 171:108727. https://doi.org/10.1016/j.cep.2021.108727

19. Malla R, Shah D, Gajendragadkar C, Vamanan V, Singh D, Gupta S, Vengovan D, Trivedi R, Weichert H, Carpio M, Chandran K (2021) Seed train process intensification strategy offers potential for rapid, cost-effective scale-up of biosimilars manufacturing. Bioprocess J 20. https://doi.org/10.12665/J20OA.Malla
20. Cytiva (2020) One-step seed culture expansion from one vial of high-density cell bank to 2000 L production bioreactor, Application Note 29160932. https://cdn.cytivalifesciences.com/api/public/content/digi-17729-original
21. Mauter M (2009) Environmental life-cycle assessment of disposable bioreactors. BioProcess Int 8:18–28
22. Pietrzykowski M, Flanagan W, Pizzi V, Brown A, Sinclair A, Monge M (2013) An environmental life cycle assessment comparison of single-use and conventional process technology for the production of monoclonal antibodies. J Clean Prod 41:150–162. https://doi.org/10.1016/j.jclepro.2012.09.048
23. Flanagan W, Brown A, Pietrzykoski M, Pizzi V, Monge M, Sinclair A (2014) An environmental life cycle assessment of single-use and conventional process technology:comprehensive environmental impacts. Biopharm Int 27
24. Whitford WG, Petrich MA, Flanagan WP (2019) Environmental impacts of single-use systems. In: Eibl R, Eibl D (eds) Single-use technology in biopharmaceutical manufacture. Wiley, pp 169–179
25. Ottinger M, Wenk I, Carvalho Pereira J, John G, Junne S (2022) Single-use technology in the biopharmaceutical industry and sustainability: a contradiction? Chemie Ingenieur Technik 94: 1883–1891. https://doi.org/10.1002/cite.202200105
26. BioPlan Associates Inc (2020) Seventeenth annual report and survey of biopharmaceutical manufacturing capacity and production, ISBN 978-1-934106-37-2
27. Hammond M, Marghitoiu L, Lee H, Perez L, Rogers G, Nashed-Samuel Y, Nunn H, Kline S (2014) A cytotoxic leachable compound from single-use bioprocess equipment that causes poor cell growth performance. Biotechnol Prog 30:332–337. https://doi.org/10.1002/btpr.1869
28. Hammond M, Nunn H, Rogers G, Lee H, Marghitoiu A-L, Perez L, Nashed-Samuel Y, Anderson C, Vandiver M, Kline S (2013) Identification of a leachable compound detrimental to cell growth in single-use bioprocess containers. PDA J Pharm Sci Technol 67:123–134. https://doi.org/10.5731/pdajpst.2013.00905
29. Peng J, Zhao Y, Hong Y, Burkhalter RS, Hogue CL, Tran E, Wei L, Romeo L, Dolley-Sonneville P, Melkoumian Z, Liang X, Fang Y (2018) Chemical identity and mechanism of action and formation of a cell growth inhibitory compound from polycarbonate flasks. Anal Chem 90:4603–4610. https://doi.org/10.1021/acs.analchem.7b05102
30. Jenke DR, Stults CLM, Paskiet DM, Ball DJ, Nagao LM (2015) Materials in manufacturing and packaging systems as sources of elemental impurities in packaged drug products: a literature review. PDA J Pharm Sci Technol 69:1–48. https://doi.org/10.5731/pdajpst.2015.01005
31. Shea J, Goodrich E, Killian P, Becket P, Broschard T (2015) Adopting a fully single-use process: a leachables and patient safety evaluation case study. In: Mahajan E, Lye GJ (eds) Single-use technologies II: bridging polymer science to biotechnology applications. ECI Sympos. Ser., New York
32. Magarian N, Lee K, Nagpal K, Skidmore K, Mahajan E (2016) Clearance of extractables and leachables from single-use technologies via ultrafiltration/diafiltration operations. Biotechnol Prog 32:718–724. https://doi.org/10.1002/btpr.2277
33. Hauk A, Jurkiewicz E, Pahl I, Loewe T, Menzel R (2018) Filtration membranes – scavengers for leachables? Eur J Pharm Sci 120:191–198. https://doi.org/10.1016/j.ejps.2018.04.044
34. Ding W, Madsen G, Ekta M, O'Connor S, Wong K (2014) Standardized extractables testing protocol for single-use systems in biomanufacturing. Pharm Eng 34:1–11

35. Scott B, Ullsten S, Wang P, Hathcock J, Madsen GL, Dale G, Wong KM (2020) BioPhorum best practice guide for: extractables testing of polymeric single-use components used in manufacturing. BioPhorum 2:1–26
36. Eibl R, Steiger N, Fritz C, Eisenkrätzer D, Bär J, Müller D, Eibl D (2014) Standardized cell culture test for the early identification of critical films for cho cell lines in chemically defined culture media. DECHEMA Biotechnol:1–24
37. ASTM (2019) Standard guide for cell culture growth assessment of single-use material. In: Book of standards
38. US Pharmacopeia (2021) <665> Plastic componnets and systems used to manufacture pharmaceutical drug products and biopharmaceutical drug substances and products. US Pharmacopeia
39. Peuker T, Eibl D (2011) Biopharmaceutical manufacturing facilities integrating single-use systems. In: Eibl R, Eibl D (eds) Single-use technology in biopharmaceutical manufacture. Wiley, Hoboken, pp 145–158
40. Eibl R, Eibl D (2019) Single-use technology in biopharmaceutical manufacture II. Wiley
41. Werner S, Kraume M, Dieter E (2019) Mixing Systems for Single-use. In: Eibl R, Eibl D (eds) Single-use technology in biopharmaceutical manufacture. Wiley, Hoboken, pp 25–35
42. de Boulard A, Kienle K (2022) Trends in single-use mixing for biomanufacturing with an insight into Lonza Ibex® solutions. Chemie Ingenieur Technik 94:1962–1967. https://doi.org/10.1002/cite.202200090
43. Rotondi M, Grace N, Betts J, Bargh N, Costariol E, Zoro B, Hewitt CJ, Nienow AW, Rafiq QA (2021) Design and development of a new ambr250® bioreactor vessel for improved cell and gene therapy applications. Biotechnol Lett 43:1103–1116. https://doi.org/10.1007/s10529-021-03076-3
44. Costariol E, Rotondi MC, Amini A, Hewitt CJ, Nienow AW, Heathman TRJ, Rafiq QA (2020) Demonstrating the manufacture of human CAR-T cells in an automated stirred-tank bioreactor. Biotechnol J 15:2000177. https://doi.org/10.1002/biot.202000177
45. Teale M, Jossen V, Eibl D, Eibl R (2021) Chemically defined, Xeno-free expansion of human mesenchymal stem cells (hMSCs) on benchtop-scale using a stirred single-use bioreactor. In: Methods in molecular biology
46. Jossen V, Kaiser SC, Schirmaier C, Herrmann J, Tappe A, Eibl D, Siehoff A, den Bos C, van Eibl R (2014) Modification and qualification of a stirred single-use bioreactor for the improved expansion of human mesenchymal stem cells at benchtop scale. Pharm Bioprocess 2:311–322. https://doi.org/10.4155/pbp.14.29
47. Schirmaier C, Jossen V, Kaiser SC, Jüngerkes F, Brill S, Safavi-Nab A, Siehoff A, van den Bos C, Eibl D, Eibl R (2014) Scale-up of adipose tissue-derived mesenchymal stem cell production in stirred single-use bioreactors under low-serum conditions. Eng Life Sci 14:292–303. https://doi.org/10.1002/elsc.201300134
48. Schirmer C, Müller J, Steffen N, Werner S, Eibl R, Eibl D (2020) How to produce mAbs in a cube-shaped stirred single-use bioreactor at 200 L scale. In: Methods in molecular biology. Humana Press Inc., pp 169–186
49. Kaiser SC, Decaria PN, Seidel S, Eibl D (2022) Scaling-up of an insect cell-based virus production process in a novel single-use bioreactor with flexible agitation. Chemie Ingenieur Technik. https://doi.org/10.1002/cite.202200103
50. Bögli NC, Ries C, Bauer I, Adams T, Greller G, Eibl R, Eibl D (2011) Bag-based rapid and safe seed-train expansion method for Trichoplusia nisuspension cells. BMC Proc 5:P124. https://doi.org/10.1186/1753-6561-5-S8-P124
51. Hundt B, Best C, Schlawin N, Kaßner H, Genzel Y, Reichl U (2007) Establishment of a mink enteritis vaccine production process in stirred-tank reactor and wave® bioreactor microcarrier culture in 1–10L scale. Vaccine 25:3987–3995. https://doi.org/10.1016/j.vaccine.2007.02.061
52. Jossen V, Pörtner R, Kaiser SC, Kraume M, Eibl D, Eibl R (2014) Mass production of mesenchymal stem cells – impact of bioreactor design and flow conditions on proliferation and differentiation. In: Eberli D (ed) Cells and biomaterials in regenerative medicine. InTech, p 119–174

53. Jossen V, Schirmer C, Mostafa Sindi D, Eibl R, Kraume M, Pörtner R, Eibl D (2016) Theoretical and practical issues that are relevant when scaling up hMSC microcarrier production processes. Stem Cells Int 2016:1–15. https://doi.org/10.1155/2016/4760414
54. Imseng N, Steiger N, Frasson D, Sievers M, Tappe A, Greller G, Eibl D, Eibl R (2014) Single-use wave-mixed versus stirred bioreactors for insect-cell/BEVS-based protein expression at benchtop scale. Eng Life Sci 14:264–271. https://doi.org/10.1002/elsc.201300131
55. Eibl R, Steiger N, Wellnitz S, Vicente T, John C, Eibl D (2013) Fast single-use VLP vaccine productions based on insect cells and the baculovirus expression vector system: influenza as case study. Disposable Bioreactors II:99–125
56. Yang J, Guertin P, Jia G, Lv Z, Yang H, Ju D (2019) Large-scale microcarrier culture of HEK293T cells and Vero cells in single-use bioreactors. AMB Express 9:70. https://doi.org/10.1186/s13568-019-0794-5
57. Eibl R, Löffelholz C, Eibl D (2014) Disposable bioreactors for inoculum production and protein expression. In: Animal cell biotechnology: methods and protocols, p 265–284
58. Xu P, Clark C, Ryder T, Sparks C, Zhou J, Wang M, Russell R, Scott C (2017) Characterization of TAP Ambr 250 disposable bioreactors, as a reliable scale-down model for biologics process development. Biotechnol Prog 33:478–489. https://doi.org/10.1002/btpr.2417
59. Šrom O, Trávníková V, Wutz J, Kuschel M, Unsoeld A, Wucherpfennig T, Šoóš M (2022) Characterization of hydrodynamic stress in ambr250® bioreactor system and its impact on mammalian cell culture. Biochem Eng J 177:108240. https://doi.org/10.1016/j.bej.2021.108240
60. Velez-Suberbie ML, Morris SA, Kaur K, Hickey JM, Joshi SB, Volkin DB, Bracewell DG, Mukhopadhyay TK (2020) Holistic process development to mitigate proteolysis of a subunit rotavirus vaccine candidate produced in *Pichia pastoris* by means of an acid pH pulse during fed-batch fermentation. Biotechnol Prog 36. https://doi.org/10.1002/btpr.2966
61. Tai M, Ly A, Leung I, Nayar G (2015) Efficient high-throughput biological process characterization: definitive screening design with the Ambr250 bioreactor system. Biotechnol Prog 31:1388–1395. https://doi.org/10.1002/btpr.2142
62. Olughu W, Galbraith D, Paget C, Ruscoe S, Smith J, Mason A (2021) Does the BioBLU 0.3f single-use scale to the BioFlo® 320 reuseable bioreactor on a matched volumetric oxygen mass transfer coefficient? World J Microbiol Biotechnol 37:11. https://doi.org/10.1007/s11274-020-02968-2
63. Silva Couto P, Rotondi MC, Bersenev A, Hewitt CJ, Nienow AW, Verter F, Rafiq QA (2020) Expansion of human mesenchymal stem/stromal cells (hMSCs) in bioreactors using microcarriers: lessons learnt and what the future holds. Biotechnol Adv 45:107636. https://doi.org/10.1016/j.biotechadv.2020.107636
64. Dufey V, Tacheny A, Art M, Becken U, de Longueville F (2016) Expansion of human bone marrow-derived mesenchymal stem cells in BioBLU 0.3c single-use bioreactors. Appl Note 305:1–8
65. Jiang Y, Langenberg K, Borgdorff V, Dunska M, Post R, Bartulos O, Doornbos M, Braam S, Reijerkerk A (2019) Controlled, large-scale manufacturing of hipsc-derived cardiomyocytes in stirred-tank bioreactors, Eppendorf Application Note 409
66. Nienow AW, Rielly CD, Brosnan K, Bargh N, Lee K, Coopman K, Hewitt CJ (2013) The physical characterisation of a microscale parallel bioreactor platform with an industrial CHO cell line expressing an IgG4. Biochem Eng J 76:25–36. https://doi.org/10.1016/j.bej.2013.04.011
67. Rafiq QA, Hanga MP, Heathman TRJ, Coopman K, Nienow AW, Williams DJ, Hewitt CJ (2017) Process development of human multipotent stromal cell microcarrier culture using an automated high-throughput microbioreactor. Biotechnol Bioeng 114:2253–2266. https://doi.org/10.1002/bit.26359/abstract
68. Kaiser S, Löffelholz C, Werner S, Eibl D (2011) CFD for characterizing standard and single-use stirred cell culture bioreactors. In: Minin IV, Minin OV (eds) Computational fluid dynamics. InTech, pp 97–122

69. Löffelholz C, Kaiser SC, Kraume M, Eibl R, Eibl D (2013) Dynamic single-use bioreactors used in modern liter- and m3- scale biotechnological processes: engineering characteristics and scaling up. In: Advances in biochemical engineering/biotechnology, p 1–44
70. Zanghi A, Lin D, Balsara T, Young L, Wolf B, Huang P (2017) Increased efficiency and product quality with the UniVessel single use bioreactor for CHO fed-batch cultures, Sartorius Application Note https://www.sartorius.com.cn/download/96790/7/appl-univessel-su-cho-sbi4014-e-data.pdf
71. Garcia L, Turon M, Comellas M, Mouriño M, Urniza A (2015) Veterinary vaccine production using bioreactors: scale up study from laboratory and pd up to commercial production. BMC Proc 9:P76. https://doi.org/10.1186/1753-6561-9-S9-P76
72. Moreno F, Lip F, Rojas H (2022) Development of an insect cell-based adeno-associated virus packaging cell line employing advanced rep gene expression control system. Mol Ther Methods Clin Dev 27:391–403. https://doi.org/10.1016/j.omtm.2022.10.015
73. Kaiser SC, Eibl R, Eibl D (2011) Engineering characteristics of a single-use stirred bioreactor at bench-scale: the Mobius CellReady 3L bioreactor as a case study. Eng Life Sci 11:359–368. https://doi.org/10.1002/elsc.201000171
74. Micheletti M, Ducci A (2019) Engineering parameters in single-use bioreactors: flow, mixing, aeration, and suspension. In: Eibl R, Eibl D (eds) Single-use technology in biopharmaceutical manufacture, 2nd edn. Wiley, pp 259–269
75. Odeleye AOO, Marsh DTJ, Osborne MD, Lye GJ, Micheletti M (2014) On the fluid dynamics of a laboratory scale single-use stirred bioreactor. Chem Eng Sci 111:299–312. https://doi.org/10.1016/j.ces.2014.02.032
76. Limke T (2009) Comparability between the Mobius CellReady 3 L bioreactor and 3 L glass bioreactors. BioProcess Int 7:122–123
77. Cierpka K, Elseberg CL, Niss K, Kassem M, Salzig D, Czermak P (2013) hMSC production in disposable bioreactors with regards to GMP and PAT. Chemie Ingenieur Technik 85:67–75. https://doi.org/10.1002/cite.201200151
78. Grein TA, Leber J, Blumenstock M, Petry F, Weidner T, Salzig D, Czermak P (2016) Multiphase mixing characteristics in a microcarrier-based stirred tank bioreactor suitable for human mesenchymal stem cell expansion. Process Biochem 51:1109–1119. https://doi.org/10.1016/j.procbio.2016.05.010
79. Lawson T, Kehoe DE, Schnitzler AC, Rapiejko PJ, Der KA, Philbrick K, Punreddy S, Rigby S, Smith R, Feng Q, Murrell JR, Rook MS (2017) Process development for expansion of human mesenchymal stromal cells in a 50L single-use stirred tank bioreactor. Biochem Eng J 120:49–62. https://doi.org/10.1016/j.bej.2016.11.020
80. Löffelholz C, Husemann U, Greller G, Meusel W, Kauling J, Ay P, Kraume M, Eibl R, Eibl D (2013) Bioengineering parameters for single-use bioreactors: overview and evaluation of suitable methods. Chemie Ingenieur Technik 85:40–56. https://doi.org/10.1002/cite.201200125
81. Pall (2018) Allegro™ STR single-use stirred tank bioreactors. Product brochure USD3117f/GN22.07961. https://cdn.cytivalifesciences.com/api/public/content/brIzQmrKqUq6YUcn-vPIGA-pdf
82. Thermo Fisher Scientific (2021) 3,000 and 5,000L HyPerforma DynaDrive single-use bioreactors. Data sheet COL014369 0221
83. Thermo Fisher Scientific (2020) 500L Hyperforma DynaDrive single-use bioreactor. Data sheet COL013983 1220
84. Thermo Fisher Scientific (2021) 50L HyPerforma DynaDrive single-use bioreactor. Data sheet COL015641 0521
85. Cobia J, Madsen B (2021) Scale-up evaluation of the HyPerforma DynaDrive S.U.B.s. Thermo Fisher Scientific Application Note EXT4705 0323
86. Steinwedel T, Dahlmann K, Solle D, Scheper T, Readon KF, Lammers F (2019) Sensors for disposable bioreactor systems. In: Eibl R, Eibl D (eds) Single-use technology in biopharmaceutical manufacture, 2nd edn. Wiley, pp 69–82

87. Webster TA, Hadley BC, Hilliard W, Jaques C, Mason C (2018) Development of generic raman models for a GS-KOTM CHO platform process. Biotechnol Prog 34:730–737. https://doi.org/10.1002/btpr.2633
88. Webster TA, Hadley BC, Dickson M, Busa JK, Jaques C, Mason C (2021) Feedback control of two supplemental feeds during fed-batch culture on a platform process using inline Raman models for glucose and phenylalanine concentration. Bioprocess Biosyst Eng 44:127–140. https://doi.org/10.1007/s00449-020-02429-y
89. Jossen V, Eibl R, Eibl D (2019) Single-use bioreactors – an overview. In: Single-use technology in biopharmaceutical manufacture. Wiley, p 37–52
90. Zhan C, Hagrot E, Brandt L, Chotteau V (2019) Study of hydrodynamics in wave bioreactors by computational fluid dynamics reveals a resonance phenomenon. Chem Eng Sci 193:53–65. https://doi.org/10.1016/j.ces.2018.08.017
91. Öncül AA, Kalmbach A, Genzel Y, Reichl U, Thévenin D (2010) Characterization of flow conditions in 2 L and 20 L wave bioreactors® using computational fluid dynamics. Biotechnol Prog 26:101–110. https://doi.org/10.1002/btpr.312
92. Bartczak M, Wierzchowski K, Pilarek M (2022) Mixing performance in a litre-scale rocking disposable bioreactor: DoE-based investigation of mixing time dependence on operational parameters. Chem Eng J 431. https://doi.org/10.1016/j.cej.2021.133288
93. Werner S, Eibl R, Lettenbauer C, Röll M, Eibl D, de Jesus M, Zhang X, Stettler M, Tissot S, Bürki C, Broccard G, Kühner M, Tanner R, Baldi L, Hacker D, Wurm FM (2010) Innovative, non-stirred bioreactors in scales from milliliters up to 1000 liters for suspension cultures of cells using disposable bags and containers – a Swiss contribution. Chimia (Aarau) 64:819–823. https://doi.org/10.2533/chimia.2010.819
94. Pilarek M, Sobieszuk P, Wierzchowski K, Dąbkowska K (2018) Impact of operating parameters on values of a volumetric mass transfer coefficient in a single-use bioreactor with wave-induced agitation. Chem Eng Res Des 136:1–10. https://doi.org/10.1016/j.cherd.2018.04.012
95. Seidel S, Maschke RW, Kraume M, Eibl R, Eibl D (2022) CFD modelling of a wave-mixed bioreactor with complex geometry and two degrees of freedom motion. Front Chem Eng 4. https://doi.org/10.3389/fceng.2022.1021416
96. Meusel W, Löffelholz C, Husemann U, Dreher T, Greller G, Kauling J, Eibl D, Kleebank S, Bauer I, Glöckler R, Huber P, Kuhlmann W, John GT, Werner S, Kaiser SC, Pörtner R, Kraume M (2016) Recommendations for process engineering characterization of single-use bioreactors and mixing systems by using experimental methods. DECHEMA:1–60
97. Eibl R, Werner S, Eibl D (2009) Bag bioreactor based on wave-induced motion: characteristics and applications. In: Disposable bioreactors. Springer, Berlin/Heidelberg, pp 55–87
98. Kalmbach A, Bordás R, Öncül AA, Thévenin D, Genzel Y, Reichl U (2011) Experimental characterization of flow conditions in 2- and 20-l bioreactors with wave-induced motion. Biotechnol Prog 27:402–409. https://doi.org/10.1002/btpr.516
99. Kaiser S, Eibl D, Eibl R (2015) Single-use bioreactors for animal and human cells. In: Al-Rubeai M (ed) Animal cell culture. Springer, pp 445–499
100. Eibl R, Kaiser S, Lombriser R, Eibl D (2010) Disposable bioreactors: the current state-of-the-art and recommended applications in biotechnology. Appl Microbiol Biotechnol 86:41–49. https://doi.org/10.1007/s00253-009-2422-9
101. Junne S, Solymosi T, Oosterhuis N, Neubauer P (2013) Cultivation of cells and microorganisms in wave-mixed disposable bag bioreactors at different scales. Chem Ing Tech 85:57–66
102. Oosterhuis NMG, Neubauer P, Junne S (2013) Single use bioreactors for microbial cultivation. Pharm Bioprocess 1:167–177
103. Hitchcock T (2009) Production of recombinant whole-cell vaccines with disposable manufacturing systems. BioProcess Int 5:36–45
104. Dreher D, Husemann U, Zahnow C (2013) High cell density Escherichia coli cultivation in different single-use bioreactor systems. Chemie Ingenieur Technik 85:162–171

105. Westbrook A, Scharer J, Moo-Young M, Oosterhuis N, Perry Chou C (2014) Application of a two-dimensional disposable rocking bioreactor to bacterial cultivation for recombinant protein production. Biochem Eng J 88:154–161. https://doi.org/10.1016/j.bej.2014.04.011
106. Werner S, Kaiser SC, Kraume M, Eibl D (2014) Computational fluid dynamics as a modern tool for engineering characterization of bioreactors. Pharm Bioprocess 2:85–99. https://doi.org/10.4155/pbp.13.60
107. Timmins NE, Kiel M, Günther M, Heazlewood C, Doran MR, Brooke G, Atkinson K (2012) Closed system isolation and scalable expansion of human placental mesenchymal stem cells. Biotechnol Bioeng 109:1817–1826. https://doi.org/10.1002/bit.24425
108. Fritsche E, Volk H-D, Reinke P, Abou-El-Enein M (2020) Toward an optimized process for clinical manufacturing of CAR-Treg cell therapy. Trends Biotechnol 38:1099–1112. https://doi.org/10.1016/j.tibtech.2019.12.009
109. van der Loo JCM, Wright JF (2016) Progress and challenges in viral vector manufacturing. Hum Mol Genet 25:R42–R52. https://doi.org/10.1093/hmg/ddv451
110. Gallo-Ramírez LE, Nikolay A, Genzel Y, Reichl U (2015) Bioreactor concepts for cell culture-based viral vaccine production. Expert Rev Vaccines 14:1181–1195. https://doi.org/10.1586/14760584.2015.1067144
111. Pieralisi I, Rodriguez G, Micheletti M, Paglianti A, Ducci A (2016) Microcarriers' suspension and flow dynamics in orbitally shaken bioreactors. Chem Eng Res Des 108:198–209. https://doi.org/10.1016/j.cherd.2015.11.020
112. Klöckner W, Diederichs S, Büchs J (2014) Orbitally shaken single-use bioreactors. Adv Biochem Eng Biotechnol 138:45–60. https://doi.org/10.1007/10_2013_188
113. Anderlei T, Lailaw D, Richter A, Bürgin T, Bürki C, Wurm F Orbital shaken bioreactors: The scale up fast track (from mL to 2500L). Poster https://kuhner.com/wAssets/docs/download/Poster/Poster-Kuehner/Orbital-Shaken-Bioreactors_The-Scale-Up-Fast-Track.pdf
114. Laidlaw D, Anderlei T, Burgin T, Richter A, Adams T, Herrmann J, Bruki C, Wurm FM, Monteil D, Kuhner M Orbital mixing as an alternative to stirred-tank bioreactors for scale-up and commercial production of cell-based therapies. Poster https://kuhner.com/wAssets/docs/download/Poster/Poster-Kuehner/Orbital-Mixing-as-an-Alternative-to-Stirred-Tank-Bioreactors.PDF
115. Klöckner W, Raven N, Anderlei T, Schillberg S, Büchs J (2014) Characterization of orbitally shaken single-use bioreactors for plant cell cultivation. Poster https://kuhner.com/wAssets/docs/download/Poster/Poster-RWTH-Aachen/Characterization-of-orbitally-shaken-single-use-bioreactors.pdf
116. Dittler I, Badertscher B, Blaschczok K (2016) A novel approach to increase efficiency of insect cell bacolovirus vector system-based production processes. In: Eibl R, Eibl D (eds) BioTech
117. Anderlei T, Cesana C, Bürki C, de Jesus M, Kühner M, Wurm F, Lohser R (2009) Shaken bioreactors provide culture alternative. Gen 29
118. Nguyen HC, Langland AL, Amara JP, Dullen M, Kahn DS, Costanzo JA (2019) Improved HCP reduction using a new, all-synthetic depth filtration media within an antibody purification process. Biotechnol J 14:1700771. https://doi.org/10.1002/biot.201700771
119. Almeida A, Chau D, Coolidge T, El-Sabbahy H, Hager S, Jose K, Nakamura M, Voloshin A (2022) Chromatographic capture of cells to achieve single stage clarification in recombinant protein purification. Biotechnol Prog 38. https://doi.org/10.1002/btpr.3227
120. Koehler KC, Jokondo Z, Narayan J, Voloshin AM, Castro-Forero AA (2019) Enhancing protein a performance in mAb processing: a method to reduce and rapidly evaluate host cell DNA levels during primary clarification. Biotechnol Prog 35. https://doi.org/10.1002/btpr.2882
121. Singh N, Arunkumar A, Peck M, Voloshin AM, Moreno AM, Tan Z, Hester J, Borys MC, Li ZJ (2017) Development of adsorptive hybrid filters to enable two-step purification of biologics. MAbs 9:350–363. https://doi.org/10.1080/19420862.2016.1267091
122. Sartorius (2022) Ksep systems. Product Datasheet

123. Saballus M, Nisser L, Kampmann M, Greller G (2021) A novel clarification approach for intensified monoclonal antibody processes with 100 million cells/mL using a single-use fluidized bed centrifuge. Biochem Eng J 167. https://doi.org/10.1016/j.bej.2020.107887
124. Kuehberger M, Biewald A, Goehmann R (2022) GEA kytero: Efficiency leap in single-use biopharma clarification. Whitepaper GEA
125. Hundley R, Königsson S (2020) Development of a single-use, hermetic centrifuge system. BioProcess Int 18(10):26-34
126. Alfa Laval (2021) CultureOne. Product brochure 100005090-1-EN 2112. https://www.alfalaval.com/globalassets/documents/products/separation/biopharma/cultureone-range_sales-leaflet_en.pdf
127. Ott V, Müller J, Schirmer C, Weiss N, Göhmann R, Biewald A, Michel A, Eibl D, Eibl R (2022) Qualification of a single-use disk stack separator for cell separation in mammalian cell-based antibody production. Chemie Ingenieur Technik. https://doi.org/10.1002/cite.202200096
128. Manser B, Glenz M, Bisschops M (2019) Single-use downstream processing for biopharmaceuticals: current state and trends. In: Eibl R, Eibl D (eds) Single-use technology in biopharmaceutical manufacture, 2nd edn. Wiley, pp 117–126
129. Pall (2016) Cadence acoustic separator. Product brochure USD3133/ GN16.6491
130. Collins M, Levison P (2016) Development of a high-performance, integrated, and disposable clarification solution for continous bioprocessing. BioProcess Int. https://bioprocessintl.com/downstream-processing/separation-purification/high-performance-integrated-and-disposable-clarification-solution
131. Pierlot C, Vanhecke A, Thompson K, Gloeckler R, Kehl D (2019) Single-use technology for formulation and filling applications. In: Eibl R, Eibl D (eds) Single-use technology in biopharmaceutical manufacture, 2nd edn. Wiley, pp 285–294
132. Konstantinov KB, Cooney CL (2015) White paper on continuous bioprocessing may 20–21 2014 continuous manufacturing symposium. J Pharm Sci 104:813–820. https://doi.org/10.1002/jps.24268
133. Gupta SK (2017) Upstream continuous process development. In: Subramanian G (ed) Cotinuous biomanufacturing – innovative technologies and methods. Wiley-VCH Verlag GmbH & Co. KGaA, Weinheim, pp 201–232
134. Broly H, Costioli MD, Guillemont-Potelle C, Mitchell-Logean C (2010) Cost of goods modeling and quality by design for developing cost-effective processes. Biopharm Int 23:26–35
135. Langer ES, Rader RA (2014) Single-use technologies in biopharmaceutical manufacturing: a 10-year review of trends and the future. Eng Life Sci 14:238–243. https://doi.org/10.1002/elsc.201300090
136. Whitford WG (2013) Single-use technology supporting the comeback of continuous bioprocessing. Pharm Bioprocess 1:249–253. https://doi.org/10.4155/pbp.13.28
137. Clincke M, Mölleryd C, Zhang Y, Lindskog E, Walsh K, Chotteau V (2013) Very high density of CHO cells in perfusion by ATF or TFF in WAVE bioreactor™. Part I. Effect of the cell density on the process. Biotechnol Prog 29:754–767. https://doi.org/10.1002/btpr.1704
138. Clincke M-F, Mölleryd C, Zhang Y, Lindskog E, Walsh K, Chotteau V (2011) Study of a recombinant CHO cell line producing a monoclonal antibody by ATF or TFF external filter perfusion in a WAVE bioreactor™. BMC Proc 5. https://doi.org/10.1186/1753-6561-5-s8-p105
139. Müller J, Teale M, Steiner S, Junne S, Neubauer P, Eibl D, Eibl R (2021) Intensified and continuous mab production with single-use systems. In: Appreciation to Professor Mohamed Al-Rubeai. Springer International Publishin, Cham, p 401–429
140. Chotteau V (2015) Perfusion processes. In: Animal cell culture. Springer, pp 407–443
141. Bisschops M, Schofield M, Grace J (2017) Two mutually enabling trends: continuous bioprocessing and single-use technologies. In: Subramanian G (ed) Continuous biomanufacturing – innovative technologies and methods. Wiley-VCH Verlag GmbH & Co. KGaA, Weinheim, pp 149–170

142. Zarrineh M, Mashhadi IS, Farhadpour M, Ghassempour A (2020) Mechanism of antibodies purification by protein A. Anal Biochem 609:113909. https://doi.org/10.1016/j.ab.2020.113909
143. Lain B, Cacciuttolo M, Zarbis-Papastoitsis G (2009) Development of ahigh-capacity MAb capture step based on cation-exchange chromatography. BioProcess Int 7:26–34
144. Martinez M, Spitali M, Norrant EL, Bracewell DG (2019) Precipitation as an enabling technology for the intensification of biopharmaceutical manufacture. Trends Biotechnol 37: 237–241. https://doi.org/10.1016/j.tibtech.2018.09.001
145. Kateja N, Agarwal H, Saraswat A, Bhat M, Rathore AS (2016) Continuous precipitation of process related impurities from clarified cell culture supernatant using a novel coiled flow inversion reactor (CFIR). Biotechnol J 11:1320–1331. https://doi.org/10.1002/biot.201600271
146. Hammerschmidt N, Tscheliessnig A, Sommer R, Helk B, Jungbauer A (2014) Economics of recombinant antibody production processes at various scales: industry-standard compared to continuous precipitation. Biotechnol J 9:766–775. https://doi.org/10.1002/biot.201300480
147. Smejkal B, Agrawal NJ, Helk B, Schulz H, Giffard M, Mechelke M, Ortner F, Heckmeier P, Trout BL, Hekmat D (2013) Fast and scalable purification of a therapeutic full-length antibody based on process crystallization. Biotechnol Bioeng 110:2452–2461. https://doi.org/10.1002/bit.24908
148. Kruse T, Kampmann M, Rüddel I, Greller G (2020) An alternative downstream process based on aqueous two-phase extraction for the purification of monoclonal antibodies. Biochem Eng J 161:107703. https://doi.org/10.1016/j.bej.2020.107703
149. Lim JAC, Sinclair A, Kim DS, Gottschalk U (2007) Economic benefits of single-use membrane chromatography in polishing – a cost of goods model. BioProcess Int 5:60–64
150. Casey C, Rogler K, Gjoka X, Gantier R, Ayturk E (2016) Cadence™ single-pass TFF coupled with chromatography steps enables continuous bioprocessing while reducing processing times and volumes. Am Pharm Rev 19(6)
151. Research and Markets (2022) Single-use bioprocessing market size, share & trends analysis report by product (work equipment, apparatus & plants), by workflow (upstream, downstream), by end-use, by region, and segment forecasts, 2022–2030, ID:4582080

Chapter 2
Recent Developments in Bioprocess Monitoring Systems

Vida Rahmatnejad, Yunqian Wei, and Govind Rao

Abstract Bioprocess monitoring systems are essential tools for achieving an optimal manufacturing process for pharmaceuticals. In recent years, there have been advancements in the development of these systems. One of the most notable developments in bioprocess monitoring systems is the use of biosensors, which can detect and measure biological molecules such as enzymes and proteins. The use of imaging technologies, such as microscopy and flow cytometry, has allowed for the non-invasive monitoring of cell culture, providing valuable information for process optimization. Other non-invasive monitoring systems are being developed with the aim of reducing the risk of contamination. Another important advancement is the use of machine learning and artificial intelligence in bioprocess monitoring systems. These technologies have the ability to learn from historical data and predict future outcomes, which can be beneficial for optimizing bioprocesses in real-time. Overall, recent advancements in sensor technology can help in improving the efficiency and effectiveness of bioprocessing. The integration of advanced sensing technologies, and machine learning can provide valuable insights and improve the ability to monitor and control bioprocesses online. In this chapter, we will discuss recent advancements in bioprocess monitoring systems, specifically for monoclonal antibody and cell and gene therapy manufacturing processes.

Keywords Monoclonal antibody (mAb) · Process analytical technology (PAT) · Single-use bioreactors · Cell and gene therapy · Noninvasive monitoring · Modeling-enhanced bioprocessing

---

V. Rahmatnejad · Y. Wei · G. Rao (✉)
Center for Advanced Sensor Technology and Department of Chemical, Biochemical and Environmental Engineering, University of Maryland Baltimore County, Baltimore, MD, USA
e-mail: vrahmat1@umbc.edu; yunqian1@umbc.edu; grao@umbc.edu

R. Pörtner (ed.), *Biopharmaceutical Manufacturing*, Cell Engineering 11,
https://doi.org/10.1007/978-3-031-45669-5_2

Abbreviations

2D-LC	Two-dimensional liquid chromatography
ACO	Ant Colony Optimization)
CEX	Cation Exchange Chromatography
cGE	Capillary Gel Electrophoresis
cGMP	Current Good Manufacturing Practices
CHO	Chinese Hamster Ovary
cIEF	Capillary Isoelectric Focusing
CoAdReS	Competitive Adaptive Reweighted Sampling
CPPs	Critical Process Parameters
CQA	Critical Quality Attributes
CSPR	Cell Specific Perfusion Rate
DCO_2	Dissolved Carbon Dioxide
DHI	Digital Holography Imaging
DNA	Deoxyribonucleic Acid
DO	Dissolved Oxygen
DO	Dissolved Oxygen
DSP	Downstream Processing
ELISA	Enzyme-Linked Immunosorbent Assay
FDA	Food and Drug Administration
FETs	Field Effect Transistors
FIA	Flow Injection Analysis
FT-MIR	Fourier Transform Mid Infrared Spectroscopy
GFP	Green Fluorescent Protein
GLC	Glucose Level
HCP	Host Cell Protein
hMSCs	Human Mesenchymal Stem Cells
HMW	High Molecular Weight
HPLC	High Performance Liquid Chromatography
HP-SEC	High Performance Size-Exclusion Chromatography
iPSCs	Induced Pluripotent Stem Cells
LC-MS	Liquid chromatography-mass spectrometry
LFI	Lens-free Imaging
mAb	Monoclonal Antibody
MIR	Mid Infrared Spectroscopy
ML	Machine learning
MLS	Multi-Angle Light Scattering
MPC	Model Predictive Control
MVDA	Multiple Variable Data Analysis
NIR	Near Infrared Spectroscopy
OUR	Oxygen Uptake Rate
PAT	Process Analytical Technology
PCA	primary component analysis

PLS	Partial least squares
PSM	Process sample manager
PTM	Posttranslational modifications
RFID	Radio Frequency Identification
RuOx	Ruthenium Oxide
SUB	Single Use Bioreactor
TPE	Two Protein Excitation
UPLC	Ultra-Performance Liquid Chromatography
USP	Upstream Processing
UV	Ultraviolet
UV-Vis	Ultraviolet–visible spectroscopy
VCD	Viable Cell Density
YFP	Yellow Fluorescent Protein

Introduction

The manufacturing of biopharmaceuticals has grown substantially in recent decades. Unlike small molecules, biopharmaceuticals offer more specific target interactions, even personalized options, and thus require increasingly complex manufacturing processes. To ensure the quality and efficacy of the product, analytical measurements are applied throughout the process for both process parameters and Critical Quality Attributes (CQAs). The Food and Drug Administration (FDA) has introduced the concept of Process Analytical Technology (PAT) to encourage the industry to incorporate product quality into the process design through on-line monitoring, which forms the foundation for Quality by Design (QbD) [1–3].

The purpose of PAT is to improve the reproducibility of the manufacturing process and ensure product quality by monitoring, analyzing, and controlling the process. PATs are utilized at various stages, such as early discovery and development to gain a better understanding of the process and enable better control at manufacturing stage. In biomanufacturing, the ideal PAT scenario is an end-to-end application that enables real-time product release, starting from raw materials, vial thaw, seed training propagation, purification to packaged and labeled drug products [4]. This review will focus on emerging PATs in manufacturing process of monoclonal antibodies (mAbs), and cell and gene therapy products.

Due to the increase in the use of mAbs, there is a growing interest in developing a reliable and scalable manufacturing process for these products. To achieve such robust process and guarantee product quality, there is a growing trend towards the application of PATs in the manufacturing process. Figure 2.1 provides an overview of the PATs utilized in the production of mAbs. Monitoring technologies are essential PAT tools as they are crucial in collecting data from the process. This information could be used for developing models and algorithms. Furthermore, sensors are helpful in designing effective methods for controlling operational

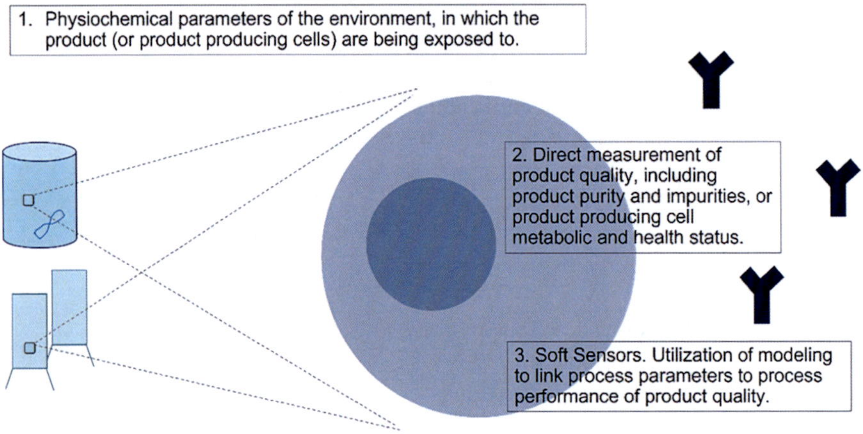

Fig. 2.1 Overview of PATs in monoclonal antibody production processes. For upstream, PATs that measure the physicochemical conditions of the cells include dissolved oxygen, pH, dissolved CO_2, osmolality, glucose, glutamine, glutamate, and lactate which does not provide direct information on product quality; PATs that measure biomass, cell status and product quality include viable cell density, viability, metabolic status, morphology, MAb titer, aggregate level, and glycosylation. Soft sensors are developed to link sensor reading on process parameters to product quality and cell status. For downstream, PATs that measure the physiochemical environment of the product to ensure its stability include pH and conductivity; PATs that measure directly product quality include protein concentration, aggregate and fragment level, particle size, charge variant, deaminization isoforms, DNA, HCP, peptide sequence and higher protein structure. Similarly, soft sensors are developed to link process parameters to product quality through modeling

variables. This contributes to providing pre-validation data and shortening the release time [5, 6].

Developing a well-defined manufacturing process is not limited to the manufacturing of mAbs. It is a crucial concept in the growing field of cell and gene therapy as well. This is specifically important as compared to mammalian cells, the culture of stem cells is associated with more challenges, and manufacturing process for cell and gene therapy products has a direct impact on the quality of the product. Enabling sufficient expansion of cells while maintaining the quality and pluripotency of these cells is of high importance for delivering safe and effective products. The manufacturing process is becoming automated by utilizing different bioreactors. Despite the available technologies, the processes are not precisely defined due to the lack of appropriate monitoring systems. Specifically, to optimize a manufacturing process for induced pluripotent stem cells (iPSCs) in large scale bioreactors (larger than 5 L) and control the differentiation of stem cells, novel monitoring techniques are required [7]. Sensors are beneficial for both autologous (cells that patient receives are derived from themselves) and allogeneic (cells that patient receives are derived from another person) therapies. In autologous therapies where quality control must be conducted for individual patients, application of sensors is more critical [8, 9]. The FDA's PAT initiative has resulted in numerous efforts to explore novel sensing and analytical tools [10]. However, industry has been slow to adopt them into routine operations.

The range of sensors and analytical methods used as PATs is extensive and can be categorized based on their relevance to product quality. The following are commonly used to categorize PATs:

1. The most widely used parameters for measuring and monitoring the physiochemical conditions of the unit operations are pH, temperature, dissolved oxygen (DO), dissolved carbon dioxide (DCO_2) and glucose, which are critical for bioreactors [11]. For chromatography and filtration steps, the most commonly used parameters are pH and conductivity [12]. These analytical tools provide information about the environment of the product or cells, but not product quality itself. Sensors and controls are in place to ensure that the cells or products are maintained in the desired physiochemical conditions for productivity or stability.
2. Product quality indicating PATs include technologies for measurement of target product concentration, purity and impurities, cell culture metabolites, and cell imaging [13–15]. The measurements are directly related to intermediate or final product quality.
3. "Soft sensors", also known as "software sensors", are another category. Machine learning and/or mechanistic modeling have been utilized to link process parameters (direct measurements) to the cell metabolic state or CQA [16, 17].

Figure 2.2 illustrates the PATs in the above-mentioned categories.

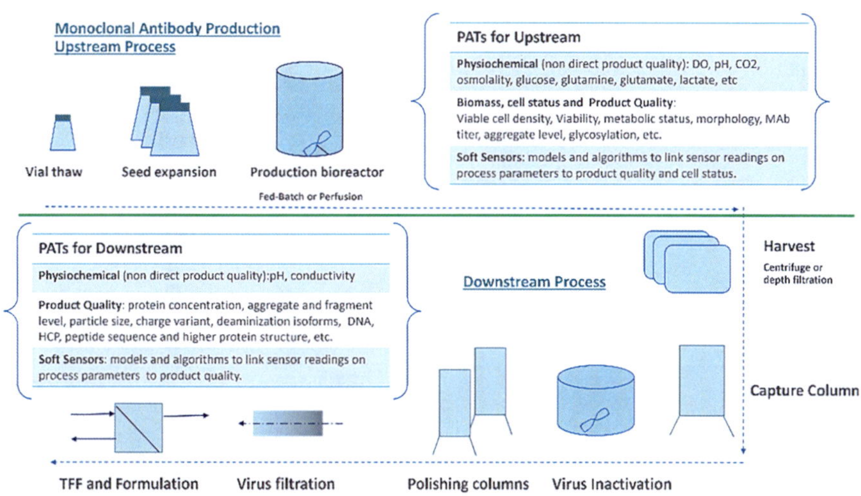

Fig. 2.2 Illustration of different categories of PATs including monitoring systems in unit operations, product quality indicating technologies, and soft sensors

Terminology Regarding Bioprocess Monitoring Systems

Given the significance of sensors, various types of monitoring systems have been developed. These systems are optimized for ease of use and cost effectiveness while being capable of detecting different parameters. Each of these techniques are used in a specific mode. To help frame this chapter, we start by discussing different modes for monitoring in the following paragraphs:

On-line or in-line: monitoring methods provide continuous and up-to-date information from bioprocesses. This information could be real-time (online method) or near real-time (in-line method). These methods are useful for continuous control in bioreactors.

Off-line or at-line: methods do not provide real-time information from the state of the process. In fact, this method demands a high degree of labor and physical separation of the sample and the process. As a result, this method is not ideal for control purposes as the data is obtained discontinuously.

Invasive: monitoring systems have probes that come into direct contact with cells or cell culture medium. In this method, sensors are located inside the bioreactor and are referred to internal sensors. A picture of internal sensors is shown in Fig. 2.3a. These sensors have drawbacks such as increasing the risk of contamination, and the adverse effects on the medium [1]. Biofouling is another issue with this type of monitoring systems, and they are coated by biocompatible materials.

Noninvasive: methods obtain information from cell culture process with no direct contact with the cell culture environment. In this method, sensors are located outside the bioreactor and are referred to as external sensors. A picture of external sensors is shown in Fig. 2.3b. An important new application of noninvasive methods is the introduction of a diffusible barrier between the sensor and the bioreactor contents,

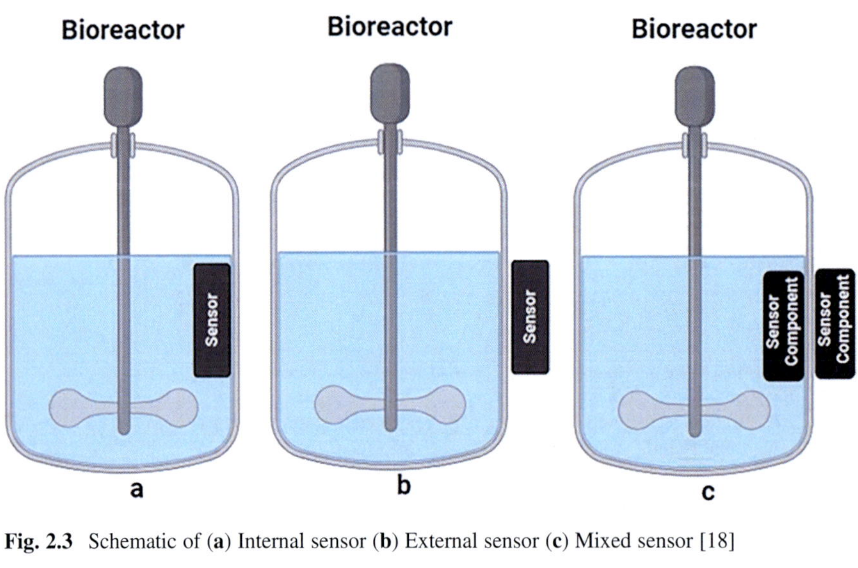

Fig. 2.3 Schematic of (**a**) Internal sensor (**b**) External sensor (**c**) Mixed sensor [18]

which eliminated the risk of contamination and minimizes the introduction of leachables and extractables into the medium. Furthermore, these sensors can be replaced or recalibrated without disrupting the process.

Indirect: monitoring methods where automated or manual sampling techniques are used to draw samples from the culture and analyze it outside of the bioreactor. For this purpose, mixed sensors are used. These sensors have components inside and outside of the bioreactor. A picture of a mixed sensor is shown in Fig. 2.3c [8, 19].

Destructive: methods necessitate the utilization of cells for analysis. In these techniques, cells cannot be used after analysis. As a result, these kinds of analysis are not appropriate for autologous therapies where a limited number of cells are available [8].

Non-Destructive: methods do not involve using cells for analysis. In fact, some of these methods enable returning sampled cells to the bioreactor after analysis is completed [8].

As previously mentioned, implementing a reliable control system in bioprocesses will result in higher process reproducibility and product quality. Among different methods, on-line and in-line monitoring techniques are preferred methods to develop a robust control system [8].

With a deeper understanding of different types of monitoring techniques in cell culture, we will now discuss the technologies recently developed for monitoring cell therapies as well as the development and manufacturing processes of mAbs.

Recently Developed Monitoring Tools

Different types of in-line sensors are used in development stage to gather process understanding as much as possible and detect the source of variabilities. Based on the operating mechanism of sensors they are defined in different categories as below:

Electrochemical Sensors

Existing bioprocesses mainly utilize electrochemical sensors for online monitoring of pH and DO. In this technique, the interaction between the sensing surface of the electrode and analytes is detected and measured, and the data is then converted to an electric signal [20]. Electrochemical sensors are known for being reliable and have higher accuracy at high oxygen concentrations. However, they have disadvantage of being bulky, and their monitoring method is completely invasive. In addition, the sensor requires recalibration after each sterilization cycle [21, 22].

Optical Sensors

Optical sensors operate by measuring changes in different properties related to light such as absorbance, fluorescence signal emission, and phase angle via an indicator dye. The measurements can be converted to the concentration of the analyte. This type of sensor is typically used for monitoring DO, DCO_2, and pH. Optical sensors require little to no contact with the cell culture medium, reducing the risk of contamination. Furthermore, they are the best candidate for processes with low volumes where large-sized sensors cannot be used. However, low long-term stability, and toxicity of the dye are concerns regarding using optical sensors [10, 23].

Spectroscopic Sensors

This type of monitoring system mainly works based on the effect of electromagnetic waves on the bonds in the molecules. In this method, the spectral range for bioprocesses varies between ultraviolet (UV) and middle infrared (MIR). Among different types of infrared spectroscopy, MIR and near-infrared (NIR) are more sensitive for bioprocessing applications. Raman spectroscopy gained the most polularity because it is cheaper, more stable and has minimal fluorescence interferences. In addition, Raman Spectroscopy can be used as an online, in-line, at-line, or offline monitoring system due to its flexible sampling method.

Spectroscopic sensors are mainly used for the measurement of biomass in upstream processing (USP) and downstream processing (DSP). However, dissolved compounds in the cell culture medium such as glucose, lactate and amino acids in the cell culture environment can be monitored using this type of sensor [7, 8, 24]. Another advantage of spectroscopic sensors is that measurements are conducted in a non-invasive way as sampling is not required. This minimizes the risk of contamination. Conducting measurements with no need for sampling makes Raman spectroscopy an advantageous method for monitoring parallel processes and reducing costs. The main drawback of Raman spectroscopy is having infrequent feedback signals. As a result, multiple efforts such as constructing linear autoregression with an exogenous input model are made to improve this method [8, 21, 22].

Chromatography steps largely employ UV absorbance (A_{280} nm) for collection and pooling criteria, which assumes that protein concentration is linearly correlated to absorbance. This method works well for collections designed from baseline, but it has limited capability to distinguish between target protein and its isoforms. If a chromatography step is designed to separate target protein and its isoforms, UV absorbance-based collection is not as ideal and could be combined with other forms of analytical testing [25]. A control algorithm was implemented to terminate column loading upon breakthrough detection as low as 0.5 mg/mL. Hakemeyer et al. reported the at line use of NIR of CHO culture supernatant with spectra being collected daily, and multivariate process fingerprints were computed. Data set span from 2.5 L bioreactor to 1000 L bioreactor and was divided into calibration set and validation set. Multiple process monitoring attributes, including product titer, glucose, lactate, glutamate, glutamine, ammonium, cell density and viability were reported from the primary component analysis (PCA). The accuracy was 15% or higher compared with the reference methods. In addition, process performance evaluation can be developed by comparing process signatures with the historical database. Multiparameter monitoring is possible for spectroscopic sensors, which makes them promising as PAT. Several authors reported the inline application of Raman Spectroscopy to monitor multiple parameters and attributes, including nutrients, metabolites, and biomass (glucose, glutamine, glutamate, lactate, ammonium and VCD) [26–28]. Partial least squares (PLS) regression was used to analyze the spectrum to calibrate and train the model. Combined with its noninvasive nature, Raman spectroscopy offers great potential for gaining process knowledge and establishing process control.

Image-Based Monitoring Systems

Image-based monitoring of various parameters in bioprocesses is a highly effective tool for visually analyzing cells. This has resulted in the development of sensors for in-situ imaging. Differential interference contrast imaging is a real-time imaging technique that measures the confluence of cells on micro-carriers, density and distribution of cells. This method provides on-line imaging from the culture and can provide information about cell size, aggregation level, and cell density. However, the invasive nature of the sampling method and having a small field of view are major disadvantages of this technique. Additionally, the results in this method depend on manual analysis making it time-consuming and prone to errors method. Digital holographic imaging is another example of image-based method that is used for measuring cell number, area, thickness and volume. This is a real-time technique with a large field of view. The main advantage of this method is being noninvasive, and the major drawback of this method is being only applicable to planar cultures [8].

Now that we are familiar with different types of monitoring systems, it is worth mentioning that in general NIR Spectroscopy, MIR Spectroscopy and Raman spectroscopy have been utilized for various parameter monitoring in upstream [29]. Upstream process refers to cell expansion process from vial thaw, seed expansion to production bioreactor, where the cells will expand to desired number and produce target protein [30]. Chinese Hamster Ovary (CHO) cell lines have been used as a workhorse for mAb production due to its mammalian nature and the benefit of post translation modifications [31–33].

In line PATs currently utilized in upstream process are limited to pH, DO and temperature. Both electrochemical and optical sensors are widely used for pH and DO detection. For media components and metabolites monitoring such as glucose and lactate, off lines samples are taken and monitored based on high-performance liquid chromatography (HPLC) or enzymatic- amperometric measurements. Viable cell density and product titer are being monitored through sampling and offline testing as well. As mentioned before, sampling could cause potential contamination and require additional labor. Extensive effort must be made to ensure the representativeness of the samples. Another disadvantage is the extended analysis time, which makes real time processing decisions impossible. For example, offline measurement-based glucose feeding strategy could introduce high variability from batch to batch, due to the timing of sampling and analysis period.

Downstream process normally refers to the process of recovering and purifying the target product from process and product related impurities and buffer exchanging into stability enabling formulation buffers. In industry setting, downstream process consists of chromatography and filtration (dead end and tangential flow) unit operations. Starting from downstream, the application of PATs focusses on measuring protein concentration, product quality and impurities. For chromatography steps, regular PATs are limited to pH, conductivity, and UV trace. While for filtration steps, only physical parameters like pressure and flow rate are being monitored. Shifting to in-line or at-line application has been observed since it enables (near) real time data analysis and process control [29].

Here we will discuss the recent developments in monitoring systems for measurement of crucial parameters in bioprocesses:

Glucose

Glucose is the most important source of energy for cells. Too high or too low glucose level in the medium can result in unusual cellular metabolism. For example, studies show that glucose level affects the mineral deposition of mesenchymal stem cells (MSCs) during osteogenesis. As a result, glucose sensors can help to provide more information about glucose level and improve our understanding of the culture condition. Monitoring systems for this parameter can function based on (a) detecting optical properties of enzymes, and other related compounds (b) monitoring the products of glucose oxidase (c) application of boronic acid (d) using glucose binding proteins.

NIR and MIR spectroscopy are available methods for monitoring glucose. Despite the availability of these methods, continuous on-line monitoring of this parameter via spectroscopy is limited, and electrochemical sensors are the only common method used for glucose monitoring [8]. NIR spectroscopy is specifically popular because of being a non-destructive and fast method.

A recent study shows the application of NIR spectroscopy system for monitoring glucose during scale-up of the mAb production process. This system has been developed by connecting a spectrometer to a BRUKER Matrix-F FT-NIR system using a 10 m long fiber optic cable. The specifications of each spectrum are 256 scans with the resolution of 8. During the cell culture processes in a shake flask, the probe was inserted in the cell culture medium. OPUS 6.5 was used to control the system. For bioreactor applications, the probe was coupled with the system with a fiber-optic cable via the vessel port. A predictive model was created using NIR-based method. The models were more accurate than other methods in batch cultures as a higher error was observed in fed-batch mode [34].

A novel study shows the development of an optical sensor which works based on an available oxygen sensor coated with glucose oxidase containing a layer of an enzyme. After the enzymatic conversion of glucose and oxygen, the oxygen diffuses through a membrane and enters the enzyme layer. This sensor consists of an optical oxygen transducer that measures the oxygen consumption inside the enzyme layer. The monitored oxygen partial pressure is inversely correlated to the glucose concentration in the cell culture. This sensor is disposable and suitable for suspension cell culture processes [35].

Raman Spectroscopy has been investigated by many to be used for online measurement of glucose and feedback control to maintain its level for (fed) batch processes [36, 37]. Most recently Graf et al. reported the inclusion of a Raman spectroscopy flow cell on the permeate side of the perfusion bioreactor and validated the glucose concentration readings against offline reference data. A control loop was designed based off Raman spectroscopy glucose readings [29].

Another method for glucose measurement works based on glucose-binding protein (GBP). This engineered protein has a single mutation that allows it to adopt a fluorophore sensitive to the environment. The protein responds to its ligand in micromolar range and can be stored at 4 °C for about 5 months. In this technique, to measure glucose concentration, GBPs were immobilized on Ni-NTA microbeads confined to the tip of an optical fiber which was connected to fluorimeter. The glucose sensing method was tested in yeast and *Ecoli* fermentations as well as mammalian culture. Throughout the cell culture process, a dialysis tubing was submerged in the cell culture medium, and glucose diffusing through this sampler was transferred to the microbeads. Glucose concentration was then measured based on the fluorescence intensity of the labeled protein. The measurements obtained from this technique were compared with measurements from YSI 2700 Chemistry Analyzer. The results show that the method was successful in measuring glucose at different concentrations. Another benefit of using this platform is that a sample volume as small as 1 μL is sufficient for glucose measurement making it suitable for small-scale cultures. Additionally, this technique shows high sensitivity to glucose which makes it a great choice for high throughput bioprocessing purposes. The only requirement in this method is the dilution of the sample. The dilution step helps in reducing the autofluorescence interference which is a common problem in other optical sensors. The technique was also proven to be successful in transdermal glucose measurement. A similar approach with a sampler appropriate to the cell culture vessel could be used for noninvasive monitoring of glucose throughout the culture process [38–41].

Lactate

Lactate is the final product of the anaerobic pathway in cell culture processes, and its concentration can increase from almost 0 to a few mM in a typical cell culture process. Lactate can inhibit cell growth and productivity in many cell types. Therefore, it is important to monitor this parameter throughout the culture. LED-based 2D-fluorescence spectroscopy method is a newly developed for online monitoring of lactate during the CHO cultivation process. The device is made of sensor and a probe. The sensor consists of an excitation and a detection part. For online measurement of lactate during the process, the probe is inserted in the bioreactor and the sample is illuminated by fibers inside the probe. Longevity and low maintenance of the device makes it an ideal method for industry purposes, and adaptable to different applications [42].

In a technology recently reported, a nanohybrid-functionalized Pt microelectrode has been designed for in-vitro detection of lactate. This technique works based on an enzyme-free electrochemical tool with a microneedle electrode. The microneedle was modified with Au nanoparticles, and polydopamine nanospheres. The results of this study show high sensitivity and selectivity in detection of lactate in cell culture medium for different cell lines and in a wide concentration range [43].

A new multiparametric sensing device is reportedly capable of real-time monitoring of different cell culture parameters including lactate. In this platform, an electrochemical sensing part is embedded in the flow-cell. The platform is electrically connected using Molex connectors. The microfluidic section is also connected to the flow cell via luer lock connectors. To monitor lactate, an amperometric enzymatic sensor is used, and the measurement happens from two reactions: first lactate is catalyzed, and hydrogen peroxide is produced. The peroxide molecules then diffuse through the enzymatic biofilm on the electrode and are further reduced. The reductive current is then used for lactate monitoring [44].

In another newly developed method, a composite of reduced graphene oxide, carbon nanotubes, and gold nanoparticles has been used to modify the platinum electrode and detection of lactate in normoxic and hypoxic bioprocesses. In this method, lactate oxidase is immobilized on the electrode which catalyzes the lactate to hydrogen peroxide. Hydrogen peroxide plays an important role in production of electrons that generate an amperometric signal. The proposed sensor has high sensitivity, stability, reproducibility, and repeatability [45].

pH

The pH of the cell culture medium can directly affect the intracellular enzymatic activity and cellular metabolism as a result. This is specifically important when culturing stem cells as some studies show that slight changes in pH of the cell culture medium in human mesenchymal stem cells (hMSCs) can inhibit osteogenic differentiation. Therefore, it is of high importance to maintain this parameter in an optimal range which is 6.8–7.4 to enhance the cell viability and productivity [46, 47].

pH is normally monitored by using pH electrodes in both USP and DSP. Therefore, electrochemical sensors are common tools for measuring pH in bioprocesses. Optical sensors are another type of pH sensors. These sensors work based on a dye immobilized on an optical fiber. This dye goes through changes in molecular structure by H+ ions, and the changes are reflected in the spectroscopic phenomena such as absorption or reflection. Wavepod II-pHOPT from GE Healthcare, iTube pH Bioreactor from PreSens, TurFluor pH from Fitnesse, and OptiSens pH from Sartorius are commercially available optical sensors. Most of these sensors require direct contact with the cell culture environment.

A noncontact optical method for pH measurement has been developed recently. The system consists of a sensor cassette, a pump, a support structure, and a flow through cuvette, where the sample is transferred to an LED cassette. The transmitted light from the LED module through the sample is measured with a sensor module. This method has a higher sensitivity, accuracy, and responsivity. The measurements are conducted more rapidly due to fast chemical equilibration of the indicator. The most important advantage of the setup is that it decreases the risk of contamination [18].

MIR and NIR are other methods used for monitoring pH in the cell culture medium [8]. A novel method is developed to measure pH via an on-chip ruthenium oxide (RuOx) electrode. This sensor operates potentiometrically, and consists of glass chip, Teflon chip holder, pogo pins connector, and Transwell insert. For monitoring pH in cell culture using this method, the cells are cultured on top of a membrane. The membrane is placed 3 mm above the glass chip that is suspended using a Teflon holder. In this sensor, nafion 117 solution has been used to coat the electrode in order to avoid biofouling. The findings show that RuOx electrodes are reliable tools for pH measurements in long cell culture processes and complex cell culture medium. In addition, the sensitivity of the sensor is not affected over time [48].

Another reported technology is sensing cell culture flask (SCCF) sensor chip for cell culture processes. In this method, a chip is embedded in the bottom of the flask and consists of amperometry and working electrode (amperometric and potentiometric electrodes), reference electrode, and indicator electrode. The potentiometric sensors successfully monitored the cellular acidification in breast cancer (T-47D) cells. The disadvantage of this method is that the technique is limited to 2D cell culture processes; however, it still provides valuable information about the details of the diffusion in the cell culture system [20].

Oxygen

Oxygen is an important factor in cell metabolism. Studies show that low levels of oxygen can enhance the differentiation of stem cells. In addition, the findings from 3D mesenchymal stem cell cultures show that a change in the oxygen level can noticeably affect the result of the culture. In fact, in these types of cultures, an appropriate level of oxygen can improve the quality of the culture. Considering the importance of oxygen, it is critical to carefully monitor and maintain this parameter at a level appropriate to the culture condition. To monitor oxygen in gas phase, different technologies such as semiconducting, electrochemical, and paramagnetic sensors are commonly used. For measuring oxygen in liquid phase, electrochemical and optical sensors are available. Electrochemical sensors, specifically Clark-type electrodes, are most commonly used method for measuring DO. However, these sensors are bulkier than optical sensors [8, 9, 49, 50].

Multiple technologies have been recently developed for oxygen measurement. One of these techniques involves the use of a single on-chip sensor that simultaneously monitors oxygen and pH during the cell culture process. This sensor conducts the measurements through a ruthenium oxide (RuOx) electrode. The basis of the sensor was described in previous section on "the novel method for pH measurement via a chip". The RuOx electrode can be used for cell metabolism studies. Throughout the culture, DO is reduced on the electrode surface, and the current resulting from that is linearly proportional to the oxygen concentration in the medium. Reducing the size of the electrode provides better time resolution for sampling. However, at current sensor size, it is not possible to conduct oxygen measurements at rates higher than once an hour [48]. SCCF technology, described earlier as a technology for pH measurement, is capable of oxygen monitoring in the cell culture medium. In this method, amperometric oxygen sensors are used to monitor respiration [20]. Another novel study reports developing a wireless DO sensor by coupling a Clark-type DO sensor with a BLE chipset. The DO sensor has a three-electrode configuration which quantifies the oxygen partial pressure. The system consists of an electronic module, electrochemical DO sensor, and 3D-printed enclosure. A microcontroller is used for performing periodic DO measurements. The data is then transferred wirelessly. In addition, a reservoir has been created by connecting a fluorinated ethylene propylene membrane to the surface of the sensor The system has been examined in a bench-scale bioreactor, and data is obtained using Bluetooth low-energy communication between the setup and a mobile phone [51].

Carbon Dioxide

As a byproduct of cell respiration, carbon dioxide can pass through the cell membrane and affect the intracellular pH, thereby impacting the metabolism and well-being of the cell. To monitor the CO_2 level in the cell culture environment and maintain it at an acceptable level, different techniques have been developed. Electrochemical, and optical sensors are commonly used for measuring CO_2 in bioprocesses. In addition, off-line analyzers provide a means for monitoring DCO_2. This technique monitors the level of this gas without direct contact with the cell culture medium. In comparison with CO_2 probes, off-gas sensors are inexpensive, remain stable in long term processes, and have no direct contact with the cell culture environment. However, this method can only provide an average value for DCO_2 [52]. commercially available off-gas sensors from companies such as YSI or Metler Toledo are used to monitor DCO_2 in bioprocesses.

Among the newly developed techniques for monitoring CO_2 in the cell culture medium, one noteworthy method is the "credit card" microsystem, which integrates microfluidics, a gas-diffusion module, and an optical flow cell. The sampling rate is 30 samples per hour in this system. During sampling, the carbon dioxide from the sample is transferred to a pH-sensitive bromothymol blue solution and the color change is optically measured. Low cost is an important advantage of using this system. However, this method has a detection limit of 83 mg/L [8, 53, 54]. Another novel approach to measure the CO_2 level in cell culture medium works based on the initial diffusion rate of CO_2 through silicone membrane. In this method, a small area of the cell culture vessel is replaced with a silicone membrane, which is permeable to CO_2. Therefore, the CO_2 dissolved in the cell culture medium diffuses through the membrane attached to the cell culture vessel, where it is then transferred to a sensor for measurement. The main advantage of this method is its noninvasive nature, as sampling is conducted without any direct contact with cell culture medium [55, 56].

Redox

The condition of intracellular redox affects the pathways influencing manufactured biologics. Different technologies developed to measure this parameter, and they all are helpful in understanding the effect of redox on the process and on improving the yield and quality of the products. Membrane pressure sensors and redox electrodes are efficient ways of measuring redox potential in the USP and DSP [21, 22, 57]. Redox-sensitive yellow fluorescent protein (rx YFP) and green fluorescent protein (GFP) (roGFPs) both are currently engineered and used for online monitoring of thiol redox changes. The cysteine residues on the surface of the protein in the β-barrel are capable of forming a disulfide bond which results in changes in the fluorescence of protein [58].

Optical technique is another method with the capability to monitor cell differentiation online based on measuring the redox level. For example, a novel technique that works based on the redox ratio value is developed to determine the differentiation state in mesenchymal stem cells (MSCs). This approach is based on the observation that the optical redox ratio in mesenchymal stem cells decreases when cells undergo differentiation. In this method, fluorescence-lifetime imaging microscopy is combined with two proton excitation (TPE) fluorescence microscopy and determines the differentiation state of MSCs by measuring the optical redox ratio and fluorescence lifetimes of NAD(P)H and FAD. However, a monitoring system for online monitoring of differentiation is not currently available. However, to develop such a system, a probe must be designed for direct contact with cells and data collection. Another newly developed method provides an efficient way to monitor aggregation of hiPSCs in a continuous stirred tank bioreactor. This device is an in situ microscopic imaging tool. Given the importance of maintaining the pluripotency of iPSCs throughout the process, the development of a tool for detection and assessment of cell pluripotency during the culture is important [7].

Amino Acids

Amino acids are important in bioprocesses especially in CHO culture processes. To design an effective feeding strategy, it is essential to understand the amino acid profile during cell culture process. This is especially important considering that the limiting amino acids vary from process to process. A recent study has reported the development of an HPLC-based system for monitoring amino acids in CHO culture. In this technique to provide online monitoring system, an HPLC autosampler is included in the set-up. The sampling happens by drawing 10 μL water and 30 μL sample. The sampling is conducted every 3 hours and measurements happen 45 minutes after the sample is drawn. The information about concentration of 18 different amino acids can be provided using this method. One of the main advantages of this system is that it provides measurements at a higher frequency compared to Raman sensors. In addition, this sensor highly reduces the measurement errors and has higher accuracy compared to monitoring systems working based on spectroscopy [59].

Product Concentration, Quality Attributes and Impurities

A new monitoring system has been designed to measure the concentration of proteins and excipients at ultrafiltration and diafiltration units by using Mid-IR platform. In this method, two Mid-IR sensor probes are placed in the bulk solution and retentate line. The signal is acquired in this step. The spectra are then automatically transferred to a process monitoring platform for real-time visualization of excipients and protein concentration. This technique displays the results in graphical and tabulated form and helps to improve process understanding, troubleshooting and eventually leads to a faster development process [60].

Sensors for detecting proteins are developed based on the unit operations and the type of protein being targeted. For example, in chromatography, sensors based on UV absorbance and fluorescence are used for the in-line monitoring of proteins and nucleic acids [5]. UV absorbance at 280 nm wavelength has been used to quantify total proteins based on absorption from aromatic amino acids. UV spectroscopy has high sensitivity and can test protein concentrations to mg/L range [61]. Host cell protein (HCP) has been quantified by labor intensive ELISA. Capito et al. report using Fourier Transform Mid Infrared Spectroscopy (FT-MIR) to quantify HCPs and the results show the applicability of the method in higher HCP ranges that varies from 20,000 to 200,000 ng/mL [62]. FT-MIR is vibrational spectrometry and requires chemometrics to establish the calibrated prediction models. Cao et al. connected a ZipChip interface with mass spectrometry to enable CE-MS based native intact analysis and peptide mapping. The microfluidic-based analysis shows comparable levels of separation for charge variants and resolution of characterization results reducing both sampling volume and run time significantly [63].

Current product release and characterization assays include multiple techniques such as Capillary Isoelectric Focusing (cIEF) for charge variance, Size Exclusion Chromatography (SEC), Capillary Gel Electrophoresis (cGE) for size variance and Enzyme-Linked Immunosorbent Assay (ELISA) for HCP characterization. However, even with the combination of these assays, molecular level insights are not provided. There is a growing trend to utilize mass spectrometry coupled with liquid chromatography (LC-MS) for peptide mapping-based analysis. This can provide characterization on a wide range of attributes, such as posttranslational modifications (PTM) and impurities. The sequence of a regular MS includes sample preparation, MS data acquisition and treatment. Each can be labor intensive and complicated. Song et al. shows automating the whole sequence and establishing the automated platform at 3 different sites. In this method, MAM has been used as a powerful tool to guide process development and stability monitoring [64]. A series of application have been reported, further demonstrating its ability to reveal molecular level structural information. Interest has been expressed to cGMP applications as well [65–68]. Yigzaw Y et al. report the application of A410 nm to monitor HCP breakthrough at the depth filtration step to determine the end of collection criteria [69].

To report product quality attributes, including aggregate levels for making pooling decisions, Ultra-Performance Liquid Chromatography (UPLC) has been used for downstream chromatography steps through a T-connector [70]. Multi-Angle Light Scattering (MLS) technology, which employs a static light scattering technique, measures the light scattering intensity from multiple angles to provide insights toward molecular weight and/or particle size. In a study by Patel et al., the coupling of a HIC purification system with inline MLS is discussed as a means for monitoring and controlling aggregate levels in the eluate. The flow-through HIC step is designed to remover aggregate, where the aggregates with higher hydrophobicity have a longer residence time than monomers. The MLS intensity is proportional to protein concentration and weight average molecular weight. When coupled with UV-Vis, real time pooling decisions are made based on molecular weight to remove aggregates from the product stream. This eliminates the need for fractionation and offline SEC testing [53]. Additionally, protocols have been developed that combine SEC with MLS, bringing together the separation power of SEC with the analysis power of MLS [71].

Temperature

Temperature is a physical parameter that is monitored in both upstream processes (USP) and downstream processes (DSP) using thermometer. Thermocouples are commonly used for in-line monitoring of temperature [21, 22].

Biomass

Impedance sensors and fluorescence sensors are used in the monitoring of biomass and cell morphology, respectively, in upstream processes [21, 22]. A novel flow-cytometry based technique has recently been developed for monitoring biomass. This technology provides a great understanding of the physiology and morphology of the cells, something that is not possible with other online and offline monitoring systems. The technology is designed specifically for measurements in medium with high particle background. In other words, unlike other conventional methods, optical density, and viable cells are distinguished from background media in this method. In addition, viability, and size distribution through the use of fluorescent staining and morphological assessment can be obtained in a single measurement [72].

To monitor cell viability during the manufacturing process using specific perfusion rate (CSPR) control system, a capacitance probe is inserted in the cell culture medium and applies an electrical field polarizing the cells passing through it. Viable cells have an intact membrane, thus become polarized, resulting in an increase in their capacitance. This increase is measured and converted to the number of viable cells. This device can be used as a probe by inserting it into the cell culture medium. The only drawback of this technique is the invasive nature of the measurement. Despite this, the probes have been shown to be robust and reliable. Therefore, it is expected that there will be an increasing trend in the usage of such PATs in the near future [73].

Another approach for assessing viable cell numbers, confluence, and distribution in stem cells cultured on microcarriers is through use of optic-based technique methods. 2D imaging techniques such as lens-free imaging (LFI) and digital holography imaging (DHI) are other technologies available [8]. A novel technique for monitoring biomass is LED-based fluorescence microscopy method. The device consists of a sensor and a probe. To monitor biomass during bioprocess, the probe is inserted into a bioreactor through a port and sterilized in place. A detection fiber, located at the middle part of the probe, is surrounded by 280 nm excitation fibers. An outer layer made of 365 and 455 nm fibers surround the inner layer fibers. The fibers will illuminate the sample homogeneously and are connected to LEDs in the sensor in the detection part. This part consists of CCD-spectrometer and obtains spectra with a spectral resolution of less than 4 nm. This 2D fluorescence spectroscopy method is a technology to distinguish between viable, dead, and lysed cells [42].

A recent study has reported a titer reporting method through the use of microfluidic chip. In this method microfluidic capillary electrophoresis is coupled with mass spectrometry for charge-based separation in samples. The titer information is obtained by comparing the light chain signal intensity after sampling with heavy labeled mAb. This method is fast and simple, and it has the potential for further improvements as an analytical tool for in-line titer monitoring through the integration of an automated sample loading option [74]. Thakur et al. implemented NIR spectroscopy to quantify the concentration of mAb and excipients in real time during the ultrafiltration step for a high concentration product [75].

NIR probe was connected to the retention flow path and PLS model was used to calibrate the predicted concentration against standard testing. By implementing a python control algorithm, the excipient drift was effectively mitigated. The PAT application was critical for ensuring product quality and stability. In a similar study, Wasalathanthri et al. utilized another spectroscopy method-Fourier transform infrared spectroscopy for the same purpose and achieved a high level of accuracy. West et al. explored UPLC for a similar application. They incorporated an auto-sampler and quantified the concentration of antibody (High Molecular Weight (HMW)) and recipients (arginine and histidine) based of absorbance at 280 nm and 214 nm respectively during the ultrafiltration and diafiltration process [76, 77].

Titer is an important indication of upstream process performance, and its measurement is important for further downstream operation. With the industry moving toward continuous mode, real-time titer reporting is becoming increasing important [78–80]. Offline HPLC (affinity chromatography) has traditionally been used for titer quantification. Chemmalil et al. reported at line/online titer reporting using ultra performance liquid chromatography (UPLC) with a retrofitted process sample manager (PSM) for both fed-batch and perfusion bioreactors [70]. Titer measurement was carried out according to a preset schedule. The same author later published work utilizing 2D-LC to report both antibody titer and product quality, with the first dimension being protein A chromatography to purify and quantify cell culture samples, and the second dimension being either SEC to detect size variants or CEX for charge variants [81]. A SegFlow autosampler was adopted to reduce the loading volume for the second dimension LC, thus reduce the injection volume and improve resolution. Raman Spectroscopy can also be used to predict titer for fed-batch supernatant samples. Li et al. reported the Raman Spectroscopy analysis of samples from 37 CHO production batches, from which product titer was predicted. Competitive adaptive reweighted sampling (CoAdReS) and Ant Colony Optimization (ACO) variable selection methods were adopted to simplify spectral information for chemometric analysis. PLS regression was used to correlate the processed Raman spectra with product titer. Both variable selection methods were proven to improve the predictability compared to full spectrum analysis [82].

Aggregates and Fragments

HP-SEC is the gold standard in the industry for aggregate and fragment detection. Aggregates are among the most monitored product related impurities due to immunogenicity concerns, and fragments provide insights toward degradation pathways. HP-SEC has its own beauty of simplicity and is widely used in off-line setting. However, the major hurdle of using this method is its turnaround time. Normally, the sampling, testing, and analyzing of HP-SEC take about 30 min. Rathore et al. evaluated the application of online HP-SEC for real time process decision making by deducing the testing time to 13.5 min and increasing the processing step time to make it sufficient for process decision making [25]. This approach has the advantage of using product quality attributes directly for forward processing decisions. HP-SEC also has drawbacks including the lack of resolution between different aggregate forms and the potential for inaccurate reports due to nonspecific interaction between the stationary phase and the protein. As the industry shifts towards continuous operation for cost reduction, higher productivity, and smaller footprint, the first challenge to be addressed is to continuous loading onto the capture column from perfusion bioreactor without sampling and offline testing for titer. Rudt et all applied PLS modeling on the load eluent UV/Vis spectra to detect antibody breakthrough. In addition, refractive x-based sensors are used for detecting lipids and carbohydrates [83–85].

Sensors for Single Use Bioreactors

Manufacturers of mAbs, and cell and gene therapy products must comply with regulations set forth by FDA and EMA. Good Manufacturing Practices (GMP) guidelines specifically require the elimination of contamination throughout the process. This can be achieved by different sterilization processes at different stages of the biomanufacturing process. However, these practices are time-consuming. Using single-use technologies is a good way for reducing the risk of contamination and the need for sterilization at the same time. This results in reducing the sterilization time, faster biomanufacturing process and higher productivity [86]. Considering the advantages, disposable bioreactor technologies are gaining popularity nowadays.

Single-use bioreactors are equipped with monitoring systems through the use of sensors with different interfaces, similar to conventional bioreactors. These systems provide valuable information on the bioprocesses and help to improve cell culture processes. A commonly used monitoring method is at-line sensors which analyze liquid samples from bioreactors. To provide cell-free sampling, in-line filtration using microfiltration or dialysis membranes as a sterile barrier is an effective way. An at-line monitoring method is Flow injection analysis (FIA). However, the major disadvantage of these systems is the risk of contamination during frequent samplings. To address this issue, sampling containers are sterilized, then connected to the bioreactors, and sealed.

Any type of off-gas sensor used in conventional bioreactors is suitable for single-use bioreactors. In at-line monitoring systems, sterility is not a limiting factor as the sensor can be placed in the non-sterile side of the filter. Ultrasonic or spectroscopic sensors are proper types of monitoring systems for disposable vessels. Field effect transistors (FETs), electrochemical pH sensors, radio-frequency identification-based sensors, chemo and biosensors are examples of available in-situ monitoring systems. Sensors with noninvasive interfaces provide noninvasive monitoring of the process in single-use bioreactors. Temperature and weight are two common parameters measured using sensors with noninvasive interfaces. This method works through components mounted on the surface of the bioreactor. Optical sensors, impedance spectroscopy, ultrasonic sensors, and free-floating sensors are other types of sensors with noninvasive interfaces. Wireless technologies are also great candidates for single-use vessels as they do not require penetration through the reactor.

A newly developed sensor for measuring CO_2 dissolved in the cell culture medium is a noninvasive monitoring technology. This method works based on the initial diffusion rate of the CO_2 through a silicone layer. The details of this technology was earlier explained among available methods for monitoring CO_2. This technology can be easily integrated with single-use vessels. The monitoring system includes a sampler attached to the vessel and a sensor placed outside the cell culture environment. The sampler is made of inexpensive material and can be discarded after usage of the bioreactor while the sensor part of the system could be reused with a new sampler and bioreactor [55, 87].

To measure cell viability while manufacturing mAbs with CSPR (cell specific perfusion rate) control systems in single-use bioreactors, a patch is welded into the film of the bioreactor. The measurement happens based on a principle similar to the viability probe described previously [73]. Radio frequency identification (RFID) tags is capable of passively interrogating the reactor headspace have been demonstrated for monitoring temperature and pressure within single-use bioreactors [51]. Another important point about single-use vessels is the cost of the monitoring system as the sensor is discarded with the bioreactor after each usage. By developing communication protocols, and standard interfaces, the road to the transition to using novel sensors could be paved as scientists improve the accuracy, reproducibility, and durability of the monitoring systems [55, 87].

Omic Tools for Monitoring High-Throughput Bioreactors

The utilization of omic tools have proven to be beneficial in enhancing our understanding of intracellular characteristics, and monitoring bioprocesses conducted in bioreactors. In this method, different biological molecules such as genes, proteins, and metabolites are quantified, and analyzed. Recently, a trend towards the application of omics technologies in the stem cell culture field has emerged. The information provided by this method is crucial in many cell culture processes in stirred-tank bioreactors such as embryonic stem cell (ESC) cultures. Moreover, it provides insights into physiological and metabolic changes in iPSC cultures as well as investigating the pluripotency of stem cells. Despite the advantages of this method, there is a need for further improvement, such as enhancing computational techniques to link between omics and phenotypes. Additionally, it must be noted that biomarkers are cell-line specific, and the development of omic to address this issue is necessary [8].

Modeling-Enhanced Bioprocessing

As discussed in previous sections, monitoring systems are utilized in upstream bioprocesses to monitor critical process parameters (CPPs) such as pH, DCO_2 and DO throughout the process. These environmental parameters can impact the outcome of the bioprocess. For instance, the pluripotency of hiPSCs is influenced by changes in the microenvironment during the culture. The information obtained from process monitoring can be used to produce products with desirable CQAs. Model-based monitoring is a strategy to link CPPs to CQAs applying a model. This method can be combined with a desired output trend to develop a model-based control system. In this system, the best control settings are predicted based on the measured data and a model. Model-based monitoring is more commonly used in mammalian cell cultures and very few studies are conducted using this method in stem cell cultures. This strategy can be specifically beneficial in processes with heterogenic input materials. However, adjustments must be conducted to cover the regulatory requirements for these cultures [7, 8].

A recent study reports developing a model predictive control (MPC) system based on machine learning algorithms. The aim was to provide optimal feeding for achieving the highest cell growth and metabolite production. A linear regression model is initially developed. A neural network model is then trained with several regressors. Based on the best model, the MPC is designed, and optimal feed strategy is determined to obtain the maximum viable cell density. Regression models are designed based on the correlation between culture variables and glucose level (GLC). The relationship between the GLC and glucose feed volume is obtained using mass balance. The mass balance and regression model result in a hybrid model representing the fed-batch process. Experimental studies were conducted for validation of the control method. The results show 2% improvement in protein production after using the developed control method [88].

Other techniques have been developed for non-destructive monitoring of apoptosis and cell viability using machine learning and deep learning algorithms [76, 89]. A model for real-time calculation of oxygen uptake rate and biomass in CHO culture was recently developed. In this method, KL_a was characterized during the culture to estimate OUR, and a segmented model showed that viable biomass is correlated with OUR. This model is a robust technique for control-based bioprocesses [90]. Wallocha et al. used off-gas oxygen measurement to determine the Oxygen Uptake Rate (OUR) and correlate it to biomass through mechanistic modeling in a single use bioreactor (SUB) in perfusion setting. The biomass prediction can be used to automatically control cell bleeding during steady state process. In addition to multiple variable data analysis (MVDA), machine learning (ML) has seen increasing use in the pharmaceutical industry over the past 5 years [91, 92].

Conclusions

Nowadays, with the increasing popularity of cell and gene therapy products and monoclonal antibodies (mAbs), there is a growing demand to develop robust and consistent manufacturing processes. Sensors provide data from environmental parameters throughout the process and are thus useful tools for gaining a better understanding of these processes and developing reliable bioprocess. Therefore, adopting monitoring technologies in processes is crucial. Furthermore, by combining data from sensors with model-based techniques, environmental parameters can be controlled in an efficient manner. The model-based technologies can also help identify risks of batch failure in stem cell bioprocessing [8]. Given the diversity of approaches and tools employed, there is an increasing need for standardization and reference benchmarks for the various sensors and instrumentation.

Conflict of Interest Statement GR has several patents related to the subject matter. VR and GR have filed additional patents.

References

1. Chan AC, Carter PJ (2010) Therapeutic antibodies for autoimmunity and inflammation. Nat Rev Immunol 10:301–316. https://doi.org/10.1038/nri2761
2. Weiner LM, Wang S (2010) Monoclonal antibodies: versatile platforms for cancer immunotherapy. Nat Rev Immunol 10(5):317–327. https://doi.org/10.1038/nri2744
3. FDA (2004) Guidance for industry, PAT-A framework for innovative pharmaceutical development, manufacturing and quality assurance. Retrieved from http://www.fda.gov/cder/guidance/published.html
4. Rathore AS, Ghare V (2010) Process analytical technology (PAT) for biopharmaceutical products. Anal Bioanal Chem 398(1):137–154. https://doi.org/10.1007/s00216-010-3781-x
5. Maruthamuthu MK, Rudge SR, Ardekani AM, Ladisch MR, Verma MS (2020) Process analytical technologies and data analytics for the manufacture of monoclonal antibodies. Trends Biotechnol 38:1169–1186. https://doi.org/10.1016/j.tibtech.2020.07.004
6. Lashkari SG (2017) Controlling recombinant Escherichia coli cultures to the boundary of oxidative and overflow metabolism (BOOM) for robust efficient. Growth. https://doi.org/10.13140/RG.2.2.34153.36966
7. Polanco A, Kuang B, Yoon S (2020) Bioprocess technologies that preserve the quality of iPSCs. Trends Biotechnol 38:1128–1140. https://doi.org/10.1016/j.tibtech.2020.03.006
8. Papantoniou I, Lambrechts T, Gupta P, de Bournonville S, Loverdou N, Geris L, Aerts J-M (2018) Bioreactor sensing and monitoring for cell therapy manufacturing. In: Cabral JMS, da Silva CL (eds) Bioreactors for stem cell expansion and differentiation, 1st edn. CRC Press, pp 243–268
9. Fuentes P, Torres MJ, Arancibia R, Aulestia F, Vergara M, Carrión F, Osses N, Altamirano C (2022) Dynamic culture of mesenchymal stromal/stem cell spheroids and secretion of paracrine factors. Front Bioeng Biotechnol 10:916229. https://doi.org/10.3389/fbioe.2022.916229
10. Rao G (2020) Sensor techknowledge: sensors for process development: a practical guide. UMBC Chemical, Biochemical & Environmental Engineering Department

11. Mitra S, Murthy GS (2022) Bioreactor control systems in the biopharmaceutical industry: a critical perspective. Syst. Microbiol. Biomanufact. 2(1):91–112. https://doi.org/10.1007/s43393-021-00048-6
12. Khanal O (2022) Mathematical modeling and process analytical technology for continuous chromatography of biopharmaceutical products. Curr Opin Biotechnol 78:102796. https://doi.org/10.1016/j.copbio.2022.102796
13. Liu Z, Zhang Z, Qin Y, Chen G, Hu J, Wang Q, Zhou W (2021) The application of Raman spectroscopy for monitoring product quality attributes in perfusion cell culture. Biochem Eng J 173:108064. https://doi.org/10.1016/j.xphs.2021.09.011
14. Reyes SJ, Durocher Y, Pham PL, Henry O (2022) Modern sensor tools and techniques for monitoring, controlling, and improving cell culture processes. Processes 10(2):189. https://doi.org/10.3390/pr10020189
15. Wang B, Bowles-Welch AC, Yeago C, Roy K (2022) Process analytical technologies in cell therapy manufacturing: state-of-the-art and future directions. J Adv Manuf Process 4(1):e10106. https://doi.org/10.1002/amp2.10106
16. Luttmann R, Bracewell DG, Cornelissen G, Gernaey KV, Glassey J, Hass VC, Kaiser C, Preusse C, Striedner G, Mandenius CF (2012) Soft sensors in bioprocessing: a status report and recommendations. Biotechnol J 7(8):1040–1048. https://doi.org/10.1002/biot.201100506
17. Randek J, Mandenius CF (2018) On-line soft sensing in upstream bioprocessing. Crit Rev Biotechnol 38(1):106–121. https://doi.org/10.1080/07388551.2017.1312271
18. Kattipparambil Rajan D, Patrikoski M, Verho J, Sivula J, Ihalainen H, Miettinen S, Lekkala J (2016) Optical non-contact pH measurement in cell culture with sterilizable, modular parts. Talanta 161:755–761. https://doi.org/10.1016/j.talanta.2016.09.021
19. Holzberg TR, Watson V, Brown S, Andar A, Ge X, Kostov Y, Tolosa L, Rao G (2018) Sensors for biomanufacturing process development: facilitating the shift from batch to continuous manufacturing. Curr Opin Chem Eng 22:115–127. https://doi.org/10.1016/j.coche.2018.09.008
20. Kieninger J, Tamari Y, Enderle B, Jobst G, Sandvik J, Pettersen E, Urban G (2018) Sensor access to the cellular microenvironment using the sensing cell culture flask. Biosensors 8:44. https://doi.org/10.3390/bios8020044
21. Li F, Vijayasankaran N, Shen A, Kiss R, Amanullah A (2010) Cell culture processes for monoclonal antibody production. MAbs 2:466–479. https://doi.org/10.4161/mabs.2.5.12720
22. Joynes L, Zhang J (2022) A feedback control strategy for a fed-batch monoclonal antibody production process utilising infrequent and irregular sampled measurements. PRO 10:1448. https://doi.org/10.3390/pr10081448
23. Dhinakaran V, Vigneswari K, Lavanya M, Varsha Shree M (2020) Point-of-care applications with graphene in human life. In: Comprehensive analytical chemistry. Elsevier, pp 235–262
24. Esmonde-White KA, Cuellar M, Uerpmann C, Lenain B, Lewis IR (2017) Raman spectroscopy as a process analytical technology for pharmaceutical manufacturing and bioprocessing. Anal Bioanal Chem 409:637–649. https://doi.org/10.1007/s00216-016-9824-1
25. Rathore AS, Yu M, Yeboah S, Sharma A (2008) Case study and application of process analytical technology (PAT) towards bioprocessing: use of on-line high-performance liquid chromatography (HPLC) for making real-time pooling decisions for process chromatography. Biotechnol Bioeng 100(2):306–316
26. Hakemeyer C, Menezes JC (2012) At-line NIR spectroscopy as effective PAT monitoring technique in Mab cultivations during process development and manufacturing. Talanta 90:12–21. https://doi.org/10.1016/j.talanta.2011.12.042
27. Abu-Absi NR, Kenty BM, Cuellar ME, Borys MC, Sakhamuri S, Strachan DJ et al (2011) Real time monitoring of multiple parameters in mammalian cell culture bioreactors using an in-line Raman spectroscopy probe. Biotechnol Bioeng 108(5):1215–1221
28. Whelan J, Craven S, Glennon B (2012) In situ Raman spectroscopy for simultaneous monitoring of multiple process parameters in mammalian cell culture bioreactors. Biotechnol Prog 28(5):55

29. Graf A, Lemke J, Schulze M, Soeldner R, Rebner K, Hoehse M, Matuszczyk J (2022) A novel approach for non-invasive continuous in-line control of perfusion cell cultivations by Raman spectroscopy. Front Bioeng Biotechnol 10:719614. https://doi.org/10.3389/fbioe.2022.719614
30. Sharker SM, Rahman A (2021) A review on the current methods of Chinese hamster ovary (CHO) cells cultivation for the production of therapeutic protein. Curr Drug Discov Technol 18(3):354–364
31. Kunert R, Reinhart D (2016) Advances in recombinant antibody manufacturing. Appl Microbiol Biotechnol 100:3451–3461. https://doi.org/10.1007/s00253-016-7388-9
32. Kelley B (2009) Industrialization of mAb production technology: the bioprocessing industry at a crossroads. MAbs 1(5):443–452. https://doi.org/10.4161/mabs.1.5.9448. Epub 2009 Sep 16. PMID: 20065641; PMCID: PMC2759494
33. Jayapal KP, Yap MGS (2007) Recombinant protein therapeutics from CHO cells – 20 years and counting. Chem Eng Prog 103(10):40–47
34. Kozma B, Salgó A, Gergely S (2019) On-line glucose monitoring by near infrared spectroscopy during the scale up steps of mammalian cell cultivation process development. Bioprocess Biosyst Eng 42:921–932. https://doi.org/10.1007/s00449-019-02091-z
35. Tric M, Lederle M, Neuner L, Dolgowjasow I, Wiedemann P, Wölfl S, Werner T (2017) Optical biosensor optimized for continuous in-line glucose monitoring in animal cell culture. Anal Bioanal Chem 409:5711–5721. https://doi.org/10.1007/s00216-017-0511-7
36. Craven S, Glennon B (2014) Glucose concentration control of a fed-batch mammalian cell bioprocess using a nonlinear model predictive controller. J Process Control 24:344–357. https://doi.org/10.1016/j.jprocont.2014.02.007
37. Berry BN, Dobrowsky TM, Timson RC, Kshirsagar R, Ryll T, Wiltberger K (2016) Quick generation of Raman spectroscopy based in- process glucose control to influence biopharmaceutical protein product quality during mammalian cell culture. Biotechnol Prog 32:224–234. https://doi.org/10.1002/btpr.2205
38. Brown S, Zambrana PN, Ge X, Bagdure D, Stinchcomb AL, Rao G, Tolosa L (2018) Minimally invasive technique for measuring transdermal glucose with a fluorescent biosensor. Anal Bioanal Chem 410:7249–7260. https://doi.org/10.1007/s00216-018-1336-8
39. Tiangco C, Fon D, Sardesai N, Kostov Y, Sevilla F, Rao G, Tolosa L (2017) Fiber optic biosensor for transdermal glucose based on the glucose binding protein. Sensors Actuators B Chem 242:569–576. https://doi.org/10.1016/j.snb.2016.11.077
40. Ge X, Rao G, Tolosa L (2008) On the possibility of real-time monitoring of glucose in cell culture by microdialysis using a fluorescent glucose binding protein sensor. Biotechnol Prog 24: 691–697. https://doi.org/10.1021/bp070411k
41. Ge X, Tolosa L, Simpson J, Rao G (2003) Genetically engineered binding proteins as biosensors for fermentation and cell culture. Biotechnol Bioeng 84:723–731. https://doi.org/10.1002/bit.10830
42. Graf A, Claßen J, Solle D, Hitzmann B, Rebner K, Hoehse M (2019) A novel LED-based 2D-fluorescence spectroscopy system for in-line monitoring of Chinese hamster ovary cell cultivations – part I. Eng Life Sci 19:352–362. https://doi.org/10.1002/elsc.201800149
43. Li Q, Zhang Y, Fan H, Gong Y, Xu Y, Lv Q, Xu Y, Xiao F, Wang S, Wang Z, Wang L (2021) In vitro and in vivo detection of lactate with nanohybrid-functionalized Pt microelectrode facilitating assessment of tumor development. Biosens Bioelectron 191:113474. https://doi.org/10.1016/j.bios.2021.113474
44. Chmayssem A, Verplanck N, Tanase CE, Costa G, Monsalve-Grijalba K, Amigues S, Alias M, Gougis M, Mourier V, Vignoud S, Ghaemmaghami AM, Mailley P (2021) Development of a multiparametric (bio)sensing platform for continuous monitoring of stress metabolites. Talanta 229:122275. https://doi.org/10.1016/j.talanta.2021.122275
45. Hashemzadeh S, Omidi Y, Rafii-Tabar H (2019) Amperometric lactate nanobiosensor based on reduced graphene oxide, carbon nanotube and gold nanoparticle nanocomposite. Microchim Acta 186:680. https://doi.org/10.1007/s00604-019-3791-0

46. Schulz E, Münzel T (2011) Intracellular pH: a fundamental modulator of vascular function. Circulation 124:1806–1807. https://doi.org/10.1161/CIRCULATIONAHA.111.061226
47. Abou-el-Enein M, Elsallab M, Feldman SA, Fesnak AD, Heslop HE, Marks P, Till BG, Bauer G, Savoldo B (2021) Scalable manufacturing of CAR T cells for cancer immunotherapy. Blood Cancer Discovery 2:408–422. https://doi.org/10.1158/2643-3230.BCD-21-0084
48. Tanumihardja E, Slaats RH, van der Meer AD, Passier R, Olthuis W, van den Berg A (2021) Measuring Both pH and O 2 with a single on-Chip sensor in cultures of human pluripotent stem cell-derived cardiomyocytes to track induced changes in cellular metabolism. ACS Sens 6:267–274. https://doi.org/10.1021/acssensors.0c02282
49. Al-Ani A, Toms D, Kondro D, Thundathil J, Yu Y, Ungrin M (2018) Oxygenation in cell culture: critical parameters for reproducibility are routinely not reported. PLoS One 13: e0204269. https://doi.org/10.1371/journal.pone.0204269
50. Miller AT (1966) The role of oxygen in metabolic regulation. Helgolander Wiss Meeresunters 14:392–406. https://doi.org/10.1007/BF01611634
51. Stine JM, Beardslee LA, Sathyam RM, Bentley WE, Ghodssi R (2020) Electrochemical dissolved oxygen sensor-integrated platform for wireless in situ bioprocess monitoring. Sensors Actuators B Chem 320:128381. https://doi.org/10.1016/j.snb.2020.128381
52. Kroll P, Herwig C, Mueller H (2019) Control of dissolved carbon dioxide in driving force limited cell culture processes. Bluesens. https://www.bluesens.com/fileadmin/user_upload/downloads-products/Control_of_dissolved_carbon_dioxide_in_driving_force_limited_cell_culture_processes.pdf, https://doi.org/10.13140/RG.2.2.36330.54723
53. Chatterjee M, Ge X, Uplekar S, Kostov Y, Croucher L, Pilli M, Rao G (2015) A unique noninvasive approach to monitoring dissolved O_2 and CO_2 in cell culture: noninvasive approach to monitoring O_2 and CO_2. Biotechnol Bioeng 112:104–110. https://doi.org/10.1002/bit.25348
54. Calvo-López A, Ymbern O, Izquierdo D, Alonso-Chamarro J (2016) Low cost and compact analytical microsystem for carbon dioxide determination in production processes of wine and beer. Anal Chim Acta 931:64–69. https://doi.org/10.1016/j.aca.2016.05.010
55. Rahmatnejad V, Tolosa M, Ge X, Rao G (2022) A novel approach to noninvasive monitoring of dissolved carbon dioxide in small-scale cell culture processes. Front Bioeng Biotechnol 10: 968294. https://doi.org/10.3389/fbioe.2022.968294
56. Rahmatnejad V (2021) Noninvasive sensor applications in cell culture. Master's thesis, University of Maryland, Baltimore County. ProQuest Dissertations Publishing. https://doi.org/10.13140/RG.2.2.14230.42569
57. Sinharoy P, McFarland KS, Majewska NI, Betenbaugh MJ, Handlogten MW (2021) Redox as a bioprocess parameter: analytical redox quantification in biological therapeutic production. Curr Opin Biotechnol 71:49–54. https://doi.org/10.1016/j.copbio.2021.06.017
58. Schwarzländer M, Dick TP, Meyer AJ, Morgan B (2016) Dissecting redox biology using fluorescent protein sensors. Antioxid Redox Signal 24:680–712. https://doi.org/10.1089/ars.2015.6266
59. Hofer A, Kroll P, Barmettler M, Herwig C (2020) A reliable automated sampling system for on-line and real-time monitoring of CHO cultures. PRO 8:637. https://doi.org/10.3390/pr8060637
60. Wasalathanthri DP, Feroz H, Puri N, Hung J, Lane G, Holstein M, Chemmalil L, Both D, Ghose S, Ding J, Li ZJ (2020) Real-time monitoring of quality attributes by in-line Fourier transform infrared spectroscopic sensors at ultrafiltration and diafiltration of bioprocess. Biotechnol Bioeng 117:3766–3774. https://doi.org/10.1002/bit.27532
61. Rolinger L, Rüdt M, Hubbuch J (2020) A critical review of recent trends, and a future perspective of optical spectroscopy as PAT in biopharmaceutical downstream processing. Anal Bioanal Chem 412:2047–2064. https://doi.org/10.1007/s00216-020-02407-z
62. Capito F, Skudas R, Kolmar H, Stanislawski B (2013) Host cell protein quantification by Fourier transform mid infrared spectroscopy (FT-MIR). Biotechnol Bioeng 110(1):252–259. https://doi.org/10.1002/bit.24611

63. Cao L, Fabry D, Lan K (2021) Rapid and comprehensive monoclonal antibody characterization using microfluidic CE-MS. J Pharm Biomed Anal 204:114251
64. Song YE, Dubois H, Hoffmann M, D́Eri S, Fromentin Y, Wiesner J, Pfenninger A, Clavier S, Pieper A, Duhau L, Roth U (2021) Automated mass spectrometry multi-attribute method analyses for process development and characterization of mAbs. J Chromatogr B 1166:122540
65. Murisier A, Duivelshof BL, Fekete S, Bourquin J, Schmudlach A, Lauber MA, Nguyen JM, Beck A, Guillarme D, D'Atri V (2021) Towards a simple on-line coupling of ion exchange chromatography and native mass spectrometry for the detailed characterization of monoclonal antibodies. J Chromatogr A 11(1655):462499. https://doi.org/10.1016/j.chroma.2021.462499
66. Füssl F, Strasser L, Carillo S, Bones J (2021) Native LC–MS for capturing quality attributes of biopharmaceuticals on the intact protein level. Curr Opin Biotechnol 71:32–40
67. Graf T, Heinrich K, Grunert I, Wegele H, Haberger M, Bulau P, Leiss M (2020 Jul) Recent advances in LC-MS based characterization of protein-based bio-therapeutics – mastering analytical challenges posed by the increasing format complexity. J Pharm Biomed Anal 15(186):113251. https://doi.org/10.1016/j.jpba.2020.113251
68. Ren D (2020) Advancing mass spectrometry technology in cGMP environments. Trends Biotechnol 38(10):1051–1053
69. Yigzaw Y, Piper R, Tran M, Shukla AA (2006) Exploitation of the adsorptive properties of depth filters for host cell protein removal during monoclonal antibody purification. Biotechnol Prog 22(1):288–296
70. Chemmalil L, Prabhakar T, Kuang J, West J, Tan Z, Ehamparanathan V, Song Y, Jianlin X, Ding J, Li Z (2020) Online/at-line measurement, analysis and control of product titer and critical product quality attributes (CQAs) during process development. Biotechnol Bioeng 117:3757–3765
71. Patel BA, Gospodarek A, Larkin M, Kenrick SA, Haverick MA, Tugcu N, Brower MA, Richardson DD (2018) Multi-angle light scattering as a process analytical technology measuring real-time molecular weight for downstream process control. MAbs 10(7):945–950. https://doi.org/10.1080/19420862.2018.1505178
72. Vees CA, Veiter L, Sax F, Herwig C, Pflügl S (2020) A robust flow cytometry-based biomass monitoring tool enables rapid at-line characterization of S. Cerevisiae physiology during continuous bioprocessing of spent sulfite liquor. Anal Bioanal Chem 412:2137–2149. https://doi.org/10.1007/s00216-020-02423-z
73. Jagschies G (2018) Biopharmaceutical processing: development, design, and implementation of manufacturing processes, 1st edn. Elsevier, Amsterdam
74. Wang Y, Feng P, Sosic Z, Zang L (2017) Monitoring glycosylation profile and protein titer in cell culture samples using ZipChip CE-MS. J Anal Bioanal Tech 08. https://doi.org/10.4172/2155-9872.1000359
75. Thakur G, Hebbi V, Rathore AS (2021) Near infrared spectroscopy as a PAT tool for monitoring and control of protein and excipient concentration in ultrafiltration of highly concentrated antibody formulations. Int J Pharm 1(600):120456. https://doi.org/10.1016/j.ijpharm.2021.120456
76. Phillip JM, Han K-S, Chen W-C, Wirtz D, Wu P-H (2021) A robust unsupervised machine-learning method to quantify the morphological heterogeneity of cells and nuclei. Nat Protoc 16:754–774. https://doi.org/10.1038/s41596-020-00432-x
77. West JM, Feroz H, Xu X, Puri N, Holstein M, Ghose S, Ding J, Li ZJ (2021) Process analytical technology for on-line monitoring of quality attributes during single-use ultrafiltration/diafiltration. Biotechnol Bioeng 118(6):2293–2300. https://doi.org/10.1002/bit.27741
78. Godawat R, Konstantinov K, Rohani M, Warikoo V (2015) End-to-end integrated fully continuous production of recombinant monoclonal antibodies. J Biotechnol 10(213):13–19. https://doi.org/10.1016/j.jbiotec.2015.06.393
79. Croughan MS, Konstantinov KB, Cooney C (2015) The future of industrial bioprocessing: batch or continuous? Biotechnol Bioeng 112(4):648–651

80. Burcham CL, Florence AJ, Johnson MD (2018) Continuous manufacturing in pharmaceutical process development and manufacturing. Ann Rev Chem Biomol Eng 9:253–281
81. Chemmalil L, Wasalathanthri DP, Zhang X, Kuang J, Shao C, Barbour R et al (2021) Online monitoring and control of upstream cell culture process using 1D. Biotechnol Bioeng 118(9): 3593–3603
82. Li B, Ray BH, Leister KJ, Ryder AG (2013) Performance monitoring of a mammalian cell based bioprocess using Raman spectroscopy. Anal Chim Acta 24(796):84–91. https://doi.org/10.1016/j.aca.2013.07.058
83. Thakur G, Hebbi V, Rathore AS (2020) An NIR-based PAT approach for real-time control of loading in protein a chromatography in continuous manufacturing of monoclonal antibodies. Biotechnol Bioeng 117(3):673–686
84. Yumioka R, Sato H, Tomizawa H, Yamasaki Y, Ejima D (2010 Feb) Mobile phase containing arginine provides more reliable SEC condition for aggregation analysis. J Pharm Sci 99(2): 618–620. https://doi.org/10.1002/jps.21857
85. Rüdt M, Brestrich N, Rolinger L, Hubbuch J (2017) Real-time monitoring and control of the load phase of a protein a capture step. Biotechnol Bioeng 114(2):368–373
86. Nogueira DES, Cabral JMS, Rodrigues CAV (2021) Single-use bioreactors for human pluripotent and adult stem cells: towards regenerative medicine applications. Bioengineering 8:68. https://doi.org/10.3390/bioengineering8050068
87. Busse C, Biechele P, de Vries I, Reardon KF, Solle D, Scheper T (2017) Sensors for disposable bioreactors. Eng Life Sci 17:940–952. https://doi.org/10.1002/elsc.201700049
88. Rashedi M, Khodabandehlou H, Demers M, Wang T, Garvin C (2022) Model predictive controller design for bioprocesses based on machine learning algorithms. IFAC-PapersOnLine 55:45–50. https://doi.org/10.1016/j.ifacol.2022.07.420
89. Bannon D, Moen E, Schwartz M, Borba E, Kudo T, Greenwald N, Vijayakumar V, Chang B, Pao E, Osterman E, Graf W, Van Valen D (2021) DeepCell Kiosk: scaling deep learning–enabled cellular image analysis with Kubernetes. Nat Methods 18:43–45. https://doi.org/10.1038/s41592-020-01023-0
90. Pappenreiter M, Sissolak B, Sommeregger W, Striedner G (2019) Oxygen uptake rate soft-sensing via dynamic kLa computation: cell volume and metabolic transition prediction in mammalian bioprocesses. Front Bioeng Biotechnol 7:195. https://doi.org/10.3389/fbioe.2019.00195
91. Wallocha T, Popp O (2021) Off-gas-based soft sensor for real-time monitoring of biomass and metabolism in Chinese hamster ovary cell continuous processes in single-use bioreactors. PRO 9(11):2073. https://doi.org/10.3390/pr9112073
92. Banner M, Alosert H, Spencer C, Cheeks M, Farid SS, Thomas M, Goldrick S (2021) A decade in review: use of data analytics within the biopharmaceutical sector. Curr Opin Chem Eng 34: 100758

Chapter 3
Advances in Characterization of Industrial Bioreactors for Cell Culture Process

Jürgen Fitschen, Sebastian Hofmann, Lotta Kursula, Ingrid Haase, Thomas Wucherpfennig, and Michael Schlüter

Abstract Aerated stirred tank reactors are widely used in the biopharmaceutical industry for cell culture applications due to their flexibility and broad range of uses. This book chapter provides an overview of the main characterization methods for aerated stirred tank reactors in the context of cell culture processes. The methods described include mechanical power input, volumetric mass transfer, flow field, and mixing characterization. Examples are given using stirred tank reactors of different scales, ranging from a 3 L laboratory scale to a 15,000 L industrial scale. The chapter also discusses recent developments in the use of mobile, encapsulated sensors known as Lagrangian Sensor Particles (LSP) for continuous data collection and transmission throughout the entire reactor volume, providing insights from the perspective of cells during a cell culture process. Additionally, the state-of-the-art in numerical flow simulation is discussed, using aerated stirred tank reactors as an example to illustrate the challenges and opportunities offered by numerical flow simulation. It is highlighted that the methods described are specifically focused on cell culture applications, which have lower specific power inputs and aeration rates compared to other types of applications such as microbial fermentations or chemical reactions. The chapter also provides a brief overview of the fundamentals of aerated stirred tank reactors, including their design and components, and emphasizes the importance of equipment characterization for successful process transfer and scale-up in cell culture processes.

Keywords Aerated stirred tank reactors · Biopharmaceutical industry · Cell culture applications · Characterization methods · Mechanical power input · Mass transfer · Mixing · Lagrangian sensor particles · Numerical flow simulation

---

J. Fitschen · S. Hofmann · L. Kursula · I. Haase · M. Schlüter (✉)
Institute of Multiphase Flows, TUHH, Hamburg, Germany
e-mail: juergen.fitschen@tuhh.de; sebastian.hofmann@tuhh.de; lotta.kursula@tuhh.de; ingrid.haase@tuhh.de; michael.schlueter@tuhh.de

T. Wucherpfennig
Boehringer Ingelheim Pharma GmbH & Co. KG, Biberach an der Riß, Germany
e-mail: thomas.wucherpfennig@boehringer-ingelheim.com

R. Pörtner (ed.), *Biopharmaceutical Manufacturing*, Cell Engineering 11,
https://doi.org/10.1007/978-3-031-45669-5_3

Abbreviation

4D PTV	4D Particle Tracking Velocimetry
BMS	Biochemical measuring system
CFD	Computational fluid dynamics
CT	Computer tomography
DFG	German Research Foundation
DO	Dissolved oxygen
DTU	Technical University of Denmark
ECT	Electrical capacitance tomography
FNR	Agency for renewable resources
GPU	Graphic processing unit
IMU	Inertial measurement unit
LB	Lattice Boltzmann
LED	Light emitting diode
LES	Large Eddy simulation
LSP	Lagrangian sensor particle
MRF	Moving reference frame
MRT	Magnetic resonance tomography
OTR	Oxygen transfer rate
PDF	Probability distribution function
PEEK	Polyether ether ketone
PLIF	Planar laser-induced fluorescence
RANS	Reynolds-averaged Navier-stokes
SUB	Single-use bioreactor

Symbols

a	m^{-1}	Specific surface area
c	$mol\ m^{-3}$	Concentration
$c*$	$mol\ m^{-3}$	Saturation concentration
D	$m^2\ s^{-1}$	Diffusion coefficient
$d_{stirrer}$	m	Stirrer diameter
D_{STR}	m	Tank diameter
$d_{orifice}$	m	Orifice diameter
F	–	Proportion factor
g	$m\ s^{-2}$	Gravity constant
h	m	Bottom clearance
H	m	Filling height
H_s^{cc}	–	Henry's law constant

H_s^{cp}	mol m^{-3} atm^{-1}	Henry's law constant
H_{total}	m	Total tank height
I	–	Intensity
k_L	m s^{-1}	Mass transfer coefficient
k_La	s^{-1}	Volumetric mass transfer coefficient
l	m	Length
M	–	Degree of homogeneity
M_{idle}	N m^{-1}	Idle torque
$M_{stirrer}$	N m^{-1}	Stirrer torque
n	s^{-1}	Stirrer frequency
$\dot{n}$	mol m^{-3} s^{-1}	Molar flux
$P_{stirrer}$	kg m^2 s^{-3}	Power input
p_{part}	N m^{-2}	Partial pressure
p_{hydro}	N m^{-2}	Hydrostatic pressure
q	m^3 s^{-1}	Aeration rate
R	Ω	Electrical resistance
s	m	Stirrer spacing
t	s	Time
T	°C	Temperature
u	m s^{-1}	Velocity component in x-direction
u_E	m s^{-1}	Eulerian velocity
u_G	m s^{-1}	Superficial gas velocity
u_L	m s^{-1}	Lagrangian velocity
v	m s^{-1}	Velocity component in y-direction
v	m s^{-1}	Gas exit velocity
V	m^3	Liquid volume
V_{fill}	m^3	Filling volume
w	m s^{-1}	Velocity component in z-direction
X	–	Blending ratio
X_L	–	Lagrangian fluid parcel
x	m	Cartesian coordinate system
y	m	Cartesian coordinate system
z	m	Cartesian coordinate system

Greek Symbols

δ	m	Boundary layer thickness
λ	m	Kolmogorov's length
θ	s	Mixing time
ν	m^3 s^{-1}	Kinematic viscosity
ε	–	Gas hold-up
ε	m^2 s^{-3}	Energy dissipation rate
ρ	kg m^{-3}	Density

Introduction

In biopharmaceutical industry, aerated stirred tank reactors are widely used due to their simplicity and comparably good mixing and mass transfer performance [1]. Aerated stirred tank reactors are characterized by their flexible and broad range of applications, including the cultivation of yeast, bacteria and mammalian cells. However, the demands on the equipment and the process vary greatly depending on the cell type selected. For example, in the cultivation of microorganisms with a high oxygen demand, oxygen as well as nutrients must be provided in a short time [2]. In the case of mammalian cells, which exhibit significantly slower metabolism compared to microorganisms, a constant supply of oxygen is just as important. Nevertheless, for mammalian cells, additional care must be taken to ensure that the required oxygen and mixing of nutrients is provided with a limited hydrodynamic stress [3]. Compared to other reactor types such as bubble columns or airlift reactors, aerated stirred tanks are characterized by the fact that mass transfer and mixing can be ensured not only via the buoyancy driven flow by the gaseous phase but additionally actively via the momentum input by the stirrer, which allows more controlled process conditions.

This book chapter describes the main characterization methods for aerated stirred tank reactors in the context of cell culture applications. In the following, the established methods for the description of the mechanical power input, the volumetric mass transfer, the flow field as well as the mixing are described. In this context, examples will be given repeatedly using stirred tank reactors on a 3 L laboratory as well as on a 15,000 L industrial scale. For the latter, an acrylic glass reactor was constructed together with the company Boehringer Ingelheim Pharma GmbH & Co. KG at the Institute of Multiphase Flows at the Hamburg University of Technology. An illustration of the available transparent stirred tank reactors is shown in Fig. 3.1.

Based on established methods and recent developments the characterization of aerated stirred tank reactors for biopharmaceutical cell culture processes is described in the following.

Already being developed by multiple research teams for over six decades, the potential utilization of mobile, encapsulated sensors in biotechnological processes is presented. Also known as Lagrangian Sensor Particles (LSP), they are able to collect and transmit continuous data throughout the entire reactor volume, and thus may provide measurement data from the point of view of cells during a cell culture process. Finally, the state-of-the-art in the field of numerical flow simulation is discussed, using the example of aerated stirred tank reactors to illustrate the challenges but also the opportunities offered by numerical flow simulation.

All methods described in the following can be used for both the characterization of aerated stirred tank reactors and also for other reactor types, such as bubble columns and airlift reactors. However, it is pointed out that the descriptions in the

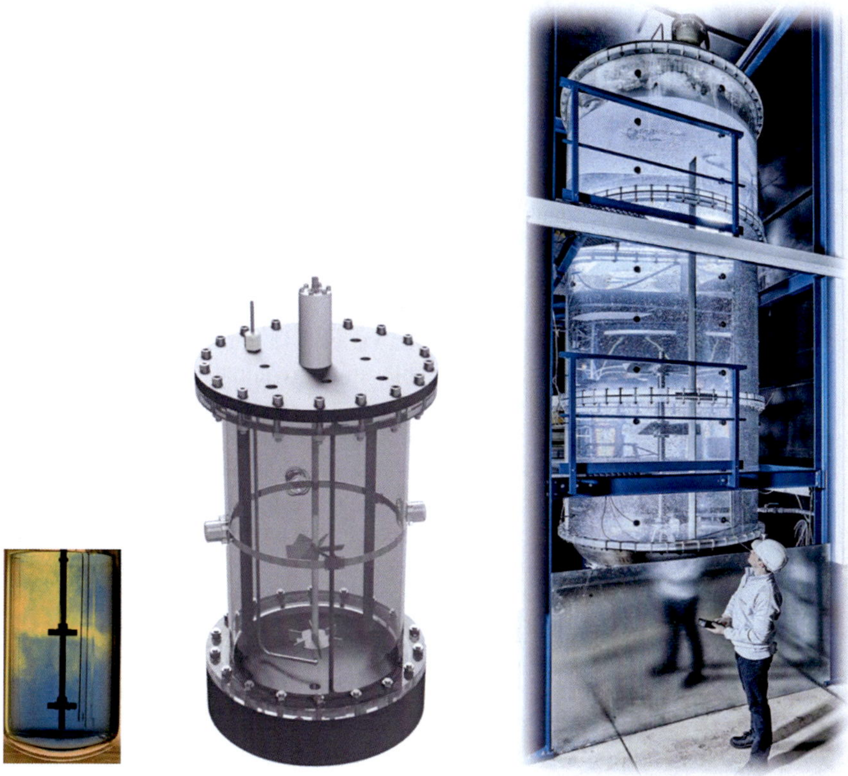

Fig. 3.1 Aerated and transparent acrylic glass stirred tank reactors at the Institute of Multiphase Flows in Hamburg. (left: 3 L Applikon, center: 30 L scale-down of the 15,000 L stirred tank reactor (right))

following are explicitly considered in the context of cell culture applications. This is due to the fact that cell culture processes are performed at significantly lower specific power inputs and aeration rates, when compared to microbial fermentations or chemical reactions. Moreover, the fermentation broth is a low-viscosity Newtonian medium. As a result, knowledge from other types of applications such as microbial fermentation or the chemical industry cannot be transferred directly.

Aerated stirred tank reactors are already intensively described in the literature. Therefore, for further information the reader is referred to the relevant books and publications by Zlokarnik, Nienow and Kraume [4–6]. In the following, the basic designs and components of stirred tank reactors are briefly discussed. This is used as a reference for the characterization methods related to cell culture application which are described in detail below.

Fundamentals of Aerated Stirred Tank Reactors

The design of stirred tank reactors is not subject to any limitations, but there are design rules and restrictions. For example, the reactor size is limited by the transportability of the components [5]. In addition, the cultivation of mammalian cells requires the highest standards of manufacturing quality regarding material and surface quality. Since stirred tank reactors are frequently used in process engineering, their design is regulated in DIN 28130. A stirred tank reactor essentially consists of a stirrer shaft with stirring elements, a motor, baffles and a temperature control unit. It is, however, the dimension of the individual components and their respective arrangement that determine the greatest design variety [4]. Furthermore, in the case of multiphase gas/liquid systems, a gas supply unit is required. This increases the degree of individual designs further. Besides classical stainless-steel tanks, single-use bioreactors (SUBs) become more popular.

Based on the basic components listed, it is clear that a large number of different designs are possible due to the free combination and positioning of the components within the system. This makes it clear that the physical relationships described in the literature contribute to understanding, but can rarely be applied directly to one's own applications. Especially, for cell culture applications, detailed equipment characterization is important for successful process transfer and process scale-up. In this case, equipment characterization is focused on the mechanical power input, volumetric mass transfer coefficient of oxygen, and mixing time. This is due to the fact that during the seed train phase, which includes reactors in the range of mL $< V_{\mathrm{fill}} <$ m^3, the same process conditions, such as oxygen level, concentration profile and maximum shear stress are always ensured. Figure 3.2 schematically shows the most

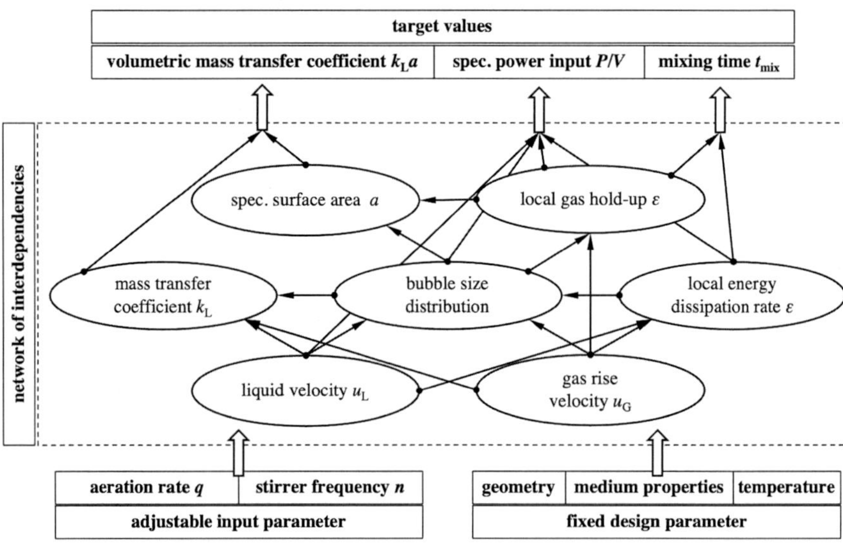

Fig. 3.2 Schematic representation of the dependencies between the adjustable input parameters and fixed design parameters on the target values

important target values characterizing the operation of aerated stirred tank reactors as a function of multiple parameters. For the latter, a distinction must be made between adjustable input parameters and fixed design parameters. Consequently, only the input parameters (aeration rate and stirrer frequency) can be used to influence the target values of mass transport and mixing during operation.

In addition, the volumetric mass transfer coefficient of carbon dioxide, the bubble size distribution, the gas hold-up, the general flow field and the maximum shear stress can be characterized. However, it is not always necessary to characterize all physical properties in detail, as not all physical properties of the system can be kept constant during scale-up [7]. Therefore, scale-up is often based on the target values listed in Fig. 3.2. Furthermore, a complete characterization of all physical properties hardly makes sense, since for example the volumetric mass transfer coefficient depends on the bubble size distribution, the local energy dissipation rate and the gas hold-up (see Fig. 3.2). However, in case of a troubleshooting, the characterization of further parameters can help to find proper solutions.

For a better understanding but also as a basis for the following explanations, the main properties of an aerated stirred tank reactor are discussed below, using the example of the transparent 15,000 L acrylic glass reactor. Figure 3.3 shows the basic components commonly used in the design of stirred tank reactors.

The number and shape of the baffles and stirrers in the system can be freely selected. In the case of low viscosity media, baffles are used to avoid the formation of a vortex and to increase the turbulence and thus the mechanical power input into the system. From four baffles upwards, the system is also called a fully baffled system [5]. A fully baffled system is characterized by the fact that the mechanical power input cannot be further increased by installing additional baffles. Furthermore, heating coils or built-in heat exchangers can also act as baffles. However, such installations are always difficult to clean due to their complex geometry and pose an increased risk of contamination, especially for cell culture processes. Therefore, in case of heat transfer for cell culture, a simple double jacket reactor can be used, in

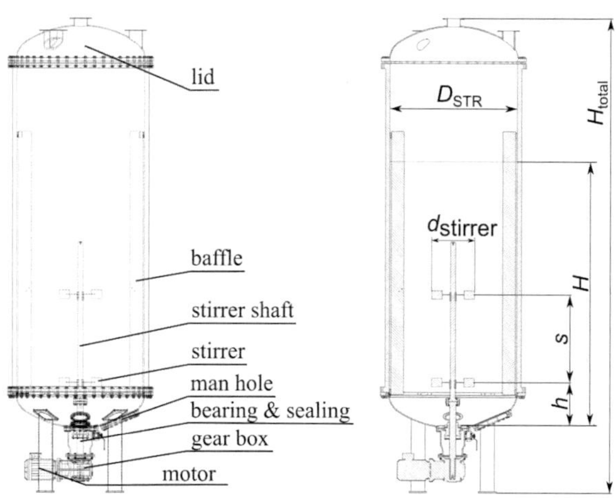

Fig. 3.3 Schematic visualization of an industrial stirred tank reactor showing the main components (left) and the main dimensions of the reactor geometry (right)

which the cold/hot medium circulates in the double jacket and the temperature in the stirred tank reactor is controlled by an external heat exchanger.

In principle, an aspect ratio of filling height to reactor diameter $\frac{H}{D_{STR}} = 1$ is recommended and can often be found in small scale applications [5]. However, the vessel diameter rapidly increases with increasing reactor volume $D_{STR} > 4.6$ m [5], which limits the vessel size due to transport constraints. For this reason, an aspect ratio of $\frac{H}{D_{STR}} > 1$ is used for large-scale stirred tank reactors with a volume $V_{STR} > 2$ m^3. This results in disadvantages, which must be solved by modifying the design. On the one hand, it is recommended to distribute several stirrers over the reactor height H. This means that the stirrer shaft must be longer and thus more stable. On the other hand, the mixing time increases accordingly with increasing aspect ratio. Particularly, when using several radially pumping stirrers, compartmentalization of the system may occur. This must either be taken into account in the design or compensated for by alternative stirrer combinations.

The choice of stirrer is based on two main operating conditions and three operating tasks. For the operating conditions, a distinction is made between axially and radially pumping stirrers. For the operating tasks, a distinction is made between homogenizing, dispersing and suspending. Figure 3.4 shows examples of the idealized single-phase flow structures of axially and radially pumping stirrers as single- or multi-stage configuration. It can be assumed that the ideal flow structures shown are independent of the stirrer frequency for Newtonian media. In contrast to a double configuration of two axially pumping agitators, multiple coherent vortex structures form in a double radially pumping configuration. Axially pumping stirrers are used for axial mixing and homogenization of the continuous phase.

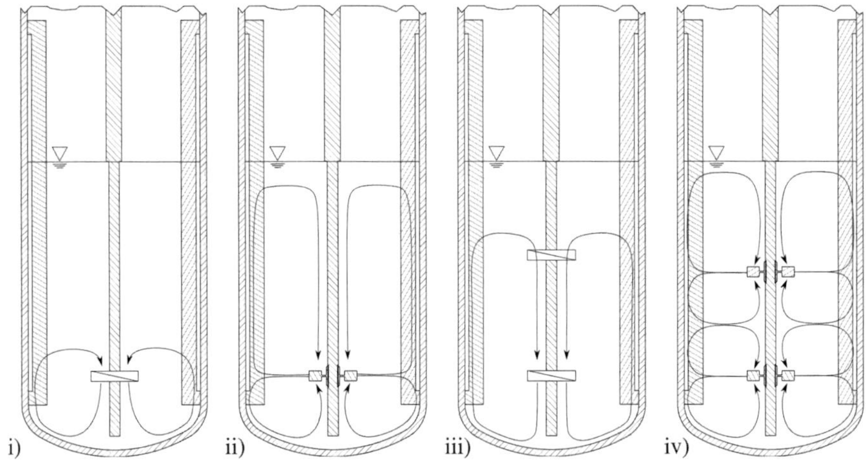

Fig. 3.4 Representation of the characteristic ideal single-phase flow behavior of a: (i) single pitched blade turbine (down pumping), (ii) single Rushton turbine, (iii) double pitched blade turbine (down pumping) and (iv) double Rushton turbine

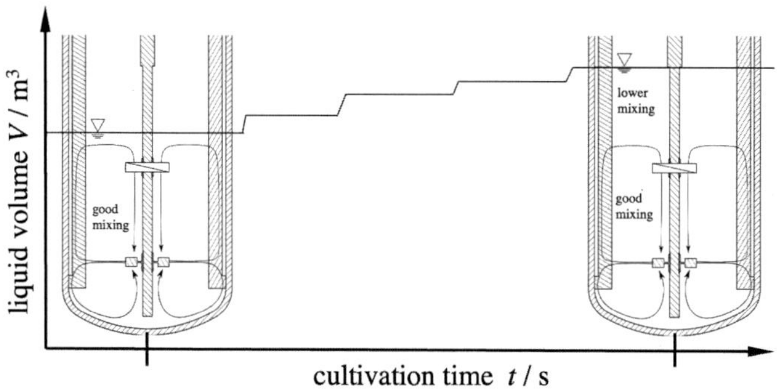

Fig. 3.5 Exemplary illustration of the dominant single-phase flow structure as a function of the cultivation time and the reactor volume, respectively

Furthermore, it must always be considered that classical cell culture applications are operated in fed batches. In this case, a minimum overlap height of the top stirrer to the start volume must be maintained. As disadvantage of this configuration, the upper reactor volume is not actively stirred. The increasing reactor volume results in a larger volume above the upper stirring element and, in the worst case, in the formation of a compartment, which is characterized by lower mixing. The influence due to the changing reactor volume is exemplarily shown in Fig. 3.5.

The compartments resulting from changing liquid volume can be broken up by the gaseous phase depending on the operating parameters [8]. Furthermore, the change in reactor volume also causes deviation in mass transfer and mixing behavior.

In the case of conventional cell culture applications, the stirrers are utilized for homogenizing the liquid phase as well as for dispersing the gaseous phase. Radially pumping stirrers are particularly suitable for dispersing, since the gaseous phase is dispersed in the vortex trails caused by the stirrer blades and distributed radially over the reactor cross-section [9]. Nevertheless, the dispersing effect depends on the stirrer type, the aeration rate as well as on the reactor scale [10]. More precisely, the size ratio of bubble size to stirrer blade height determines how much the stirrer can effectively contribute to gas dispersion, since cell culture processes in particular are limited in maximum power input. Depending on the aeration rate and the stirrer frequency, a distinction is made between three conditions, which are shown as examples in Fig. 3.6.

The most efficient gas dispersion takes place in the loading regime. However, if the stirrer frequency is reduced at constant aeration rate below a critical minimum stirrer frequency or if the aeration rate rises above a critical value at constant stirrer frequency, the stirrer may be flooded by the gaseous phase. The flooding regime is characterized by a reduced mechanical power consumption and a significantly reduced gas dispersion. This is usually accompanied by reduced mass transfer performance as well. To describe the important two-phase flow structure, that dominates the mixing and mass transfer performance, the dimensionless Flow number

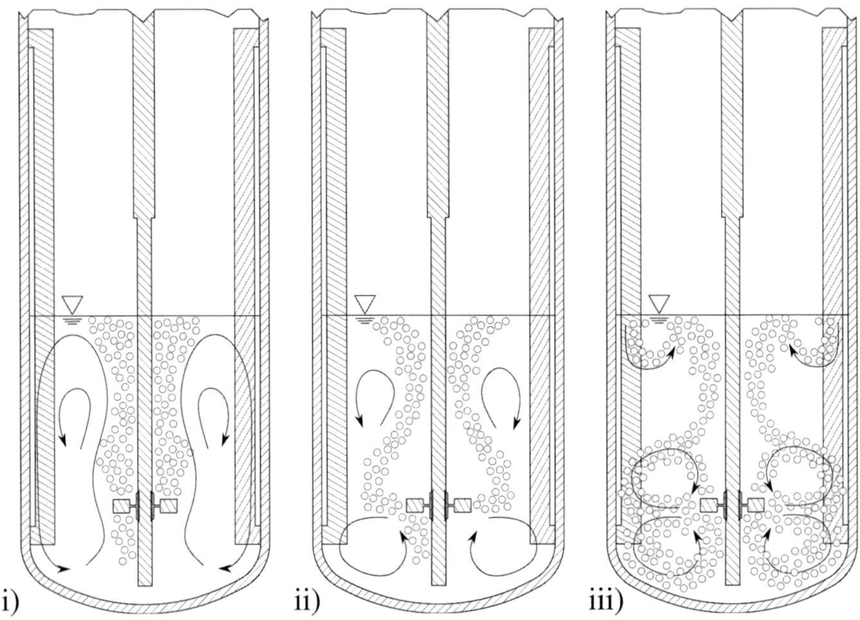

Fig. 3.6 Visualization of the characteristic stirrer gas/liquid behavior of an aerated stirred tank reactor with a single radially pumping stirrer: (i) stirrer flooding, (ii) stirrer loading, (iii) stirrer loading with gas recirculation, according to Warmoeskerken and Smith [11]

$$Fl = \frac{q}{n\,d^3} \tag{3.1}$$

with the aeration rate q, the stirrer frequency n and the stirrer diameter d as well as the stirrer Froude number

$$Fr = \frac{n^2\,d}{g} \tag{3.2}$$

with the gravity constant g can be used. Figure 3.7 shows the typical dependence of the critical gas flow rate on the Fl number for aerated stirred tank reactors according to the literature. The dependencies of the transition point from the flooding to the loading regime shows a similar dependence on the Fr number for the different correlations. The differences can be explained in particular by the fact that different stirrers or reactor geometries have been used. The data points in Fig. 3.7 show measurement data from Rosseburg et al. [8] obtained from the transparent acrylic glass reactor in Hamburg, which agree well with the literature data from Mikulcova et al. [12]. Additinally, Fig. 3.7 show typically operation windows for different reactor scales. Based on the operating parameters in Fig. 3.7, it is evident that a switch between loading and flooding on an industrial scale is to be expected in cell culture applications.

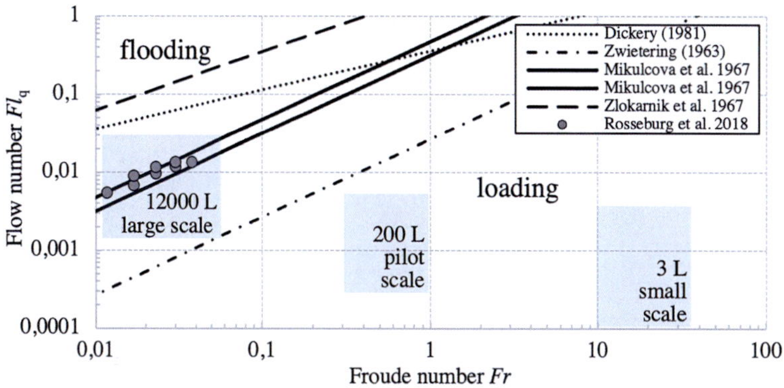

Fig. 3.7 Transition between loading and flooding regime in aerated stirred tank reactors of different geometry and size [13], blue boxes mark typical operation windows

While the choice of the stirrer has an influence on the gas dispersion, the choice of the sparger type is not less important. With increasing stirred tank reactor scale (filling volume mL < V_{fill} < m^3), the ratio of bubble size to stirrer blade height decreases. Thus, the influence of the stirrer becomes more important with increasing stirred tank reactor scale [8, 14]. Common sparger geometries are simple open tubes, tubes with several holes, sintered sparger or even membranes for bubble-free aeration. However, membranes are not described here, since their use is a special case in the operation of stirred tank reactors.

An important parameter for the description of sparger is the Weber number

$$We = \frac{d_{\text{orifice}}\,\rho\,v^2}{\sigma} \tag{3.3}$$

with the orifice diameter d_{orifice}, the liquid density ρ, the surface tension σ and the gas exit velocity v [14]. The initial bubble size depending on the Weber number has a direct influence on their rising velocities and therefore the two-phase flow structure and the mixing as well as mass transfer performance. For this reason, knowledge of the bubble size distribution and two-phase flow structure within the reactor is crucial for efficient and successful scale-up and process transfer. For this reason, the most important measurement methods for the characterization of aerated stirred tank reactors are described below and their practical implementation is discussed.

Experimental Characterization Methods for Aerated Stirred Tank Reactors

The characterization methods for aerated stirred tank reactors in context of cell culture applications are described below.

Characterization of Power Input

The mechanical power input of the stirrer is an important characterization parameter for aerated stirred tank reactors, which is frequently used not only for the design but also for the scale-up of the reactor setup [15].

With regard to stirred tank reactors, the volumetric power input P/V or the mean energy dissipation rate

$$\varepsilon = \frac{P}{V} * \rho^{-1} \tag{3.4}$$

with the mechanical power P, the filling volume V and the density ρ is often used. The mean energy dissipation rate ε is furthermore used to quantity the hydrodynamic stress. Especially for cell culture processes with mammalian cells, knowledge of the maximum hydrodynamic shear stress and shear stress distribution are important, since the maximum shear stress has an influence on important cell parameters such as viability, cell growth or even cell death [3, 16]. The exact knowledge of the power input is therefore a decisive characterization parameter that determines the success or failure of aerated stirred tank reactor operations. For this reason, the most important fundamentals and methods for determining the mechanical power input are discussed in detail below. The characterization of mechanical power input is distinguished between the single-phase and two-phase operation. Due to the fact that oxygen demand in cell culture processes is comparatively low (compared to microbial applications), a characterization in single-phase operation is usually sufficient. Nevertheless, the influence of the gaseous phase on the power input is also described below.

The most important dimensionless numbers for the description of aerated stirred tank reactors are the Reynolds number

$$Re = \frac{n\, d^2}{\nu} \tag{3.5}$$

with the stirrer frequency n, the stirrer diameter d and the liquid viscosity ν as well as the power number

$$Po = \frac{P_{\mathrm{stirrer}}}{\rho\, n^3\, d^5} \tag{3.6}$$

with the liquid density ρ and the mechanical power

$$P_{\mathrm{stirrer}} = 2\, \pi\, M_{\mathrm{stirrer}}\, n \tag{3.7}$$

with the stirrer frequency n and the stirrer torque M_{stirrer}.

If the stirrer torque M_{stirrer} cannot be measured directly because the torque sensor is located on the motor or behind the shaft bearing, for example, the idle torque of the bearing has to be measured in air first. A variety of measuring techniques are known for measuring mechanical stresses on shafts. Analog measuring techniques include classical displacement and angle indicators, such as the linear moving coil movement [17] or pendulum machines. Analog instruments are based on laws of classical mechanics. Digital measuring instruments are usually based on the use of electrical torque transducers. Strain gauges are among the best-known torque transducers [18]. In this case, the electrical resistance R_0 of the strain gauge changes proportionally to the path length $R = R_0 \pm \Delta R$ when the length Δl changes. In order to achieve the greatest possible absolute changes in path length with small component dimensions, strain gages are often designed in a meandering shape [17]. Furthermore, torque sensors are available which measure the torsional force on the shaft by means of piezoelectric sensors. Piezoelectric sensors generate a voltage proportional to the applied force when pressure is applied. Compared to strain gauges, piezoelectric sensors are characterized by a larger measuring range, dynamics and a higher overload capacity.

Regardless of the measurement method, torque is proportional to frequency squared. This physical relationship can be used to check the torque measurement for plausibility. In the case of a fully baffled system, no aeration and operation in the turbulent regime, the torque

$$M_{\text{stirrer}} = a\, n^2 + b \tag{3.8}$$

is proportional to the stirrer frequency squared [19]. The linear deviation b from the origin of the regression includes static errors of the measurement method like total measurement offset. This error must be subtracted from the measured torque. Therefore, for the turbulent regime, the unaerated power number

$$Po = \frac{P_{\text{stirrer}}}{\rho\, n^3\, d^5} = \frac{2\,\pi\, n\, M_{\text{stirrer}}}{\rho\, n^3\, d^5} = \frac{2\,\pi\, n\,(a\, n^2)}{\rho\, n^3\, d^5} = a \frac{2\,\pi}{\rho\, d^5};\; Re > 10^4 \tag{3.9}$$

is calculated based on the slope a of the fitted torque. The determination of the power number for the turbulent range has the decisive advantage that the uncertainties in the determination of the idle torque are reduced. This is clearly shown in Fig. 3.8, which shows the stirrer torque and the idling torque plotted against the stirrer frequency.

For the transition regime and the laminar regime this relationship does not apply and the power number is calculated directly with the measured torques with respectively higher uncertainties.

The power input characteristics of different stirrers and stirrer combinations are shown in Fig. 3.9 for single-phase operation. In single-phase operation, the power number is a function of the Reynolds number and can be divided into three regions. The laminar, the transition and the turbulent region. For a fully baffled system, the power number in the laminar region is inversely proportional to the Reynolds number. Whereas in the turbulent region the power number is more or less constant.

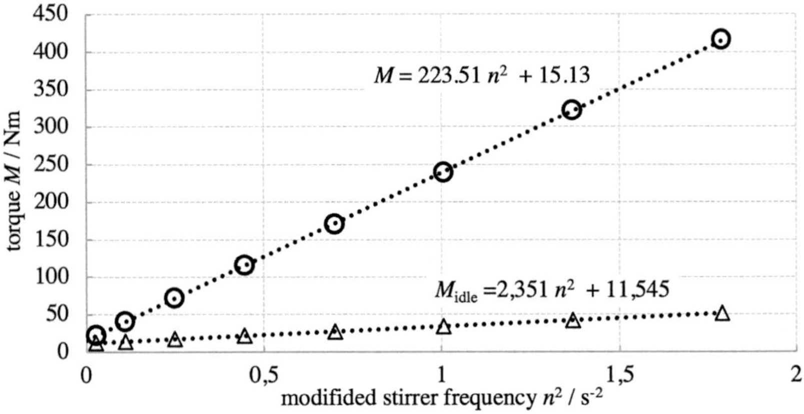

Fig. 3.8 Schematic visualization of the measured total and idle torque depending on the modified stirrer frequency n^2

In systems without baffles, the power number will decrease constantly even in the turbulent region, which is especially problematic for scale-up. Finally, in fully baffled systems, it can be assumed that as long as the stirred tank reactor is operated in the turbulent regime, the Newton number will be constant. Furthermore, the power number is a stirrer-specific number and is further dependent on the overall reactor geometry (installation height, aspect ratio, number of baffles) and must therefore always be determined for the stirred tank to be described.

Figure 3.9 illustrates that, in order to determine the power input characteristics of stirrers for all flow regimes (laminar to turbulent), a range of Reynolds coefficients have to be covered over several decades. The simplest variant for this will be the variation of the stirrer frequency, whereby very low stirrer frequencies must be set for the laminar range. The resulting very low required torques are extremely difficult to measure. Furthermore, a torque measuring device will be necessary, which can measure in the range of a few Nmm to several Nm. For this reason, it has proven more practical to vary the viscosity using glycerol. This has the advantage that neither the setup nor the measuring system has to be changed.

Furthermore, the power input characteristics under aeration conditions have to be investigated, since the mechanical power input of the stirrer decreases depending on the stirrer type and the stirrer configuration.

In the literature, a large number of different empirical correlations exist which describe the effect of aeration on the mechanical power input [19]. However, the fact that the proposed correlations cannot always be applied without restrictions can be seen in Fig. 3.10. It shows the relative drop in power input by aeration as a parity plot compared to literature. The measured drop in power number under aeration conditions for a 30 L and an industrial 12,500 L stirred tank reactor are plotted against those calculated according to Calderbank [20]. The further parameter refers also to the dimensionless gas flow number Fl_G.

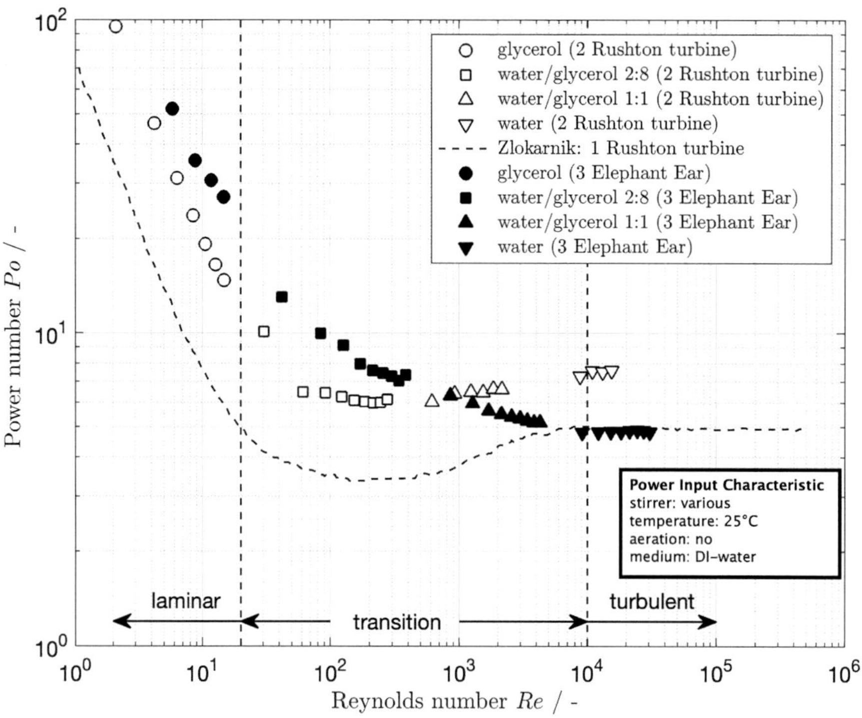

Fig. 3.9 Power input characteristics for different stirrers in a single phase operated baffled stirred tank reactor

Figure 3.10 shows further the influence of the stirrer configuration on the power number under aeration. For this purpose, the stirrer clearance *s* between the stirrers has been varied in relation to the stirrer diameter *d*. It can be seen that the influence of aeration on the mechanical power input for the 30 L stirred tank reactor can be predicted well by the correlation, whereas the prediction for the industrial stirred tank fails.

Measurement of Global and Local Mixing Time

An important key parameter, which is often used to characterize aerated stirred tank reactors, is the mixing time. Especially for multiphase operation, where two or more phases have to be continuously brought into contact over a long period of time. In case of cell culture processes, supplied constant nutrient and dissolved oxygen concentration have to be ensured in the entire reactor volume. Efficient mixing ensures sufficient nutrient supply in the entire system over the entire process time. Furthermore, efficient mixing minimizes local concentration gradients (e.g. after pH adjustment). In the following, fundamentals of mixing and the characterization of mixing time and mixing performance in stirred tank reactors for cell culture are

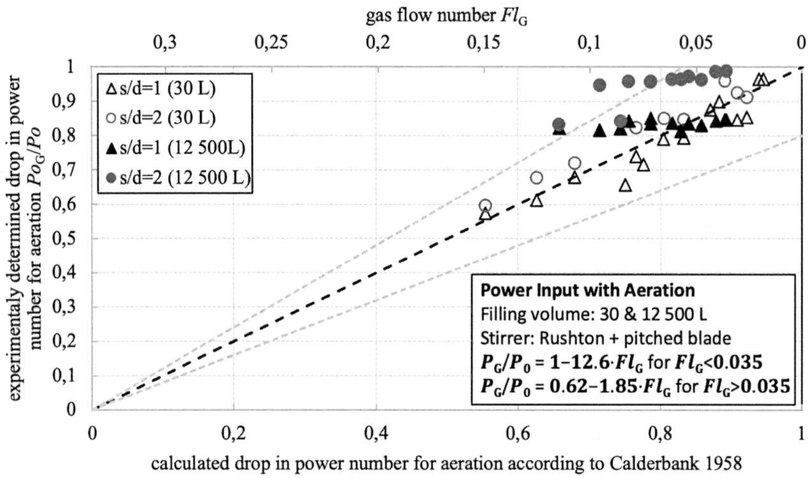

Fig. 3.10 Comparison of the influence of aeration on the mechanical power input for different stirred tanks and stirrer combinations (d: stirrer diameter, s: stirrer clearance) with literature data

described. Furthermore, important terms such as micro, meso, and macro mixing but also local and global mixing times are described. At this point, the physical fundamentals of mixing are discussed. Based on the fundamentals, typical mixing behaviors of aerated stirred tank reactors are described. Finally, the common measurement methods for the determination and characterization of the mixing time are described.

Definition of Mixing

The mixing time describes the time until a full homogenization of two or more different phases (liquid/liquid or solid/liquid) is reached. For this reason, mixing time is often defined by the degree of homogeneity in a system. To determine the mixing time

$$\theta_{0.05} = t(I(t) = 1 \pm 0.05) \tag{3.10}$$

usually the 95 % criterion is used [19, 21, 22] which describes the time at which the measured variable I remains within a range of 5 % around the final value after a step response. Figure 3.11 shows an example of the time course of the measured variable I after a step response measurement. The system is considered homogenized when 95 % of the final value is reached. The 95 % criterion is a statistical tool, which is used to allow better validity and comparability, because it can take forever to reach a stable 100 % homogeneous state. This can be due to the measurement position, the response characteristics of the probe but also due to static experimental errors.

Furthermore, it is necessary to distinguish in the description of the mixing time between macro, meso-, and micro mixing [23]. Due to the fact that the mixing time is defined by the compensation of concentration differences, the concentration

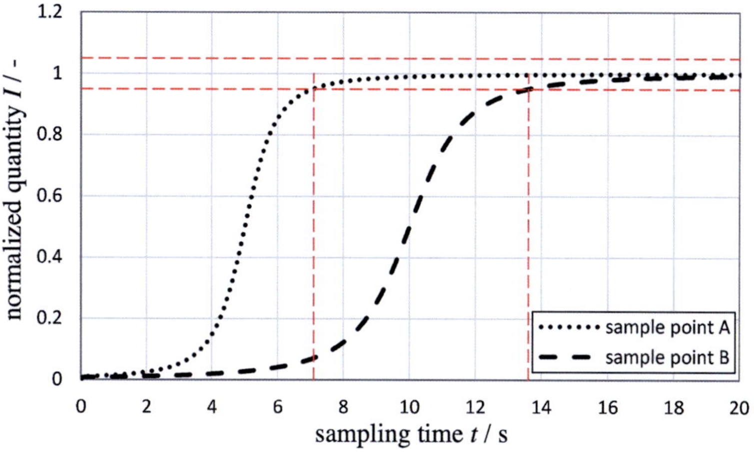

Fig. 3.11 Illustration of the development of a measurement quantity I after a pulse response measurement

compensation must be equalized on a micro scale. This means that the stirrer is responsible for mixing on the macro scale, but without diffusion on the micro scale, complete homogeneity cannot be achieved [5]. An important parameter describing the length scale of micro mixing is the Kolmogorov's length

$$\lambda = \left(\frac{\nu^3}{\varepsilon}\right)^{1/4} \tag{3.11}$$

with the volume average energy dissipation rate ε and the liquid viscosity ν, which describes the minimum expected vortex size. Whereas macro mixing depends on the reactor geometry and the stirrer, the time for micro mixing

$$\theta_{micro} = \frac{\lambda^2}{D} \tag{3.12}$$

depends only on the diffusion coefficient D and the smallest vortex size λ in the reactor and is thus independent of the reactor geometry [5]. Because macro mixing is largely dependent on the reactor geometry and thus the equipment used. Especially for equipment characterization as well as process transfer and scale-up, knowledge of the macro mixing is crucial. ed. Furthermore, in case of cell culture cultivation the overall mixing depends mainly on the macro mixing. For this reason, an experimental method to distinguish between micro and macro mixing has been developed [24]. The method is based on the optical observation of the mixing time and the specific control of the excess of hydrochloric acid used for the neutralization reaction with sodium hydroxide solution. Figure 3.12 shows an example of the measurement procedure for determining micro and macro mixing. To determine the micro mixing,

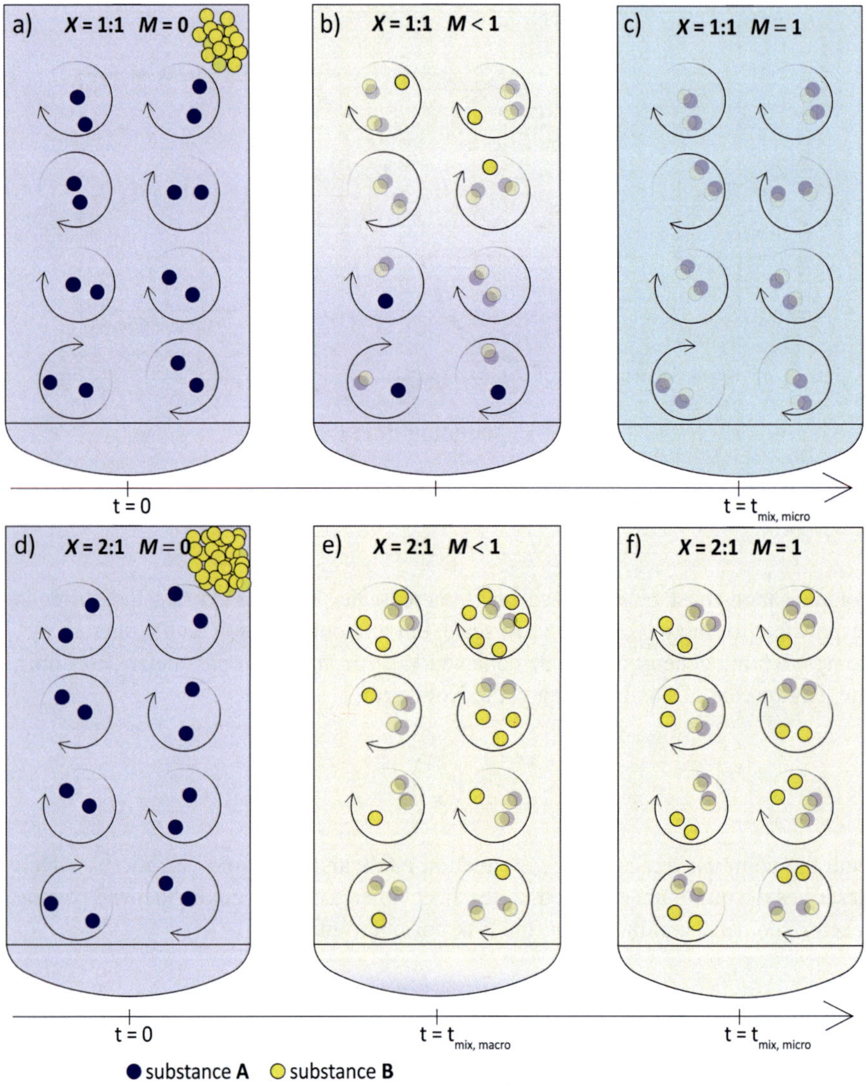

Fig. 3.12 Micro vs. macro mixing – Schematic representation of macro and micro mixing of two substances A and B with the same substance mixture (upper row) and a double excess of substance B (lower row); M: degree of homogeneity

a 1:1 mixing ratio of acid and alkali must be used, since complete color change can only be achieved at homogeneity with equimolar mixing. The pH tracer used to observe mixing is bromothymol blue. Bromothymol blue has the advantage of being able to differentiate between the acidic (yellow), neutral (green) and basic (blue) pH regions. For the determination of macro-mixing, the consideration of the degree of homogeneity is not important, since on a microscopic scale it is observed where in the system the tracer has already spread. To determine macro mixing, the concentrations do not necessarily have to be balanced on the microscopic scale. The distribution of acid in the reactor volume leads to a color change from blue to yellow as soon as an excess of acid prevails in each sub volume.

Besides the definition of micro and macro mixing, the definition of local and global mixing time is described in more detail below. The term 'local' and 'global' mixing time is used in respect to the framework for which the mixing time is described. The global mixing time of a system is a mean value for the entire system and, therefore, a mean of all locally determined mixing times. A subdivision into macro or micro mixing can be made further for a local as well as global description [24].

Two-Phase Mixing

For the mixing characterization of stirred tank reactors, a distinction is made between mixing in aerated and unaerated operation. The distinction is important, since the dispersed gaseous phase has a significant influence on the mixing behavior. Commonly, the mixing time decreases with increasing stirrer frequency respectively power input without aeration. In case of aeration, the interaction of the stirrer with the dispersed gaseous phase can intensify or even significantly reverse this effect. This can be primarily explained by the prevailing operating condition of the stirrer. Thus, the mixing time of the system can be significantly reduced when the stirrer is flooded (see Fig. 3.6). However, if the stirrer frequency is increased at a constant aeration rate and the stirrer regime changes from flooding to loading, the mixing time will increase even though the mechanical power input has been increased.

The two-phase mixing time characteristic is shown in Fig. 3.13. The mixing time characteristic is plotted against the stirrer Reynolds number *Re*, which has been measured and correlated for an industrial scale aerated stirred tank reactor by Rosseburg et al. [8]. In the range of small Reynolds numbers and aeration rates, the global multiphase mixing time is much smaller than the single-phase mixing

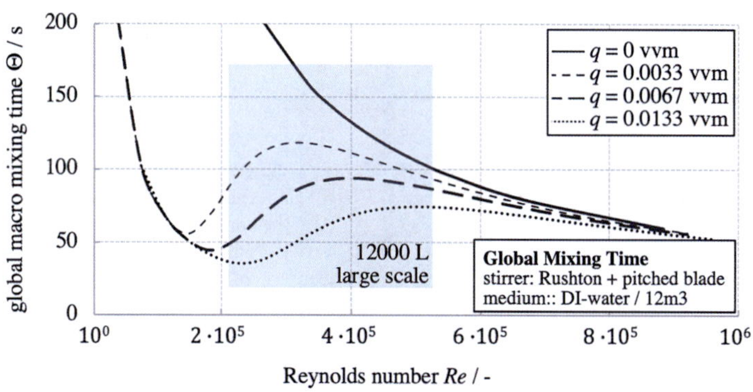

Fig. 3.13 Multiphase mixing time characteristic of an industrial scale aerated stirred tank reactor [8], blue box marks typical operation windows of the investigated STR

time. In the range of $2 \cdot 10^5 < Re < 4 \cdot 10^5$ the multiphase mixing time increases with increasing Reynolds number as a function of the aeration rate until the multiphase and single-phase mixing time values collapse at high Reynolds numbers.

The determination of the time required for mixing is generally based on the measurement of an impulse response. For example, a local peak in conductivity is used to measure the change in conductivity over time. For this purpose, a small volume of increased conductivity, for example, is added locally at time t_0 in order to measure the conductivity over time. Instead of a step response by means of increased conductivity, a temperature or pH step can also be used or any other process variable which can be measured reliably. To explain the different experimental measurement methods, the most important methods are listed below and discussed in the context of cell culture application.

Measurement Methods

In general, the measuring methods for determining the mixing time can be divided into invasive and non-invasive methods. Figure 3.14 provides a general overview of invasive and non-invasive measurement methods for determining the mixing time of stirred tank reactors, which are discussed in more detail below.

Non-invasive methods are based on the optical or tomographic observation of the mixing process, which can be visualized by means of different methods like colometry, fluorescence or density differences. Invasive methods are usually related to concentration, temperature or pH measurements by means of local probes.

Furthermore, a distinction between chemical and physical methods has to made. The chemical or physical methods differ in the mechanism by which the mixing is quantified. Basically, the mixing is quantified by measuring a step response. How the step response is generated is specific to each method. For example, in the chemical method, the neutralization reaction is used to measure the pH invasively or to visualize it non-invasively using a colorimetric pH tracer. In physical methods, the

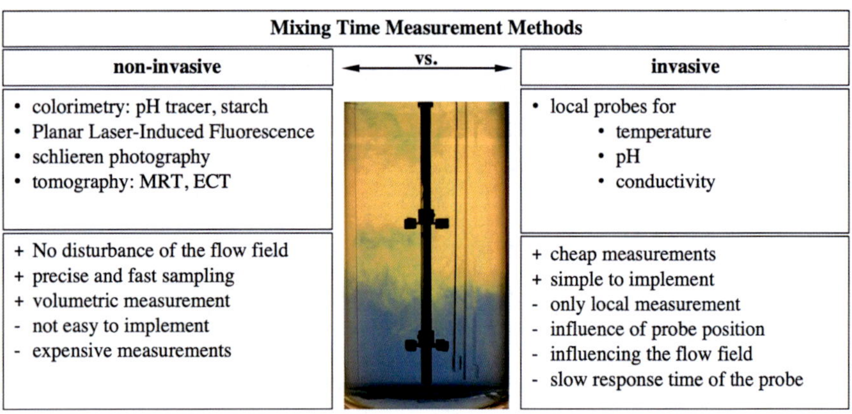

Fig. 3.14 Overview of invasive and non-invasive measurement methods for determining the mixing time in stirred tank reactors. Image shows the blending of an acid-base reaction visualized with bromothymol blue

conductivity, temperature, density or pH is influenced by adding a small amount of liquid with respectively higher physical property, which can be measured invasively using local probes. The change in physical properties can be also measured by non-invasive imaging methods such as PLIF (Planar Laser-Induced Fluorescence) or tomographic methods [21].

The chemical methods are also often called "optical methods" because the color of the medium in the stirred tank reactor changes based on a chemical reaction. With the chemical method, a distinction is made between single and multiple transition systems. The single transition systems include the redox reaction of sodium thiosulfate with iodine or the neutralization of sodium hydroxide solution with hydrochloric acid [25]. The multi-transition system includes the neutralization reaction visualized by a multicolor pH tracer like bromothymol blue.

Furthermore, it is important to differentiate between local and global measurements. Local measurements refer to the measurement of a physical parameter of the system (e.g. conductivity) at a specific location by means of a single or multiple probes. Global measurements mean the observation of, for example, a neutralization reaction by means of a pH-sensitive tracer. In case of local probes, the response time of the probe is particularly important and has to be significantly faster than the expected mixing time. Optical methods usually do not have a problem with the response time of the measurement method, since, for example, in the neutralization reaction the pH shift can be visualized almost instantaneously. Furthermore, in case of invasive measurements, the measuring position has several influences on the result depending on the size of the stirred tank for the measurement of the mixing time. Whereas in small stirred tank reactors, the probe influences the flow field due to its relative size to the reactor diameter and thus also the mixing time. In larger stirred tank reactors, for example, the probe may be installed in a small, poorly mixed area, which is not representative of the entire stirred tank reactor [26, 27].

Global measurements can be made either by PLIF or by other optical visualization methods. PLIF is an established method to measure concentration fields on a single slice in the system. The method can also be used to observe the concentration over time, and thus to describe the mixing. However, since the method requires expensive lasers and optical lenses, as well as highly sensitive and fast cameras, it is not easy to implement. A limitation of PLIF is that the mixing can only be observed on one plane. This means that the resulting mixing time is thus only valid for the evaluated slice, which is, however, considerably more significant than in the case of a local measurement at only one point. Measurement methods using a simple color change are also referred to as the decolorization method. The change in color can either made by redox reaction of sodium thiosulfate with iodine or by the neutralization of sodium hydroxide solution with hydrochloric acid. The advantage of the decolorization method is that possible dead zones are clearly visible, but an excess of iodine or hydrochloric acid cannot be detected [19]. This is due to the fact that the threshold for the color change is not necessarily associated with complete homogeneity in the system. In order to obtain a broader spectrum of "decolorization", for example in the neutralization reaction, multi color pH indicators such as bromothymol blue can be used instead of commonly used phenolphthalein.

Visualization of the Flow

The knowledge about the flow fields in aerated stirred tank reactors contributes decisively to the understanding of the process. Based on the exact knowledge of the flow fields, mixing processes, for example, can be better understood. Furthermore, flow field measurements can be used to identify possible existing gradients (temperature, oxygen or nutrient concentration, pH) and to optimize the process based on them. Finally, experimentally determined flow fields help to validate numerical simulations, which in turn can provide much deeper insights into the process. In the following, the most important theoretical principles are described, as well as an overview of the available measurement techniques.

For the description of flow fields, a distinction between the Eulerian and the Lagrangian approach is made. In the Eulerian approach, the flow field is primarily regarded as a continuous phase, and the Navier-Stokes equations for the flow field are solved using the finite volume scheme method. Mostly utilized in the three-dimensional cartesian coordinates, this method involves establishing a fixed measurement volume in space, through which the fluid is moving. The Lagrangian approach treats the flow as a discrete phase consisting of particles moving in space and carrying specific physical properties. The trajectory of the fluid particle is plotted and monitored based on its position relative to its surrounding. In the fluid element's viewpoint, it remains stationary while the fluid around it is in motion [28]. By means of the combining of the Eulerian and Lagrangian approach, it becomes possible to couple a microorganism's reaction (within the Lagrangian framework) with the environment it encounters (within the Eulerian framework) [29]. In the following, a closer look at the Eulerian and Lagrangian flow specifications, and how they are connected to each other, is taken.

The Eulerian specification of a flow field defines the instantaneous distribution of the flow characteristic, such as velocity or pressure, over the control volume. Thus, the Eulerian velocity u_E

$$u_E(x, y, z, t) = (u(x, y, z, t), v(x, y, z, t), w(x, y, z, t)) \tag{3.13}$$

is defined as a function of space and time. For each time step, the three components u , v and w describe the velocities at a certain point in space parallel to the respective axes x , y and z.

The Lagrangian specification of a flow defines the characteristics of a flow as a function of the positions of a fluid parcel and time. The three components of the velocity give the instantaneous velocity of the parcel in the directions of the respective axes. The Lagrangian view is understood as not defining a distribution, but rather, as looking at individual small fluid parcels moving in the Eulerian flow field. These parcels must be sufficiently small, meaning that they should not affect the Eulerian velocity field in any way, but just passively flow with it. The Lagrangian velocity u_L

$$u_L(X_L, t) = u_E(X_L(t) = (x_L, y_L, z_L), t) = \frac{dX_L(t)}{dt} \tag{3.14}$$

of a parcel X_L at a given time t is equal to the Eulerian velocity at the position of the parcel at the same time t. The derivative of the Lagrangian velocity, material derivative or acceleration of the fluid parcel, can be expressed as:

$$\frac{du_L(X_L, t)}{dt} = \frac{\partial u_L}{\partial t} + \frac{\partial u_L}{\partial x_L} u_L + \frac{\partial u_L}{\partial y_L} v_L + \frac{\partial u_L}{\partial z_L} w_L \tag{3.15}$$

The Lagrangian parcel follows its trajectory, which is defined as the initial position of the parcel and the integral of the Lagrangian velocity as shown in Eq. 3.15. Integration of the trajectory yields the position of the Lagrangian parcel as shown in Eq. 3.16 [30].

$$X(t) = X(t_0) + \int u_L(X(t), t)dt \tag{3.16}$$

$$X(t) = X(t_{n-1}) + \frac{1}{2}\Delta t(u_L(t) + u_L(t_{n-1})) \tag{3.17}$$

Various measurement methods can be used to describe velocity fields in aerated stirred tank reactors. However, when selecting the appropriate method for the particular system to be characterized, some aspects have to be taken into account. The major difficulty to be overcome is the general access to the system, since usually the system to be characterized is permanently installed, which means that for existing plants either a transparent replica has to be made or, for new plants, appropriate measurement time has to be scheduled for characterization before commissioning. Furthermore, fermentation broths are usually opaque media and thus poorly accessible for optical measurement methods, which further limits the choice of suitable measurement methods. As a result, characterization studies for cell culture applications are often carried out in a PBS-Pluronic medium, which has comparable rheological properties and is also much easier to use.

An overview of the most common measurement techniques for describing the flow fields in case of cell culture application is given in Fig. 3.15. In addition, further measurement techniques as well as variations of the measurement techniques listed here can be found in the literature [31–33].

Development and Application of Lagrangian Sensor Particles

The past decades have witnessed growth in biotechnological processes due to significant advancements in microbiology, genetic engineering, animal cell cultures, genomics and proteomics. Thereby, process monitoring and process control are fundamental keys in order to guarantee reproducibility, a high-quality standard

Flow Field Measurement Methods

opaque media	vs.	transparent media
invasive probes: • hot wire anemometer • impeller anemometer + *easy to handle* - *disturbs the flow*		**invasive probes:** • hot wire anemometer + *easy to use* - *disturbs the flow*
non-invasive probes: • positron emissions particle tracking • X-ray CT • Magnetic Resonance Tomography + *3D flow field* - *expensive* + *no interference* - very complex		**non-invasive probes:** • laser doppler anemometer + *very accurate* - *local & average* • particle image velocimetry + *state of the art* - *complex evaluation* • particle tracking velocimetry + *precise 3D data* - *complex setup* + *simple evaluation* - *expensive*

Fig. 3.15 Overview of invasive and non-invasive measurement methods for determining the flow field in stirred tank reactors. Image shows a flow field measurement combined with an image of the blending of an acid-base reaction visualized by bromothymol blue

and to increase efficiency and productivity. To do so, locally fixed environment sensors are utilized, such as temperature, pressure, (dissolved) gas, pH or biomass density sensors [34]. However, especially industrial-scale bioreactors would require a spatio-temporal acquisition of process parameters, and thus those static probes may not constitute a representative picture of what is really happening inside. Their single-point measurements near the reactor wall depict only a time dependent mean value for the respective volume fraction. Spatial and temporal gradients in form of dead zones and compartments, which are experienced by microorganisms in the course of the cultivation need to be estimated. Most recent studies on quorum-sensing and cell-cell interactions show the impact during dynamically changing environments [35, 36], which is why microbial *lifelines* [29, 37] must be investigated thoroughly to understand population heterogeneity during a fermentation [38]. Multiple research groups have addressed this issue from a numerical standpoint [39–42] or investigate large-scale impacts in multi-compartment scale-down reactors [38, 43]. However, experimental data are still necessary in order to validate Lagrangian particle simulations [44, 45], calibrate compartment models and eventually enhance the understanding of this topic.

That is the reason why a unique development in the healthcare industry served as an example for the process industry to overcome aforementioned obstacles: Mobilizing local sensors. Due to the commercial development of *radio pills* described by Rowlands and Wolff in 1961 [46] and Watson et al. [47] a year later, those cylindrically shaped particles with a length of approx. 20 mm were subsequently utilized for measurements of internal stresses of granular material in bunkers [48]. This direct linkage between a transducer and a transmitter in an encapsulated shell can be considered as one of the first wireless free-floating sensor particle system used for investigations in process engineering. In 2005 and 2007, Wadke et al. demonstrated with their "smart" sphere, which has a size of 25 mm in diameter, how to receive commands wirelessly and store temperature data continuously in its internal memory while being introduced into a rotary drum mixer [49, 50]. Moreover, mobilized environmental sensor devices were presented by Deng et al. to assess the risk of fish injury during spillway passage before and after the installation of baffle blocks at a hydroelectric dam [51–53]. Another team employs an instrumented particle, measuring 70 mm in diameter, to evaluate hydrogeomorphological processes in various environments, such as fluvial, coastal, and aeolian settings to understand the incipient entrainment of coarse particles on a bed surface [54, 55].

With respect to biotechnology, in order to measure the liquid circulation time inside a stirred tank reactor during a fermentation, first implementations took place in 1969 by means of a hollow sphere with a total size of 20 mm in diameter, equipped with a radio transmitter working at a carrier frequency of about 2 MHz. Basic evaluations were successfully obtained in stirred tanks of 450 L capacity [56]. Afterwards, further developments were conducted regarding measuring circulation times in stirred and aerated tanks by means of small encapsulated radio transmitters [57–60]. In the following section, nine recent research groups are briefly presented (see Fig. 3.16), which have adopted this approach to gain an internal view into a bioreactor.

Five of those designs may utilize wireless data transmission, whereas the remaining four designs are intended to store the recorded data internally. By means of applying and merging multiple sensor systems into a mobile, encapsulated,

Team	Size/mm	T	p	pH	C	c_{gas}	IMU	Further sensors
1	ø 7.9 - 8.9	X						
2	ø 25	X	X	X	X			
3	ø 44			X	X			Potassium, sodium, glucose
4	ø 60					X		
5	ø 40 - 90	X	X		X		X	Ultra-wide-band module
6	ø 39						X	Ultrasound transducer
7	ø 45 - 55	X	X	X		(X)	X	
8	ø 36							Acceleration, angular velocity
9	ø 40		X				X	

Fig. 3.16 Lagrangian Sensor Particle (LSP) developers for biotechnological applications – **1**: “Sens-o-Sphere”, Lauterbach et al. (Technische Universität Dresden, Germany); **2**: “smartCAPS”, Zimmermann et al. (smartINST SAS, France); **3**: Todtenberg et al. (Innovations for High Performance Microelectronics & Fraunhofer IBMT, Germany); **4**: “BPOD”, Stine et al. (University of Maryland, USA); **5**: Reinecke & Buntkiel (Helmholtz-Zentrum Dresden-Rossendorf, Germany); **6**: “Xploring WiseMotes”, Duisterwinkel et al. (Eindhoven University of Technology & INCAS3, The Netherlands); **7**: “Fermsense 3D”, Bisgaard et al. (Freesense ApS, Denmark); **8**: Cabrera & Cobelli (CNRS, France & Universidad de Buenos Aires and IFIBA, Argentina); **9**: Hofmann et al. (Hamburg University of Technology, Germany); Sensor types – T: temperature; p: pressure; pH: pH value; C: conductivity; c_{gas}: dissolved gas; IMU: Inertial Measurement Unit

free-floating Lagrangian Sensor Particle (LSP), their common goal is to understand cell-cell and cell-bioreactor interactions and to mimic the movement of a cell. For a thorough and critical review that evaluates existing numerical and experimental ideas and obstacles related to determining *lifelines* and applying a *lifeline* analysis, the reader is referred to Blöbaum et al. [29].

1. With a diameter of just 7.9 mm, Lauterbach et al. developed the smallest LSP for a biotechnological application, called *Sens-o-Spheres*. The inductively rechargeable spheres contain an antenna for sending measured temperature data (fastest transmission frequency of 0.1 Hz) via a carrier frequency of 433 MHz to a receiver outside of the reactor. Up to 24 LSPs can be utilized simultaneously [61–63]. Experimental trials show successful implementations not only in laboratory shake flasks, but also in a 200 L and 2000 L single-use bioreactor filled with deionized water.
2. The second team, which implemented *in situ* wireless measurements with different types of LSPs with the size of 25 mm in diameter is smartINST. Based on research findings by Zimmermann et al., either a temperature, pH, conductivity or pressure sensor was integrated in a shell made of polyether ether ketone (PEEK) [64–66].
3. Todtenberg et al. implemented as a third research team a radio communication system between their LSP and a base station with a carrier frequency of 433 MHz. The team's objective was to design a sensor capsule capable of remotely monitoring the growth of an algae culture inside a photobioreactor by means of potentiometric, amperometric and conductivity sensors. The first prototype showed a converged packet error rate of 26 % in nutrient solution while flowing through the reactor. Although a successful signal processing chain was demonstrated for the second prototype for potentiometric sensor values, further testing of the proprietary Biochemical Measuring System (BMS)-chip is still pending to the author's knowledge [67, 68].
4. A promising *in situ* oxygen measurement is demonstrated by Stine et al. Their LSP is called "bPod", it has a size of 60 mm in diameter and was already tested in a 2 L glass vessel and a 10 L lab-scale bioreactor. It includes a potentiostat analog-front-end, a Bluetooth Low Energy microcontroller, and a Clark-type dissolved oxygen (DO) sensor. The bPod was tested and found to have a linear electrochemical current response corresponding to DO saturation levels. Compared to the aforementioned research teams, the data transmission works via Bluetooth Low Energy Wireless Communication [69, 70].
5. In collaboration with the Agency for Renewable Resources (FNR) within the project "NeoBio" and the Münster University of Applied Sciences, the Helmholtz-Zentrum Dresden-Rossendorf develops LSPs, which were tested in an air-water column reactor, multiple biogas reactors and waste water treatment plants. Due to its integrated buoyancy control unit, achieved by a piston and the possibility to fill the void with payload, the first shape was rather cylindrical with a diameter of 45 mm and a height of approx. 85 mm. In 2021, a spherical design was presented with an outer diameter of 90 mm. A conductivity sensor was added

to an already integrated IMU, temperature and pressure sensor. Furthermore, a module based on ultra-wide-band radio technology performs localization and data transmission at the fluid surface. The third design is also spherical, but only 40 mm in diameter, and thus equipped with less sensors due to the limited space. Current studies show promising results in a 15,000 L bioreactor to characterize hydrodynamic behavior with axial pumping impellers regarding circulation time distributions and residence times [71–75].

6. The research group Duisterwinkel et al. introduced *Xploring WiseMotes*, which measure 39 mm in diameter, in order to conduct measurements of an operational pipeline *in situ*. They are equipped with an Inertial Measurement Unit and an ultrasound emitter and were introduced into a 260 m pipeline section with water flow. The data is analyzed offline subsequently. Another application takes place in a bubble column to analyze the LSP's interaction with gas bubbles to determine the void fraction and bubble size distribution. The concept with the ultrasound emitter is based on their study in 2013, where they present a technique in which a swarm of sensor motes conduct ultrasonic ranging measurements to identify the location of the motes within the swarm while simultaneously mapping the surrounding environment traversed by the swarm [76–78].
7. Established in 2016, the company Freesense developed LSPs as a collaboration between the Danish pharma industry and the Technical University of Denmark (DTU). Basically, two types of LSPs were published, one with a size of 43 mm and the other with 52 mm in diameter. The first ("FermSense 3D") was equipped with an ISFET pH sensor, a temperature sensor, a pressure sensor and an Inertial Measurement Unit (IMU). For the latter ("FermSense DO"), the pH sensor is substituted by a dissolved oxygen sensor. The first design has been implemented in a 600 m^3 bubble column to monitor temperature and pH levels. The data was correlated to the axial position derived from pressure measurements and further utilized to develop an automated compartment model [79–81].
8. The LSP design of Cabrera & Cobelli is presented as a method to characterize turbulent free-surface flows. The particle is capable of measuring its 3D translational acceleration and angular velocity and recording them on a removable memory card. The particle's shell is formed using Acrylonitrile Butadiene Styrene (ABS) 3D printing and has a spherical shape with a diameter of 36 mm. Although the device is not utilized immersed particularly, its basic structure being built from commercial off-the-shelf components serves as an example for an accessible, reproducible and open-source LSP development project [82].
9. Within the framework of the priority program *InterZell* (SPP 2170), funded by the German Research Foundation (DFG), Hofmann et al. developed two LSP designs with 40 mm in diameter, one equipped with a pressure sensor and one with an IMU. Since the housing is made of machined Poly(methyl methacrylate) (acrylic glass), the differently colored installed LEDs inside the shell can be detected from the outside and utilized for the validation of recorded position data. The evaluation of IMU and pressure data was performed in acrylic glass bioreactors with volumes of 200 L and 15,000 L. This analysis enabled the identification of compartments by determining circulation times and residence times.

Multi-scale experiments with the utilization of LSPs in combination with numerical approaches can aid in understanding and the successful scaling-up of processes. By providing guidance for scale-down experiments and identifying targets for strain engineering or bioreactor optimization, the development and application of LSPs is fundamental for biotechnology and upcoming challenges related to climate change.

Oxygen and Carbon Dioxide Mass Transfer

For the fermentation of mammalian cells, a sufficient amount of dissolved oxygen must always be made available to the cells. Especially due to its low solubility in water, knowledge about the ideal supply of oxygen to the fermentation broth is crucial for a successful operation [2]. Furthermore, the carbon dioxide resulting from cell respiration must be removed from the fermentation broth to prevent over-acidification. To describe gas-liquid mass transfer, the two-film theory and the penetration theory are the most commonly used. Since this chapter aims to summarise hands-on information, only the two-film theory is reviewed, which is the most widely applied [83]. For more information on the remaining theories the reader is referred to [4, 84]. In the following, the mass transfer theory is discussed first, followed by a description of the determination of the volumetric mass transfer coefficient for oxygen and carbon dioxide.

For a two-phase system, the two-film theory proposes two laminar concentration boundary layers on both sides of the interface, as depicted in Fig. 3.17.

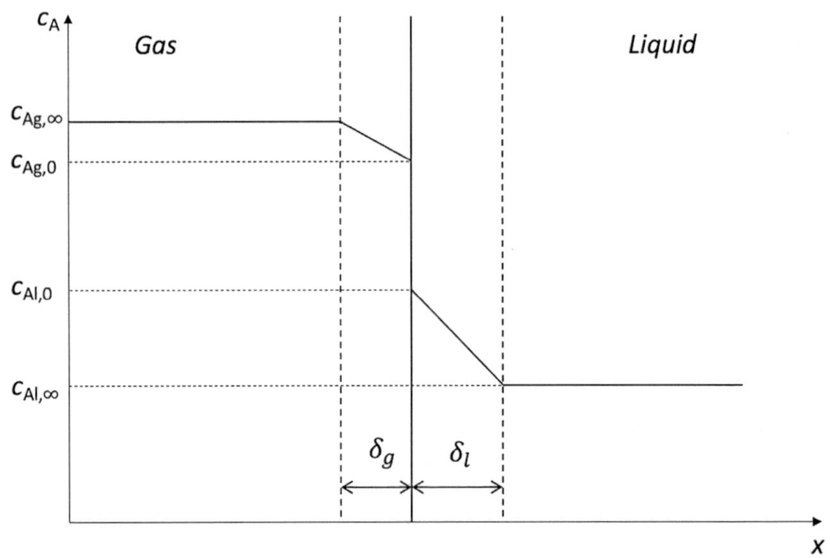

Fig. 3.17 Two-film theory with gas and liquid boundary layers with respective thicknesses δ_G and δ_L corresponding to k_G and k_L at both sides of the interface

At a sufficient distance from the interface, the concentration of each species *A* and *B* is considered to stay constant. Within the boundary layers, the concentration gradient between the bulk and the interface is described. The molar flux of the gaseous species A through the gaseous boundary layer is described by

$$\dot{n} = k_G \left(c_{Ag,\infty} - c_{Ag,0} \right) \tag{3.18}$$

whereas the flux of species *A* through the liquid boundary layer is defined as:

$$\dot{n} = k_L \left(c_{Al,0} - c_{Al,\infty} \right) \tag{3.19}$$

At the interface, an equilibrium state between the two phases persists, where the Henry's law constant describes the discontinuity in the concentrations:

$$H_s^{cc} = \frac{c_{Al,0}}{c_{Ag,0}} = RTH_s^{cp} \tag{3.20}$$

For high values of the Henry's law constant, meaning low solubility of the gaseous phase in the liquid, the resistance on the liquid side boundary layer is much higher than the one in the gaseous boundary layer, yielding in much higher k_L values than k_G [2]. Hence, the mass transfer rate k_G is often considered negligible and, therefore, the concentration of the gaseous species *A* is considered constant in the gaseous phase, as shown in Fig. 3.18.

In two-phase gas-liquid reactors, the mass transfer is quantified utilising the volumetric mass transfer rate $k_L a$. Since experimentally it is often not possible to determine the liquid side mass transfer coefficient k_L and the volume specific surface area a separately, they are determined together as one coefficient. To calculate the volumetric mass transfer coefficient,

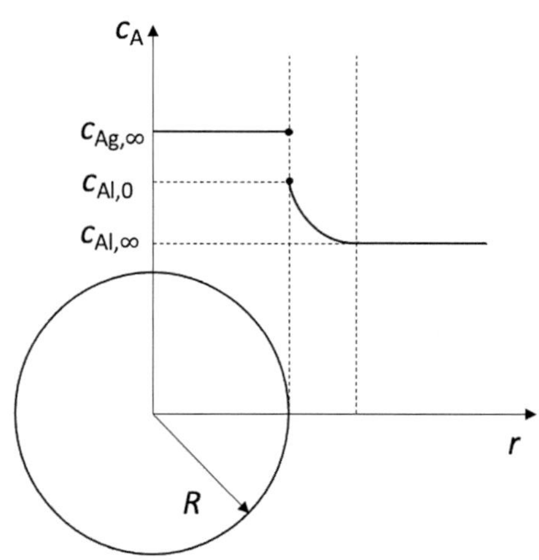

Fig. 3.18 Mass transfer from a bubble into the liquid bulk, according to Kraume [4]

$$\frac{dc}{dt} = -k_{\mathrm{L}}a(c^* - c(t)) \tag{3.21}$$

is solved yielding

$$k_{\mathrm{L}}a = \frac{\ln\left(\frac{c^* - c(t)}{c^* - c(t_0)}\right)}{t - t_0} \tag{3.22}$$

For determining the volumetric mass transfer coefficient in a reactor, the concentration of dissolved gas is measured over time. This can be conducted by dynamic degassing, the dynamic pressure method or by chemical methods. However, at this point the focus is based on the first one, because in case of mammalian cell cultivation and the respectively low oxygen uptake rate, the dynamic degassing is the most efficient measurement method. For more information on the latter the reader is referred to Garcia-Ochoa and Gomez [2].

Determining the Volumetric Oxygen Mass Transfer Rate $k_{\mathrm{L}}a_{\mathrm{O2}}$

The volumetric oxygen mass transfer coefficient $k_{\mathrm{L}}a_{\mathrm{O2}}$ is an import process specific parameter, which can be used to calculate the oxygen transfer rate (OTR). The most common method to determine the volumetric oxygen mass transfer coefficient is the dynamic method [2]. In general, the dynamic method is based on the subsequent absorption and desorption of the oxygen under real process conditions. Since most oxygen probes cannot measure absolute values but only relative oxygen saturations (0–100 %), we will not make a conversion into absolute values in the following. A conversion between relative and absolute values is not necessary under the assumption that during the measurement the fluid properties, system parameters and, thus, the Henry's law constant remain constant. This simplifies the evaluation considerably, since the Henry's law constant does not have to be determined beforehand.

For the determination of the volumetric oxygen mass transfer coefficient the concentration of dissolved oxygen in the liquid bulk is measured over time. Therefore, first the oxygen concentration is decreased to zero by stripping with nitrogen. Next, by aeration the air or pure oxygen concentration is increased again until saturation with oxygen is reached. The volumetric mass transfer coefficient $k_{\mathrm{L}}a$ can then be calculated in accordance with Eq. 3.22. Practically, it is sufficient to work within a smaller range of oxygen saturation concentration to save time as well as resources. Therefore, it is recommended to strip with nitrogen until the dissolved oxygen concentration is below a threshold of 10 %. To ensure that all nitrogen bubbles have left the system in the evaluated concentration range, data above 20 % of saturation concentration can be used. For a better understanding, Fig. 3.19 shows an ideal concentration pattern after complete degassing of oxygen as well as a typical measured concentration curve. Especially in the first minutes, the measured oxygen signal lags behind the real course.

A further reduction of the oxygen concentration is not reasonable, since on the one hand the mass transfer rate decreases significantly due to the decreasing concentration gradient (liquid to gas). This leads to very long gassing times of the

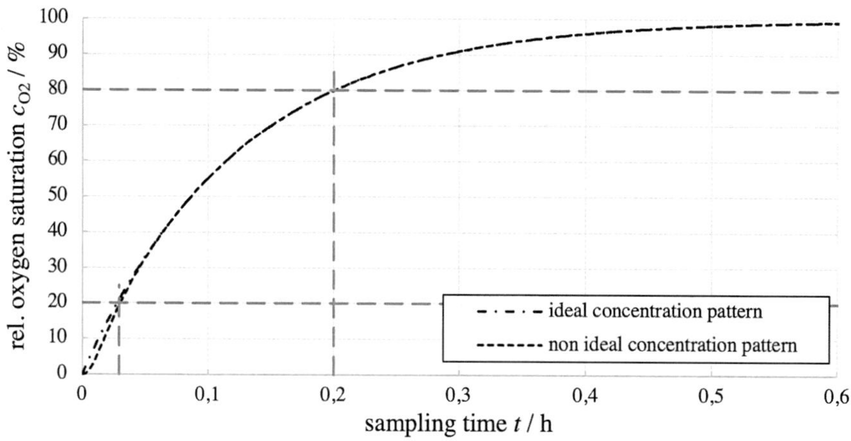

Fig. 3.19 Schematic illustration of the relative oxygen concentration pattern of a dynamic degassing measurement

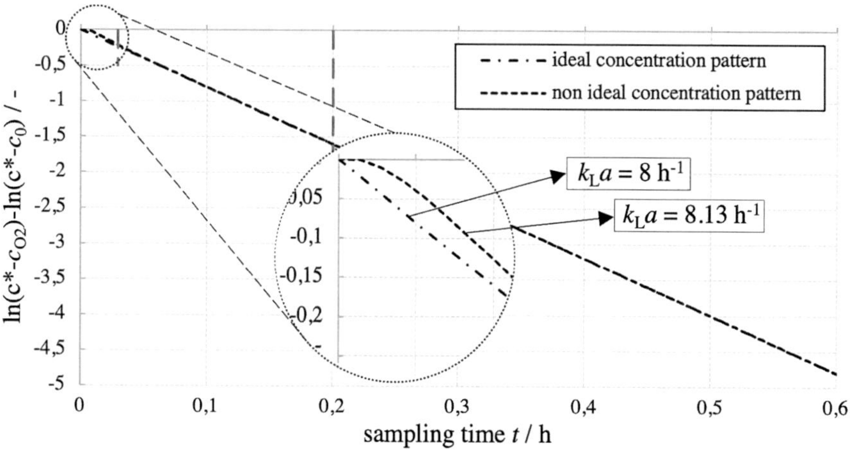

Fig. 3.20 Schematic visualization of the influence of a non-ideal measurement on the volumetric mass transfer coefficient

system with nitrogen being necessary. On the other hand, due to the response time of the oxygen probe, a time delay in the measurement signal occurs at the beginning of the oxygen aeration anyway, which leads to an erroneous measurement and, therefore, wrong volumetric oxygen mass transfer coefficient values. The influence of the initial ramp up phase in particular on the final volumetric oxygen mass transfer coefficient is illustrated in Fig. 3.20, which shows the logarithmized oxygen concentration curve for an ideal and a non-ideal pattern. The volumetric oxygen mass transfer coefficients shown in Fig. 3.20 are evaluated in the range of 20–80 % oxygen saturation.

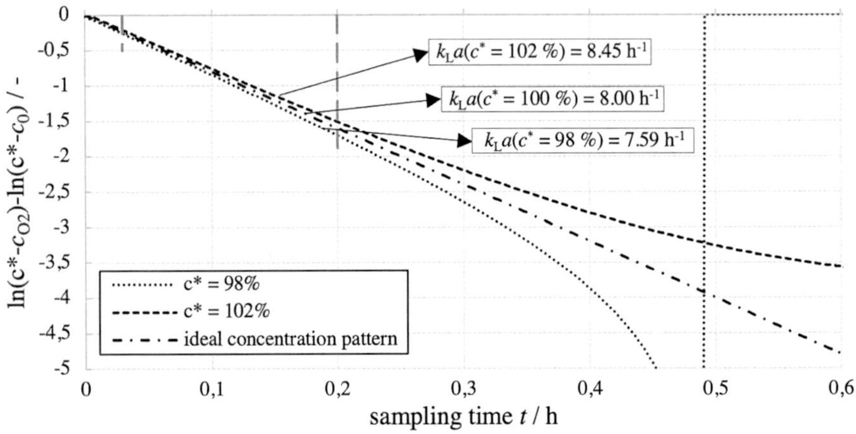

Fig. 3.21 Schematic visualization of the influence of a wrong calibrated oxygen sensor on the volumetric mass transfer coefficient

Figure 3.20 clearly shows that even small deviations from the ideal concentration curve lead to a noticeable difference in the resulting volumetric mass transfer coefficients (8.13 compared to 8 h^{-1}). It should be noted that the case shown in Fig. 3.20 is clearly exaggerated to illustrate the influence of a non-ideal measurement. In reality, the influence is significantly smaller, depending on the mass transfer performance and the specific oxygen sensor. However, every measurement should always be reviewed in this regard.

Further, to keep the experimental duration within an acceptable time range as well as the error due to a wrong sensor calibration, analyses until 80 % of saturation are hence recommended. Measurements significantly above 80 % oxygen saturation are not reasonable, because the static and statistical measurement error of the respective probe leads to significantly greater error, which complicates the evaluation further. Figure 3.21 shows the logarithmized concentration curve for the case where the probe was calibrated to an incorrect saturation concentration (±2 %). It becomes clear that the error in the calculation of the volumetric oxygen mass transfer coefficient value increases when values above 80 % are included in the evaluation.

Therefore, the saturation concentration c^* has to be measured under real process conditions (temperature, over pressure, salt concentration, surfactants) by aeration until the liquid is saturated by air, and the oxygen concentration stays constant. The evaluation ranges (20–80 %) given here are based on the experience of the authors. It should be noted that, taking the measurement errors listed above into account, the evaluation range can be adapted but may of course also be varied.

Determining the Volumetric Carbon Dioxide Mass Transfer Rate k_La_{O2}

Besides a sufficient dissolved oxygen supply into the system, also the efficient stripping of carbon dioxide is of large importance for the cultivation of mammalian cells [85]. The carbon dioxide produced by the metabolism of mammalian cells is transported out of the cells, and thus is dissolved in the culture medium [2]. If the culture medium consists mainly of water, a small portion of the dissolved carbon dioxide reacts with water (H_2O) to form carbonic acid (H_2CO_3)

$$CO_2 + H_2O \rightleftharpoons H_2CO_3. \tag{3.23}$$

Carbonic acid is a strong, biprotonic acid and dissociates first to hydrogen carbonate (HCO_3^-) and a proton (H^+)

$$H_2CO_3 \rightleftharpoons HCO_3^- + H^+. \tag{3.24}$$

The hydrogen carbonate then dissociates to carbonate (CO_3^{2-}) and another proton

$$HCO_3^- \rightleftharpoons CO_3^{2-} + H^+. \tag{3.25}$$

The concentration of protons can be described by the pH value for diluted solutions. As the three carbonic acid species carbon dioxide, hydrogen carbonate and carbonate are related to each other by the law of mass action, the proportion F of each species can be determined at a given pH value. In Fig. 3.22, the logarithmized fraction is presented as a plot against pH value [86, 87].

At pH value of less than 8.3, the proportion of carbonate is below 1 % (lg $F =$ – 2). The second dissociation step (Eq. 3.24) is therefore not significant and can be neglected. However, below pH value of 8.3, the proportion of carbon dioxide is

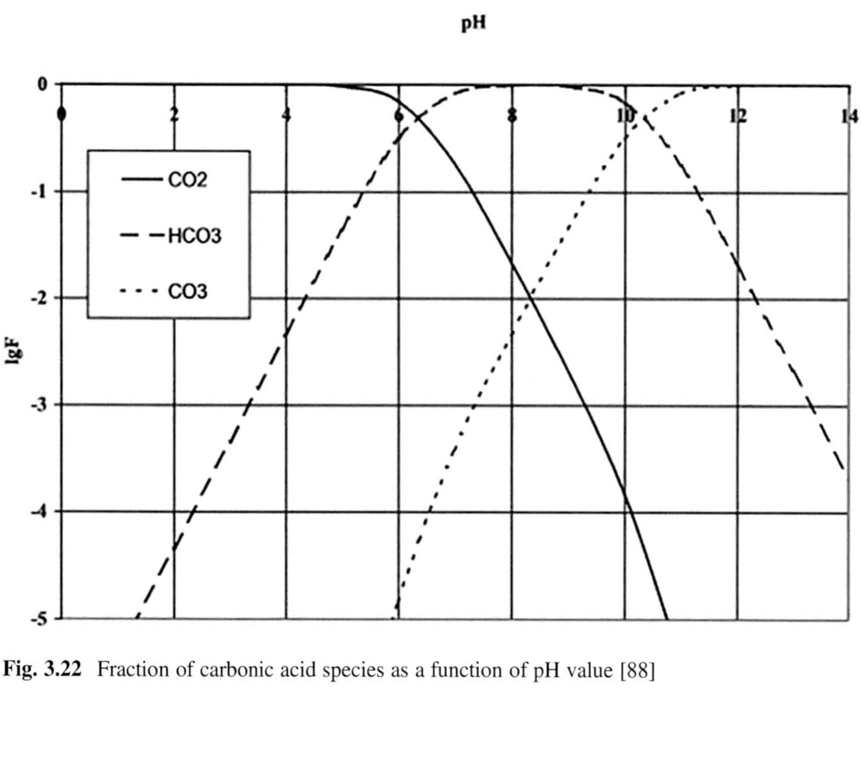

Fig. 3.22 Fraction of carbonic acid species as a function of pH value [88]

critical and the first dissociation step (Eq. 3.23) is relevant in pH range of 4.3–8.3. Carbon dioxide is a weak acid that does not lower the pH value of water below 4.5. In addition, the pH value influences the solubility of carbon dioxide [87–89]. In bioprocess engineering, productions are performed under the conditions that the concentration of dissolved carbon dioxide does not seriously affect the growth or productivity of mammalian cells. The processes are operated in a pH environment of 4–8 because mammalian cells are also sensitive to pH value. The pH range corresponds to that of the first dissociation step (Eq. 3.23). However, the hydrogen carbonate concentration has no influence on the mass transfer of carbon dioxide under pseudo-stationary conditions. Pseudo-stationary implies stationary or slowly varying pH value and carbon dioxide formation rate, which are applicable for most production processes [90].

For the removal of the produced carbon dioxide, volumetric mass transfer rate is a decisive parameter. To determine the volumetric mass transfer rate of carbon dioxide removal, the carbon dioxide concentration during aeration is measured. Therefore, the saturation concentration c^* is set to zero, and Eq. 3.22 modifies to

$$k_L a = \frac{-\ln\left(\frac{c(t)}{c(t_0)}\right)}{(t - t_0)} \tag{3.26}$$

In principle, the measurement errors shown and discussed in Figs. 3.20 and 3.21 can also occur when measuring the volumetric carbon dioxide mass transfer coefficient. However, due to the significantly greater solubility of carbon dioxide in water compared to oxygen, the concentration gradients over time are significantly smaller compared to oxygen. Because of this, the errors described in Figs. 3.20 and 3.21 are much less significant in case of the volumetric carbon dioxide mass transfer coefficient determination.

Influence of Hydrostatic Pressure

As indicated by the Henry's law,

$$H_s^{cp} = \frac{c^*}{p_{part}} \tag{3.27}$$

the saturation concentration for oxygen depends on the partial pressure of oxygen within the bubbles. In large reactors, the partial pressure p_{part} increases linearly with increasing hydrostatic pressure p_{hyd} and, therefore, the achievable saturation concentration changes as well. In small reactors with a liquid volume of a few litres, this change in saturation concentration is negligible, however, not in reactors of industrial size or pilot scale. At the top of such reactors, partial pressures at atmospheric conditions can be used. However, the hydrostatic pressure increases by approximately 0.1 bar with each metre of depth below the surface and, therefore, also the partial pressure of the gaseous phase increases. This linear change in partial pressure as well as in achievable saturation concentrations respectively is depicted in

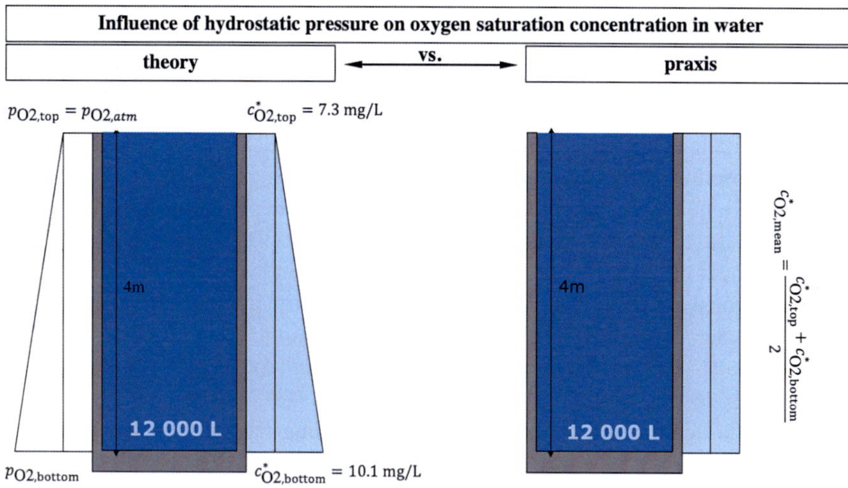

Fig. 3.23 Influence of hydrostatic pressure on the oxygen saturation concentration and partial pressure in an industrial-scale reactor according to [8]

Fig. 3.23 for a temperature of $T = 37$ °C, liquid volume of 12 m^3 and a height of 4 m [8].

Hence, the assumption of one global saturation concentration in a reactor may lead the erroneous conclusions. To overcome this limitation and obtain reasonable experimental data, it is good practise to place multiple probes over the height into the reactor and determine the saturation concentration for each.

In a well-mixed, buoyancy driven system, Rosseburg et al. measured one global saturation concentration in an industrial-scale reactor. The probes at the top of the reactor, as well as on the bottom, both measured the mean value between the theoretical saturation concentrations at the reactor top and the bottom [8]. For the exemplary reactor in Fig. 3.23, the global saturation concentration would therefore yield in $c^* = 8.7\,\text{mg L}^{-1}$. Nevertheless, it is recommended to determine the saturation concentration for the system experimentally by using multiple probes.

Numerical Flow Simulation Methods for Characterization of Aerated Stirred Tank Reactors

Equipment characterization is nowadays as dependent on computational methods as on experimental methods. Computational Fluid Dynamics (CFD) is often used as a tool to model the multiphase flow inside stirred tanks [91]. It can provide a high level of detail of single-phase and multiphase variables, from the very small scale [92], to the very large scale [93]. To calculate the fluid velocity and pressure CFD works by solving the Navier-Stokes equations. Additionally, conservation of mass, momentum, and energy equations can be used to calculate concentration or temperature distribution of a fluid. These equations are typically solved numerically using finite volume or finite element methods, a Reynolds Averaged Navier-Stokes Equations (RANS) turbulence model is used. As turbulence model the k$-\varepsilon$ two-equation is often used for the simulation of turbulent flows in stirred tanks [94–97]. Two transport equations for the turbulent kinetic energy and the dissipation rate are solved to close the RANS equations. In combination with the steady-state Multiple Reference Frame (MRF) approach to account for impeller rotation, this setup has been shown to accurately predict power numbers in stirred tank reactors [98–100]. The method averages the equations over time and space, which simplifies the calculations but will result in some loss of accuracy. Since the Reynolds-averaged Navier–Stokes (RANS) $k - \varepsilon$ model is a computationally rather inexpensive method, it is often applied and has been proven to be a valuable tool for equipment characterization [101, 102].

Another, but computationally much more demanding method for simulating the flow inside of a stirred tank is the Large Eddy Simulation (LES). This approach divides the flow into large and small scales and solves the Navier–Stokes equations for the large scales, while modeling the effect of the small scales using a subgrid-scale model [103]. As a result, the LES method can capture more detailed effects of turbulence and unsteady flow, which are important in stirred tanks. Haringa et al., were able to show clearly that an LES approach or RANS sliding mesh showed better performance, than the static RANS MRF method [104, 105].

A recently popular alternative approach are Lattice Boltzmann simulations. The Lattice Boltzmann method is based on the Boltzmann equation, which describes the evolution of a probability distribution function (PDF). The technique discretizes the Bolzmann equation in velocity space, physical space and time on a lattice, where each node of the lattice represents a possible velocity direction of the particles. An advantage is its computational efficiency compared to traditional CFD techniques [106]. The technique can simulate the fluid flow at a relatively coarse grid resolution, which reduces the computational cost and allows for faster simulations. Further, the Lattice Bolzmann method is a highly local technique, which can be easily parallelized by using graphic processing units (GPU), which further reduces the simulation time [107].

Standard simulation procedures for stirred tanks in industrial applications are still lacking, making it hard to compare results from different publications or companies [108], even if some recommendations to the industry on the formatting, organization, and content of reports for CFD are provided by the U.S. Food and Drug Administration [109].

The accuracy of CFD simulations depends on several factors, such as the quality of the mesh, the turbulence model, and the boundary conditions. The fact that CFD models can often be predictive without estimating parameters has enabled the directional use of CFD without rigorous comparison to experimental data, many of the models have an unknown credibility [110]. Therefore, it is essential to validate CFD simulations with experimental data before using them for design and optimization purposes.

For lab scale bioreactors several studies exist were experimental and simulated flow patterns are compared [111, 112]. For the LB LES methods, such a study for a 3 L bioreactor system was recently published by Kuschel et al.. The simulation results were compared to novel 4D particle tracking velocimetry (4D PTV) experiments, showing good accordance of time averaged flow field from LB LES to experimental measurements [45]. A comparison of RANS and LB LES methods for a 54 m^3 bioreactor and a single impeller case is available from Haringa [113].

The quality of the mesh, including its size, shape, and distribution, can significantly affect the accuracy and computational efficiency of the CFD simulation. To ensure the validity of the CFD model, meshing studies are conducted to optimize mesh parameters and ensure that the mesh is fine enough to capture the details of the fluid flow in the bioreactor, but coarse enough to ensure short calculating times. This involves evaluating different meshing techniques, mesh element sizes, and cell shapes, as well as assessing the impact of the mesh on the simulation results. Since many biopharmaceutical processes are transferred based on specific power input, identification of accurate power numbers *Po* is a major goal of CFD simulations. Regarding the mesh study, power numbers are often determined at different mesh sizes. Usually, the power number is unstable at coarser meshes and converges once the mesh gets finer [45]. Convergence for the power number at a relatively coarse grid size has been demonstrated in the literature [98]. Besides the mesh, the power number is influenced by the rotating reference frame around the agitator. The influence of the size and aspect ratio of the reference frame should be studied in course of the validation of the CFD model [45, 97, 102, 114].

Results from a classical one phase simulation of a 2000 L disposable bioreactor are shown in Fig. 3.24. The turbulent eddy dissipation rate (A) gives an indication on introduced power, whereas velocity streamlines (B) demonstrate the fluid flow in the bioreactor. These are the classical pictures typically associated with CFD

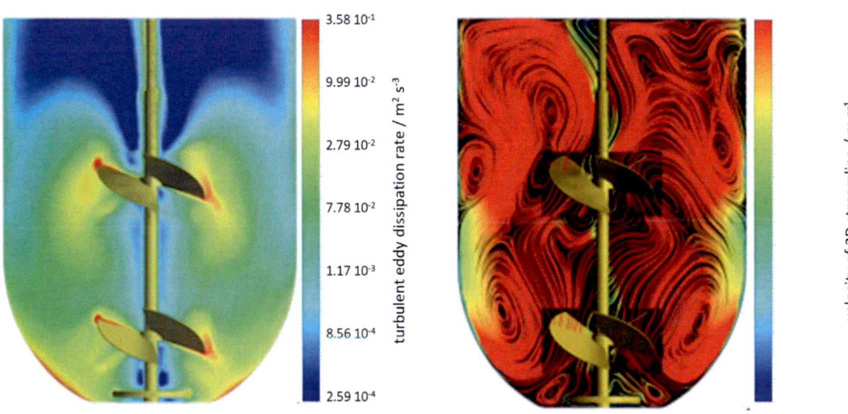

Fig. 3.24 (left) Turbulent dissipation rate for Sartorius STR 2000 L disposable bioreactor at 70 rpm; (right) Streamlines for the same simulation; The minimum and maximum values are cut-offs in both contour plots

simulations, which allow some insight into the flow field and might give an indication on dead zones and optimal impeller configuration.

The simulation of mixing times with CFD is an established procedure in the literature [91, 115–117], which is often in good agreement with experimental data [19]. For better accuracy sliding mesh or LES should be used rather than the MRF approach [104, 105]. Since mixing is strongly dependent on the flow field, it is important, especially for turbulent flow conditions in stirred tank reactors, that mixing is calculated based on the transient velocity data. To simulate the mixing process typically, a sphere at the top of the liquid volume is patched with a tracer and the tracer is tracked at defined positions. Mixing time simulations are commonly performed with the probe positions from the experiments [118]. The mixing time is the last timepoint when the tracer concentration signal is smaller than 5 % variation of the average final concentration. This method can also be beneficial in the scale-up of more complex processes like perfusion systems [119].

Shear stress is another hydrodynamic parameter which can be well predicted by CFD [115]. Even if the severity of shear stress for animal cell cultivation is sometimes in doubt, a stress threshold for animal cell culture was recently confirmed by Šrom et al. [3]. A detailed discussion on the simulation of stress by CFD can be found in the literature [120, 121].

Modeling of mass transfer by CFD requires multiphase simulations and is more challenging. Two general approaches are commonly used. The Euler-Euler model assumes that all phases are continuous and are described by the same set of governing equations. The properties of each phase are averaged over a small volume of the flow [103]. The Euler-Lagrange model treats one phase as a continuous phase and the other phase as a set of discrete particles. The continuous phase is modeled using the standard CFD equations, while the discrete phase is computed using Lagrangian particle tracking [103]. Often, good models can be established for a single bioreactor but have to be adapted when using it for another vessel [122]. Typically, the surface renewal theory form Higbie is applied to calculate volumetric mass transfer coefficient $k_L a$ values [106, 122]. Bubble size distributions are seldomly incorporated in the models, even though some models for breakage and coalescence are available [123]. Good overall agreement of simulated and measured volumetric mass transfer coefficients were shown in recent Lattice Boltzmann studies [106, 107], in both cases there were problems with larger scales and a very high number of bubbles. A new publication with well predicted bubble size distributions in small scale is available from Krýsa and Šoóš [124]. Such studies for industrial scale reactors, especially with experimental values for the initial bubble sizes are crucial to establish versatile multiphase models.

Successful application examples in the prediction of mixing times, power input, and mass transfer coefficients k_La have led to its use for other unit operations in the industry including: centrifugation, chromatography, ultrafiltration, membrane systems, microfiltration, spray drying, and freeze drying [108]. Overall, CFD methods are an important tool for characterizing stirred tanks. By providing detailed information on the flow patterns, shear rates and mixing characteristics, CFD simulations can help to optimize the design and operation of these systems, leading to improved performance and efficiency.

References

1. Kraume M (2014) Die Entwicklung der Rührtechnik von einer empirischen Kunst zur Wissenschaft. Chemie Ingenieur Technik 86(12):2051–2062. https://doi.org/10.1002/cite.201400124
2. Garcia-Ochoa F, Gomez E (2009) Bioreactor scale-up and oxygen transfer rate in microbial processes: an overview. Biotechnol Adv 27(2):153–176. https://doi.org/10.1016/j.biotechadv.2008.10.006
3. Šrom O et al (2022) Characterization of hydrodynamic stress in ambr250® bioreactor system and its impact on mammalian cell culture. Biochem Eng J 177:108240. https://doi.org/10.1016/j.bej.2021.108240
4. Kraume M (2012) Mischen und Rühren. In: Kraume M (ed) Transportvorgänge in der Verfahrenstechnik: Grundlagen und apparative Umsetzungen, VDI-Buch. Springer, Berlin/Heidelberg, pp 555–601. https://doi.org/10.1007/978-3-642-25149-8_18
5. Zlokarnik M (2001) Stirring: theory and practice. Wiley, Weinheim
6. Harnby N, Edwards MF, Nienow MW (1997) Mixing in the process industries, 2nd edn. Butterworth-Heinemann, Oxford
7. Zehner P (2002) Modelltechnik und Maßstabsübertragung. In: Kraume M (ed) Mischen und Rühren, 1st edn. Wiley, pp 405–423. https://doi.org/10.1002/3527603360.ch16
8. Rosseburg A, Fitschen J, Wutz J, Wucherpfennig T, Schlüter M (2018) Hydrodynamic inhomogeneities in large scale stirred tanks – influence on mixing time. Chem Eng Sci 188: 208–220. https://doi.org/10.1016/j.ces.2018.05.008
9. Tatterson GB (1991) Fluid mixing and gas dispersion in agitated tanks. McGraw-Hill, New York
10. Rosseburg A (2019) Influence of heterogeneous bubbly flows on mixing and mass transfer performance in stirred tanks for mammalian cell cultivation – a study in transparent 3 L and 12 000 L reactors, PhD thesis. Hamburg University of Technology
11. Warmoeskerken MMCG, Smith JM (1985) Flooding of disc turbines in gas-liquid dispersions: a new description of the phenomenon. Chem Eng Sci 40(11):2063–2071
12. Mikulcova E, Kudrna V, Vlcek J (1967) Bestimmung der Grenzbedingungen für die Überflutung des Rührers K1, Scientific paper of the Institute of Chemical Technology, Prague, pp 167–183
13. Zwietering TN (1963) Discussion contribution in Dutch without title. De Ingenieur 75:60–61
14. Räbiger N, Schlüter M (2013) L4.1 Bildung und Bewegung von Tropfen und Blasen. In: VDI-Wärmeatlas, 11th edn. Springer, Berlin/Heidelberg, pp. 1413–1428. Accessed 30 Mar 2020. [Online]. http://link.springer.com/10.1007/978-3-642-19981-3
15. Platas Barradas O (2014) Process and cultivation strategies for the human industrial cell line AGE1.HN, thesis. Technische Universität Hamburg. Accessed 26 Mar 2023. [Online]. https://tore.tuhh.de/handle/11420/1205
16. Neunstoecklin B et al (2016) Pilot-scale verification of maximum tolerable hydrodynamic stress for mammalian cell culture. Appl Microbiol Biotechnol 100(8):3489–3498. https://doi.org/10.1007/s00253-015-7193-x
17. Tränkler H-R, Fischerauer G (2014) Das Ingenieurwissen: Messtechnik. In: Hütte: Das Ingenieurwissen, 33rd edn. Springer, Berlin/Heidelberg. https://doi.org/10.1007/978-3-662-44030-8
18. Fischer R (2016) Elektrotechnik für Maschinenbauer. Springer Fachmedien Wiesbaden, Wiesbaden. https://doi.org/10.1007/978-3-658-12515-8
19. Fitschen J et al (2021) Novel evaluation method to determine the local mixing time distribution in stirred tank reactors. Chem Eng Sci 10:100098. https://doi.org/10.1016/j.cesx.2021.100098
20. Calderbank PH (1958) Physical rate processes in industrial fementation. Part 1: the interfacial area in gas-liquid contacting with mechanical agitation. Trans Inst Chem Eng 36: 443–463. Accessed 26 Mar 2023. [Online]. https://cir.nii.ac.jp/crid/1571980075271358848

21. Kraume M, Pahl MH, VDI-Gesellschaft Verfahrenstechnik und Chemieingenieurwesen (2002) Mischen und Rühren: Grundlagen und moderne Verfahren. Weinheim, Wiley
22. Hiby JW (1979) Definition und Messung der Mischgüte in flüssigen Gemischen. Chemie Ingenieur Technik 51(7):704–709. https://doi.org/10.1002/cite.330510705
23. Bockhorn H, Mewes D, Peukert W, Warnecke H-J (eds) (2010) Micro and macro mixing, Heat and mass transfer. Springer, Berlin/Heidelberg. https://doi.org/10.1007/978-3-642-04549-3
24. Fitschen J (2021) Hydrodynamic Characterization of heterogeneities in aerated stirred tank reactors – from an Eulerian to a Lagrangian perspective, PhD thesis. Hamburg University of Technology, Hamburg
25. Sieblist C, Jenzsch M, Pohlscheidt M, Lübbert A (2011) Insights into large-scale cell-culture reactors: I. Liquid mixing and oxygen supply. Biotechnol J 6(12):1532–1546. https://doi.org/10.1002/biot.201000408
26. Manna L (1997) Comparison between physical and chemical methods for the measurement of mixing times. Chem Eng J 67(3):167–173. https://doi.org/10.1016/S1385-8947(97)00059-4
27. Zhang A, Tsang VL, Korke-Kshirsagar R, Ryll T (2014) Effects of pH probe lag on bioreactor mixing time estimation. Process Biochem 49(6):913–916. https://doi.org/10.1016/j.procbio.2014.03.005
28. Azarpira M, Zarrati AR, Farrokhzad P (2021) Comparison between the Lagrangian and Eulerian approach in simulation of free surface air-core vortices. Water 13(5):5. https://doi.org/10.3390/w13050726
29. Blöbaum L, Haringa C, Grünberger A (2023) Microbial lifelines in bioprocesses: from concept to application. Biotechnol Adv 62:108071. https://doi.org/10.1016/j.biotechadv.2022.108071
30. Elghobashi S, Truesdell GC (1992) Direct simulation of particle dispersion in a decaying isotropic turbulence. J Fluid Mech 242:655–700. https://doi.org/10.1017/S0022112092002532
31. Schlüter M (2011) Lokale Messverfahren für Mehrphasenströmungen. Chemie Ingenieur Technik 83(7):992–1004. https://doi.org/10.1002/cite.201100039
32. Nitsche N, Brunn A (2006) Strömungsmesstechnik, VDI-Buch, 2nd edn. Springer, Berlin/Heidelberg. https://doi.org/10.1007/3-540-32487-9
33. Merzkirch W (2007) Flow visualization. In: Tropea C, Yarin AL, Foss JF (eds) Springer handbook of experimental fluid mechanics, Springer handbooks. Springer, Berlin/Heidelberg, pp 857–870. https://doi.org/10.1007/978-3-540-30299-5_11
34. Busse C, Biechele P, de Vries I, Reardon KF, Solle D, Scheper T (2017) Sensors for disposable bioreactors. Eng Life Sci 17(8):940–952. https://doi.org/10.1002/elsc.201700049
35. Hoang MD, Doan DT, Schmidt M, Kranz H, Kremling A, Heins A-L (2022) Application of an Escherichia coli triple reporter strain for at-line monitoring of single-cell physiology during L-phenylalanine production. Eng Life Sci 23(1):15. https://doi.org/10.1002/elsc.202100162
36. Mukherjee S, Bassler BL (2019) Bacterial quorum sensing in complex and dynamically changing environments. Nat Rev Microbiol 17(6):371–382. https://doi.org/10.1038/s41579-019-0186-5
37. Lapin A, Müller D, Reuss M (2004) Dynamic behavior of microbial populations in stirred bioreactors simulated with Euler−lagrange methods: traveling along the lifelines of single cells. Ind Eng Chem Res 43(16):4647–4656. https://doi.org/10.1021/ie030786k
38. Nadal-Rey G, McClure DD, Kavanagh JM, Cornelissen S, Fletcher DF, Gernaey KV (2021) Understanding gradients in industrial bioreactors. Biotechnol Adv 46:107660. https://doi.org/10.1016/j.biotechadv.2020.107660
39. Haringa C et al (2016) Euler-Lagrange computational fluid dynamics for (bio)reactor scale down: an analysis of organism lifelines. Eng Life Sci 16(7):652–663. https://doi.org/10.1002/elsc.201600061
40. Kuschel M, Siebler F, Takors R (2017) Lagrangian trajectories to predict the formation of population heterogeneity in large-scale bioreactors. Bioengineering 4(4):27. https://doi.org/10.3390/bioengineering4020027

41. Nadal-Rey G et al (2021) Development of dynamic compartment models for industrial aerobic fed-batch fermentation processes. Chem Eng J 420:130402. https://doi.org/10.1016/j.cej.2021.130402
42. Sarkizi Shams Hajian C, Haringa C, Noorman H, Takors R (2020) Predicting by-product gradients of Baker's yeast production at industrial scale: a practical simulation approach. PRO 8(12):12. https://doi.org/10.3390/pr8121554
43. Gaugler L, Mast Y, Fitschen J, Hofmann S, Schlüter M, Takors R (2022) Scaling-down biopharmaceutical production processes via a single multi-compartment bioreactor (SMCB). Eng Life Sci elsc.202100161. https://doi.org/10.1002/elsc.202100161
44. Hofmann S, Weiland C, Fitschen J, von Kameke A, Hoffmann M, Schlüter M (2022) Lagrangian sensors in a stirred tank reactor: comparing trajectories from 4D-particle tracking velocimetry and Lattice-Boltzmann simulations. Chem Eng J 449:22. https://doi.org/10.1016/j.cej.2022.137549
45. Kuschel M, Fitschen J, Hoffmann M, von Kameke A, Schlüter M, Wucherpfennig T (2021) Validation of novel Lattice Boltzmann large Eddy simulations (LB LES) for equipment characterization in biopharma. PRO 9(6):950. https://doi.org/10.3390/pr9060950
46. Rowlands EN, Wolff HS (1961) The radio pill: telemetering from the digestive tract. Br Commun Electron 7:598
47. Watson BW, Ross B, Kay AW (1962) Telemetering from within the body using a pressure-sensitive radio pill. Gut 3(2): 181–186. Accessed 10 Aug 2020. [Online]. https://www.ncbi.nlm.nih.gov/pmc/articles/PMC1413324/
48. Handley MF, Perry MG (1965) Measurements of stresses in flowing granular materials. Rheol Acta 4(3):225–235. https://doi.org/10.1007/BF01969259
49. Wadke PM, Hounslow MJ, Salman AD (2005) The 'smart' sphere. Chem Eng Res Des 83(11): 1298–1302. https://doi.org/10.1205/cherd.05091
50. Wadke PM, Salman AD, Hounslow MJ (2008) The 'smart' temperature sphere: application in rotary drum mixers. Powder Technol 185(3):274–279. https://doi.org/10.1016/j.powtec.2007.10.029
51. Deng D et al (2017) Evaluation of boundary dam spillway using an autonomous sensor fish device. J Hydro Environ Res 14:85–92. https://doi.org/10.1016/j.jher.2016.10.004
52. Pflugrath BD, Boys CA, Cathers B, Deng ZD (2019) Over or under? Autonomous sensor fish reveals why overshot weirs may be safer than undershot weirs for fish passage. Ecol Eng 132: 41–48. https://doi.org/10.1016/j.ecoleng.2019.03.010
53. Salalila A, Deng ZD, Martinez JJ, Lu J, Baumgartner LJ (2019) Evaluation of a fish-friendly self-cleaning horizontal irrigation screen using autonomous sensors. Mar Freshw Res 70(9): 1274. https://doi.org/10.1071/MF19194
54. AlObaidi K, Valyrakis M (2021) Linking the explicit probability of entrainment of instrumented particles to flow hydrodynamics. Earth Surf Processes Landf 46(12): 2448–2465. https://doi.org/10.1002/esp.5188
55. Curley EAM, Valyrakis M, Thomas R, Adams CE, Stephen A (2021) Smart sensors to predict entrainment of freshwater mussels: a new tool in freshwater habitat assessment. Sci Total Environ 787:147586. https://doi.org/10.1016/j.scitotenv.2021.147586
56. Mitchell FRG (1969) Low power telemetry for circulation studies. J Phys E Sci Instrum 2(9): 812–812. https://doi.org/10.1088/0022-3735/2/9/416
57. Bryant J, Sadeghzadeh S (1979) Circulation rates in stirred and aerated tanks. In: Proceedings of the 3rd European conference on mixing: held at the University of York. BHRA Fluid Engineering, Cranfield, pp 325–336
58. Middleton JC (1979) Measurement of circulation within large mixing vessels. In: Proceedings of the 3rd European conference on mixing: held at the University of York. BHRA Fluid Engineering, Cranfield, pp 15–36
59. Van Barneveld J, Smit W, Oosterhuis NMG, Pragt HJ (1987) Measuring the liquid circulation time in a large gas-liquid contactor by means of a radio pill. Part 1. Flow pattern and mean circulation time. Ind Eng Chem Res 26(11):2185–2192. https://doi.org/10.1021/ie00071a003

60. Van Barneveld J, Smit W, Oosterhuis NMG, Pragt HJ (1987) Measuring the liquid circulation time in a large gas-liquid contractor by means of a radio pill. Part 2. Circulation time distribution. Ind Eng Chem Res 26(11):2192–2195. https://doi.org/10.1021/ie00071a004
61. Lauterbach T et al. (2017) Sens-o-Spheres – mobile, miniaturisierte Sensorplattform für die ortsungebundene Prozessmessung in Reaktionsgefäßen, 13. Dresdner Sensor-Symposium 2017, pp 89–93. https://doi.org/10.5162/13dss2017/2.6
62. Lauterbach T et al (2019) Measurements on the fly– introducing mobile micro-sensors for biotechnological applications. Sensors Actuators A Phys 287:29–38. https://doi.org/10.1016/j.sna.2019.01.003
63. Lauterbach T, Ziebart N, Bley T, Walther T, Lenk F (2019) Mobile Sensoren für die Biotechnologie – Ortsunabhängige, miniaturisierte Prozessmessung. Chemie Ingenieur Technik 91(12):1827–1832. https://doi.org/10.1002/cite.201900114
64. smartINST (2015) smartCAPS: in situ wireless measurements. In: smartCAPS: in situ wireless measurements. https://webma9021.wixsite.com/smartinstnew/smartcaps-in-situ-wireless-measurem. Accessed 19 Mar 2020
65. Zimmermann R, Fiabane L, Gasteuil Y, Volk R, Pinton J-F (2013) Characterizing flows with an instrumented particle measuring Lagrangian accelerations. New J Phys 15(1):015018. https://doi.org/10.1088/1367-2630/15/1/015018
66. Zimmermann R, Fiabane L, Gasteuil Y, Volk R, Pinton J-F (2013) Measuring Lagrangian accelerations using an instrumented particle. Phys Scr T155:014063. https://doi.org/10.1088/0031-8949/2013/T155/014063
67. Todtenberg N, Klatt J, Schmitz-Hertzberg S-T, Jorde F, Schmalz K (2013) Wireless sensor capsule for bioreactors. In: 2013 IEEE MTT-S international microwave workshop series on RF and wireless technologies for biomedical and healthcare applications (IMWS-BIO), IEEE, Singapore, pp 1–3. https://doi.org/10.1109/IMWS-BIO.2013.6756236
68. Todtenberg N et al (2015) Autonomous sensor capsule for usage in bioreactors. IEEE Sensors J 15(7):4093–4102. https://doi.org/10.1109/JSEN.2015.2412652
69. Stine JM et al (2020) Wireless sensor-integrated platform for localized dissolved oxygen sensing in bioreactors. J Microelectromech Syst 29(5):713–719. https://doi.org/10.1109/JMEMS.2020.2999089
70. Stine JM, Beardslee LA, Sathyam RM, Bentley WE, Ghodssi R (2020) Electrochemical dissolved oxygen sensor-integrated platform for wireless in situ bioprocess monitoring. Sensors Actuators B Chem 320:128381. https://doi.org/10.1016/j.snb.2020.128381
71. Buntkiel L, Reinecke S, Richtungsaufgelöste Messung von Beschleunigungen mit Sensorpartikeln in industriellen Prozessbehältern. In: 6. Sensoren für Verfahrenstechnik, 16. Dresdner Sensor-Symposium 2022, Dresden, 2022, pp 51–56. https://doi.org/10.5162/16dss2022/6.3
72. Reinecke S, Deutschmann A, Jobst K, Kryk H, Friedrich E, Hampel U (2012) Flow following sensor particles—validation and macro-mixing analysis in a stirred fermentation vessel with a highly viscous substrate. Biochem Eng J 69:159–171. https://doi.org/10.1016/j.bej.2012.09.010
73. Reinecke SF, Buntkiel L, Kipping R, Hampel U (2022) Process characterization in industrial vessels by flow-following sensor particles. Meas Sci Technol 33(9):095106. https://doi.org/10.1088/1361-6501/ac75af
74. Thiele S, Schöne S, Voigt F, Da Silva MJ, Hampel U (2009) Design of a neutrally buoyant self-powered multi-parameter sensor for data logging in flow applications. In: 2009 IEEE Sensors, pp 1927–1930. https://doi.org/10.1109/ICSENS.2009.5398357
75. Buntkiel L, Reinecke S, Heller A, Budelmann C, Hampel U (2021) Mit UWB-Lokalisierung gekoppelte inertiale Lage- und Bewegungsverfolgung für instrumentierte Strömungsfolger. In: Vorträge, Online: AMA Service GmbH, Von-Münchhausen-Str. 49, 31515 Wunstorf, pp 22–27. https://doi.org/10.5162/15dss2021/2.2
76. Duisterwinkel E, Demi L, Dubbelman G, Talnishnikh E, Wörtche HJ, Bergmans JW (2014) Environment mapping and localization with an uncontrolled swarm of ultrasound sensor

motes, presented at the 166th meeting of the Acoustical Society of America, San Francisco, California, p 030001. https://doi.org/10.1121/1.4879264
77. Duisterwinkel EHA, Talnishnikh E, Krijnders D, Wortche HJ (2018) Sensor motes for the exploration and monitoring of operational pipelines. IEEE Trans Instrum Meas 67(3): 655–666. https://doi.org/10.1109/TIM.2017.2775404
78. Duisterwinkel EHA, Dubbelman G, Talnishnikh E, Bergmans JJWM, Wortche HJ, Linnartz J-PMG (2020) Go-with-the-flow swarm sensing in inaccessible viscous media. IEEE Sensors J 20(8):4442–4452. https://doi.org/10.1109/JSEN.2019.2960448
79. Bisgaard J et al (2020) Flow-following sensor devices: a tool for bridging data and model predictions in large-scale fermentations. Comput Struct Biotechnol J 18:2908–2919. https://doi.org/10.1016/j.csbj.2020.10.004
80. Bisgaard J et al (2021) Characterization of mixing performance in bioreactors using flow-following sensor devices. Chem Eng Res Des 174:471–485. https://doi.org/10.1016/j.cherd.2021.08.008
81. Bisgaard J, Zahn JA, Tajsoleiman T, Rasmussen T, Huusom JK, Gernaey KV (2022) Data-based dynamic compartment model: modeling of *E. coli* fed-batch fermentation in a 600 m3 bubble column. J Ind Microbiol Biotechnol 49(5):kuac021. https://doi.org/10.1093/jimb/kuac021
82. Cabrera F, Cobelli PJ (2021) Design, construction and validation of an instrumented particle for the Lagrangian characterization of flows: application to gravity wave turbulence. Exp Fluids 62(1):19. https://doi.org/10.1007/s00348-020-03121-3
83. Whitman WG (1962) The two film theory of gas absorption. Int J Heat Mass Transf 5(5): 429–433. https://doi.org/10.1016/0017-9310(62)90032-7
84. Baehr HD (2013) Wärme- und Stoffübertragung, Springer-lehrbuch, 8th edn. Springer Vieweg, Berlin. [Online]. http://www.gbv.de/dms/weimar/toc/746715293toc.pdf
85. Nienow AW (2015) Mass transfer and mixing across the scales in animal cell culture. In: Al-Rubeai M (ed) Animal cell culture, Cell engineering. Springer, Cham, pp 137–167. https://doi.org/10.1007/978-3-319-10320-4_5
86. Hobiger G (2015) Kohlendioxid in Wasser mit Alkalinität: Berechnung und grafische Darstellung der chemischen Gleichgewichte. Springer, Berlin/Heidelberg. https://doi.org/10.1007/978-3-662-45466-4
87. Boyd CE (2019) Water quality: an introduction. Springer Nature
88. Martens-Menzel R (2011) Physikalische Chemie in der Analytik: eine Einführung in die Grundlagen mit Anwendungsbeispielen, 2., Aktualisierte Aufl. in Chemie in der Praxis. Vieweg + Teubner, Wiesbaden
89. Hu W-S (2017) Engineering principles in biotechnology. Wiley, Hoboken
90. Royce PNC, Thornhill NF (1991) Estimation of dissolved carbon dioxide concentrations in aerobic fermentations. AICHE J 37(11):1680–1686. https://doi.org/10.1002/aic.690371111
91. Nadal-Rey G et al (2022) Computational fluid dynamics modelling of hydrodynamics, mixing and oxygen transfer in industrial bioreactors with Newtonian broths. Biochem Eng J 177: 108265. https://doi.org/10.1016/j.bej.2021.108265
92. Wutz J, Steiner R, Assfalg K, Wucherpfennig T (2018) Establishment of a CFD-based kLa model in microtiter plates to support CHO cell culture scale-up during clone selection. Biotechnol Prog 34(5):1120–1128. https://doi.org/10.1002/btpr.2707
93. Siebler F, Lapin A, Hermann M, Takors R (2019) The impact of CO gradients on C. Ljungdahlii in a 125 m3 bubble column: mass transfer, circulation time and lifeline analysis. Chem Eng Sci 207:410–423. https://doi.org/10.1016/j.ces.2019.06.018
94. Kerdouss F, Bannari A, Proulx P (2006) CFD modeling of gas dispersion and bubble size in a double turbine stirred tank. Chem Eng Sci 61(10):3313–3322. https://doi.org/10.1016/j.ces.2005.11.061
95. Laakkonen M, Alopaeus V, Aittamaa J (2006) Validation of bubble breakage, coalescence and mass transfer models for gas–liquid dispersion in agitated vessel. Chem Eng Sci 61(1): 218–228. https://doi.org/10.1016/j.ces.2004.11.066

96. Montante G, Horn D, Paglianti A (2008) Gas–liquid flow and bubble size distribution in stirred tanks. Chem Eng Sci 63(8):2107–2118. https://doi.org/10.1016/j.ces.2008.01.005
97. Sommerfeld M, Decker S (2004) State of the art and future trends in CFD simulation of stirred vessel hydrodynamics. Chem Eng Technol 27(3):215–224. https://doi.org/10.1002/ceat.200402007
98. Coroneo M, Montante G, Paglianti A, Magelli F (2011) CFD prediction of fluid flow and mixing in stirred tanks: numerical issues about the RANS simulations. Comput Chem Eng 35(10):1959–1968. https://doi.org/10.1016/j.compchemeng.2010.12.007
99. Ebrahimi M, Tamer M, Villegas RM, Chiappetta A, Ein-Mozaffari F (2019) Application of CFD to analyze the hydrodynamic behaviour of a bioreactor with a double impeller. PRO 7(10):10. https://doi.org/10.3390/pr7100694
100. Cortada-Garcia M, Dore V, Mazzei L, Angeli P (2017) Experimental and CFD studies of power consumption in the agitation of highly viscous shear thinning fluids. Chem Eng Res Des 119:171–182. https://doi.org/10.1016/j.cherd.2017.01.018
101. Murthy BN, Joshi JB (2008) Assessment of standard k–ε, RSM and LES turbulence models in a baffled stirred vessel agitated by various impeller designs. Chem Eng Sci 63(22):5468–5495. https://doi.org/10.1016/j.ces.2008.06.019
102. Brucato A, Ciofalo M, Grisafi F, Micale G (1998) Numerical prediction of flow fields in baffled stirred vessels: a comparison of alternative modelling approaches. Chem Eng Sci 53(21):3653–3684. https://doi.org/10.1016/S0009-2509(98)00149-3
103. Crowe CT, Schwarzkopf JD, Sommerfeld M, Tsuji Y (2011) Multiphase flows with droplets and particles, 2nd edn. CRC Press. https://doi.org/10.1201/b11103
104. Haringa C, Vandewijer R, Mudde RF (2018) Inter-compartment interaction in multi-impeller mixing: part I. Experiments and multiple reference frame CFD. Chem Eng Res Des 136:870–885. https://doi.org/10.1016/j.cherd.2018.06.005
105. Haringa C, Vandewijer R, Mudde RF (2018) Inter-compartment interaction in multi-impeller mixing. Part II. Experiments, sliding mesh and large Eddy simulations. Chem Eng Res Des 136:886–899. https://doi.org/10.1016/j.cherd.2018.06.007
106. Witz C, Treffer D, Hardiman T, Khinast J (2016) Local gas holdup simulation and validation of industrial-scale aerated bioreactors. Chem Eng Sci 152:636–648. https://doi.org/10.1016/j.ces.2016.06.053
107. Thomas JA et al (2021) A mechanistic approach for predicting mass transfer in bioreactors. Chem Eng Sci 237:116538. https://doi.org/10.1016/j.ces.2021.116538
108. Roush D et al (2020) Toward in silico CMC: an industrial collaborative approach to model-based process development. Biotechnol Bioeng 117(12):3986–4000. https://doi.org/10.1002/bit.27520
109. U.S. Food & Drug Administration (2016) Reporting of computational modeling studies in medical device submissions. FDA. Accessed 30 Mar 2023. [Online]. Available: https://www.fda.gov/regulatory-information/search-fda-guidance-documents/reporting-computational-modeling-studies-medical-device-submissions
110. Babi DK et al (2022) Opportunities and challenges for model utilization in the biopharmaceutical industry: current versus future state. Curr Opin Chem Eng 36:100813. https://doi.org/10.1016/j.coche.2022.100813
111. Delafosse A, Line A, Morchain J, Guiraud P (2008) LES and URANS simulations of hydrodynamics in mixing tank: comparison to PIV experiments. Chem Eng Res Des 86(12):1322–1330. https://doi.org/10.1016/j.cherd.2008.07.008
112. Sharp KV, Adrian RJ (2001) PIV study of small-scale flow structure around a Rushton turbine. AICHE J 47(4):766–778. https://doi.org/10.1002/aic.690470403
113. Haringa C (2023) An analysis of organism lifelines in an industrial bioreactor using Lattice-Boltzmann CFD. Eng Life Sci 23(1). https://doi.org/10.1002/elsc.202100159
114. Concha-Gómez ADDL, Ramírez-Muñoz JJ, Márquez-Baños VE, Haro C, Alonso-Gómez AR (2019) Effect of the rotating reference frame size for simulating a mixing straight-blade

impeller in a baffled stirred tank. Revista Mexicana de Ingeniería Química 18(3):3. https://doi.org/10.24275/uam/izt/dcbi/revmexingquim/2019v18n3/DeLaConcha
115. Villiger TK et al (2018) Experimental and CFD physical characterization of animal cell bioreactors: from micro- to production scale. Biochem Eng J 131:84–94. https://doi.org/10.1016/j.bej.2017.12.004
116. Delafosse A et al (2014) CFD-based compartment model for description of mixing in bioreactors. Chem Eng Sci 106:76–85. https://doi.org/10.1016/j.ces.2013.11.033
117. Bach C et al (2017) Evaluation of mixing and mass transfer in a stirred pilot scale bioreactor utilizing CFD. Chem Eng Sci 171:19–26. https://doi.org/10.1016/j.ces.2017.05.001
118. Wutz J, Waterkotte B, Heitmann K, Wucherpfennig T (2020) Computational fluid dynamics (CFD) as a tool for industrial UF/DF tank optimization. Biochem Eng J 160:107617. https://doi.org/10.1016/j.bej.2020.107617
119. Kuschel M, Wutz J, Salli M, Monteil D, Wucherpfennig T (2023) CFD supported scale up of perfusion bioreactors in biopharma. Front Chem Eng 5:1076509. https://doi.org/10.3389/fceng.2023.1076509
120. Soos M, Ehrl L, Bäbler MU, Morbidelli M (2010) Aggregate breakup in a contracting nozzle. Langmuir 26(1):10–18. https://doi.org/10.1021/la903982n
121. Soos M, Kaufmann R, Winteler R, Kroupa M, Lüthi B (2013) Determination of maximum turbulent energy dissipation rate generated by a rushton impeller through large eddy simulation. AICHE J 59(10):3642–3658. https://doi.org/10.1002/aic.14206
122. Wutz J et al (2016) Predictability of kLa in stirred tank reactors under multiple operating conditions using an Euler–Lagrange approach. Eng Life Sci 16(7):633–642. https://doi.org/10.1002/elsc.201500135
123. Sungkorn R, Derksen JJ, Khinast JG (2012) Euler–Lagrange modeling of a gas–liquid stirred reactor with consideration of bubble breakage and coalescence. AICHE J 58(5):1356–1370. https://doi.org/10.1002/aic.12690
124. Krýsa P, Šoóš M (2022) Modelling of bubble breakage and coalescence in stirred and sparged bioreactor using the Euler-Lagrange approach. Int J Heat Mass Transf 199:123466. https://doi.org/10.1016/j.ijheatmasstransfer.2022.123466

Chapter 4
Continuous Bioprocessing for Upstream Process: Perfusion Process

Veronique Chotteau, Hubert Schwarz, and Meeri Mäkinen

Abstract Perfusion processes are gaining higher interest in the industry for the production of glycoproteins. The intensification that this operation mode can bring is attractive and inclusion of perfusion process is nowadays often envisioned for a switch towards continuous processing including downstream, for an intensification only in the upstream production process or for the intensification at the level of the seed bioreactor only. In all cases, the intensification is based on operation at high or very high cell density. The present chapter presents an overview of the recent knowledge about perfusion processes, their main concepts and their development, including medium development, scale-up, scale-down, life cycle and cost analysis.

Keywords Recombinant protein biomanufacturing · Perfusion process · Continuous bioprocessing · Perfusion medium optimization · Cell specific perfusion rate · CSPR · Monoclonal antibody · Chinese hamster ovary cells · Human embryonic kidney cells · Microbioreactors · Tubespin · Pseudo-perfusion · Design of Experiments

Abbreviations and Notations

Abbreviations

ATF	Alternating tangential flow filtration
CapEx	Capital expenditures
CHO	Chinese hamster ovary cells

V. Chotteau (✉) · H. Schwarz · M. Mäkinen
Cell Technology Group CETEG, Department of Industrial Biotechnology, School of Engineering Sciences in Chemistry, Biotechnology, and Health, KTH – Kungliga Tekniska Högskolan (Royal Institute of Technology), Stockholm, Sweden

AdBIOPRO, Competence Centre for Advanced BioProduction by Continuous Processing, Stockholm, Sweden
e-mail: veronique.chotteau@biotech.kth.se; chotteau@kth.se; hschwarz@kth.se; meerim@kth.se

R. Pörtner (ed.), *Biopharmaceutical Manufacturing*, Cell Engineering 11,
https://doi.org/10.1007/978-3-031-45669-5_4

COGS	Cost of goods
CSPR	Cell specific perfusion rate
$CSPR_{min}$	Minimum cell specific perfusion rate
DHFR	Dihydrofolate reductase
DO	Dissolved oxygen
DoE	Design of Experiments
EDR	Energy dissipation rate
GS	Glutamine synthetase
HCD	High cell density process
HEK293	Human embryonic kidney 293 cells
LCD	Low cell density process
mAb	Monoclonal antibody
MBR	Microbioreactor
MSX	Methionine sulfoximine
OD	Optical density
PoI	Product of interest
RV	Reactor volume
SS	Steady-state
SSPP	Steady-state perfusion process
STR	Stirred-tank bioreactor
SUB	Single-use bioreactor
TAFE	Targeted feeding
TFF	Tangential flow filtration
UF	Ultra-filter

Notations

B	Bleed rate
CO_2	Oxygen concentration in liquid phase
CO_2*	Saturation oxygen concentration in liquid phase
C_p	Concentration of the product p in the culture
$C_{p,in}$	Concentration of the product p in the feed medium
$C_{PoI,\ Bioreactor}$	Product of interest concentration in the bioreactor
$C_{PoI,\ Harvest}$	Product of interest concentration in the harvest
C_s	Concentration of component s in the bioreactor
$C_{s,\ in}$	Concentration of component s in the feed
D	Perfusion rate
H	Harvest rate
H_{POI}^{tot}	Accumulated harvested product of interest production per bioreactor volume
k_La	Volumetric mass transfer coefficient
N-1	N-1 refers to the seed bioreactor
P/V	Power/Volume
$q_{glucose}^{target}$	Targeted cell specific uptake rate of glucose

$q_{s_i}^{target}$	Targeted cell specific uptake rate of sugar s_i, where s_i = glucose, mannose or galactose
$q_{sugar,T}^{target}$	Sum of the targeted cell specific uptake rate of the sugars
q_{O2}	Cell specific oxygen uptake rate
q_{PoI}	Cell specific productivity of product of interest
$Q_{PoI,\ Harvest}$	Volumetric productivity of product of interest
q_s	Cell specific consumption/production rate of component s
t	Time
V	Bioreactor volume
$V_{harvest}$	Harvest volume
V_{in}	Volume entering the bioreactor per time unit
V_{out}	Volume exiting the bioreactor per time unit
X	Cell density
μ	Specific growth rate
σ_H	Sieving coefficient

Introduction

The biopharmaceutical field has been dominated by fed-batch processes during decades and this operation mode continues to be predominant. However the industry is progressively also adopting continuous processing based on perfusion process due to its potential of process intensification, detailed below. This interest is present at different levels, depending on the company production capacity, its experience with continuous systems and its desire to drive newer approaches, which can potentially be associated to higher risks. In perfusion mode, the culture medium is continuously renewed, allowing to continuously provide new nutrients while eliminating the toxic by-products. The culture medium is renewed while the cells are retained in the bioreactor thanks to a cell separation device. Section "Basics of Perfusion Processes" introduces the basics of perfusion processes. *Cell separation devices* have been previously reviewed [1] and section "Cell Separation Devices" gives an update of this report. We refer as well to the review of Castilho and Medronho [2] for detailed features of cell separation devices not based on filters. The perfusion mode can provide also a *shorter residence* time of the product of interest (PoI) in the bioreactor, which is favorable for the *product quality* and is a requirement in case of sensitive glycoproteins such as enzymes or coagulation factors [3–6]. Beyond the production of these type of molecules, perfusion mode enables achieving high and very high cell densities while maintaining a high viability as published for the first time more than 10 years ago [7, 8]. These last years, the field has strived after *intensification*, where increasing the cell density is a key factor. Figure 4.1 recapitulates the main ways the field envisions process intensification based on increased cell density supported by medium renewal. One way, in the *steady-state perfusion process* (SSPP), is to increase the cell density to a given high level, such as 100×10^6

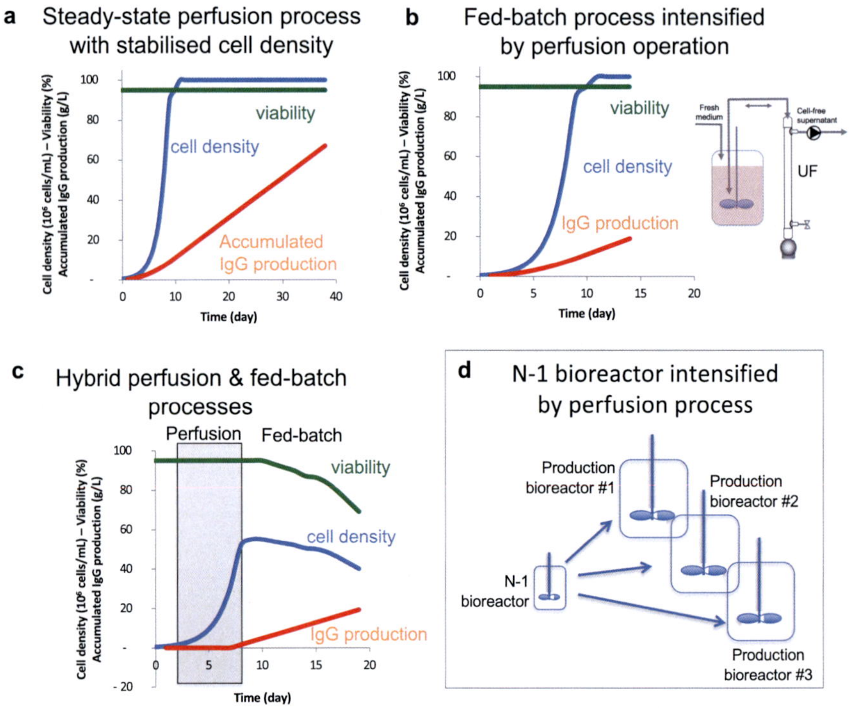

Fig. 4.1 Schematic representation of the main ways that perfusion operation is included in biopharmaceutical processes nowadays; (**a**) steady-state perfusion process showing that the cell density increases to a given high level, where it is maintained by cell bleed, and showing as well a high viability and the accumulated produced PoI; (**b**) concentrated fed-batch, operated with an ultra-filter in the cell separation device, and in which the cell density is continuously increased, while the PoI accumulates in the bioreactor retained by the UF; (**c**) example of hybrid solution allying a period of perfusion to increase the cell density, followed by fed-batch phase for the production; (**d**) implementation of perfusion operation in the seed bioreactor, N-1, to increase the inoculum cell density of one or several production bioreactors

cells/mL, and to maintain it at this level by continuously discarding cells in *cell bleed* at the same rate as the cells are growing. In Fig. 4.1a, one can see that after a growth phase, the cell density is maintained at a stable level of 100×10^6 cells/mL around day 10 until the culture end. The product is continuously produced by the cells and directed into the harvest tank downstream of the cell separation device. A high cell density can generate a very large total accumulated production, which is the daily volumetric productivity integrated over time. The favorable environment of medium renewal can support a high viability, given that the cell separation device does not generate mechanical stress affecting the cells. Such cultures can be maintained during several weeks or even months. Concerning the process length, the longest published commercial process was 6 month long and produced the coagulation factor VIII [3]. An actual tendency is cultures lasting 1 or 2 months. Maintaining the cell stability for the PoI production is of course a major constraint in these systems. A drawback of the SSPP approach is that the harvest volume is very high and needs then to be processed in the capture step, creating a heavy burden for this operation. An alternative way to take benefit of the medium renewal favoring very high cell densities without necessitating to capture the PoI from large harvest volumes is to include an *ultra-filter* (UF) in the cell separation device. This is represented in Fig. 4.1b, where the hollow fiber cartridge used in the cell separation device is a UF filter. The cut-off of the filter is selected such that the PoI is retained in the bioreactor while the molecules of smaller sizes are renewed by the perfusion operation. In other words, during this continuous renewal of the small molecules, the nutrients are provided from the fresh medium, while the by-products present in the spent medium are removed from the bioreactor, generating favorable conditions for cell growth and production while the PoI stays in the bioreactor. This process associates the advantages of the perfusion and fed-batch operations and is therefore called *concentrated fed-batch* or *enhanced fed-batch* and sometimes *intensified fed-batch* [7, 9–12]. The main benefit is that this process requires a single harvest instead of harvesting several times during the cultivation run or implementing a continuous capture step. In this case, no cell bleed is implemented and the cell density increases until the UF filter gets clogged, why these processes last a couple of weeks, i.e. typically much shorter than SSPP. Noticeably, their single harvest is challenging due to the very high cell density achieved of more than 100×10^6 cells/mL. Hybrid solutions have also been proposed such as depicted in Fig. 4.1c where both perfusion and fed-batch modes are used with aimed of using the perfusion to increase the cell density and the fed-batch for the production phase [13]. Finally implementation of perfusion operation in the seed bioreactor, N-1, is the most adopted intensification by the industry [14–21]. The inoculum cell density of the production bioreactor can easily be increased, representing a substantial gain of time for the cell growth phase in the production bioreactor operated in fed-batch or in perfusion mode. Potentially several production bioreactors can be inoculated from the same N-1 seed bioreactor, as schematically illustrated in Fig. 4.1d.

As mentioned above, the main benefit of perfusion mode compared to batch and fed-batch operation is the medium renewal, which can be associated with very high cell density. In comparison with these, this can provide higher daily volumetric production, shorter residence time of the PoI in the production bioreactor, a longer culture period, healthier cells, better and importantly reversible control of the cell environment. The short residence time provides a product quality superior to fed-batch due to reduced enzymatic action of proteases and sialidases [22–24]. Higher cell density associated with higher volumetric productivity together with long cultivation run, represent a significant intensification for commercial production in bioreactors of size much smaller, e.g. 2000 L, in comparison with fed-batch bioreactors [25]. Consequently such intensified processes are highly compatible with disposable / single-use equipment. Due to the smaller bioreactor size, the investment cost (capital expenditures, CapEx) required to build new facilities equipped with continuous intensified processes is significantly lower than building plants based on fed-batch bioreactors ensuring the same capacity. This has been one of the major factor for the rising of perfusion processes these last years in industry. Another factor is the compatibility with single-use equipment providing as well a high flexibility needed in multi-purpose plant [26], which is a trend even more important nowadays with the increase of new modalities, potentially requiring different type of equipment, i.e. using the concept of ball room. Furthermore, in case of sudden increase of the demand for a biologics, the production capacity of an intensified process realized in disposable equipment can potentially rapidly be enlarged by scaling-out the process, with also the possibility of transfer to another site. These last arguments might be of low importance for companies where the production capacity is already ensured by plants equipped with large bioreactors, and for which the intensification brought by perfusion processes is not necessary.

Concerning the N-1 step, a general trend, followed by many companies, is the adoption of high cell density perfusion process to obtain a high inoculum in the production bioreactor, generating a reduced growth period in the production bioreactor and a more efficient seed train in comparison with legacy batch culture.

Another popular industrial application of perfusion process is the manufacturing of cell bank. A cell suspension grown in bioreactor in perfusion mode at high cell density, e.g. 100×10^6 cells/mL in wave-induced bioreactor, can be directly cryopreserved by addition of a cryopreservant (i.e. dimethyl sulfoxide DMSO) without centrifugation and with excellent recovery at thaw [7, 27, 28].

A final aspect, which has nowadays taken a large importance, is that intensified processes relying on disposable equipment consumes less energy and clean water, and are smaller in size in comparison with large legacy fed-batch plants and thus have a more favorable carbon dioxide footprint [25, 29, 30]. *Life cycle and cost analysis* are presented in section "Life Cycle and Cost Analysis".

The reasons why perfusion processes have been avoided until the years 2000–2010's except for unstable PoI, were the lack of robust and simple cell separation device, a higher process complexity and, consequently, a higher challenge for the sterility, in comparison with legacy fed-batch processes. A corollary to this has been the lack of competence and knowledge in biopharma field dominated by fed-batch processes, including as well Contract Manufacturing Organizations (CMOs). This is also clear from the literature more abundant for fed-batch processes in comparison to perfusion since 2007, a trend however, which has recently changed, i.e. since 2020 (Fig. 4.2). Furthermore, the large volumes of harvest and fresh medium generate as well challenges for the logistics and the downstream team. One of the key factors, which has enabled the perfusion technology has been the hollow fiber-based cell separation devices, and in particular the alternating tangential flow filtration (ATF) introduced by Shevitz [31]. A history has been recently published [32].

The present chapter is partially based on [33].

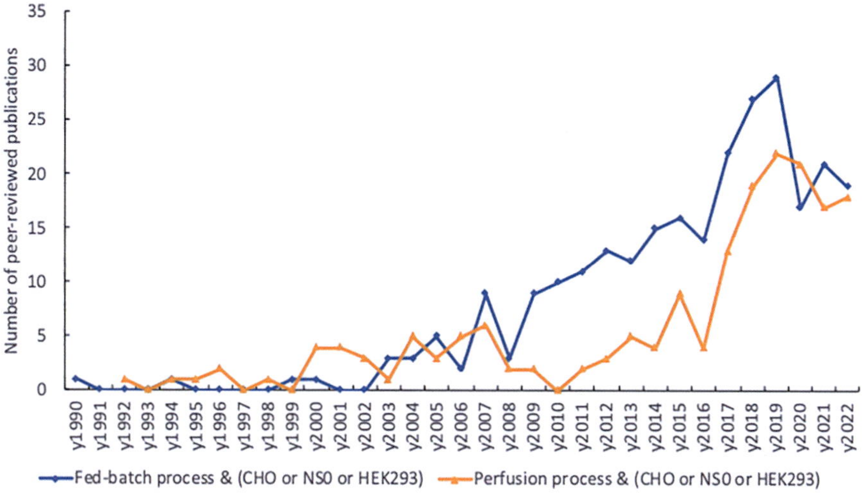

Fig. 4.2 Number of peer-reviewed articles published between 1990 and 2022, related to fed-batch or to perfusion processes with CHO, NS0 or HEK293 cells obtained in https://pubmed.ncbi.nlm.nih.gov (search Fed-batch process AND (CHO or NS0 or HEK293); search Perfusion process AND (CHO or NS0 or HEK293))

Basics of Perfusion Processes

Operation and Control of a Perfusion Bioreactor

Perfusion Concept

A schematic representation of a typical set-up for a perfusion process can be found in Fig. 4.3. The culture medium is continuously renewed by a cell separation device, retaining the cells in the bioreactor. In SSPP, a growth phase until a given cell density is followed by a production phase at a cell concentration stably maintained at steady-state (SS) by cell bleed at a rate equal to the growth rate. The cell bleed is discarded and represents thus a loss of material. It can be attractive to exploit this material however this provides a higher complexity to the process, increasing the risk of failure and the validation, and is therefore not a common practice. During the growth phase, the medium and feed(s) are optimized to maximize the growth rate, while at SS, a cell growth is desired because it ensures healthy cells but should not be excessive since the PoI in the cell bleed is lost. The cell growth can be slowed down by decreasing the temperature, which should be optimized. Notice that an alternative continuous culture process is the chemostat in which the cell suspension is continuously removed at the same rate as the growth rate without retention of the cells in the bioreactor. Chemostat operation is not used for production using mammalian cells but is sometimes used to develop perfusion processes.

The rate at which the medium is renewed in the bioreactor is characterized by the perfusion rate (D), with unit number of reactor volumes per day or RV/day or day^{-1}.

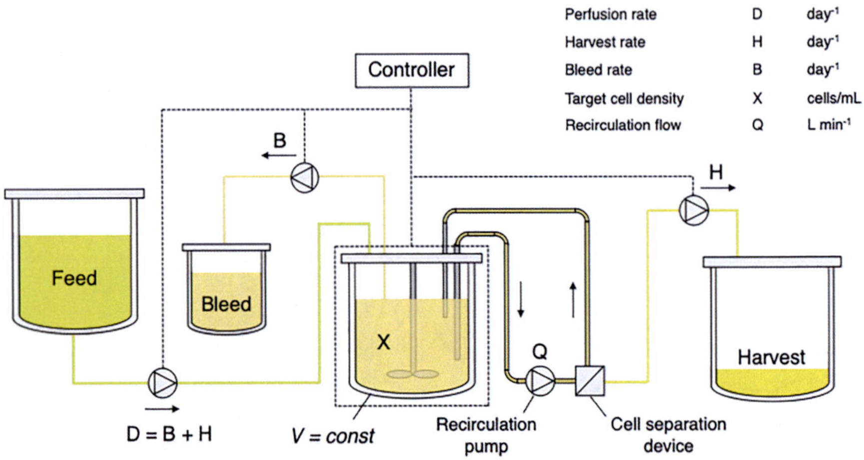

Fig. 4.3 Example of perfusion process showing the bioreactor, the feed tank containing the fresh medium (Feed), the harvest tank (Harvest) cooled at 4 °C, the cell bleed tank (Bleed) collecting the cell suspension to be discarded, the cell separation device connected, in this example, to the bioreactor via a recirculation pump actuated at rate Q, and the controller, which actuates the flow rates to maintain the working volume V constant, the cell density X, the perfusion rate, D, at desired targets. Thanks to the cell separation device, the supernatant free of cells is continuously pumped into the harvest tank at harvest rate H while the cells, more concentrated, are returned from the cell separation device into the bioreactor; fresh medium is automatically added into the bioreactor from the feed tank at rate D such that the bioreactor volume is in average constant; the excess of cells is pumped from the bioreactor into the bleed tank at rate B. (Reproduced from [33])

This continuous medium renewal provides a constant environment to the cells, which is favorable for the cell metabolism, growth and health. The storage of the PoI in the harvest tank at 4 °C, where enzymatic cleavage is highly reduced, is highly favorable for the PoI quality, a property indispensable for labile glycoproteins [34–37]. Interestingly, even stable proteins produced by fed-batch, can present a superior quality profile in SSPP, e.g. a high level of glycosylation maturation [12, 24]. An exception is of course the case of concentrated fed-batch, in which the PoI is kept in the bioreactor during the whole run, so the enzymatic cleavage is present. Nutrients are provided from the fresh medium continuously added to avoid their depletion, and concurrently the removal of spent medium eliminates the toxic by-products, i.e. lactate, ammonia, or others as listed in [38] or [39], cellularly produced CO_2 and HCO_3^- issued from the alkali added to control the pH. Other fields than glycoprotein manufacturing use as well perfusion operation such as cultures of stem cells, immune or primary cells/tissue [40, 41].

Perfusion Process Operation

Different flows are coming in- or leaving the bioreactor in a perfusion process. The perfusion rate D, introduced in the previous paragraph, is the overall renewal flow rate. Its setpoint is given by the operator and is selected during the process development. The harvest rate, H [RV/day or day^{-1}] is the rate of the cell-free supernatant pumped out of the cell separation device. To maintain a process in steady-state, cells are removed with a bleed rate, B [RV/day or day^{-1}]. No bleed rate is not applied when it is desired to increase the cell concentration, i.e. B is then equal to 0. Each of these flow rates is typically controlled by a pump. The perfusion rate is the sum of the harvest rate and the bleed rate;

$$D = H + B \tag{4.1}$$

The removal rate of these exiting liquids is equal to the rate of fresh medium, denoted as '*feed medium*', '*perfusion medium*' or simply 'feed') pumped into the bioreactor, and the bioreactor volume is constant. As previously mentioned, B is equal to the cell growth rate, μ. For a given D, H is given by the difference between D and B, i.e. D − B. Since B represents a loss of product, it is desirable to minimize the cell growth at steady-state, so that H is maximized for the given D. The perfusion set-up includes dedicated tanks or single-use bags for the harvest, the bleed and the feed (Fig. 4.3). Notice that the volumes of feed medium and harvest are much larger than the bioreactor volume.

In the practical implementation of the perfusion process, a feedback control is put in place with the objective to maintain the bioreactor at a constant volume. Typically H and B are measured and are the inputs of this feedback control which automatically controls the feed rate pump. This automatic addition of fresh medium from a cooled medium tank should be operated at a slow rate to avoid cooling of the culture. It is also thinkable to implement a control of the harvest rate, H, with the fresh medium addition as input. However from an operational safety point of view, in case of failure, the first alternative leads to an empty bioreactor, while the second one will cause overflooding, which represents a more severe issue. In lab scale, it is possible to use calibrated pumps for the rates D, H and B without implementation of a feedback control, given that the process is well monitored to ensure a constant bioreactor volume. Peristaltic pumps do not always generate exactly the same flows. In particular after long time period in use, tube fatigue occurs, affecting the performances. Monitoring of the bioreactor volume is carried out by weigh (i.e. load cell) or level.

Cell Specific Perfusion Rate

The cell specific perfusion rate, CSPR, [pL/cell/day or nL/cell/day] is a well-recognized parameter to characterize and select the perfusion rate [42, 43]. It is equal to the perfusion rate divided by the cell density;

$$\mathrm{CSPR} = \frac{D}{X} \tag{4.2}$$

In short, if a certain perfusion rate D_1 has been well selected for a given cell density X_1, i.e. providing adequately the nutrient to the culture and by-product elimination, for another higher cell density X_2, the perfusion rate D_2 should be linearly increased with the same factor as the cell density increase, X_2 / X_1, keeping the CSPR constant;

$$\mathrm{CSPR} = \frac{D_1}{X_1} = \frac{D_2}{X_2}$$

It is advantageous to minimize the CSPR since the smaller the CSPR, the smaller the consumption of medium, the smaller the generation of harvest, and the smaller the PoI dilution, i.e. the higher the PoI titer in the harvest line. A common approach is to select a CSPR for a given feed medium, including potential additives, and to deduce to perfusion rate for a given cell density from the CSPR value. The bleed rate and the perfusion rate for a given CSPR can be automatically controlled based on the information of the viable cell density measured by e.g. a bioimpedance probe [44, 45] or the total cell density in case of high viability given by optical density [46].

The concept of CSPR is illustrated here by two examples;

A CSPR = 10 pL/(cell day) is very low, and corresponds to D = 1 RV/day at cell density = 100×10^6 cells/mL.

A CSPR = 50 pL/(cell day) is not unusual, in particular for older processes, and corresponds to D = 1 RV/day at cell density = 20×10^6 cells/mL, or D = 5 RV/day at cell density = 100×10^6 cells/mL.

Performance Parameters and Mass Balance Equations in Perfusion Process

By-products and Substrates

The mass balance equation of a process states that the variation of the concentration of a component with time (*t*) is equal to the consumption or production of this component, plus what enters in the bioreactor, minus what is removed from the bioreactor. Let us take the example of ammonia, the mass balance for this product (p) is

$$\frac{d(VC_{p)}}{dt} = q_p\, X\, V + V_{in} C_{p,\ in} - V_{out} C_p \tag{4.3}$$

where C_p is the concentration of p, q_p is the cell specific production rate of p, V_{in} and V_{out} are respectively the volumes entering and exiting the bioreactor per time unit, $C_{p,in}$ is the concentration of the product p in the feed medium. For instance many media include a concentration of 2 mM ammonium.

$V_{in} = V_{out}$ and the perfusion rate, D, is the ratio of V_{in} or V_{out} by V;

$$D = \frac{V_{in}}{V} = \frac{V_{out}}{V} \tag{4.4}$$

In a perfusion process, the volume of the bioreactor is constant, which brings a simplification of the mass balance eq.

$$V\frac{d\,C_p}{dt} = V\left(q_p X + \frac{V_{in}}{V} C_{p,\ in} - \frac{V_{out}}{V} C_p\right)$$
$$\Rightarrow \frac{d\,C_p}{dt} = q_p X + \frac{V_{in}}{V} C_{p,\ in} - \frac{V_{out}}{V} C_p \tag{4.5}$$

Furthermore in case of established steady-state, the concentrations of the different components are constant, so the variation with time of these concentrations is 0. Introducing as well Eqs. 4.4 and 4.5 becomes

$$\frac{d\,C_p}{dt} = 0 = q_p X + D\left(C_{p,\ in} - C_p\right) \tag{4.6}$$

It follows that the concentration of p in the bioreactor can be expressed at steady-state as

$$C_p = q_p \frac{X}{D} + C_{p,\ in} \tag{4.7}$$

and taking Eq. 4.2 into account, Eq. 4.7, expressing the concentration of a product in steady-state perfusion process, becomes

$$C_p = \frac{q_p}{CSPR} + C_{p,\ in} \tag{4.8}$$

Equation 4.8 can be used to understand which concentrations of the product p can occur, given that estimations of the values of q_p and $C_{p,in}$ are given. Notice that although Eq. 4.8 is rather simple, the value of q_p is highly depending on the cell environment. For instance, the cell specific production rates of the by-products lactate and ammonia depend on the concentrations of their substrates, i.e. sugar, glutamine and other amino acids. Figure 4.4 illustrates which concentrations are obtained for different values of the cell specific production rates, the CSPR and the presence (or not) of the by-product in the feed medium on the concentrations of these toxic by-products in SSPP. It is obvious that the cell specific production rates of these by-products should be minimized as much as possible. From Fig. 4.4a, it can be observed that a too high lactate concentration, e.g. ≥25 mM, can rather easily occur when the CSPR is low, even for low values of its cell specific production rate, $q_{lactate}$. Therefore providing too large amount of glucose to the culture generating a high $q_{lactate}$, will cause a high lactate concentration. A strategy to address this issue is given in section "Process Optimization". Interestingly, a concentration of ammonia considered to be too high, e.g. ≥8 mM, occurs for a range of rather high cell specific production rates of ammonia, $q_{ammonia}$. This can be visualized by drawing an imaginary horizontal line at 25 mM lactate concentration and another one at 8 mM

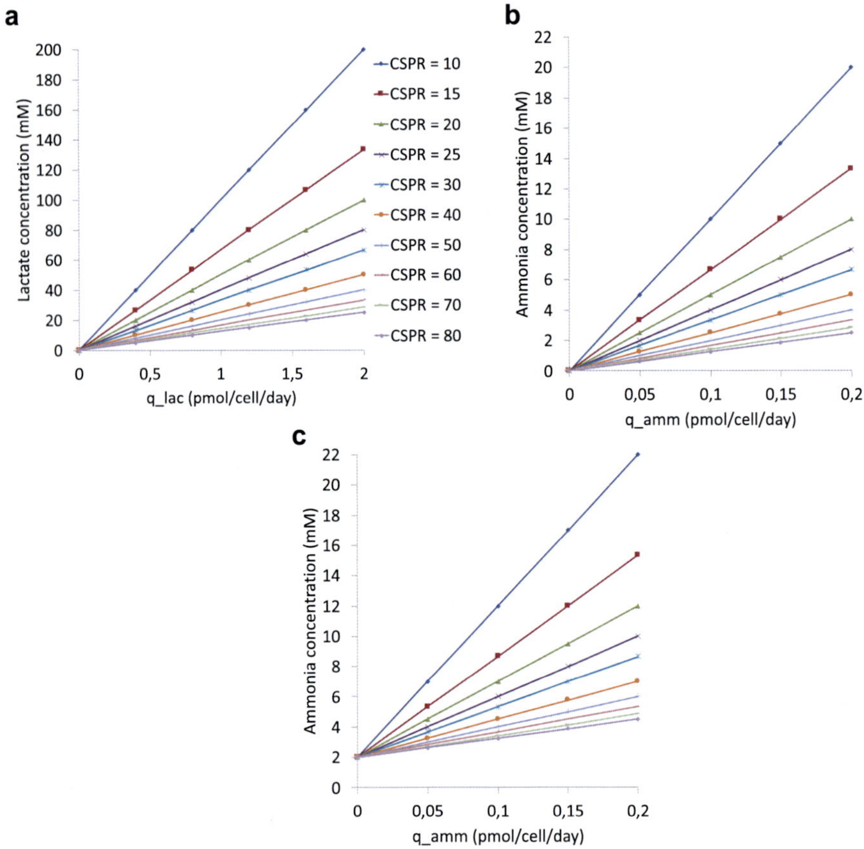

Fig. 4.4 Simulation of lactate and of ammonia concentrations in the bioreactor as a function of the cell specific production rates of these by-products when the CSPR varies (CSPR unit = pL/cell/day), using Eq. 4.8; a simulation of lactate concentration; b simulation of ammonia concentration in case of no ammonia is present in the fresh medium; c simulation of ammonia concentration in case of 2 mM ammonia is present in the fresh medium

ammonia concentration in Fig. 4.4a, b, respectively. Comparing these imaginary lines, the line at 25 mM lactate concentration crosses all the CSPR cases except the two highest values, while the line at 8 mM ammonia crosses only the three lowest values of CSPR. It is therefore easier to maintain the ammonia concentration at a low level even for low CSPR values. In other words, the detrimental effect of a high concentration of lactate is potentially more likely to occur than too high ammonia concentration, not mentioning the use of cell lines able to synthase their own glutamine synthetase, and thus not prone to produce ammonia.

Equation 4.8 and Fig. 4.4 can be used during the process development to understand the interaction of these different parameters and to identify what should be the cell specific production rate tolerated of a by-product for a given CSPR.

Likewise the mass balance for a substrate s, e.g. glucose, glutamine, can be expressed as follows;

$$\frac{d\,C_s}{dt} = -q_s\,X + C_{s,\ in} - \frac{V_{out}}{V}C_s \tag{4.9}$$

where C_s is the concentration of s, q_s is the cell specific consumption rate of s, $C_{s,in}$ is the concentration of the substrate s in the feed medium. In a similar way as Eq. 4.8 was derived, the following can be expressed for steady-state perfusion culture;

$$C_s = -\frac{q_s}{CSPR} + C_{s,\ in} \tag{4.10}$$

Equation 4.10 can be used during the process development to identify the concentration of the substrates in the feed medium or additives. This will be developed below in section "Process Optimization".

Product of Interest

A mass balance equation for the PoI can be derived similarly to Eq. 4.8, expressing the variation with time the PoI concentration, C_{PoI} [g/L]. Notice that units/L are often used for enzymes but the text refers to g/L only for clarity. However two other factors are also important: the loss of a certain amount of PoI in the cell bleed and the fact that the cell separation device can potentially partially retain the PoI in the bioreactor. This latter is, in particular, an issue in filter-based device, see section "Filtration-based Cell Separation Devices".

The mass balance equation for the PoI is

$$\frac{d\,C_{PoI}}{dt} = q_{PoI}\,X - \frac{V_{out}}{V}C_{PoI} \tag{4.11}$$

where q_{PoI} [pg/cell/day] is the cell specific production rate of the PoI. Taking into account the harvest rate H [RV/day] and the bleed rate B [RV/day], such that $D = H + B$, Eq. 4.11 becomes

$$\frac{d\,C_{PoI}}{dt} = q_{PoI}\,X - H\,C_{PoI,Harvest} - B\,C_{PoI,Bioreactor} \tag{4.12}$$

where $C_{PoI,Harvest}$ and $C_{PoI,Bioreactor}$ are the concentrations of the PoI in the harvest line and in the bioreactor respectively. After rearrangement of Eq. 4.12, the cell specific production rate of the PoI can be expressed as

$$q_{PoI} = \frac{1}{X}\left(\frac{d\,C_{PoI}}{dt} + H\,C_{PoI,Harvest} + B\,C_{PoI,Bioreactor}\right) \tag{4.13}$$

and the volumetric productivity per bioreactor volume, $Q_{\mathrm{PoI,Harvest}}$ [g/L/day], which is effectively collected, is accounting for the product harvested as follows

$$Q_{\mathrm{PoI,Harvest}} = \mathrm{H}\ C_{\mathrm{PoI,Harvest}} \tag{4.14}$$

The total productivity per volume of bioreactor of a run is the accumulated harvested PoI production, H_{POI}^{tot} [g/L], and is given by the integration of the volumetric productivity with time *t*;

$$\mathrm{H}_{POI}^{tot} = \int_0^t \mathrm{H}\ C_{\mathrm{PoI,Harvest}} dt = \int_0^t Q_{\mathrm{PoI,Harvest}} dt \tag{4.15}$$

As previously mentioned, the PoI in the cell bleed is most often discarded due to the complexity of harvesting this material.

Biomass

The mass balance equation for the biomass can be expressed as

$$\frac{\mathrm{dX}}{\mathrm{dt}} = \mu\ \mathrm{X} - \mathrm{B}\ \mathrm{X} \tag{4.16}$$

In steady-state perfusion culture, the cell density is constant, so Eq. 4.16 becomes

$$0 = \mu\ \mathrm{X} - \mathrm{B}\ \mathrm{X} \quad \text{or} \quad \mu = \mathrm{B} \tag{4.17}$$

i.e. the bleed rate is equal to the growth rate to main the cell density setpoint.

Cell Separation Devices

The cell retention device separates the cells, which are returned to the bioreactor, from the cell-free supernatant transferred to the harvest tank. The device should generate mechanical stress tolerable for the cells, so that the cell viability is affected as little as possible. The residence time of the cells in the devices should not be long. The operation should be stable over long period of time, weeks or potentially months. It should be suitable for commercial scale operation. The retention of the PoI should be noll or minimal. It should be possible to exchange parts or the whole system during a culture run in case of failure, such as clogging. For this reason external separation devices are preferred. For process intensification, the cell retention device should be able to operate at very high cell densities. More detailed reviews have been previously reported [1, 2, 37, 47]. The present text gives a short overview including as well an update of report [1].

The separation devices used for animal cell perfusion processes are based on different main physical principles: (i) filtration, (ii) acceleration including gravity, (iii) electric field, reviewed by Castilho and Medronho [2].

Filtration-based Cell Separation Devices

Cell separation devices based on tangential flow filtration have received an increasing interest and have seen new developments this last decade. In contrast, rotating spin filters easily prone to clogging have not received the same attention. Cell separation can be obtained by a filter mounted in the bioreactor, such as the internal floating filter of the wave-induced bioreactor. Due to its easiness and compactness, this system has seen a growing interest for applications such as cell cryopreservation [48] or seed bioreactor [20, 21]. Cell densities up to 250×10^6 cells/mL have been reported with this device [48].

In the tangential flow filtration, the cell suspension circulates tangentially to the filter surface, which avoids pressing particles in the filter pores and has proven to be highly advantageous to avoid filter clogging. A favored filter system consists in a cartridge of hollow fiber filter, in the lumen of which the cell suspension passes while the cell-free permeate is pumped out of the cartridge. In the tangential flow filtration (TFF), the cell suspension is pumped outside of the bioreactor, circulated in the filter cartridge and sent back to the bioreactor. The pump of this device has been traditionally a peristaltic pump, nowadays potentially replaced by a levitating centrifugal pump. Introduced by Shevitz [31], the alternating tangential flow (ATF) includes a diaphragm pump at one end of the filter cartridge, which alternatively pulls and pushes the cell broth in the hollow fiber filter with a cycle of $\approx$60 seconds. The main advantages of the cell separation devices TFF and ATF are the cell low residence time in the separation device, rather simple and flexible operation, good scalability and demonstrated ultra-high cell density $>200 \times 10^6$ cells/mL [8]. The TTF and ATF have been compared in studies realized in stirred bioreactors [49–51], Wave bioreactors [7] and orbital shake bioreactor [52]. The cell damage has been reported to be sometimes higher when using the TFF in comparison with the ATF. An identified cause is the shear created in the peristaltic pump, causing the so-called 'pump effect'. If the pump setup is adequately selected with a large tube, both systems generate similar cell growth and viability for CHO cells [8]. Furthermore using a smooth inner surface in silicone reduces the pump effect even more [53]. Replacing the peristaltic pump by a levitating centrifugal pump provides an even better solution due to the low shear of this type of pump [51].

A drawback of the ATF or the TFF filtration methods is the retention of the PoI in the bioreactor due to the formation of a gel layer at the filter surface. The gel layer formation in filters used for mammalian cell culture is due to particles of 20–200 nm size, constituted of DNA and other cell debris [35, 54–56], and can be affected as well by antifoam [57]. It is formed after a couple of days to 1 week and can increase with time. This causes a reduced transmission of the PoI through the filter membrane. The yield of product recovery is characterized by the sieving coefficient $\sigma_H = c_{PoI,\ harvest}/c_{PoI,\ bioreactor}$. The sieving decay of the PoI through the filter membrane often leads to an exchange of the filter cartridge during a run to avoid PoI loss. A noticeable difference for the performances of the ATF vs. TFF is the tendency for more severe PoI retention for the latter [7]. The superior product recovery observed with the ATF is attributed to a backflush movement of the permeate through the filter pores due to the alternating directions of the flow in the ATF, preventing the membrane fouling. The decay of PoI sieving is a major issue for which solutions have been sought. The product sieving can be improved with the usage of hollow fiber filter with a pore size of ≥2 μm, i.e. larger than the usual size of 0.1–0.45 μm [58, 59]. This improvement however requires an additional step of filtration since the permeate from the hollow fiber cartridge has a too high turbidity for a direct loading of the capture chromatography step [60]. Interestingly, the same group [56] observed that the nominal pore size of the filter was not the only criterion to take into account but the structure of the filter was a major factor for a possible sieving decay. Using wide-surface pore membranes with conical shaped pores of 0.2 μm filter pore size (Asahi Kasei Microza membrane) in a TFF setting generated a lower sieving decay of the produced IgG and of DNA, which, consequently, did not cause a PoI sieving deterioration, in comparison with a 0.65 μm filter pore size of the same manufacturer but with a different pore shape. Another solution was proposed by [61] in a TFF setting using two levitating centrifugal pumps, used alternatively to create an alternating movement of the cell suspension in the hollow fiber cartridge, leading to a sieving decay similar to the one observed in an ATF setting with the same filter cartridge. Beyond the choice of the membrane, such as the pore size, shape, material and filter area, the operational setup of the ATF or TFF have a significant impact on the product retention. Walther et al. recommended permeate flux rates ≤60 L m^{-2} day^{-1} for ATF operation however these were the cell density and run length had to be taken into account [23]. Furthermore, the shear rate inside the hollow fiber lumen, which is proportional to the recirculation flow rate, is also different in the ATF and the TFF. In the recirculation loop of the ATF, due to the alternating flow, the cell suspension velocity varies from 0 to a maximum value, which is the speed constantly applied in the TFF. The average of the velocity in the ATF is 0.64 lower than in the TFF [53]. Consequently the shear stress in the ATF is lower by the same factor 0.64 in the TFF. This difference potentially does not impact CHO cells but it was showed that for the more sensitive HEK293 cells, this difference affected the cells, independently of the pump effect. Models of the sieving obtained in ATF or in TFF have been developed bringing some insides into the involved phenomena [57, 62, 63].

Acceleration- and Gravity-based Cell Separation Devices

Based on gravity, cell sedimentation can be used to separate the cells from the supernatant. The sedimentation time of single cells is approximatively 40 min, depending on the cell size. This time is rather long, potentially jeopardizing the delivery of oxygen and nutrients to the cells, as well as the homogeneity of the cell environment and can affect the viability. It has however been successfully developed and applied for the small-scale bioreactor Ambr 15 [64]. Never the less, several features have been created to facilitate the sedimentation. Inclined settlers are made of several inclined plates (or lamellae) where the sedimentation of the cells takes place. Due to the inclination, the sedimentation is faster than in the cell suspension, thanks to a convection phenomenon, i.e. Boycott effect [65, 66]. The cell suspension is admitted close to the bottom of the inclined settler by pumping. The cells sediment on the inclined plates, from where they are sent back to the bioreactor. At the top of the settler, a pump retrieves the supernatant, which is transferred to the harvest tank. Inclined settlers are used in large-scale cultures, as reported e.g. by Pohlscheidt et al. [14] for a perfusion process at 16×10^6 cells/mL and >85% of cell retention efficiency with flow rates of up to 3000 L/day. To facilitate the sedimentation, an ultrasonic resonance field (ultrasonic waves) is created provoking zones of concentrated cells in the cell separation chambre mounted on top of the bioreactor. When the waves are interrupted, the concentrated cells easily fall into the bioreactor. Cell densities up to 42×10^6 hybridoma cells/mL have been reported [67] for this system able nowadays to exchange 200 L medium/day. Centrifugation is also used for commercial scale perfusion processes, as realized in the Centritech centrifuge, able to exchange 2880 L medium/day [1, 68]. Finally, the hydrocyclone generates a centrifugal field applied to the cell suspension without rotational mechanical part, and is the most recently introduced technique [69]. Recently a cell density of 50×10^6 cells/mL with high separation efficiency has been reported at pilot scale [70].

These different systems can generate high separation efficiency of the cells after optimization. They also have the advantage to segregate the cells by their size, so dead cells, which are smaller, can be removed from the bioreactor. Another advantage is that, by physical principle, they have a lower tendency to retain the PoI in the bioreactor in comparison with filters. A drawback is that the residence time of the cell suspension in these separation devices can be high. Furthermore the highest reported cell densities have been lower than using tangential flow filtration.

Life Cycle and Cost Analysis

The comparison of fed-batch and perfusion processes for the costs and for the environmental impact is not an easy task due to the variety of involved parameters and the necessary assumptions, not to mention the specificity of different biopharma companies [25, 29, 30] and the studies of such comparisons do not always converge. Bunnack et al. concluded that the perfusion mode has similar cost of goods, but implied increased water consumption in comparison with fed-batch and thus had a larger environmental impact [30]. They also reported a reduction of the environmental footprint by increasing the laps time between harvests of a perfusion process before pooling of the material. In comparison with fed-batch, the costs of goods are reduced when the achieved cell density is five-fold higher in perfusion [71]. The capital expenditures were shown to be 42% smaller for a perfusion-based operation with the assumptions that the production is carried out in a bioreactor fivefold smaller than fed-batch, with the corollary that the seed train for the perfusion bioreaction are reduced as well [29]. The large volume of culture medium in perfusion process is an important contributor to the operation costs, so low CSPR are sought and should be part of the assumption of such evaluations. Interestingly, Xu et al. compared the performances and medium costs of fed-batch, perfusion carried out at 14 pL/cell/day and concentrated fed-batch in a practical case study [10]. The cell densities were respectively 22, 74 and 117 × 10^6 cells/mL and the media were similar but adapted for the perfusion processes. They concluded that the q_{PoI} was around 30 pg/cell/day and comparable for all the processes, that the volumetric productivity, 2.0–2.3 g/L/day, of the concentrated fed-batch and the perfusion were higher than the fed-batch 0.34 g/L/day, while the medium costs per mAb were comparable for all the processes, and highest for the concentrated fed-batch.

The environmental impact is another aspect, which is gaining importance for the industry. The comparison of perfusion and fed-batch processes from this point of view heavily depend if single-use equipment is used or not. A consensus of the impact of single-use equipment *vs.* stainless steel-based equipment converges to point out that this latter is high consumer of energy, i.e. electricity to produce steam, and clean water, i.e. water for injection, while the plastic-based disposable equipment is not [72, 73]. Another fact is that perfusion processes are more compatible with disposable equipment due to the bioreactor size. However Cataldo et al. compared a fed-batch single-use process to a perfusion process based on stainless steel equipment and concluded that this latter had a higher water consumption per mass of antibody produced [74]. As previously mentioned, the economic and environmental assessments are complex and highly depending on the assumptions. Such studies illustrate however the increased interest and acceptance of perfusion process in commercial production for different types of biologics presenting an attractive alternative to legacy fed-batch operation.

Characteristics, Challenges and Scale-up Considerations for Perfusion Processes

Shear Stress

Different issues have to be considered in the intensification by high cell density perfusion. The cell retention devices have been discussed in section "Cell Separation Devices", pointing out mainly the mechanical stress of these systems, the cell density limitation, the PoI sieving decay and the robustness as well as the complexity of the equipment. An important factor is the targeted high (or very high) cell density, to be obtained at high cell viability. Low cell viability, potentially worsening with time, limit the perfusion process length and can jeopardize the cell separation from overload of cell DNA and debris. The cell viability can be affected by different factors in these process. The cell separation device generates a mechanical stress, which affects or not the cells depending of the stress amplitude and the cell robustness. Processes at high cell density need a high oxygen transfer capacity, brought by high agitation, adequate aeration strategy together with an efficient homogenization of the bioreactor. These factors can also increase the shear stress to the cells. An efficient homogenization relies on the design of the bioreactor and the impellers. The mechanical stress and residence time of the recirculation loop, transporting the cell suspension back and forward from the bioreactor to the cell separation device, should be minimized. During the time spent outside of the controlled environment of the bioreactor, the dissolved oxygen concentration drops very rapidly when the density of cells is high. Walther et al. reported that the likelihood of process failure increased when using an ATF device with a residence times longer than 75 seconds, apparently creating cell stress, also associated with lactate production increase, and decay of cell growth and PoI productivity [75]. This implies to optimize the dimensions of the tubes and pumping equipment of this loop. The tube diameter should be dimensioned so that it is as large as possible to reduce the wall effect however is should include a volume as low as possible to reduce the residence time in the loop, requiring to find an optimal compromise between these opposites. As previously mentioned, low-shear pumps, such as the diaphragm pump used in the ATF or the levitating centrifugal pump used in the TFF, are preferred in comparison with a peristaltic pump (see section "Filtration-based Cell Separation Devices"). However a well dimensioned peristaltic pump can also be used for TFF-based perfusion process of CHO cells, as used, for instance, in [8] for more than 40 days in a run which reached 214×10^6 cells/mL. Chalmers and collaborators have recapitulated the energy dissipation rate (EDR) generated by the different parts of a bioreactor environment, e.g. bioreactor agitation, flow through pipe, bubble rupture at the culture surface, that can damage the cells as well as the shear stress tolerance of different cell systems [76]. Using the 'torture chamber' to study the sensitivity of a cell, it is possible to evaluate its robustness to hydrodynamic forces [77, 78]. This type of tests could be carried out during the cell clone selection, helping to select more robust phenotype, better adapted for perfusion operation. In the ATF and TFF systems, the cells passing in the lumen of the hollow fibers are exposed to the shear stress. A study of the effect of increasing the recirculation flow rate in these fibers was studied for HEK293 cells, more sensitive than CHO cells [53], which demonstrated experimentally a higher damage from the shear stress in the TFF than the ATF, as presented in section "Filtration-based Cell Separation Devices". The study highlighted as well that a high shear stress caused apoptosis, identified to occur through endoplasmic reticulum stress, cytoskeleton reorganization and extrinsic signaling pathways. High shear stress was also associated to a higher expression of genes associated to the inflammation and genes associated to the unfolded protein response, which can halt the protein translation, increase chaperones involved in protein folding, degrade misfolded proteins, and eventually can lead to apoptosis. Interestingly, the application of a low shear stress had a tendency to be beneficial for the glucose/lactate metabolism and the PoI mRNA expression. In another study, the same group observed an increase in α-actinin, which is a sign of stress fiber assembly, at density higher than 165×10^6 cells/mL [79]. These fibers are typically occurring in case of exposure to mechanical forces, and are important for the cell morphogenesis, providing cell contractility. The phenomenon of assembly of stress fibers is probably the mechanism explaining the cell diameter decrease, which can be observed at very high cell densities, which creates an environment very dense in cells implying high frequency of cell collisions and contacts with the equipment walls.

Bioreactor Aeration and Very High Cell Density

The oxygen demand increases with the cell density so that the oxygenation can become a limiting factor in intensified processes. The oxygen gas transfer, characterized by the volumetric mass transfer coefficient of oxygen, k_La, required for different cell densities can be estimated based on the cell specific oxygen uptake rate (q_{O2}) as $k_La = X \frac{q_{O2}}{C^*_{O2} - C_{O2}}$, where C_{O2} and C^*_{O2} are the dissolved oxygen concentration and the dissolved oxygen concentration at air saturation in the liquid phase. For CHO cells, values between 0.12 and 0.33 pmol $cell^{-1}$ $hour^{-1}$ have been reported [80, 81]. Xu et al. reported a strategy for scaling-up fed-batch cultures up to a cell density of 30×10^6 cells/mL [80]. This approach and the reported data can be applied as well for higher densities. Using the q_{O2} reported in [80], it can be observed in Fig. 4.5c that k_La values of 13 or 34 $hour^{-1}$ are needed for CHO cell lines with low (0.125 pmol/cell/day) or high oxygen consumption (0.333 pmol/cell/day) at 100×10^6 cells/mL, 26 or 69 $hour^{-1}$ for 200×10^6 cells/mL, and 39 or 103 for $hour^{-1}$ if the density would reach 300×10^6 cells/mL. Although unpublished, this density is theoretically possible for cells of small diameter, such as 15 μm, as these would represent a solid fraction of 53% of the total volume, see Fig. 4.5.

Recall that the cells diameter is potentially smaller at very high cell density due to enhanced mechanical stress environment (see section "Shear Stress"). Notice as well that red blood cells occupy 50% of the total volume in the blood, to compare this system to another known context. Figure 4.6a provides the k_La measured in different bioreactors, it can be observed that using a frit sparger, k_La values up to 65 $hour^{-1}$ in 500 L bioreactor and 25 $hour^{-1}$ in 2000 L bioreactor have been measured. Frit sparger are necessary to deliver the oxygenation required at high cell density in large-scale, while drilled hole sparger are most often not sufficient. Furthermore, beyond using oxygen enriched air, pure oxygen for the aeration is also often used to ensure a sufficient aeration. In conclusion, it seems that nowadays bioreactors envisioned for commercial scale SSPP can sustain a sufficient oxygenation even at very high cell densities.

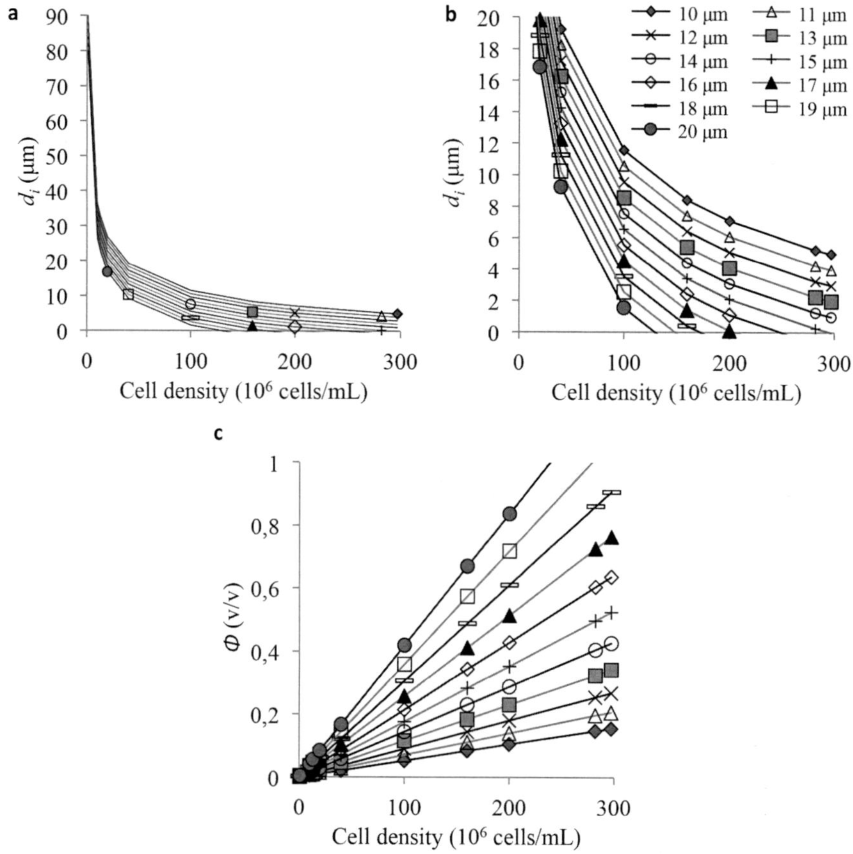

Fig. 4.5 (**a**) Theoretical distance between the cells or intercellular distances, for uncompressible and spherical cells as a function of the cell density for different cell diameters; (**b**) zoom of (**a**) including legend for (**a**–**c**); (**c**) fraction of solid (or cells) in the cell suspension. (Reproduced from [8] with permission from John Wiley and Sons)

On the other hand, the CO_2 accumulation can also represent an issue at high density, already at partial pressures >14 kPa or 98 mmHg [82], necessitating a strategy for the CO_2 stripping in large-scale. It can be seen in Fig. 4.6b that the mass gas transfers of carbon dioxide by drilled hole sparger or by frit sparger are quite close to each other's for a given gas flow rate. Another factor, not deductible from the mass gas transfer measurement, is that, in comparison to large bubbles, pure oxygen bubbles of micro size have a lower probability to reach the culture surface since their oxygen diffuses out of the bubbles during a longer residence time potentially leading to their vanishment, and thus worsening the effect of CO_2 removal by the micro bubbles. Finally, large bubbles are much less harmful that micro bubbles when reaching the surface and bursting and their foam is less stable. For these reasons, at large-scale a dual sparging system is often adopted, is recommended for high cell density and is available for single-use equipment

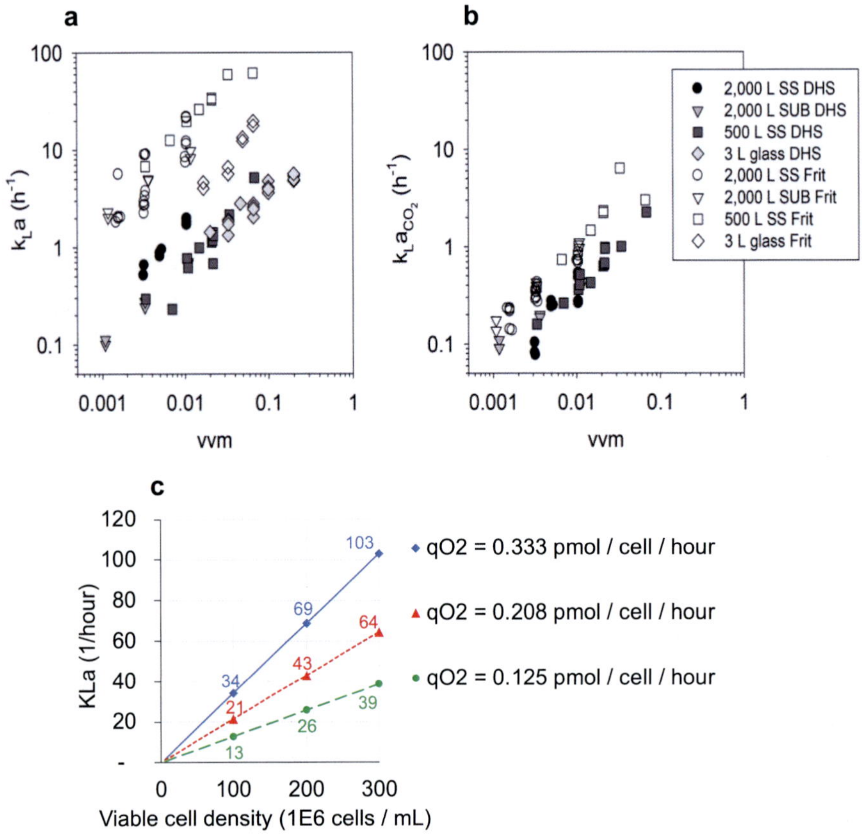

Fig. 4.6 Mass gas transfer of oxygen and carbon dioxide in different bioreactor scales and for different cell densities: (**a**) oxygen (k_La) and (**b**) carbon dioxide volumetric mass transfer coefficients as function of the gas flow in frit sparger (Frit) or drilled hole sparger (DHS) in bioreactors (stainless steel SS, single-use SUB or glass) at different scales with power input in the range of 20–40 W/m^3 for all the scales; (**c**) k_La requirement as a function of the cell density for different CHO cell specific oxygen consumption rates. (Panels a and b reproduced from Fig. 4.1 in [80] with permission from John Wiley and Sons)

nowadays. The dual sparging system includes one sparger generating micro bubbles (frit), aimed for the oxygenation and potentially using pure oxygen, and one sparger generating larger size bubbles, dedicated to the CO_2 stripping using nitrogen or air. Furthermore, this strategy enables the separation of the oxygenation and the CO_2 stripping, which alleviates both controls. Different approaches to perform the CO_2 stripping, reported for fed-batch processes, can be applied [80, 83–86], and potentially models can give better predictive tools [87]. A lower CO_2 concentration can also be partially supported by replacing the medium bicarbonate buffer by other chemicals [88]. Due to the higher requirements for the oxygenation and the CO_2 stripping, the bioreactor homogenization is very important. Notice that for small-scale bioreactors, the surface aeration is sufficient to remove the CO_2. It is however critical to optimize the bioreactor homogenization and the oxygenation even at small-scale to achieve very high cell densities. As a matter of fact, an oxygenation strategy satisfying for a cell density of 20 × 10^6 cells/mL might appear to be sub-optimal when the oxygen demand increases with the cell density and lead to foam excess, jeopardizing the culture. An example of such optimization is presented in [89] where the impeller system of a bioreactor of 200 mL working volume is optimized by including a Rushton impeller and a marine impeller, so that it can sustain culture of ≥100 × 10^6 HEK293 cells /mL, despite a higher sensitivity of these cells in comparison with CHO cells.

Viscosity

Similarly to the fact that the viscosity increases quadratically when the number of particles increases in a liquid, the cell suspension viscosity increases quadratically with the cell density [8]. One needs to be aware of this increased viscosity in comparison with medium for cultures of ≥50 × 10^6 cells/mL, and ensures an efficient homogenization in the bioreactor. Furthermore, the viscosity can be further increased by the occurrence of dead cells in the culture generating cell debris and DNA. In cultures with ATF and TFF cell separations based on filter with pore size ≤0.65 um, these debris mostly accumulate in the bioreactor as there are only partially eliminated through the cell bleeds. It might be necessary to adjust the agitation rate at high cell density to maintain the desired turbulent flow across all regions in the bioreactor, to avoid gradient formation and sub-optimal mixing time. This latter is quite critical at high cell density since the oxygen and nutrients are very rapidly consumed. This implies as well that the DO and pH sensors and controllers need to be well-calibrated/tuned, robust, and provide a fast response.

Medium and Harvest

As previously mentioned, the large volumes of medium and of harvest are a main drawbacks of perfusion processes. The reduction of the number of harvests to only one is done in case concentrated fed-batch is adopted, however with other drawbacks such as clarification of the cells at very high concentration (see section "Introduction"). Steady-state perfusion process brings a higher production but requires harvests once (or more) per week, or an integrated capture step. Obviously, efforts to reduce the cell specific perfusion rate are a main focus for the field and this will be presented in section "Process Optimization". In particular for SSPP, it is important to reduce the CSPR during the production phase at steady-state while such a reduction during the growth phase is less critical since it is a shorter time and potentially the material is not harvested to ensure a PoI uniformity [90]. In a report, joining several biopharma's, a CSPR ≤ 17 pL/cell/day has been identified as a target adequate for commercial production [91]; this represents for instance a perfusion rate of ≤1.7 RV/day for a density of 100 × 10^6 cells/mL. Such a low CSPR has been reported by several groups for CHO cells [21, 92, 93] and even for HEK293 cells [89]. The culture media and feed concentrates for CHO cells are nowadays very performant and, although developed for fed-batch, they can provide very good results for perfusion process as well, given that the glucose concentration is adapted for the perfusion mode otherwise the excessive lactate production can jeopardize the culture, see section "Process Optimization". Furthermore, nowadays media are also developed for perfusion operation.

Cell Line Stability

A cell line needs to be stable during a period starting from the cell expansion of the working cell bank to the production bioreactor run end, supplemented by the end-of-production cells culture. In SSPP, the long culture time needs to be taken into account for the cell selection. Genetic rearrangements can spontaneously occur due to the genome plasticity and loss of productivity can occur with time. The chemicals ensuring the selection pressure such as methotrexate for the dihydrofolate reductase$^-$ system or methionine sulfoximine (MSX) for the glutamine synthetase system are usually not kept in the production bioreactor due to their toxicity. Strategies to tackle the cell line stability have been recently reviewed by Chen et al. [94], concluding that omics has helped the field to identify metabolism [95] and phenotype [96], characteristics of stable and productive cell lines, which can help cell engineering to identify relevant gene integration sites.

Overview of Factors Related to High Cell Density Perfusion Processes

In summary for section "Characteristics, Challenges and Scale-up Considerations for Perfusion Processes", Fig. 4.7 summaries the different factors requiring attention in high cell density perfusion processes.

Process Development

A perfusion process needs to be optimized to obtain the desired PoI production and quality profile, and has to ensure satisfying performances in terms of reproducibility, robustness and cost of goods. This includes two main parts, the culture process in it-self and the cell separation operation. While it is necessary to optimize the process

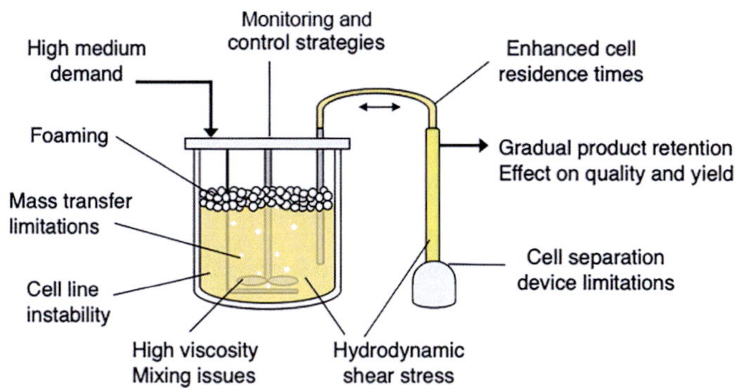

Fig. 4.7 Characteristics, challenges and scale-up considerations for perfusion processes intensified by increasing the cell density. (Reproduced from [33])

for some process parameters, other parameters are not optimized as such but the process development should bring a good understanding of their effect on the cells and the process need, in particular for large scales. Therefore the development of perfusion process can be schematically described as follows;

(a) optimization of the culture process for

(a.1)
media, feed concentrates and additives, in terms of composition and delivery
(a.2)
strategy including the medium renewal rate, i.e. CSPR
(a.3)
cell density
(a.4)
physical parameters, i.e. temperature, pH, DO, pCO_2
(a.5)
osmolality
(a.6)
cell arrest if applied
(a.7)
anti-foam addition

(b) understanding of the culture process for

(b.1) cell shear sensitivity
(b.2) impact of foam formation

(c) selection or optimization of the cell separation operation with parameters specific to the used technology, such as

(c.1) re-circulation rate in the loop passing the cell separation device
(c.2) settings of the pumps and their speed

(c.3) dimensions of the system, i.e. selection of the model or specific dimensions
(c.4) filter surface area, pore size, number of fibers, membrane type and material for the ATF and TFF
(c.5) rotor speed for the centrifuge Centritech

(d) knowledge or understanding of the impact of the cell separation device for

(d.1) mechanical damage to the cells
(d.2) retention of the PoI (i.e. PoI sieving decay)

The parameters, a.3-a.6, b.1, b.2, are optimized for any operation mode, perfusion or fed-batch, although their implications might be somehow different in perfusion in comparison with fed-batch process while a.1, a.2, c1-c.4 and d1, d.2 are specific for perfusion processes.

Scale-down Systems

Scale-down systems are used to support the process and cell line development to reduce the work load and increase the efficiency. For process development, ideally, they should behave exactly as pilot- and large-scale systems. However this is quite challenging for small and very small culture volumes, so scale-down systems can give tools to understand/predict a process and to optimize its parameters, but it is very important to be aware of their limitations. The development of the culture process, items a and b, and of the cell separation operation, items c and d, can be addressed separately if the constraints and limitations of each other's are taken into account. There are no systems of both the bioreactor and cell separation system of very small size mimicking exactly the behavior at pilot or large scale, and it is, therefore, not uncommon that the process development in bioreactor scale is performed in 5 L or 10 L bioreactor size, however this requires quite some resources in terms of labor, medium and the associated logistics. The field has therefore developed both small-size controlled bioreactor systems with perfusion ensured by a cell separation – although potentially not a real scale-down model of the large-scale device-, and high-throughput systems giving information helpful for the culture process tuning without being a real perfusion bioreactor system. A combination of both types of systems is often used to improve the efficiency by getting information obtained using high-through-put systems, followed by further tuning/development in small-scale perfusion bioreactor, and then confirmation at pilot-scale. Of course iterations of this latter description are also used.

High-throughput in Semi-perfusion Systems

Semi-perfusion cultures, also called *pseudo-perfusion*, are carried out with medium exchange performed at a given frequency, such as once per day, several times a day or every second day, while the cell density is kept at a selected level (or selected levels) at each medium exchange. It generates a culture mimicking steady-state perfusion operation and can be used for medium development or better understanding of the effect of various stimuli on the perfusion process. A popular system is the tubespin, a centrifuge tube of 50 mL with vented cap enabling both the culture and the centrifugation in the same vessel, reducing the manual operations. It is often used for high-throughput as steady-state perfusion model [97–104], and similarly deep-well plates can be used or the ambr15 system. The operations are however manual and labor intensive, in particular when one or several medium replacements per day are applied. Alternatively, automation has been introduced in the ambr15 system, where the cell separation for the medium exchange is obtained by sedimentation [64, 105]. In this latter, feedback controls of the environmental parameters, pH, temperature and DO, are carried out providing higher robustness and individualization in comparison with the tubespin. Notice however that these systems do not really achieve the same steady-state high cell densities as bioreactors operated in perfusion mode. A drawback of semi-perfusion is that the culture components make an excursion between their level after the medium exchange, e.g. fresh medium, and the next medium exchange. In the tubespin system, this represents a rather large variation [46], while the automation in the ambr15 enables more frequent medium exchange, e.g. partial medium exchange every 4 hours, which mimics quite well a perfusion process. A drawback of this latter system is that the culture mixing and the pH and DO controls are interrupted during 30–45 minutes for the cell sedimentation, which creates sub-optimal DO and pH variations, limiting these cultures to 25–30 $\times$ 10^6 cells/mL [64, 105].

Small-scale Bioreactor Systems for Perfusion

Using bioreactor sizes of 5–10 L can provide a satisfying scale-down model, in particular for scaling-down not only the culture process but the cell separation operation as well, however, as previously mentioned, these systems require quite some resources in terms of labor and medium. Bench top bioreactors are often used for small-scale process development, typically with working volumes of 1–3 L. These can offer a good compromise of mimicking the bioreactor culture process and giving good information about the cell separation, while being of rather small size. Small sizes of the ATF (ATF1 and ATF2), TFF, inclined settler, acoustic settler and Centritech centrifuge, are compatible with this bioreactor types, potentially using an intermittent mode. In intermittent mode, a cell separation device, dimensioned for pilot-scale, is used during a short time at a given frequency, e.g. the Centritech centrifuge can be used every second hour during 5 minutes. Advantageously, these systems have similar monitoring and process control as pilot- or large-scale with good scalability of the parameters a. and b. above-listed in section "Process Development" introduction. To reduce the labor and medium demand, bioreactor operated in perfusion mode with a smaller working volume of ~200 mL have been developed, such as based on the Dasbox (Eppendorf) connected to ATF or TFF with cell densities stably maintained at 100×10^6 cells/mL and also higher peaks of maximal densities [61, 89, 106–108], or based on the ambr250 (Sartorius), ambr250 perfusion operated either in ATF or TFF modes, with peaks at 100×10^6 cells/mL [109]. As previously mentioned, an issue for these small bioreactor sizes operated at very high cell densities is the risk of foam formation and the shear stress introduced by the cell separation device. Schwarz et al. optimized the agitation of the Dasbox bioreactor, i.e. impeller configuration, to obtain a high homogenization, generating a system suitable not only for CHO cells but also for the more sensitive HEK293 cells [89]. Alternative perfusion systems have also been explored for bioreactor in working volumes of 70–350 mL by spiral microfluidic cell retention chips, reaching 20×10^6 cells/mL density [110, 111]. Finally, a bioreactor of 2 mL working volume operated in perfusion thanks to a membrane-based system has been recently introduced, the ERBI (Merck KGaA, Darmstadt, Germany) able to maintain the cell density at 100×10^6 cells/mL [46]. The system does not mimic large-scale cell separation device but can advantageously be used for development of the process parameters a. listed above. It includes a fully automated perfusion, medium addition and harvest, as well as cell bleed based on online monitoring of the cell density, in addition to feedback controls of the pH, DO and temperature.

Key Performance Indicators in Perfusion Process

Cell Density, Growth Rate, Cell Bleed and Viability

The potential maximal cell density and viability in a process depend on several factors, such as the culture medium, the perfusion rate, the cell retention device, the bioreactor k_La at commercial scale together with the q_{O2}, the volumetric mass transfer coefficient for CO_2 and the bioreactor. Taking these factors into account, the cell density is selected during the process development, which should ensure robustness of the process, e.g., a steady-state cell density setpoint is defined for SSPP. In this latter, a control of the cell density at a constant level is put place, in which a cell bleed rate is applied to balance the growth rate. This is preferably carried out by a feedback control based on online monitoring of the cell density using e.g. a bioimpedance probe. Notice that the cell density in the ERBI bioreactor is monitored and control thanks to an optical density probe, i.e. giving estimation of the total cell density, and that this generates a reliable control of the SSPP. Alternatively, during the process development, the cell bleed can be performed by a continuous pump manually adjusted. It is of course important to ensure that the cell separation device has a reliable and robust function during the culture run.

Since the PoI is discarded in the cell bleed, it is important to ensure that the growth rate is reduced at steady-state while a high viability is maintained. The strategies to cause a partial cell arrest are similar to fed-batch processes, such as hypothermia or addition of a chemical such as butyrate, valproic acid or valeric acid, and have been often implemented for the production of glycoproteins in perfusion processes [5, 45, 89, 112–120]. Cell arrest can be associated to an increase of the cell specific productivity of the PoI, q_{PoI}. When hypothermia is applied, a viability higher than at 37 °C temperature is very often observed, so that this can also potentially prolong the length of a run. Butyrate, valproic acid or valeric acid supplementations are however causing cell viability decrease (in a dose dependent way), so these chemicals should be evaluated, and adopted only in case they increase the q_{PoI} [112, 120, 121]. Another approach is to exploit the fact that hyperosmolarity can cause a reduced cell growth. Hyperosmolality can be achieved by tuning the sodium and potassium levels in the culture medium, a strategy, which can also potentially increase the q_{PoI} [122, 123].

It can be attractive as well to reduce the growth rate by a reduction of the nutrients to slowdown the main metabolism of the cell, for instance by reducing the CSPR. One should however be aware that essential substrates, such as essential amino acids or vitamins, should not be depleted since this can cause apoptosis. Another risk is that a reduced CSPR causing a reduced growth rate can as well provoke a deterioration of the product productivity or quality.

The impact of the cell bleed rate increases when the perfusion rate decreases. This is easily illustrated in the following example. In a SSPP performed at perfusion rate 3 RV/day, a cell bleed of 0.3 RV/day represents a loss of 10% of PoI, while the same cell bleed of 0.3 RV/day represents a loss of 30% of PoI if the perfusion rate is 1 RV/day. A simulation of this effect, including the sieving decay, is given in Fig. 4.8, representing the product recovery in the harvest, indicative of the harvest volumetric productivity, as function of the perfusion rate and the bleed rate in SSPP. The PoI loss in the bleeding is higher for low perfusion rates (<1.5 day^{-1}) in comparison with higher perfusion rates. It can be seen as well that very high bleed rates cause severe reduction of the volumetric productivity eventually leading to annihilation of the harvested PoI in an extreme case of both the bleed rate and the perfusion rate being 0.5 RV/day. The product retention by the cell separation device accentuates this effect as shown in Fig. 4.8b, 50% sieving, in comparison with Fig. 4.8a, 100% sieving.

Production of the Product of Interest

The volumetric productivity $Q_{PoI,\ Harvest}$ of a process is determined by the cell density, the cell specific productivity q_{PoI} at that cell density and the product sieving. The total productivity of a run, H_{POI}^{tot}, is the accumulated harvested PoI production per bioreactor volume, $Q_{PoI,\ Harvest}$, over time, see section "Product of Interest".

Rearranging Eqs. 4.12 and 4.14, it follows at, steady-state, that

$$Q_{PoI,Harvest} = H\ C_{PoI,Harvest} = q_{PoI}\ X - B\ C_{PoI,Bioreactor} \tag{4.18}$$

and recall Eq. 4.15

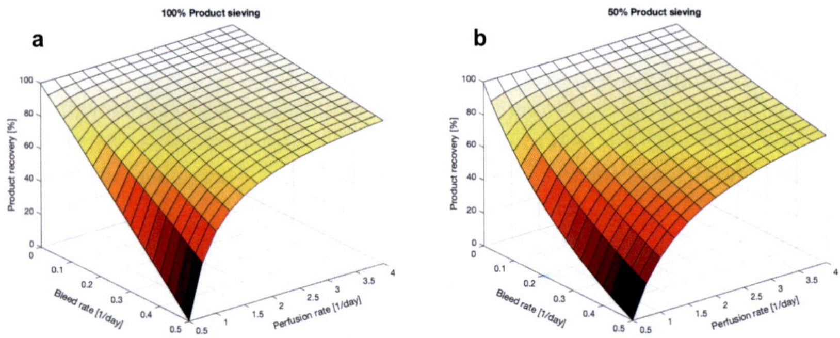

Fig. 4.8 Simulation of product recovery in the harvest as a function of the perfusion rate and the bleed rate for a sieving coefficient of (**a**) 100% or (**b**) 50%, in SSPP under the assumptions that the cell density and q_{PoI} are constant. (Reproduced from [33])

$$H_{POI}^{tot} = \int_0^t H\, C_{PoI,Harvest} dt = \int_0^t Q_{PoI,Harvest} dt$$

$Q_{PoI,\ Harvest}$ and H_{POI}^{tot} increase with the cell density, given that q_{PoI} is constant. The PoI cell specific productivity q_{PoI} can be influenced by the process conditions, i.e. medium, feed, additives, CSPR, by-products and physical environment. We have observed that q_{PoI} was maintained up to cell density of 160×10^6 CHO cells/mL, and that this was also true for the product quality [106]. It is important to recall that the cell bleed, which is equal to the growth rate, represents product loss. Fig. 4.9 illustrates by a simulation, the influence of the cell density, the sieving and the bleed rate on H_{POI}^{tot}. The harvested PoI amount is higher when the cell density is increased, or the sieving is increased, or the bleed rate is lowered, and this is of course additional. As already illustrated in Fig. 4.8, very high bleed rate can lead to a proportion of discarded product in the bleed, higher than collected in the harvest tank.

Process Optimization

CSPR Optimization

As previously mentioned, using the CSPR is a favorite approach based on the principle that the renewal of the medium, i.e. including additive(s) and feed concentrate(s), selected at a given cell density is applicable to another cell density. It was reported that the metabolomic and proteomic profiles were maintained from 5 to 120×10^6 cells/mL in a CHO cell culture at constant CSPR producing mAb [79]. The CSPR minimization is an objective in order to reduce the volumes of

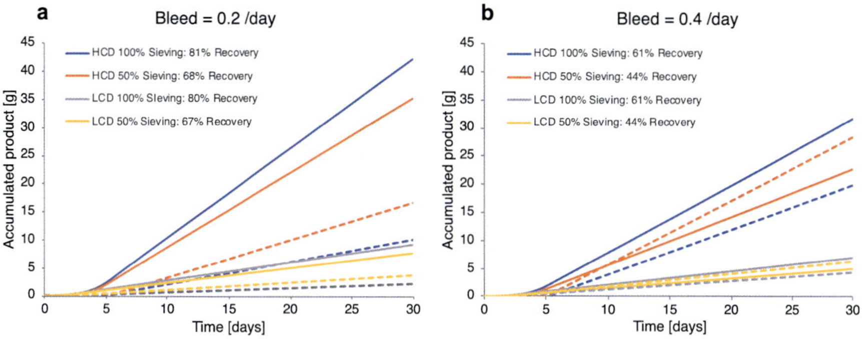

Fig. 4.9 Simulation of the PoI accumulated with time in the harvest tank (straight lines) and loss of PoI material in the cell bleed (dashed lines) in 1 L cultures at perfusion rate 1 RV/day with q_{PoI} of 20 pg/cell/day, inoculated at 5 × 10^6 cells/mL, represented for steady-state cell densities of 100 × 10^6 cells/mL (HCD) or 20 × 10^6 cells/mL (LCD), with sieving coefficients of 50% or 100% and bleed rates of 0.2 /day in (**a**) and 0.4 /day in (**b**). Reproduced from [33]

fresh medium and harvest. A constant CSPR can be applied during the cell growth, at steady-state or for a concentrated fed-batch. For steady-state perfusion processes, potentially different CSPR values can be adopted for the cell growth phase and for the production phase at steady-state, with (or without) medium switch between these phases. During the growth phase, the purpose is to increase the cell density as fast as possible and it can be decided to discard the product since the quality can be different. During the production phase, the cell growth is reduced to minimize the product loss in the cell bleeds, as mentioned in section "Production of the Product of Interest". The growth and production phases can thus benefit of different media and different medium renewals. During the production phase, the identified production medium renewed at the minimum CSPR ($CSPR_{min}$) has to robustly support high PoI productivity, high cell health and satisfying metabolism profile to provide the desired product quality attributes, e.g. the production of the by-products should not exceed a limit above which the PoI quality is deteriorated. Of course, the medium can be modified/enriched together with the efforts of decreasing the CSPR. As mentioned in section "Medium and Harvest", a $CSPR_{min}$ of 17 pL/cell/day has been named as target for an industrial process [91]. Furthermore, the perfusion rate can also be dictated by the residence time of the PoI in the bioreactor, a factor very important for labile glycoproteins. A strategy often adopted to identify the $CSPR_{min}$ is the "*push-to-low*" optimization [43]. In this approach, the perfusion rate is stepwise decreased while, for each step, the cell density is maintained at a stable steady-state level during several days, while the culture performances are monitored. A time of 7 days for each step allows to gather data of 3–4 days of steady-state while the transient period is about 3–4 days [107]. A gene expression analysis showed that most transcripts reached a steady-state within ~4 days [124]. In case, the perfusion rate has been decided due to operational constraints, e.g. cell separation device, harvesting capacity, or for labile glycoproteins for which the limit of the PoI residence time in the bioreactor is known, the perfusion rate is maintained at the

Table 4.1 Examples of perfusion processes from literature published in 2017–2022

Product / cells	Cell separation device	Cell density [10^6 cells/mL]	CSPR [pL/cell/ day]	Length at steady-state [days]	References
mAb / CHO	ATF	20	50	14	[128]
mAb / CHO	ATF	20–30	~50	21	[129]
mAb / CHO	ATF/TFF	40	50–80	14	[122]
mAb / CHO	ATF	40–60	~20	15	[130]
mAb / CHO	ATF	40–60	20–25	8	[104]
mAb / CHO	ATF	60	17	26	[131]
mAb / CHO	ATF	100	15	20	[108]
mAb / CHO	ATF	100	17	21	[21]
mAb / CHO	ATF	74	14	20	[10]
Enzyme / CHO	ATF	100	32	14	[12]
Fusion protein / CHO	ATF	80–120	~ 15	30	[132]
EPO / HEK293	ATF	80–100	10–15	37	[89]

desired level and the cell density is stepwise increased, exploring the effect of reduced CSPR values. Table 4.1 lists reported examples with different CSPRs and target cell densities, where it can be seen that processes with very low CSPR, smaller than 17 pL/cell/day, have been reported, even for cell densities larger than 100×10^6 cells/mL.

Furthermore, to save medium consumption, recycling of the spent medium, explored already in the 90's, has been studied recently in high cell density perfusion [109].

Medium Optimization

As mentioned in the previous section, the range of CSPR and its minimal value, $CSPR_{min}$, depend on the medium, which has been described as a dependency on the *medium depth.* The medium used for the perfusion operation can be derived from the knowledge existing for batch/fed-batch processes. Fed-batch media and feed concentrates can show very good performances when applied to perfusion mode if a suitable strategy is applied, including awareness of the different dynamics of nutrient consumption and by-product accumulation in perfusion and fed-batch modes. The reduction of the CSPR can advantageously benefit of an enrichment of the medium in key nutrients such as amino acids, vitamins or lipids. Ideally these should be measured when the CSPR is decreased during a push-to-low optimization experiment and be supplemented in the medium to compensate their depletion. A problem is actually the risk of medium precipitation when the medium is enriched. This can be solved by reducing the salt and buffer concentrations in the medium as reported by [125] who repurposed fed-batch medium for a perfusion process. Another approach, potentially complementary, is to use several feeds [90].

In perfusion mode, the glucose concentration in the medium or feed concentrate can lead to too high level of lactate, which is toxic for the cell. As theoretically analyzed in section "By-products and Substrates", a high concentration of lactate is more likely to be a threat for the cell health than high ammonia concentration. An illustration is given in Fig. 4.10, where it can be observed that the lactate concentration reached more than 50 mM at day 10 while the residual glucose concentration in the bioreactor was almost 0. One has to remember that the medium renewal in perfusion process physically eliminates a fraction of the lactate from the culture, however the fresh medium reintroduce glucose. This implies that the glucose availability can be very high and generate a high production of lactate. It is important to also underline that the residual glucose concentration is very close to 0 after day 7. Hence a strategy based on adjusting the perfusion rate based on the glucose concentration in the bioreactor, as an inspiration from fed-batch process would suggest (i.e. more medium is added when the residual glucose concentration is too low) is not adequate. Instead, the strategy should be to consider the delivery of the glucose (or sugars) separately from the rest of the medium during the process optimization. The results shown in Fig. 4.10 are a previous work that we have published before we developed a systematic approach to deliver glucose and to avoid too high formation of lactate. This approach, the *targeted feeding strategy* denoted as *TAFE*, is generally applicable to the substrates in a perfusion process and is described below for glucose, followed by a generalization to other sugars. It is also applicable for the delivery of the amino acids.

Another strategy to limit the lactate production, the high-end pH-control perfusion, HIPCOP, has been developed by Hiller and collaborators at Pfizer, derived from their strategy developed for fed-batch process, HIPDOG [126]. In this method, the high-end pH is controlled by automatic addition of medium containing glucose, where this glucose-based addition causes production of lactate and thus pH decrease [90]. The perfusion medium contains as well lactate as a carbon source. This strategy results in an overall perfusion rate driven by the cell density and reduced delivery of glucose.

Targeted Feeding TAFE for Sugars

The principle of the targeted feeding TAFE is that the cell specific uptake rate of the substrate, here glucose, is selected by the operator and enforced on the cells by the applied delivery of this substrate, calculated based on the perfusion rate and the cell density in the perfusion culture. The selected cell specific uptake rate is referred to as *targeted cell specific uptake rate*, $q_{glucose}^{target}$, and chosen to obtain an adequate availability of this substrate to the cells according to their needs, leading to reduced production of the by-product, lactate. It allows as well to tune the glycosylation of the product, as presented below. Thanks to the TAFE design, the concentration of glucose in the perfusion medium or in a separate feed, $C_{glucose,\ in}$, is determined based on the targeted cell specific uptake rate, as follows. Let us rearrange Eq. 4.10 as follows;

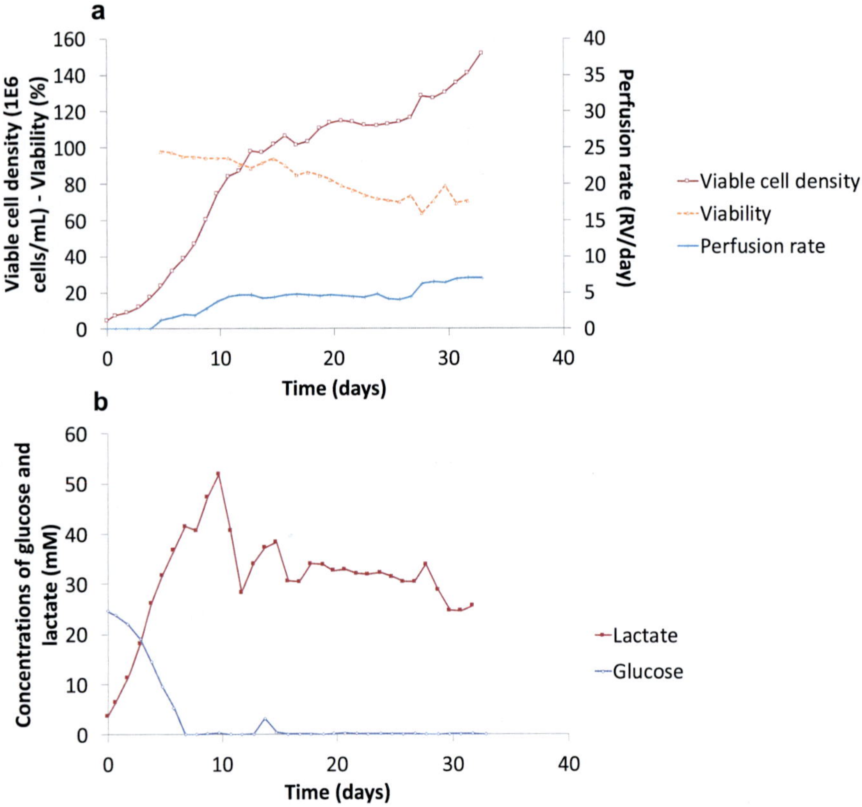

Fig. 4.10 Profiles of (**a**) the viable cell density, viability, perfusion rate, and (**b**) the concentrations of lactate and glucose in a perfusion culture performed in a system where the CHO cells producing IgG were entrapped in a non-woven polyester matrix. The lactate concentration was high despite a residual glucose concentration almost 0 from day 7 and a high perfusion rate. (Extracted from [45] under terms of Creative Commons CC-BY license)

$$C_{\text{glucose, in}} = \frac{q_{\text{glucose}}}{\text{CSPR}} + C_{\text{glucose}} \tag{4.19}$$

In case glucose is delivered to exactly meet the cell need according to the targeted cell specific uptake rate, $q_{glucose}^{target}$, but without excess, C_{glucose} can be assumed to be 0, and $C_{\text{glucose, in}}$ can be expressed from eq. 4.19 as follows;

$$C_{\text{glucose, in}} = \frac{q_{glucose}^{target}}{\text{CSPR}} \tag{4.20}$$

The choice of $q_{glucose}^{target}$ can be determined from prior knowledge of the cells and can also be experimentally tuned by studying the impact of different $q_{glucose}^{target}$ values on the PoI productivity, the byproduct formation, the cell growth and the product quality.

For this, different $C_{glucose,\ in}$ corresponding to different $q_{glucose}^{target}$ are applied at steady-state during a week and the culture performances are studied. This concept was developed with HEK293 cells perfusion culture up to 80 × 10^6 cells/mL in 2016 in Chotteau's lab, where it has been successfully applied with four HEK293 cell lines up to 120 × 10^6 cells/mL and four CHO cell lines up to 180 × 10^6 cells/mL for many perfusion cultures, and up to pilot scale at 100 × 10^6 cells/mL. For both CHO cells and HEK293 cells, a $q_{glucose}^{target}$ of 1.0 pmol/cell/day is fully satisfying in terms of cell growth and PoI productivity and results in $q_{lactate}$ well below 0.5 pmol/cell/day, between 0.2 and 0.5 pmol/cell/day. $q_{glucose}^{target}$ of 1.2 pmol/cell/day is associated with $q_{lactate}$ ~ 0.6 pmol/cell/day. An illustration is given in Fig. 4.11 for a HEK293 cell culture, where it can be seen that applying $q_{glucose}^{target}$ of 1.2 pmol/cell/day followed by 1 pmol/cell/day generated concentrations of lactate under 10 mM and 8 mM, respectively, while the residual concentration of glucose was close to 0, for a perfusion culture maintained at different steady-state levels up to ~72 × 10^6 cells/mL at very high viability well above 90%.

The TAFE concept can be generalized to other sugars and also to other substrates such as the amino acids. Zhang et al. applied this approach to combined feeding of glucose, mannose and galactose [107]. Although these sugars are not really interchangeable, CHO cells can use them as energy sources and as substrates for the glycosylation, however their respective uptake rates are different. The target consumption rates of these sugar, $q_{s_i}^{target}$ for s_i = glucose, mannose or galactose, can be a priori decided taking into account the known maximal uptake rates, e.g. maximal uptake rate of galactose of 0.4 pmol/cell/day [102]. When different sugars are provided simultaneously, the sum $q_{sugar,T}^{target}$ of the $q_{s_i}^{target}$ is selected by the operator. Considering a generalization of Eq. 4.20 for these sugars, the concentration for each sugar in the feed medium can be determined as follows;

$$c_{\mathrm{glucose,in}} = \frac{q_{glucose}^{target}}{\mathrm{CSPR}} \tag{4.21a}$$

$$c_{\mathrm{mannose,in}} = \frac{q_{mannose}^{target}}{\mathrm{CSPR}} \tag{4.21b}$$

$$c_{\mathrm{galactose,in}} = \frac{q_{galactose}^{target}}{\mathrm{CSPR}} \tag{4.21c}$$

$$\sum_{i=1}^{3} q_{s_i}^{target} = q_{glucose}^{target} + q_{mannose}^{target} + q_{galactose}^{target} = q_{sugar,T}^{target} \tag{4.21d}$$

This was applied in [107], for values of $q_{sugar,T}^{target}$ of 1.2 and 0.8 pmol/cell/day, and various feedings of glucose, galactose and mannose in CHO cells producing mAb perfusion cultures. The strategy enabled the simultaneously consumption of these sugars, while usually glucose is taken up in priority due to its faster uptake rate, see

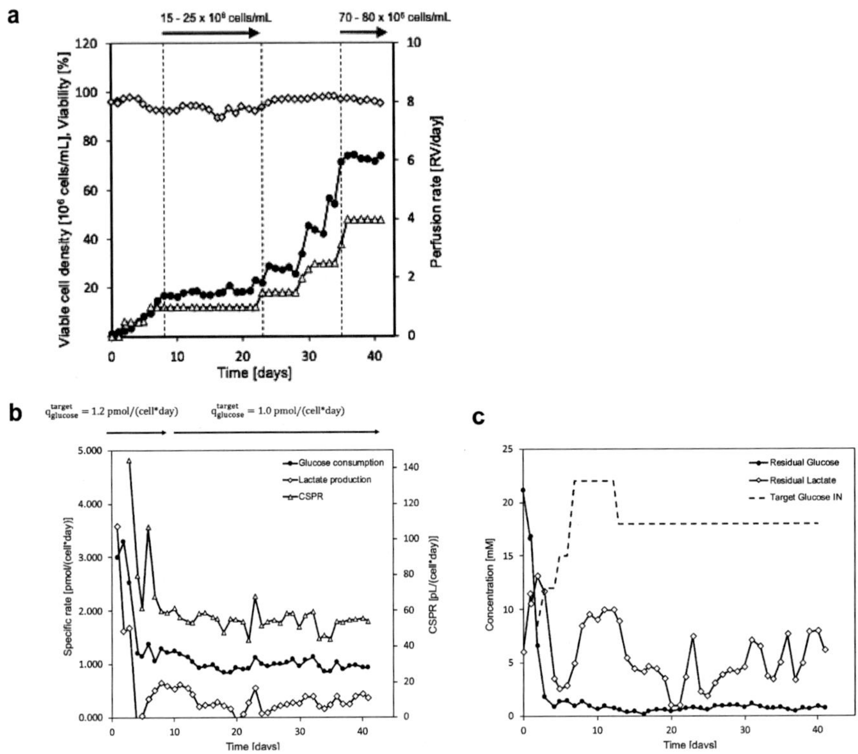

Fig. 4.11 Application of TAFE in steady-state perfusion process of HEK293 cells; (**a**) cell density (circle), perfusion rate (diamond) and viability (triangle); (**b**) cell specific glucose consumption and lactate production rate, and CSPR; **c** residual concentrations of glucose and lactate in the bioreactor, and glucose concentration in the feed medium. The perfusion was started at day 2 at rate 0.5 RV/day, resulting in a transient high CSPR until day 7 due to the low cell density, after which the CSPR stayed around 50–60 pL/cell/day. From day 4, the residual glucose concentration was ≤1.5 mM and mostly between 0 and 1 mM. The TAFE feeding of glucose with $q_{glucose}^{target}$ of 1.2 pmol/cell/day until day 12, followed by 1.0 pmol/cell/day provided a low lactate concentration under 10 mM and 8 mM, respectively. The cell density was progressively increased until a steady-state of 72 × 10^6 cells/mL, while the specific rates of glucose and lactate were stable, and the viability very high. (Reproduced from [89] Schwarz 2020 with permission of Elsevier and from [33])

Fig. 4.12. Interestingly, it resulted as well in different glycosylation profiles, providing a tool to tune and control this quality attribute (Fig. 4.12).

The TAFE strategy is a tool for process development. When a process has been optimized thanks to TAFE, it can be 'translated' into using media or feed concentrates including glucose, where its concentration has been selected based on the TAFE. It is also possible to select a q_{sugar}^{target} range and balance the CSPR and the cell density, while using commercial media and feed concentrates already including glucose at a given concentration. This was successfully realized in [21] at

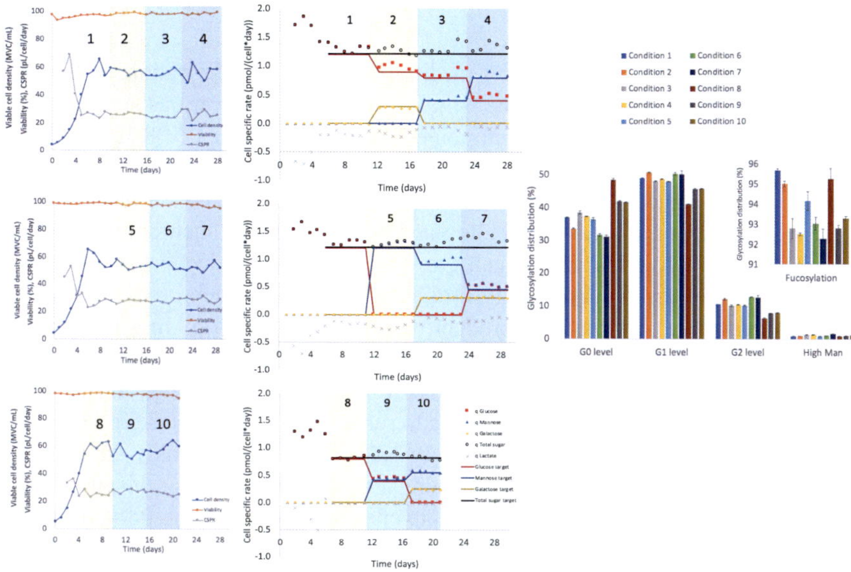

Fig. 4.12 Application of ten feeding regimes (conditions 1–10) in three CHO cell perfusion cultures producing mAb. **Left panels** cell density, viability and CSPR; **Middle panels** measured cell specific consumption rates of glucose (square), mannose (triangle), galactose (diamond), total sugar (circle), lactate (cross), target rates of glucose (red), mannose (blue), galactose (yellow) and total sugar (black); **Right panel** Glycoform distribution of the produced mAb in the ten feeding conditions, G0, G1 and G2 levels include respectively G0 and G0F, G1 and G1F, G2 and G2F, high mannose contains M5 and M6. The cell cultures were successfully maintained close to the cell density setpoint with high viability (>94%), and the cell specific consumption rates of glucose, mannose and galactose were close to the respective targets, resulting in different profiles of glycosylation. (Reproduced from [107] with permission of Elsevier)

pilot-scale in SSPP at 100 × 10^6 cells/mL in a perfusion process integrated with a continuous purification process.

Targeted Feeding TAFE for Amino Acids

The approach developed above for sugars can be used for amino acids. A high cell density perfusion process has been rapidly developed by adequately balancing the base medium and the feed concentrate of a fed-batch process producing an enzyme in CHO cells [12]. For this, the balancing has aimed at obtaining cell specific consumption rates of key amino acids and glucose selected from the fed-batch culture behavior using the TAFE strategy for a SSPP and a concentrated fed-batch, both reaching 100 × 10^6 cells/mL. The TAFE strategy was also used in a study aimed at amino acid optimization using a Design of Experiment approach, see section "Sesign of Experiments".

Design of Experiments

Design of experiment can be used for medium and process optimization. Screening experiments can be set up where the culture performances at various steady-states under the effect of different medium components, mixtures or CSPR are studied. Likewise, the effect of different pH or temperature can be studied, given that the different steady-states are reached during several days for each condition [94]. The objectives of the optimization can be to maximize the PoI volumetric productivity, q_{PoI}, cell viability, to minimize the cell growth rate, the by-product production rate, and/or to achieve the desired quality profile. As previously mentioned, at steady-state, a maximum PoI volumetric productivity is obtained both by maximizing the q_{PoI} but also by ensuring that the cell growth rate is not too high, so potentially the highest q_{PoI} is not the most optimal in case it is associated with high growth rate [46]. A fractional factorial design was used to screen feed supplements for perfusion culture [93]. This was followed by focusing on selected feeds to study the interactions and achieve medium optimization using a quadratic model. For this, the study was carried out in batch experiments in shake flasks, followed by pseudo-perfusion in tubespins and finally bioreactor runs. Kuiper [127] had a similar approach based on fed-batch medium and feeds repurposed for perfusion by a DoE. In [46], the medium optimization focused on a simplex-centroid DoE-based approach for a prototype medium with reduced amino acid composition, allowing screening for optimal amino acid cocktails. The simplex-centroid DoE is based on mixing different candidate mixtures in different proportions, always summing up to 100%, so that approach has the advantage to reduce the issues of too high osmolality or medium precipitation. The study was based on a first screening in tubespin, followed by optimization of selected amino acids by DoE in ERBI microbioreactors and confirmation in stirred tank bioreactors. The amino acid concentrations were decided based on a TAFE strategy.

Perspectives

Perfusion processes respond to the demand for intensification of the biopharma field in front of increasing demand for biopharmaceuticals. The antibodies are largely represented in this category, however multi-specific molecules rise as well new challenges where perfusion can be helpful for the volumetric productivity and the product quality. The intensification is leaning on increased cell densities, which are also supporting legacy fed-batch processes in the N − 1 step. The small footprint of intensified perfusion processes is very attractive for their compatibility with disposable equipment, and the advantages that this represents in terms of flexibility, scaling-out potential, environmental impact, etc. The integration of perfusion process with purification process represents a supplementary possibility that some companies are exploring. Perfusion mode is also interesting for the cell health and process performances in the Advanced Therapy Medicinal Products (ATMPs), an area where the learnings from biomolecule production processes can advantageously be leveraged.

Some 10 years ago, a concern was the lack of competence and knowledge for the perfusion processes and their development. This has changed and it is clear that this type of intensification will stay in the biopharma industry at different levels.

Acknowledgements This work has received the support of the Sweden's Innovation Agency VINNOVA (diarie nr. 2022-03170), Competence Centre AdBIOPRO for Advanced BioProduction by Continuous Processing, Sweden.

References

1. Chotteau V (2015) Perfusion processes. In: Al-Rubeai M (ed) Animal cell culture. Springer International Publishing, Cham, pp 407–443
2. Castilho LR, Medronho RA (2002) Cell retention devices for suspended-cell perfusion cultures. Adv Biochem Eng Biotechnol 74:129–169. https://doi.org/10.1007/3-540-45736-4_7
3. Boedeker BGD (1992) The manufacturing of the recombinant factor VIII, Kogenate. Transfus Med Rev 6:256–260. https://doi.org/10.1016/S0887-7963(92)70177-1
4. Heidemann R, Zhang C, Qi H et al (2000) The use of peptones as medium additives for the production of a recombinant therapeutic protein in high density perfusion cultures of mammalian cells. Cytotechnology 32:157–167. https://doi.org/10.1023/A:1008196521213
5. Chotteau V, Björling T, Boork S et al (2001) Development of a large scale process for the production of recombinant truncated factor VIII in CHO cells under cell growth arrest conditions. Anim Cell Technol From Target to Mark 287–292. https://doi.org/10.1007/978-94-010-0369-8_67
6. Sandberg H, Lütkemeyer D, Kuprin S et al (2006) Mapping and partial characterization of proteases expressed by a CHO production cell line. Biotechnol Bioeng 95:961–971. https://doi.org/10.1002/BIT.21057
7. Clincke MF, Mölleryd C, Samani PK et al (2013) Very high density of Chinese hamster ovary cells in perfusion by alternating tangential flow or tangential flow filtration in WAVE bioreactor™-part II: applications for antibody production and cryopreservation. Biotechnol Prog 29:768–777. https://doi.org/10.1002/btpr.1703
8. Clincke MF, Mölleryd C, Zhang Y et al (2013) Very high density of CHO cells in perfusion by ATF or TFF in WAVE bioreactor™: part I: effect of the cell density on the process. Biotechnol Prog 29:754–767. https://doi.org/10.1002/btpr.1704
9. Yang WC, Minkler DF, Kshirsagar R et al (2016) Concentrated fed-batch cell culture increases manufacturing capacity without additional volumetric capacity. J Biotechnol 217:1–1656
10. Xu S, Gavin J, Jiang R, Chen H (2017) Bioreactor productivity and media cost comparison for different intensified cell culture processes. Biotechnol Prog 33:867–7938
11. Brechmann NA, Schwarz H, Eriksson PO et al (2021) Antibody capture process based on magnetic beads from very high cell density suspension. Biotechnol Bioeng 118:3499–3510. https://doi.org/10.1002/BIT.27776
12. Särnlund S, Jiang Y, Chotteau V (2021) Process intensification to produce a difficult-to-express therapeutic enzyme by high cell density perfusion or enhanced fed-batch. Biotechnol Bioeng 118:3533–3592

13. Hiller GW, Ovalle AM, Gagnon MP et al (2017) Cell-controlled hybrid perfusion fed-batch CHO cell process provides significant productivity improvement over conventional fed-batch cultures. Biotechnol Bioeng 114:1438–1447. https://doi.org/10.1002/bit.26259/abstract
14. Pohlscheidt M, Jacobs M, Wolf S et al (2013) Optimizing capacity utilization by large scale 3000 L perfusion in seed train bioreactors. Biotechnol Prog 29:222–229. https://doi.org/10.1002/BTPR.1672
15. Padawer I, Ling WLW, Bai Y (2013) Case study: an accelerated 8-day monoclonal antibody production process based on high seeding densities. Biotechnol Prog 29:829–832. https://doi.org/10.1002/BTPR.1719
16. Yang WC, Lu J, Kwiatkowski C et al (2014) Perfusion seed cultures improve biopharmaceutical fed-batch production capacity and product quality. Biotechnol Prog 30:616–625. https://doi.org/10.1002/BTPR.1884
17. Jordan M, Kinnon NM, Monchois V, et al (2018) Intensification of large-scale cell culture processes. https://doi.org/10.1016/j.coche.2018.11.008
18. Stepper L, Filser FA, Fischer S et al (2020) Pre-stage perfusion and ultra-high seeding cell density in CHO fed-batch culture: a case study for process intensification guided by systems biotechnology. Bioprocess Biosyst Eng 43:1431–7591
19. Xu J, Xu X, Huang C et al (2020) Biomanufacturing evolution from conventional to intensified processes for productivity improvement: a case study. Taylor & Francis
20. Xu J, Rehmann MS, Xu M et al (2020) Development of an intensified fed-batch production platform with doubled titers using N-1 perfusion seed for cell culture manufacturing. Bioresour Bioprocess 7:1–16
21. Schwarz H, Gomis-Fons J, Isaksson M et al (2022) Integrated continuous biomanufacturing on pilot scale for acid-sensitive monoclonal antibodies. Biotechnol Bioeng 119:2152–2166. https://doi.org/10.1002/BIT.28120
22. Sinharoy P, Aziz AH, Majewska NI et al (2020) Perfusion reduces bispecific antibody aggregation via mitigating mitochondrial dysfunction-induced glutathione oxidation and ER stress in CHO cells. Sci Rep 10:16620. %@ 2045–2322
23. Walther J, Lu J, Hollenbach M et al (2019) Perfusion cell culture decreases process and product heterogeneity in a head-to-head comparison with fed-batch. Biotechnol J 14:1700733. %@ 1860–6768
24. Gomez N, Lull J, Yang X, et al (2020) Improving product quality and productivity of bispecific molecules through the application of continuous perfusion principles. https://doi.org/10.1002/btpr.2973
25. Pollock J, Coffman J, Ho SV, Farid SS (2017) Integrated continuous bioprocessing: economic, operational, and environmental feasibility for clinical and commercial antibody manufacture. Biotechnol Prog 33:854–866. https://doi.org/10.1002/btpr.2492
26. Konstantinov KB, Cooney CL (2015) White paper on continuous bioprocessing May 20–21 2014 continuous manufacturing symposium. J Pharm Sci 104:813–3549
27. Tao Y, Shih J, Sinacore M et al (2011) Development and implementation of a perfusion-based high cell density cell banking process. Biotechnol Prog 27:824–829. https://doi.org/10.1002/BTPR.599
28. Seth G, Hamilton RW, Stapp TR et al (2013) Development of a new bioprocess scheme using frozen seed train intermediates to initiate CHO cell culture manufacturing campaigns. Biotechnol Bioeng 110:1376–1385. https://doi.org/10.1002/BIT.24808
29. Lim AC, Washbrook J, Titchener-Hooker NJ, Farid SS (2005) A computer-aided approach to compare the production economics of fed-batch and perfusion culture under uncertainty. https://doi.org/10.1002/bit.20757
30. Bunnak P, Allmendinger R, Ramasamy SV et al (2016) Life-cycle and cost of goods assessment of fed-batch and perfusion-based manufacturing processes for mAbs. Biotechnol Prog 32:1324–1335. https://doi.org/10.1002/BTPR.2323
31. Shevitz J (2000) Fluid filtration system. patent US6544424B1. https://patents.google.com/patent/US6544424B1/en

32. Wong HE, Chen C, Le H, Goudar CT (2022) From chemostats to high-density perfusion: the progression of continuous mammalian cell cultivation. J Chem Technol Biotechnol 97:2297–2575
33. Schwarz H (2022) Tools for the development of intensified perfusion processes for mammalian cell culture. KTH Royal Institute of Technology, Stockholm
34. Ryll T, Dutina G, Reyes A et al (2000) Performance of small-scale CHO perfusion cultures using an acoustic cell filtration device for cell retention: characterization of separation efficiency and impact of perfusion on product quality. Biotechnol Bioeng 69:440–3592
35. Mercille S, Johnson M, Lanthier SP et al (2000) Understanding factors that limit the productivity of suspension-based perfusion cultures operated at high medium renewal rates. Biotechnol Bioeng 67:435–450. https://doi.org/10.1002/(SICI)1097-0290(20000220)67:4<435::AID-BIT7>3.0.CO;2-Y
36. Tang YJ, Ohashi R, Hamel JFP (2007) Perfusion culture of hybridoma cells for hyperproduction of IgG 2a monoclonal antibody in a wave bioreactor-perfusion culture system. Biotechnol Prog 23:255–264. https://doi.org/10.1021/BP060299A
37. Voisard D, Meuwly F, Ruffieux PA et al (2003) Potential of cell retention techniques for large-scale high-density perfusion culture of suspended mammalian cells. Biotechnol Bioeng 82: 751–765. https://doi.org/10.1002/BIT.10629
38. Mulukutla BC, Kale J, Kalomeris T et al (2017) Identification and control of novel growth inhibitors in fed-batch cultures of Chinese hamster ovary cells. Biotechnol Bioeng 114:1779–1790. https://doi.org/10.1002/bit.26313/abstract
39. Pereira S, Kildegaard HF, Andersen MR (2018) Impact of CHO metabolism on cell growth and protein production: an overview of toxic and inhibiting metabolites and nutrients. Biotechnol J 13:1700499. %@ 1860–6768
40. Serra M, Brito C, Correia C, Alves PM (2012) Process engineering of human pluripotent stem cells for clinical application. Trends Biotechnol 30:350–359. https://doi.org/10.1016/J.TIBTECH.2012.03.003
41. Rodrigues CAV, Fernandes TG, Diogo MM et al (2011) Stem cell cultivation in bioreactors. Biotechnol Adv 29:815–829. https://doi.org/10.1016/J.BIOTECHADV.2011.06.009
42. Ozturk SS (1996) Engineering challenges in high density cell culture systems. Cytotechnology 22:3–16. https://doi.org/10.1007/BF00353919
43. Konstantinov K, Goudar C, Ng M et al (2006) The "push-to-low" approach for optimization of high-density perfusion cultures of animal cells. Adv Biochem Eng Biotechnol 101:75–98. https://doi.org/10.1007/10_016/COVER
44. Dowd JE, Jubb A, Kwok KE, Piret JM (2003) Optimization and control of perfusion cultures using a viable cell probe and cell specific perfusion rates. Cytotechnology 42:35–45. https://doi.org/10.1023/A:1026192228471
45. Zhang Y, Stobbe P, Silvander CO, Chotteau V (2015) Very high cell density perfusion of CHO cells anchored in a non-woven matrix-based bioreactor. J Biotechnol 213. https://doi.org/10.1016/j.jbiotec.2015.07.006
46. Schwarz H, Lee K, Castan A, Chotteau V (2023) Optimization of medium with perfusion microbioreactors for high density CHO cell cultures at very low renewal rate aided by design of experiments. Biotechnol Bioeng 120(9):2523–2541. %@ 0006-3592
47. Kompala DS, Ozturk SS (2005) Optimization of high cell density perfusion bioreactors. In: Cell culture technology for pharmaceutical and cell-based therapies. CRC Press, pp 405–434. %@ 042911379X
48. Müller J, Ott V, Eibl D, Eibl R (2022) Seed train intensification using an ultra-high cell density cell banking process. Processes 10:911. %@ 2227-9717
49. Karst DJ, Serra E, Villiger TK et al (2016) Characterization and comparison of ATF and TFF in stirred bioreactors for continuous mammalian cell culture processes. Biochem Eng J 110: 17–26. https://doi.org/10.1016/j.bej.2016.02.003

50. Karst DJ, Steinhoff RF, Kopp MRG et al (2016) Intracellular CHO cell metabolite profiling reveals steady-state dependent metabolic fingerprints in perfusion culture. Am Inst Chem Eng Biotechnol Prog 33:879–890. https://doi.org/10.1002/btpr.2421
51. Wang S, Godfrey S, Ravikrishnan J et al (2017) Shear contributions to cell culture performance and product recovery in ATF and TFF perfusion systems. J Biotechnol 246:52–1656
52. Coronel J, Behrendt I, Bürgin T et al (2019) Influenza A virus production in a single-use orbital shaken bioreactor with ATF or TFF perfusion systems. Vaccine 37:7011–7018. https://doi.org/10.1016/J.VACCINE.2019.06.005
53. Zhan C, Bidkhori G, Schwarz H et al (2020) Low shear stress increases recombinant protein production and high shear stress increases apoptosis in human cells. iScience 23. https://doi.org/10.1016/j.isci.2020.101653
54. Esclade LRJ, Carrel S, Péringer P Influence of the screen material on the fouling of spin filters. https://doi.org/10.1002/bit.260380208
55. Mercille S, Johnson M, Lemieux R, Massie B (1994) Filtration-based perfusion of hybridoma cultures in protein-free medium: reduction of membrane fouling by medium supplementation with DNase I. Biotechnol Bioeng 43:833–846. https://doi.org/10.1002/BIT.260430902
56. Pinto NDS, Brower M (2020) Wide-surface pore microfiltration membrane drastically improves sieving decay in TFF-based perfusion cell culture and streamline chromatography integration for continuous bioprocessing. Biotechnol Bioeng 117:3336–3344. https://doi.org/10.1002/bit.27504
57. Kelly W, Scully J, Zhang D et al (2014) Understanding and modeling alternating tangential flow filtration for perfusion cell culture. Am Inst Chem Eng Biotechnol Prog 30:1291–1300. https://doi.org/10.1002/btpr.1953
58. Wang SB, Godfrey S, Radoniqi F et al (2019) Larger pore size hollow fiber membranes as a solution to the product retention issue in filtration-based perfusion bioreactors. Biotechnol J 14:1800137. %@ 1860–6768
59. Stressmann M, Moresoli C (2008) Effect of pore size, shear rate, and harvest time during the constant permeate flux microfiltration of CHO cell culture supernatant. Biotechnol Prog 24: 890–7938
60. Pinto NDS, Napoli WN, Brower M (2020) Impact of micro and macroporous TFF membranes on product sieving and chromatography loading for perfusion cell culture. Biotechnol Bioeng 117:117–3592
61. Pappenreiter M, Schwarz H, Sissolak B et al Product sieving of mAb and its high molecular weight species in different modes of ATF and TFF perfusion cell cultures. J Chem Technol Biotechnol 98(7): https://doi.org/10.1002/jctb.7386 %@ 0268-2575
62. Bolton GR, Apostolidis AJ (2017) Mechanistic modeling of the loss of protein sieving due to internal and external fouling of microfilters. Biotechnol Prog 33:1323–1333. https://doi.org/10.1002/BTPR.2514
63. Hadpe SR, Sharma AK, Mohite VV, Rathore AS (2016) ATF for cell culture harvest clarification: mechanistic modelling and comparison with TFF. https://doi.org/10.1002/jctb.5165
64. Kreye S, Stahn R, Nawrath K et al (2019) A novel scale-down mimic of perfusion cell culture using sedimentation in an automated microbioreactor (SAM). Biotechnol Prog 35. https://doi.org/10.1002/BTPR.2832
65. Boycott AE (1920) Sedimentation of blood corpuscles. Nature 104:532. https://doi.org/10.1038/104532b0
66. Koreo K (1949) Sedimentation in tilted vessels. Mem Fac Eng Nagoya Univ 1:9–19. https://doi.org/10.18999/MEMFENU.1.1.9
67. Dalm MCF, Cuijten SMR, Van GWMJ et al (2004) Effect of feed and bleed rate on hybridoma cells in an acoustic perfusion bioreactor: Part I. Cell density, viability, and cell-cycle distribution. Biotechnol Bioeng 88:547–557. https://doi.org/10.1002/BIT.20287

68. Johnson M, Lanthier S, Massie B et al (1996) Use of the centritech lab centrifuge for perfusion culture of hybridoma cells in protein-free medium. Biotechnol Prog 12:855–864. https://doi.org/10.1021/BP960072N
69. Jockwer A, Medronho RA, Wagner R et al (2001) The use of hydrocyclones for mammalian cell retention in perfusion bioreactors. Springer
70. Bettinardi IW, Castan A, Medronho RA, Castilho LR (2020) Hydrocyclones as cell retention device for CHO perfusion processes in single-use bioreactors. Biotechnol Bioeng 117:1915–1928. https://doi.org/10.1002/BIT.27335
71. Pollock J, Ho SV, Farid SS (2013) Fed-batch and perfusion culture processes: economic, environmental, and operational feasibility under uncertainty. Biotechnol Bioeng 110:206–219. https://doi.org/10.1002/BIT.24608
72. Budzinski K, Constable D, D'Aquila D et al (2022) Streamlined life cycle assessment of single use technologies in biopharmaceutical manufacture. New Biotechnol 68:28–36. https://doi.org/10.1016/J.NBT.2022.01.002
73. Pietrzykowski M, Flanagan W, Pizzi V et al (2013) An environmental life cycle assessment comparison of single-use and conventional process technology for the production of monoclonal antibodies. J Clean Prod 41:150–6526
74. Cataldo AL, Sissolak B, Metzger K et al (2020) Water related impact of energy: cost and carbon footprint analysis of water for biopharmaceuticals from tap to waste. Chem Eng Sci X 8:100083. https://doi.org/10.1016/J.CESX.2020.100083
75. Walther J, McLarty J, Johnson T (2019) The effects of alternating tangential flow (ATF) residence time, hydrodynamic stress, and filtration flux on high-density perfusion cell culture. Biotechnol Bioeng 116:320–3592
76. Hu W, Berdugo C, Chalmers JJ (2011) The potential of hydrodynamic damage to animal cells of industrial relevance: current understanding. Cytotechnology 63(5):445–460. https://doi.org/10.1007/s10616-011-9368-3
77. Ma N, Koelling KW, Chalmers JJ (2002) Fabrication and use of a transient contractional flow device to quantify the sensitivity of mammalian and insect cells to hydrodynamic forces. Biotechnol Bioeng 80:428–437. https://doi.org/10.1002/bit.10387
78. Mollet M, Godoy-Silva R, Berdugo C, Chalmers JJ (2007) ARTICLE acute hydrodynamic forces and apoptosis: a complex question. https://doi.org/10.1002/bit.21476
79. Zamani L, Lundqvist M, Zhang Y et al (2018) High cell density perfusion culture has a maintained exoproteome and metabolome. Wiley Online Libr 13. https://doi.org/10.1002/biot.201800036
80. Xu S, Hoshan L, Jiang R et al (2017) A practical approach in bioreactor scale-up and process transfer using a combination of constant P/V and vvm as the criterion. Biotechnol Prog 33:1146–7938
81. Goudar CT, Piret JM, Konstantinov KB (2011) Estimating cell specific oxygen uptake and carbon dioxide production rates for mammalian cells in perfusion culture. Biotechnol Prog 27:1347–1357. https://doi.org/10.1002/BTPR.646
82. Gray DR, Chen S, Howarth W et al (1996) Special issue CO2 in large-scale and high-density CllO cell perfusion culture. Cytotechnology 22:65–78
83. Mostafa SS, Gu X (2003) Strategies for improved dCO2 removal in large-scale fed-batch cultures. https://doi.org/10.1021/bp0256263
84. Xing Z, Kenty BM, Li ZJ, Lee SS (2009) Scale-up analysis for a CHO cell culture process in large-scale bioreactors. Biotechnol Bioeng 103:733–3592
85. Hoshan L, Jiang R, Moroney J, et al (2018) Effective bioreactor pH control using only sparging gases. https://doi.org/10.1002/btpr.2743
86. Doi T, Kajihara H, Chuman Y, et al (2020) Development of a scale-up strategy for Chinese hamster ovary cell culture processes using the k L a ratio as a direct indicator of gas stripping conditions. https://doi.org/10.1002/btpr.3000
87. He C, Ye P, Wang H et al (2019) A systematic mass-transfer modeling approach for mammalian cell culture bioreactor scale-up. Biochem Eng J 141:173–181. %@ 1369-703X

88. Goudar CT, Matanguihan R, Long E, et al (2007) Decreased pCO2 accumulation by eliminating bicarbonate addition to high cell-density cultures. Wiley Intersci (www.interscience.wiley.com) Biotechnol Bioeng 96:1107–1117. https://doi.org/10.1002/bit.21116
89. Schwarz H, Zhang Y, Zhan C et al (2020) Small-scale bioreactor supports high density HEK293 cell perfusion culture for the production of recombinant erythropoietin. J Biotechnol 309. https://doi.org/10.1016/j.jbiotec.2019.12.017
90. Gagnon M, Nagre S, Wang W, Hiller GW (2018) Shift to high-intensity, low-volume perfusion cell culture enabling a continuous, integrated bioprocess. Biotechnol Prog 34: 1472–1481. https://doi.org/10.1002/btpr.2723
91. Coffman J, Brower M, Connell-Crowley L et al (2021) A common framework for integrated and continuous biomanufacturing. Biotechnol Bioeng 118:1721–1735. https://doi.org/10.1002/bit.27690
92. Xu S, Chen H (2016) High-density mammalian cell cultures in stirred-tank bioreactor without external pH control. J Biotechnol 231:149–1656
93. Mayrhofer P, Kunert R (2020) Screening of media supplements for high-performance perfusion cultures by design of experiment. Methods Mol Biol 2095:27–39. https://doi.org/10.1007/978-1-0716-0191-4_3/FIGURES/3
94. Zheng C, Zhuang C, Chen Y et al (2018) Improved process robustness, product quality and biological efficacy of an anti-CD52 monoclonal antibody upon pH shift in Chinese hamster ovary cell perfusion culture. Process Biochem 65:123–5113
95. Korke R, Gatti MDL, Lei A et al (2004) Large scale gene expression profiling of metabolic shift of mammalian cells in culture. J Biotechnol 107:1–17. https://doi.org/10.1016/j.jbiotec.2003.09.007
96. Kang S, Ren D, Xiao G et al (2014) Cell line profiling to improve monoclonal antibody production. Biotechnol Bioeng 111:748–760. https://doi.org/10.1002/bit.25141/abstract
97. Villiger-Oberbek A, Yang Y, Zhou W, Yang J (2015) Development and application of a high-throughput platform for perfusion-based cell culture processes. J Biotechnol 212:21–29. https://doi.org/10.1016/J.JBIOTEC.2015.06.428
98. Hagrot E, Oddsdóttir HÆ, Hosta JG et al (2017) Poly-pathway model, a novel approach to simulate multiple metabolic states by reaction network-based model – application to amino acid depletion in CHO cell culture. J Biotechnol 259:235–247. https://doi.org/10.1016/j.jbiotec.2017.05.026
99. Gomez N, Ambhaikar M, Zhang L et al (2017) Analysis of Tubespins as a suitable scale-down model of bioreactors for high cell density CHO cell culture. Biotechnol Prog 33:490–499. https://doi.org/10.1002/BTPR.2418
100. Gagliardi TM, Chelikani R, Yang Y et al (2019) Development of a novel, high-throughput screening tool for efficient perfusion-based cell culture process development. Biotechnol Prog 35:e2811. https://doi.org/10.1002/BTPR.2811
101. Wolf MKF, Lorenz V, Karst DJ et al (2018) Development of a shake tube-based scale-down model for perfusion cultures. Biotechnol Bioeng 115:2703–2713. https://doi.org/10.1002/BIT.26804
102. Zhang L, Wang M, Castan A et al (2020) Glycan residues balance analysis – GReBA: a novel model for the N-linked glycosylation of IgG produced by CHO cells. Metab Eng 57:118–128. https://doi.org/10.1016/j.ymben.2019.08.016
103. Mayrhofer P, Reinhart D, Castan A, Kunert R (2020) Rapid development of clone-specific, high-performing perfusion media from established feed supplements. Biotechnol Prog 36. https://doi.org/10.1002/BTPR.2933
104. Mayrhofer P, Castan A, Kunert R (2021) Shake tube perfusion cell cultures are suitable tools for the prediction of limiting substrate, CSPR, bleeding strategy, growth and productivity behavior. J Chem Technol Biotechnol 96:2930–2939. https://doi.org/10.1002/JCTB.6848
105. Jin L, Wang ZS, Cao Y et al (2021) Establishment and optimization of a high-throughput mimic perfusion model in ambr® 15. Biotechnol Lett 43:423–433. https://doi.org/10.1007/S10529-020-03026-5

106. Chotteau V (2015) Perfusion process for very high cell density of CHO cells. In: Continuous processing in biopharm manufacturing, CHI's 7th annual bioprocessing summit Aug 3–7, 2015, Boston, MA, USA
107. Zhang L, Schwarz H, Wang M et al (2021) Control of IgG glycosylation in CHO cell perfusion cultures by GReBA mathematical model supported by a novel targeted feed, TAFE. Metab Eng 65:135–145. https://doi.org/10.1016/j.ymben.2020.11.004
108. Schwarz H, Mäkinen ME, Castan A, Chotteau V (2022) Monitoring of amino acids and antibody N-glycosylation in high cell density perfusion culture based on Raman spectroscopy. Biochem Eng J 182:108426. %@ 1369–703X
109. Madabhushi SR, Huang C-J, Wang X, et al (2022) An innovative strategy to recycle permeate in biologics continuous manufacturing process to improve material efficiency and sustainability. https://doi.org/10.1002/btpr.3262
110. Yin L, Au WY, Yu CC et al (2021) Miniature auto-perfusion bioreactor system with spiral microfluidic cell retention device. Biotechnol Bioeng 118:1951–1961. https://doi.org/10.1002/BIT.27709
111. Kwon T, Prentice H, De OJ et al (2017) Microfluidic cell retention device for perfusion of mammalian suspension culture. Sci Rep 7. https://doi.org/10.1038/S41598-017-06949-8
112. Wolf MKF, Closet A, Bzowska M et al (2019) Improved performance in mammalian cell perfusion cultures by growth inhibition. Biotechnol J 14. https://doi.org/10.1002/BIOT.201700722
113. Meuwly F, Papp F, Ruffieux PA et al (2006) Use of glucose consumption rate (GCR) as a tool to monitor and control animal cell production processes in packed-bed bioreactors. J Biotechnol 122:122–129. https://doi.org/10.1016/J.JBIOTEC.2005.08.005
114. Ducommun P, Rueux P-A, Kadouri A et al (2002) Monitoring of temperature effects on animal cell metabolism in a packed bed process. Biotechnol Bioeng 77:838–842. https://doi.org/10.1002/bit.10185
115. Chen Z-L, Wu B-C, Liu H et al (2004) Temperature shift as a process optimization step for the production of pro-urokinase by a recombinant Chinese hamster ovary cell line in high-density perfusion culture. J Biosci Bioeng 97:239–243. https://doi.org/10.1016/S1389-1723(04)70198-X
116. Kim JS, Ahn BC, Lim BP et al (2004) High-level scu-PA production by butyrate-treated serum-free culture of recombinant CHO cell line. Biotechnol Prog 20:1788–1796. https://doi.org/10.1021/BP025536Y
117. Angepat S, Gorenflo VM, Piret JM (2005) Accelerating perfusion process optimization by scanning non-steady-state responses. Biotechnol Bioeng 92:472–478. https://doi.org/10.1002/BIT.20635
118. Han KO, Moon KS, Yang J et al (2005) Effect of N-acetylcystein on butyrate-treated chinese hamster ovary cells to improve the production of recombinant human interferon-β-1a. Biotechnol Prog 21:1154–1164. https://doi.org/10.1021/BP050057V
119. Woo SA, Jeon JJ, Jeong YR et al (2008) Effect of culture temperature on erythropoietin production and glycosylation in a perfusion culture of recombinant CHO cells. Biotechnol Bioeng 101:1234–1244. https://doi.org/10.1002/BIT.22006
120. Kwon T, Sung HK et al (2020) Continuous online protein quality monitoring during perfusion culture production using an integrated micro/nanofluidic system. Anal Chem 92:5275. https://doi.org/10.1021/acs.analchem.9b05835
121. Coronel J, Klausing S, Heinrich C et al (2016) Valeric acid supplementation combined to mild hypothermia increases productivity in CHO cell cultivations. Biochem Eng J 114:101–109. https://doi.org/10.1016/J.BEJ.2016.06.031
122. Wang SB, Lee-Goldman A, Ravikrishnan J et al (2018) Manipulation of the sodium-potassium ratio as a lever for controlling cell growth and improving cell specific productivity in perfusion CHO cell cultures. Biotechnol Bioeng 115:921–3592
123. Qin J, Wu X, Xia Z et al (2019) The effect of hyperosmolality application time on production, quality, and biopotency of monoclonal antibodies produced in CHO cell fed-batch and

perfusion cultures. Appl Microbiol Biotechnol 103:1217–1229. https://doi.org/10.1007/S00253-018-9555-7
124. Bertrand V, Karst DJ, Bachmann A et al (2019) Transcriptome and proteome analysis of steady-state in a perfusion CHO cell culture process. Biotechnol Bioeng 116:1959–1972. https://doi.org/10.1002/BIT.26996
125. Lin H, Leighty RW, Godfrey S, Wang SB (2017) Principles and approach to developing mammalian cell culture media for high cell density perfusion process leveraging established fed-batch media. Biotechnol Prog 33:891–901. https://doi.org/10.1002/BTPR.2472
126. Gagnon M, Hiller G, Luan Y-T et al (2011) ARTICLE high-end pH-controlled delivery of glucose effectively suppresses lactate accumulation in CHO fed-batch cultures. Biotechnol Bioeng 108:1328–1337. https://doi.org/10.1002/bit.23072
127. Kuiper M, Spencer C, Fäldt E, et al (2019) Repurposing fed-batch media and feeds for highly productive CHO perfusion processes. https://doi.org/10.1002/btpr.2821
128. Wolf MKF, Müller A, Souquet J et al (2019) Process design and development of a mammalian cell perfusion culture in shake-tube and benchtop bioreactors. Biotechnol Bioeng 116:1973–1985. https://doi.org/10.1002/BIT.26999
129. Feidl F, Vogg S, Wolf M et al (2020) Process-wide control and automation of an integrated continuous manufacturing platform for antibodies. Biotechnol Bioeng 117:1367–3592
130. Bielser JM, Domaradzki J, Souquet J et al (2019) Semi-continuous scale-down models for clone and operating parameter screening in perfusion bioreactors. Biotechnol Prog 35. https://doi.org/10.1002/BTPR.2790
131. Kamga M-H, Cattaneo M, Yoon S (2018) Integrated continuous biomanufacturing platform with ATF perfusion and one column chromatography operation for optimum resin utilization and productivity. Prep Biochem Biotechnol 48:383–6068
132. Arnold L, Lee K, Rucker-Pezzini J, Lee JH (2019) Implementation of fully integrated continuous antibody processing: effects on productivity and COGm. Biotechnol J 14: 1800061. %@ 1860–6768

Chapter 5
Continuous Bioprocessing for Downstream

Lukas Gerstweiler

Abstract Continuous biomanufacturing is increasingly considered as a way to intensify production of biopharmaceuticals and to reduce manufacturing costs by decreasing equipment footprint and increase equipment utilization. For the interested reader this chapter provides an overview over the main developments in continuous downstream processing of biopharmaceuticals and introduces the main unit operations such as continuous chromatography, continuous filtration, and continuous precipitation, as well as discusses process integration and continuous manufacturing processes. Process and design principles are explained highlighting advantages and challenges of continuous downstream processing.

Keywords Continuous downstream processing · Periodic counter current chromatography · Single-pass tangential flow filtration · Integrated processing · Purification

Abbreviations

CEX	Cation Exchange
CFIR	Coild Flow Inversion Reactor
CHO	cells Chinese Hamster Ovary cells
CSTR	Continuous Stirred Tank Reactor
DBC	Dynamic Binding Capacity
FDA	Food and Drug Administration
GMP	Good Manufacturing Practices
HPLC	High Performance Liquid Chromatography
MCSGP	Multi Column Solvent Gradient Purification
PEG	Polyethylenglycol
PCC	Periodic Counter Current Chromatography

L. Gerstweiler (✉)
School of Chemical Engineering, The University of Adelaide, Adelaide, SA, Australia
e-mail: Lukas.gerstweiler@adelaide.edu.au

R. Pörtner (ed.), *Biopharmaceutical Manufacturing*, Cell Engineering 11,
https://doi.org/10.1007/978-3-031-45669-5_5

SMB Simulated Moving Bed
SPTFF Single Pass Tangential Flow Filtration
TFF Tangential Flow Filtration

Introduction

As industries mature and products become more standardised the production process usually undergoes a shift from batch to continuous operation. Many industries producing a wide range of petrochemicals, food and beverages, metals, fertilizers and various organic chemicals have applied continuous or semi-continues production approaches decades ago [1]. Continuous manufacturing usually offers many advantages over batch production, such as much smaller plants for a similar output, lower capital investments, and manufacturing flexibility. Another major benefit in particular for biopharmaceutical production is that a stable operation leads to a constant product quality with no, or less batch-to-batch variations. This is also one of the key aspects why regulatory agencies such as the FDA support the transition to continuous manufacturing [2–4]. Despite continuous manufacturing offers obvious advantages, which can lower the overall production costs, the biopharmaceutical industry lacks behind this development. However, interest in this topic has risen in recent years and many companies and universities actively pursue research in this field now [2, 4–6]. For example, Sanofi opened a continuous manufacturing facility, Enzene Bioscience and BiosanaPharma now have integrated continuous production processes with GMP status, and BiosanaPharma successfully completed Phase 1 clinical trials of a continuous produced omalizumab biosimilar [7–9].

There have been great advancements in upstream processing of biopharmaceuticals and continuous perfusion cultures are well described and utilized [2]. Significant productivity gains of more than 5 times compared to conventional fed-batch processing have been reported applying continuous perfusion culture for producing monoclonal antibodies with CHO cell lines [10]. Furthermore, critical quality parameters such as N-glycan distribution and product aggregation have been reported to be constant in continuous perfusion cultures over multiple weeks [11, 12]. To exploit the full potential of continuous manufacturing the upstream production needs to be integrated with continuous downstream purification in a seamless integrated process. As purification can account for up to 80% of the production costs of modern biopharmaceuticals, continuous downstream processing is a great way to reduce the overall costs of goods [13]. Continuous downstream processing technologies can help to reduce the cost of goods by increase equipment utilization, reducing facility and equipment footprints and shorten the production cycle [14–17].

Designing continuous and integrated downstream process, however, is challenging, as purification of modern biotherapeutics heavily rely on chromatography as a key unit operation, which is inherently a batch-wise unit operation that cannot equilibrate into a constant state. Several design principles utilising a combination of multiple chromatographic columns have been developed to circumvent some of the limitations and to further intensify downstream processing by simulating a counter-current flow. Also, alternative processes such as filtration and precipitation are investigated in their use in continuous downstream processing, either as a replacement to chromatography or for process integration purposes. This chapter aims to summarize the main developments of continuous downstream processing used to purify biopharmaceuticals derived from cell cultures and highlights some of the main design consideration of continuous downstream processing.

Continuous Chromatography

Liquid chromatography can be considered the most important purification technique in downstream processing of biopharmaceuticals. As a highly versatile process, chromatography enables purification based on a variety of principles, including hydrophobicity, charge, specific binding interactions and size at comparable mild conditions, necessary to remain biological activity of the protein [18]. Most described downstream processes to produce commercial relevant biological products, such as antibodies and vaccines are predominantly based on several steps of chromatographic purification. As chromatography is inherently a batch process, developing continuous chromatographic processes is challenging and usually requires the use of multi-column systems [19]. In its simplest way, chromatographic unit operations can just run for multiple cycles, and a quasi-continuous flow can be achieved by running multiple processes in parallel. This approach is straightforward to implement and can reduce equipment size and overall footprint, however no further advantages in terms of productivity and buffer consumption can be achieved. Therefore some more advanced multi-column approaches have been developed which also aim to improve the overall process performance by mimicking a counter current process flow of which the main concepts will be described in following subchapters [5].

Simulated Moving Bed

Developed in the 1950s simulated moving bed (SMB) mimics a counter current flow of the stationary phase and the mobile phase by combining multiple fixed bed columns with cyclic valve switching [20, 21]. SMB can be applied for the separation of binary mixtures of a weakly and a strongly adsorbing component. The basic set-up outlined in Fig. 5.1 utilizes four columns referred to as Zone I-IV, two inlets for eluent and feed intake and two outlets to withdraw either the weakly binding component, which is travelling quicker through the columns (raffinate stream), or the stronger binding component, which remains longer in the columns (extract stream)

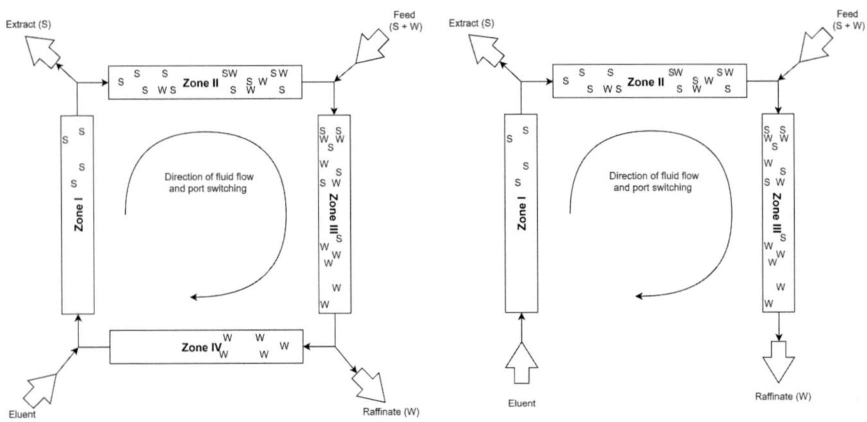

Fig. 5.1 Closed loop 4 zone SMB process and open loop 3 zone SMB process missing zone IV to separate a weakly (W) and a strongly (S) binding component. The weakly binding component travels quicker with the fluid flow through the columns. Upon port switching the zones rotate counter-clockwise and former Zone II becomes Zone I and so on, to simulate a counter-current flow of stationary and mobile phase. (Reprint from Ref. [5])

[22]. The design of SMB processes has been extensively reviewed in literature [22–24].

While a 4-zone SMB allows for buffer recycling and highly economical processes, it bears the risk of cross contamination in case non-ideal process design and does not allow for buffer modulation in each zone, which strongly limits its use in biopharmaceutical processing. Thus, the concept of a 3-zone open loop SMB, without zone IV has been developed, in which the buffer is not recycled and hence buffer properties can be changed in each zone to modulate molecule interactions. Salt- and pH gradients can be applied, which allows the use of IEX and affinity resins. Histidine tagged single chained antibody fragments have been purified continuously using 3-zone SMB with greatly improved productivities [25], as well as influenza A virus particles from cell culture supernatant with anion exchange monoliths [26].

SMB processes can also be used to transfer size exclusion chromatography and chromatographic desalting into continuous processes, in which a pseudo binary mixture of large and small molecules with different travel speeds through the columns is assumed. The feasibility of this approach has been demonstrated for the purification of plasmid DNA, influenza viruses and adenovirus particles, and desalting of BSA solution, gaining up to 6 times higher productivity compared to batch processes [27–31].

Although SMB is an extremely powerful approach for continuous downstream processing, often increasing productivities multiple times, it is not widely adapted in continuous downstream processing of biopharmaceuticals yet. A major drawback is that SMB is only feasible for separating binary mixtures, while often a ternary separation is required. It is possible to integrate multiple SMBs into a cascade in which one outlet stream is further purified in a second SMB and so on, however, this approach is highly complex to develop and operate, and therefore impracticable. Another challenge is the design of such processes, as the transfer from a batch process is not straightforward [32]. Nevertheless, open loop three zone SMB is a suitable process for biopharmaceutical downstream processing in particular for SEC chromatography for which no other continuous approach exists.

Continuous Capturing Applying Periodic Counter Current Chromatography

Periodic counter current chromatography (PCC) is the most widely described process for continuous chromatography in bind-and-elute mode but can also be used in flow through mode in which the impurities are captured while the product remains in the column flow through. It is a cyclic process which is characterized by processing a continuous feed stream, while the output is a semi-continuous process stream. PCC tackles the trade-off between resin utilization and productivity in batch chromatography, in which a lower flow rate results in a higher dynamic binding capacity (DBC), however also decreases the productivity of the process [33, 34]. The main idea of PCC is that a column is loaded beyond its DBC and the leaking product in the flow through is being captured on a second column in the loading train. In its simplest set-up it applies three columns, in which column one and two are connected in series and being loaded, while at the same time the third column is being eluted and regenerated. Once the first column is fully loaded and the third column regenerated, the third column it is being connected to the second loading column, while the formerly first column of the loading train is being disconnect from the loading and regenerated [35, 36]. This column switching is being repeated and results in a cyclic process described in Fig. 5.2 with a simulated counter current

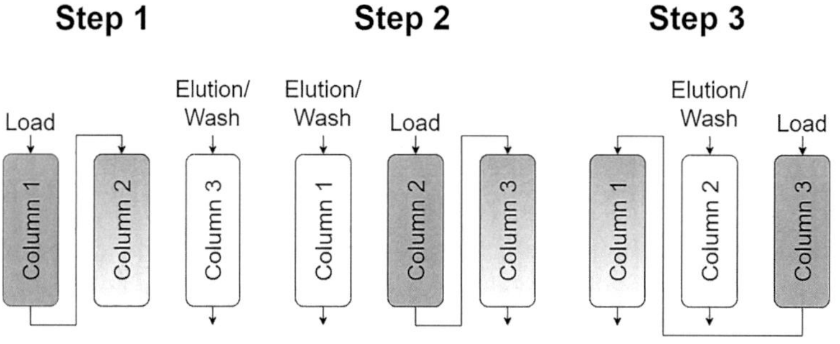

Fig. 5.2 Idealized three column PCC process. In each step two columns are being connect in series and loaded. The first column is getting overloaded, and the breakthrough is captured on the second column. Simultaneously the third column is being eluted and regenerated. In the second step the first column in the former loading is being eluted and regenerated. The former second column will be become the first column in the loading train and overloaded onto column 3

flow in the loading train. This allows for much higher column utilization, lower buffer consumption, smaller column volumes, and higher product concentrations, and usually also increases the overall productivity, compared to batch chromatography [5, 35–40]. Another benefit of using smaller columns in a repetitive manner, compared to one large column in batch mode, is a reduced residence time of the protein on the column, which can lower the burden for unstable molecules [40, 41]. Commercial PCC set-ups can differ in the total number of columns used in the loading and regeneration train, ranging from only two columns in total in the CaptureSMB process to 10 columns in the BioSMB process, and the exact column switching procedure [40, 42, 43]. Detailed description of commercially available systems and the specific set-ups are revied in literature [42, 44].

Designing an optimal PCC process is not straightforward as the performance, like in batch chromatography, is dependent on the elution-recover-regeneration time τ_R for one column, the feed stream concentration C_f, the binding capacity K_B and the residence time R_t (flow rate), which can also influence each other. Generally speaking, the PCC process can be divided into a loading zone in which the protein is binding to the columns and a regeneration zone in which the protein is eluted, and the columns are regenerated. Both zones can consist of multiple columns in which individual columns can be connected and disconnected from the train [35]. To enable a continuous inlet feed stream the loading time of one column τ_L must be larger or equal than the elution-recovery-regeneration time of one column ($\tau_L \geq \tau_R$). If the regeneration time is larger than the loading time the regeneration can be split over multiple columns and run in parallel. Godawat et al. therefore proposed Eq. (5.1) to estimate the number of necessary columns N in a PCC set-up with two columns in the loading zone [35]

$$N > 2 + \frac{\tau_R C_f}{R_t K_B} \tag{5.1}$$

From Eq. (5.1) it can be concluded that a short residence time and low binding capacities, as well as high feed stream concentrations and long elution-recover-regeneration times increase the number of required columns in the regeneration zone to keep up with the loading zone. A high number of columns in the process is however problematic as it also increases the total number of valves, pumps and generally complicates the process. A long loading time, compared to the elution-recovery-regeneration time is therefore desirable, which can be achieved by lowering the flow rate and decreasing the loading concentration [35, 40].

Another major challenge in designing a PCC process is to determine the length of the loading cycle. On the one hand lengthy loading cycles with high overloading of the first column will increase the resin utilization, however, might also lead to product loss as not all product can be captured on the second column. On the other hand, short loading cycles will result in a non-optimal resin utilization and may decrease productivity. Assuming no product loss is acceptable during loading and all columns are equal in size and capacity the optimal loading time in each cycle can be derived from the loading breakthrough curve of a single column. As shown in

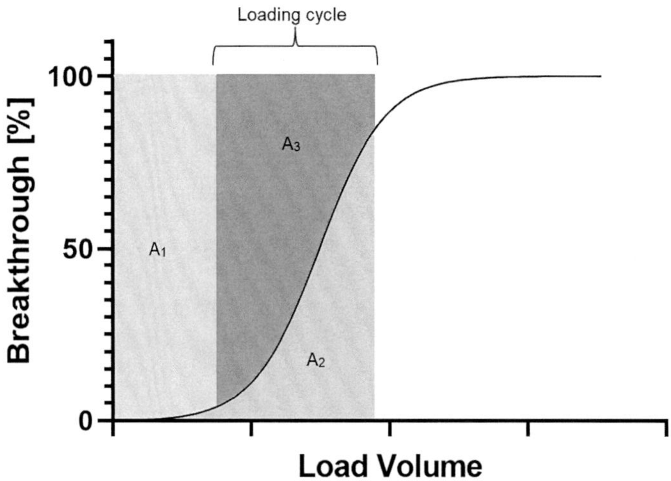

Fig. 5.3 Breakthrough curve to design loading in a PCC process. A_1 corresponds to the DBC and A_2 to the breakthrough which is being captured on the second column in the loading train. A_3 equals to the increase in binding capacity which can be achieved by overlaoding. In an ideal process A_1 equals A_2 and the volume of each loading cycle depends on the time needed to overload completley

Fig. 5.3 the area underneath the breakthrough curve A_2 represents the amount of product breaking through column 1, which is captured by the second column. The area A_1 represents the dynamic binding capacity and A_3 the increase of product binding that can be achieved by overloading [35, 45]. In an optimal process with maximum overloading without product loss A_1 equals A_2, which means that the first column at the beginning of each loading cycle is already loaded to its DBC and therefore the cycle time is dependent on how long it takes to overload the column as outlined in Fig. 5.3. This also means that PCC processes are in particular beneficial for shallow breakthrough curves and low DBCs compared to the maximum banding capacity, which can be often observed by processing large biomolecules such as mAbs, as otherwise the gain in resin utilization is minimal.

Multi-column PCC can be considered the state-of-the-art for capturing of mAbs from CHO cell supernatant using protein A affinity columns, and a seamless integration with perfusion reactors has been widely described in literature [46–49]. However, it is not limited to protein A affinity chromatography and can be applied for all bind-and-elute chromatographic steps, although less described in literature. Viral capsomeres to produce virus-like particles have been purified continuously using a mixed-mode cation exchanger in a 3-column PCC set-up [50] and both cation and anion exchange chromatography has been described with increased productivity up to 735% compared to batch processing [51, 52]. Productivities up to $50\ mg_{antibody} ml_{resin}^{-1} h^{-1}$ have been reported using an eight column BioSMB system for mAbs capturing on protein A columns [53]. Morbidelli's group developed a PCC process called CaptureSMB which applies a semi continuous inlet stream, and therefore only requires 2 columns in total [33, 34, 40]. Theoretical productivities of the twin-column have been modelled to be between 14 and 47 $mg_{antibody}ml_{resin}^{-1}h^{-1}$, and experimentally a 40% higher productivity and a 2.5 times higher resin utilization, compared to batch processing have been described [33, 34, 54].

While continuous PCC processes usually outperform a batch process, the optimal process design, and the optimal number of used columns, is challenging to find, and very little research is addressing this issue. Increasing the number of columns in the loading train generally increases the resin utilization and can lead to higher productivities. However more columns also increase the complexity of the process, which can lower the cost effectiveness [55]. Furthermore, changes in the feed stream concentration can favour different PCC set-ups and therefore this needs to be decided on a base by case basis [38].

Continuous Polishing

The beforementioned approaches utilize counter-current principles to increase resin utilization and productivity and are mostly used for initial capturing using costly affinity resins or for intermediate purification applying ion exchangers. During final polishing a high resin utilization is of less concern, as the main focuses are high yields and the removal of trace impurities or charged variants. Continuous flow through chromatography in which the product remains in the flow through and impurities bind to the resin is a straightforward approach to implement continuous polishing that usually show very high recovery rates above 95% [50, 56, 57]. A batch process can be easily transferred into a continuous process with a constant feed flow by alternating two or more columns [58] and several integrated continuous processes for the purification of mAbs apply flow through chromatography as final purification steps [46, 47]. A main advantage of flow through chromatography for continuous processing is that a feed stream can be continuously processed and opposed to bind-and-elute chromatography, the outlet stream shows only minimal fluctuations in concentration and buffer composition, which enables the seamless integration with a subsequent chromatographic step without buffer adjustment [50, 59]. Flow through chromatography can be a powerful approach for continuous downstream processing if impurities can efficiently be removed, however finding optimal buffer conditions and resins that allow a high impurity clearance might be challenging. This challenge can be overcome by combining multiple different resins such as cation with anion exchanger or even activated carbon in a seamless connected process, which can be capable of removing host cell protein, DNA and aggregates efficiently from antibodies if right buffer conditions are chosen [56, 57].

To continuously process hard to separate species that require a centre-cut approach due to overlapping peaks, such as charged mAbs variants, a process called multi-column counter current solvent gradient purification (MCSGP) has been developed [60, 61]. MCSGP combines between two and six columns and is derived from an open loop SMB with buffer adjustments, in which insufficient purified regions of the overlapping peaks are recycled and reapplied to the column (Fig. 5.4).

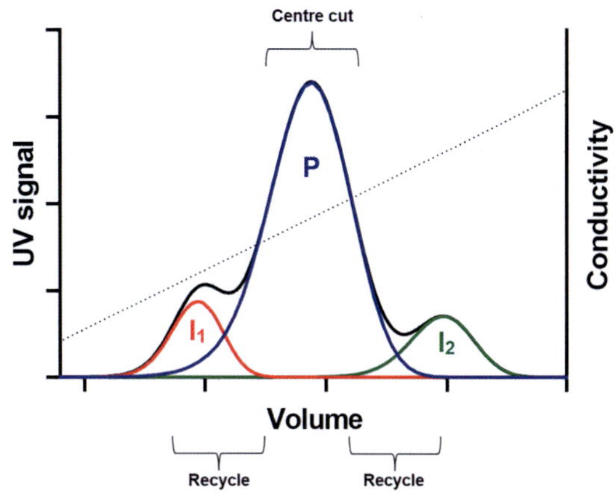

Fig. 5.4 Idealised elution profile of a product P with two overlapping impurities I_1 and I_2 that requires a centre cut for to obtain high purity product. In MSCGP the two overlapping areas, which contain both product and impurity are being recycled and reapplied to the column in a subsequent cycle

Depending on the number of columns MCSGP can run continuously or semi-continuously [60, 62–65]. The main purpose of MCSGP is to solve the yield-purity trade-off if a centre-cut is needed to achieve very high purities and higher yields compared to a batch process. For a given purity, an increase of yield of 23% has been reported and the process has successfully been tested for polishing a variety of molecules such as charged mAbs variants, peptides and other proteins [19, 61, 66].

Continuous Non-chromatographic Purification

Industrial downstream processes are usually based on multiple chromatographic purification steps, and although methods have been developed to transfer those steps into continuous unit operations, a fully equilibrated continuous flow cannot be achieved with chromatographic based methods, due to its inherent batch characteristics. Therefore, some researchers explored how alternative purification methods such as precipitation, filtration, and liquid-liquid extraction can be applied in continuous downstream processing. As these unit operations might be easier to be operated in a fully continuous manner in future. However, these approaches are still in its infancy and require further research. For a fully integrated and continuous downstream process it is also important to explore possibilities to operate other common unit operations such as buffer exchange and viral inactivation continuously.

Continuous Filtration

Filtration is a crucial unit operation in downstream processing of biopharmaceuticals, which can serve multiple purposes. Diafiltration is commonly used for buffer exchange between unit operations and concentration, nanofiltration for viral removal, and microfiltration to separate precipitates and other solids [67]. Multiple filtration units can be combined to achieve continuous filtration, an in-depth overview over possible set-ups can be found in literature [67]. For dead-end filters this can be easily realized by running them in parallel, in which the flow path is diverted to the next filter as soon as the current filter starts to block. Continuous tangential flow filtration (TFF) requires the establishment of a cascade in which each reservoir has a continuous inlet and outlet stream named feed-and-bleed set-up. The outlet of one TFF unit is processed on a subsequent TFF unit similar to cascade of continuous stirred tank reactors (CSTR). As the filter pore size can also change within the cascade, this also allows for the separation into multiple product streams, which has been shown for the purification of polysaccharides [68, 69]. It has been demonstrated that basic recirculating TFF units can be combined to continuously separate precipitated antibodies from clarified supernatant. In the described set-up two TFF were connected with constant inlet and outlet streams from each reservoir and eventually an equilibrated state established in the final outlet stream [70]. A main disadvantage of connecting multiple recirculating TFF units is a broad residence time distribution for low number of stages, comparable to a cascade of CSTRs [71]. This drawback can be overcome by single pass tangential flow filtration (SPTFF), in which filter area and residence time in the filter module is increased to a point in which a single pass is sufficient to achieve the desired filtration. SPTFF is in particular useful as an in-line concentrator for e.g., cell culture harvest to intensify subsequent chromatographic unit operations and to reduce storage volumes [72, 73]. By applying in-line dilution between multiple SPTFF modules continuous single pass diafiltration (SPDF) can be realized (Fig. 5.5) and buffer exchange rates above 99.75% have been reported in continuous antibody purification and formulation [74, 75]. SPDF requires approximately double as much buffer compared to batch diafiltration [75]. Buffer consumption can be reduced by applying a counter-current flow by recycling the permeate of a filter step in a prior step as the dilution buffer, however the buffer consumption is still around 30% higher compared to batch diafiltration [75]. The concentration profile of an ideal non-retaining small molecule (sieving factor S = 1) using SPDF differs from TFF. In TFF the concentration of component *i* in the retentate ($c^i_{retentate}$) for a constant final and initial volume can be calculated as

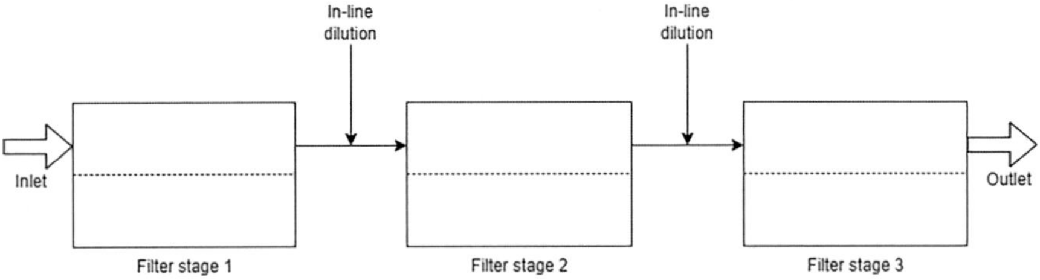

Fig. 5.5 Single pass diafiltration. In each filter stage the volume of the feed stream is being reduced. The buffer exchange is being achieved by adding new buffer to reach the initial volume after each stage, which is repeated multiple times

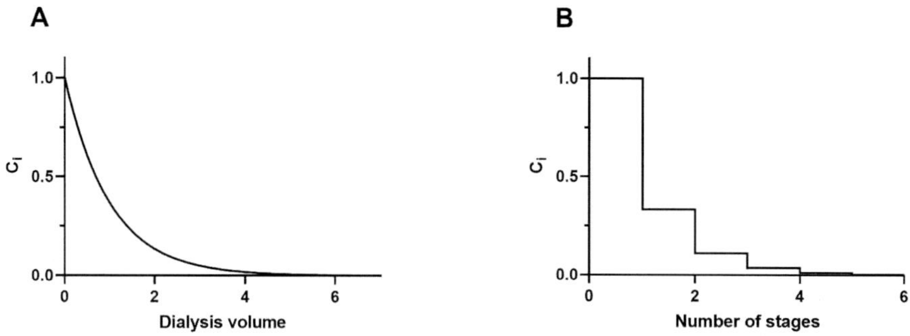

Fig. 5.6 Idealized concentration profile of a non-retaining salt during either diafiltration (**A**) or SPDF (**B**) for a consistent volume. Dilution factor for SPDF calculation X = 3

$$c^i_{retentate} = c^i_{feed} e^{-V} \tag{5.2}$$

in which V equals the required buffer volume. This leads to an exponential decrease in the concentration of the small molecule. In SPDF the concentration $c^i_{retentate}$ is dependent on the number of stages n as well as the dilution factor X in each stage and can be expressed as

$$c^i_{retentate} = c^i_{feed} \left(\frac{1}{X}\right)^n \tag{5.3}$$

Which leads to a step wise decrease in the concentration profile. An idealised concentration profile of TFF and SPDF is illustrated in Fig. 5.6.

Another approach for continuous buffer exchange is the use of hollow fibre dialysis membranes in a counter-current set-up. Buffer exchange rates of IgG solutions above 99.9% have been achieved. As the buffer exchange is mainly driven by diffusion and not by a pressure gradient over the membrane, membrane fouling effects are expected to be minimal during the process [75, 76].

Continuous Precipitation

Precipitation has been proposed as an alternative to protein A affinity chromatography as an initial capturing of monoclonal antibodies as it can effectively reduce the process volume, remove host cell protein and DNA and demonstrated high yields above 90% [70, 77]. Economic analysis unveiled that continuous antibody purification processes based upon precipitation are cheaper than the current platform approaches based on Protein A affinity chromatography [78]. Different precipitation agents such as caprylic acid, PEG, CaCl2 and ethanol have been examined in continuous precipitation of antibodies using a tubular reactor with in-line mixers with equal results compared to batch wise precipitation in a stirred tank vessel [77, 78]. The impurity clearance can be further improved by washing the precipitated product in multiples stages as shown by Burgstaller et al. [70] reaching an antibody purity of 97%, while still achieving over 95% product recovery. Continuous precipitation requires continuous mixing of the cell harvest with precipitation agent and a drawback of the widely used tubular reactors with in-line mixers is that precipitates can build up inside the mixing units over time challenging appropriate mixing in the tubular reactor [79]. An alternative are coiled-flow inversion reactors (CFIR), which enhance mixing in the tube by inducing secondary flow pattern called Dean vortices, without using static mixers. A CFIR combines helically-coiled tubing, which is bended by 90° multiples times, thus inducing centrifugal forces and flow inversion which lead to the favourable flow pattern and a plug-flow like behaviour [80–82]. A basic CFIR with four bends is illustrated in Fig. 5.7. CFIRs have been successfully used to continuously precipitate impurities from cell culture supernatant and integrated into continuous downstream processes [83, 84]. While initiating precipitation is relatively simple by mixing the precipitation agent with the supernatant in either a tubular reactor or a CFIR, a major challenge is the separation of the precipitate from the liquid. Centrifugation will compact the precipitates and thus impede a subsequent solubilization, limiting the use in continuous manufacturing [85]. For that reason continuous filtration is recommended to recover precipitated product, allowing resolubilisation within seconds, however high viscosities caused by PEG can slow down filtration [85]. Both continuous TFF in a feed-and-bleed set-up, as well hollow fibre membranes in flow through mode have been successfully implemented in continuous recovery of precipitated antibodies, achieving high yields between 70% and 95% and constant flux [70, 86, 87].

Both tubular reactors, packed bed reactors and CFIRs are also suitable to transfer time dependent reactions, which require a narrow residence time distribution and accurate mixing, such as low-pH viral inactivation into continuous mode, which show comparable results to batch viral inactivation [88–93].

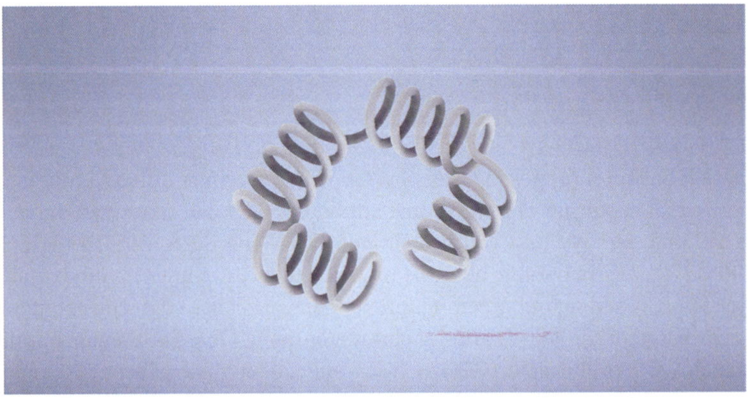

Fig. 5.7 Idealized coiled flow inversion reactor with four 90° bends.

Process Integration and Control

Continuous downstream processing can intensify the production of biopharmaceuticals and main advantages are a steady product quality but moreover reduced cost due to much lower footprints and higher productivities. Most standard unit operations in downstream processing of biopharmaceuticals have been well described in literature and hold most of the promised advantages. Stand-alone continuous unit operations however, cannot fulfill the whole potential of continuous manufacturing unless they are integrated into an end-to-end closed production line.

Several end-to-end integrated downstream processes for the production of mAbs and also VLPs have been described in recent years [46–48, 50, 53, 84, 90, 94, 95]. Some of the described processes were able to run for an entire month and the product quality of a continuous process does not differ negatively to a batch process [90]. Figure 5.8 outlines the first described fully integrated process comprising of a PCC system for initial protein A capturing, a continuous stirred tank reactor for low pH viral inactivation, a second PCC system for CEX polishing followed by AEX membrane in flow through mode as the final polishing [46]. This set-up could run continuously for 30 days with a 6 times higher productivity and a smaller footprint compared to batch processing. A resin volume reduction of 95% and 44% lower buffer consumption have been reported for a similar process applying two PCC systems for the main purification steps [53]. Another process combining PCC with multiple SPTFF and SPDF units and achieved an overall cost reduction of 15% [48]. A main concern of continuous processing is the high number of used columns, requiring extensive numbers of valves and complicated flow paths, opening up

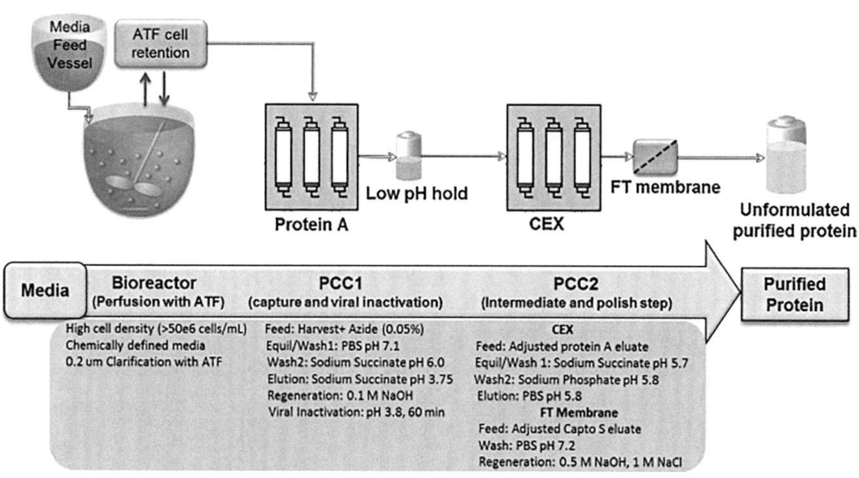

Fig. 5.8 Possible platform process for continuous and integrated processing of mAbs from cell culture, consisting of a perfusion reactor, coupled with Protein A affinity chromatography using PCC, low pH viral inactivation in stirred tank, a second PCC process for intermediate purification and a flow-through membrane for final polishing. (Reprint from Ref. [46])

possibilities for process failure. A reduction to only 5 columns in total has been shown to be feasible combining a 2 column PCC with a 2 column MCSGP and flow through polishing [47]. Another challenge is the question how different unit operations in an integrated downstream process can be efficiently connected. Outlet streams from continuous chromatographic set-ups are usually semi-continuous and fluctuate in composition and product concentration. Currently, CSTRs are usually used between unit operations, which is handy as it decouples the flow and homogenize the process fluid and dampens fluctuations in the concentration profile [96]. A major downside for continuous processing however, is the broad RTD of CSTRs [97]. A possible alternative is the use of two alternating CSTRs, in which one CSTR feeds the subsequent unit operation while the other one collects and homogenizes the outlet of the prior unit operation [97]. This approach can also be run in series in which the volume of one CSTR is periodically transferred into a second one, which allows for a defined residence time, and was used for example for viral inactivation in an integrated continuous process [53]. New process designs, which do not only transfer the established batch process into a continuous one, but rethink the whole process from an integration perspective can help to minimize the use CSTRs. Flow through chromatography offers favourable characteristics for process integration and the outlet can be seamless processed on a second flow through step or directly captured on a subsequent PCC process [50, 56]

Fluctuations in the process streams caused by the periodic nature of continuous chromatographic unit operations and deviations in titre and impurities are also a big challenge in developing control and automation strategies for continuous processing. PAT and dynamic control strategies for continuous downstream processing are in its infancy and often fail to adequately address fluctuations in inlet stream concentrations which can lead to non-optimal process conditions and product loss [45, 96]. Relative values of impurity and product instead of absolute values as a base for control signal calculations have been shown to be a feasible approach to control PCC loading with fluctuating inlet stream concentrations [96]. The implementation of dynamic control strategies require real-time online data of complex feed streams, which is challenging and a main area of investigation in current research. Spectroscopic methods based on UV-, near infrared-, and Raman spectroscopy are investigated in combination with multi wavelength analysis and multivariate data analysis to predict mAb concentrations in the fluid streams, which can be used as control signals in PCC capturing [98–100]. On-line and at-line HPLC can give further information on a wide range of properties such as product concentration, aggregate content and impurities, although there is inevitable delay in data acquisition. Feidel et al. successfully implemented a process wide control of a continuous downstream process for mAbs, capable of adjusting the process to decreasing cell-specific productivities based on data generated from multiple at-line HPLCs [94]. Fully automated control of integrated continuous downstream processing is however far from reality and one of the major challenges that need to be solved to enable wide acceptance in academia and industry.

Conclusion

Interest in continuous downstream processing of biopharmaceuticals has risen in recent years. Switching to continuous production can not only decrease overall production costs, but also speed up manufacturing, lower batch-to-batch variations, and significant lower capital investments. Continuous manufacturing has the potential of being a game-changer for the biopharmaceutical industry, in particular in cost-competing markets such as biosimilars [6]. Several studies have shown the feasibility of continuous and integrated processes to produce antibodies from cell cultures. Key unit operations such as PCC and continuous polishing are well described in literature and can greatly intensify the process. However, continuous processing is still not widely accepted and implemented. One main reason is that continuous downstream processing is relatively complex. Multi-column systems require extensive tubing, complicated flow paths and special equipment. Integration of multiple unit operations, changing flow rates and flow compositions, which all interact with each other, further complicate processing. Another reason is that control and design of those processes is still in its infancy and experts in this field are rare. Wide acceptance will require simple and robust systems that are easy to implement and operate, and new process designs specifically for continuous processing need to be developed. Developments in model-based process design, PAT and artificial intelligence will play a crucial role to solve some of the before mentioned challenges and to bring biomanufacturing to Twenty-First Century Advanced Manufacturing status.

This chapter aimed to give an introduction into continuous downstream processing of biopharmaceuticals and describes the main approaches currently applied. It is inevitable incomplete, and the interested reader is referred to the latest literature [5, 6, 101]. As most required tools for continuous downstream processing are now being available this is a great topic to explore and develop new, creative approaches and to unlock the potential of continuous biomanufacturing.

Conflict of Interest The author declares no conflict of interest in writing this chapter.

References

1. Kent JA (2007) Kent and Riegel's handbook of industrial chemistry and biotechnology. Springer, New York
2. Bielser JM et al (2018) Perfusion mammalian cell culture for recombinant protein manufacturing – a critical review. Biotechnol Adv 36(4):1328–1340
3. FDA (2019) Quality considerations for continuous manufacturing guidance for industry
4. Karst DJ, Steinebach F, Morbidelli M (2018) Continuous integrated manufacturing of therapeutic proteins. Curr Opin Biotechnol 53:76–84
5. Gerstweiler L, Bi J, Middelberg APJ (2021) Continuous downstream bioprocessing for intensified manufacture of biopharmaceuticals and antibodies. Chem Eng Sci 231

6. Rathore AS, Thakur G, Kateja N (2022) Continuous integrated manufacturing for biopharmaceuticals: a new paradigm or an empty promise? Biotechnol Bioeng
7. Ltd. EB (2019) 25/08/2022. Available from: https://www.prnewswire.com/news-releases/enzene-opens-its-first-continuous-biologics-manufacturing-facility-with-a-promise-to-disrupt-the-mab-manufacturing-cost-300953759.html?tc=eml_cleartime
8. BiosanaPharma (2020) 25/08/2022. Available from: https://www.biosanapharma.com/news/Press-releases/4
9. Sanofi (2019) Sanofi opens its first digitally-enabled, continuous manufacturing facility; ushers in next generation of biotech manufacturing. 12/12/2022. Available from: https://www.sanofi.com/en/media-room/press-releases/2019/2019-10-15-05-00-00-1929399
10. Xu S et al (2017) Bioreactor productivity and media cost comparison for different intensified cell culture processes. Biotechnol Prog 33(4):867–878
11. Karst DJ et al (2017) Modulation and modeling of monoclonal antibody N-linked glycosylation in mammalian cell perfusion reactors. Biotechnol Bioeng 114(9):1978–1990
12. Xu S, Chen H (2016) High-density mammalian cell cultures in stirred-tank bioreactor without external pH control. J Biotechnol 231:149–159
13. Farid SS (2007) Process economics of industrial monoclonal antibody manufacture. J Chromatogr B Anal Technol Biomed Life Sci 848(1):8–18
14. Woodcock J (2014) Modernizing pharmaceutical manufacturing – continuous manufacturing as a key enabler. In: International symposium on continuous manufacturing of pharmaceuticals. FDA, Cambridge, MA
15. Konstantinov KB, Cooney CL (2015) White paper on continuous bioprocessing May 20–21 2014 continuous manufacturing symposium. J Pharm Sci 104(3):813–820
16. Hummel J et al (2019) Modeling the downstream processing of monoclonal antibodies reveals cost advantages for continuous methods for a broad range of manufacturing scales. Biotechnol J 14(2):e1700665
17. Cataldo AL et al (2020) Economics and ecology: modelling of continuous primary recovery and capture scenarios for recombinant antibody production. J Biotechnol 308:87–95
18. Carta G, Jungbauer A (2010) Protein chromatography
19. De Luca C et al (2020) Modern trends in downstream processing of biotherapeutics through continuous chromatography: the potential of multicolumn countercurrent solvent gradient purification. Trends Anal Chem 132:116051
20. Szepesy L et al (1975) Continuous liquid chromatography. J Chromatogr A 108(2):285–297
21. Gerhold DBBCG (1961) Continuous sorption process employing fixed bed of sorbent and moving inlets and outlets. Universal Oil Products
22. Aniceto JPS, Silva CM (2014) Simulated moving bed strategies and designs: from established systems to the latest developments. Sep Purif Rev 44(1):41–73
23. Kim K-M et al (2017) Advanced operating strategies to extend the applications of simulated moving bed chromatography. Chem Eng Technol 40(12):2163–2178
24. Seidel-Morgenstern A, Keßler LC, Kaspereit M (2008) New developments in simulated moving bed chromatography. Chem Eng Technol 31(6):826–837
25. Martinez Cristancho CA, Seidel-Morgenstern A (2016) Purification of single-chain antibody fragments exploiting pH-gradients in simulated moving bed chromatography. J Chromatogr A 1434:29–38
26. Fischer LM, Wolff MW, Reichl U (2018) Purification of cell culture-derived influenza A virus via continuous anion exchange chromatography on monoliths. Vaccine 36(22):3153–3160
27. Krober T et al (2013) Continuous purification of influenza virus using simulated moving bed chromatography. J Chromatogr A 1307:99–110
28. Paredes G et al (2005) Simulated moving bed operation for size exclusion plasmid purification. Chem Eng Technol 28(11):1335–1345
29. Nestola P et al (2014) Adenovirus purification by two-column, size-exclusion, simulated countercurrent chromatography. J Chromatogr A 1347:111–121

30. Nestola P et al (2015) Robust design of adenovirus purification by two-column, simulated moving-bed, size-exclusion chromatography. J Biotechnol 213:109–119
31. Hashimoto K, Adachi S, Shirai Y (1988) Continuous desalting of proteins with a simulated moving-bed adsorber. Agric Biol Chem 52(9):2161–2167
32. Rajendran A, Paredes G, Mazzotti M (2009) Simulated moving bed chromatography for the separation of enantiomers. J Chromatogr A 1216(4):709–738
33. Baur D et al (2016) Optimal model-based design of the twin-column CaptureSMB process improves capacity utilization and productivity in protein A affinity capture. Biotechnol J 11(1): 135–145
34. Steinebach F et al (2016) Model based adaptive control of a continuous capture process for monoclonal antibodies production. J Chromatogr A 1444:50–56
35. Godawat R et al (2012) Periodic counter-current chromatography – design and operational considerations for integrated and continuous purification of proteins. Biotechnol J 7(12): 1496–1508
36. Mahajan E, George A, Wolk B (2012) Improving affinity chromatography resin efficiency using semi-continuous chromatography. J Chromatogr A 1227:154–162
37. Heeter GA, Liapis AI (1995) Perfusion chromatography: performance of periodic countercurrent column operation and its comparison with fixed-bed operation. J Chromatogr A 711(1): 3–21
38. Baur D et al (2016) Comparison of batch and continuous multi-column protein A capture processes by optimal design. Biotechnol J 11(7):920–931
39. Baur D et al (2018) Model assisted comparison of protein A resins and multi-column chromatography for capture processes. J Biotechnol 285:64–73
40. Angarita M et al (2015) Twin-column CaptureSMB: a novel cyclic process for protein A affinity chromatography. J Chromatogr A 1389:85–95
41. Warikoo V et al (2012) Integrated continuous production of recombinant therapeutic proteins. Biotechnol Bioeng 109(12):3018–3029
42. Steinebach F, Muller-Spath T, Morbidelli M (2016) Continuous counter-current chromatography for capture and polishing steps in biopharmaceutical production. Biotechnol J 11(9): 1126–1141
43. Bisschops M (2014) BioSMB technology as an enabler for a fully continuous disposable biomanufacturing platform. In: Subramanian G (ed) Continuous processing in pharmaceutical manufacturing. Wiley-VCH Verlag GmbH & Co. KGaA, Weinheim
44. Somasundaram B et al (2018) Progression of continuous downstream processing of monoclonal antibodies: current trends and challenges. Biotechnol Bioeng 115(12):2893–2907
45. Lofgren A et al (2021) An integrated continuous downstream process with real-time control: a case study with periodic countercurrent chromatography and continuous virus inactivation. Biotechnol Bioeng 118(4):1664–1676
46. Godawat R et al (2015) End-to-end integrated fully continuous production of recombinant monoclonal antibodies. J Biotechnol 213:13–19
47. Steinebach F et al (2017) Design and operation of a continuous integrated monoclonal antibody production process. Biotechnol Prog 33(5):1303–1313
48. Arnold L et al (2019) Implementation of fully integrated continuous antibody processing: effects on productivity and COGm. Biotechnol J 14(2):e1800061
49. Klutz S et al (2015) Developing the biofacility of the future based on continuous processing and single-use technology. J Biotechnol 213:120–130
50. Gerstweiler L et al (2022) An integrated and continuous downstream process for microbial virus-like particle vaccine biomanufacture. Biotechnol Bioeng
51. Bramer C et al (2019) Optimization of continuous purification of recombinant patchoulol synthase from Escherichia coli with membrane adsorbers. Biotechnol Prog 35(4):e2812
52. Kateja N et al (2017) Integrated continuous processing of proteins expressed as inclusion bodies: GCSF as a case study. Biotechnol Prog 33(4):998–1009

53. Gjoka X, Gantier R, Schofield M (2017) Transfer of a three step mAb chromatography process from batch to continuous: optimizing productivity to minimize consumable requirements. J Biotechnol 242:11–18
54. Sun YN et al (2020) Model-based process development and evaluation of twin-column continuous capture processes with protein A affinity resin. J Chromatogr A 1625:461300
55. Pagkaliwangan M et al (2019) Optimized continuous multicolumn chromatography enables increased productivities and cost savings by employing more columns. Biotechnol J 14(2): e1800179
56. Ichihara T, Ito T, Gillespie C (2019) Polishing approach with fully connected flow-through purification for therapeutic monoclonal antibody. Eng Life Sci 19(1):31–36
57. Orbecido A et al (2019) Connected flow-through chromatography processes as continuous downstream processing of proteins. In: MATEC web of conferences, p 268
58. Walch N, Jungbauer A (2017) Continuous desalting of refolded protein solution improves capturing in ion exchange chromatography: a seamless process. Biotechnol J 12(6)
59. Gerstweiler L et al (2021) Comparative evaluation of integrated purification pathways for bacterial modular polyomavirus major capsid protein VP1 to produce virus-like particles using high throughput process technologies. J Chromatogr A 1639:461924
60. Krattli M, Muller-Spath T, Morbidelli M (2013) Multifraction separation in countercurrent chromatography (MCSGP). Biotechnol Bioeng 110(9):2436–2444
61. Vogg S et al (2019) Experimental evaluation of the impact of intrinsic process parameters on the performance of a continuous chromatographic polishing unit (MCSGP). Biotechnol J 14(7):e1800732
62. Ulmer N, Ristanovic D, Morbidelli M (2019) Process for continuous fab production by digestion of IgG. Biotechnol J 14(10):e1800677
63. Steinebach F et al (2017) Experimental design of a twin-column countercurrent gradient purification process. J Chromatogr A 1492:19–26
64. Aumann L, Morbidelli M (2007) A continuous multicolumn countercurrent solvent gradient purification (MCSGP) process. Biotechnol Bioeng 98(5):1043–1055
65. Aumann L, Morbidelli M (2008) A semicontinuous 3-column countercurrent solvent gradient purification (MCSGP) process. Biotechnol Bioeng 99(3):728–733
66. Luca C et al (2020) From batch to continuous chromatographic purification of a therapeutic peptide through multicolumn countercurrent solvent gradient purification. J Chromatogr A 1625:461304
67. Jungbauer A (2013) Continuous downstream processing of biopharmaceuticals. Trends Biotechnol 31(8):479–492
68. Patil NV et al (2015) Separation of an inulin mixture using cascaded nanofiltration. Sep Purif Technol 146:261–267
69. Rizki Z et al (2019) Oligosaccharides fractionation cascades with 3 outlet streams. Sep Purif Technol 221:183–194
70. Burgstaller D, Jungbauer A, Satzer P (2019) Continuous integrated antibody precipitation with two-stage tangential flow microfiltration enables constant mass flow. Biotechnol Bioeng 116(5):1053–1065
71. Toson P, Doshi P, Jajcevic D (2019) Explicit residence time distribution of a generalised cascade of continuous stirred tank reactors for a description of short recirculation time (bypassing). Processes 7(9)
72. Arunkumar A et al (2017) Investigation of single-pass tangential flow filtration (SPTFF) as an inline concentration step for cell culture harvest. J Membr Sci 524:20–32
73. Casey C et al (2011) Protein concentration with single-pass tangential flow filtration (SPTFF). J Membr Sci 384(1-2):82–88
74. Rucker-Pezzini J et al (2018) Single pass diafiltration integrated into a fully continuous mAb purification process. Biotechnol Bioeng 115(8):1949–1957
75. Jabra MG, Yehl CJ, Zydney AL (2019) Multistage continuous countercurrent diafiltration for formulation of monoclonal antibodies. Biotechnol Prog 35(4):e2810

76. Yehl CJ, Jabra MG, Zydney AL (2019) Hollow fiber countercurrent dialysis for continuous buffer exchange of high-value biotherapeutics. Biotechnol Prog 35(2):e2763
77. Hammerschmidt N et al (2015) Continuous precipitation of IgG from CHO cell culture supernatant in a tubular reactor. Biotechnol J 10(8):1196–1205
78. Hammerschmidt N et al (2014) Economics of recombinant antibody production processes at various scales: industry-standard compared to continuous precipitation. Biotechnol J 9(6): 766–775
79. Zelger M et al (2016) Real-time monitoring of protein precipitation in a tubular reactor for continuous bioprocessing. Process Biochem 51(10):1610–1621
80. Saxena AK, Nigam KDP (1984) Coiled configuration for flow inversion and its effect on residence time distribution. AICHE J 30(3):363–368
81. Kumar V, Nigam KDP (2005) Numerical simulation of steady flow fields in coiled flow inverter. Int J Heat Mass Transf 48(23-24):4811–4828
82. Mridha M, Nigam KDP (2008) Coiled flow inverter as an inline mixer. Chem Eng Sci 63(6): 1724–1732
83. Kateja N et al (2016) Continuous precipitation of process related impurities from clarified cell culture supernatant using a novel coiled flow inversion reactor (CFIR). Biotechnol J 11(10): 1320–1331
84. Kateja N et al (2018) Non-protein A purification platform for continuous processing of monoclonal antibody therapeutics. J Chromatogr A 1579:60–72
85. Hammerschmidt N, Hobiger S, Jungbauer A (2016) Continuous polyethylene glycol precipitation of recombinant antibodies: sequential precipitation and resolubilization. Process Biochem 51(2):325–332
86. Dutra G et al (2020) Continuous capture of recombinant antibodies by ZnCl (2) precipitation without polyethylene glycol. Eng Life Sci 20(7):265–274
87. Li Z et al (2019) Continuous precipitation for monoclonal antibody capture using countercurrent washing by microfiltration. Biotechnol Prog 35(6):e2886
88. Martins DL et al (2020) Truly continuous low pH viral inactivation for biopharmaceutical process integration. Biotechnol Bioeng 117(5):1406–1417
89. David L et al (2019) Virus study for continuous low pH viral inactivation inside a coiled flow inverter. Biotechnol Bioeng 116(4):857–869
90. David L et al (2020) Side-by-side comparability of batch and continuous downstream for the production of monoclonal antibodies. Biotechnol Bioeng 117(4):1024–1036
91. David L et al (2020) Simulation of pH level distribution inside a coiled flow inverter for continuous low pH viral inactivation. Biotechnol Bioeng 117(2):429–437
92. Sencar J et al (2020) A narrow residence time incubation reactor for continuous virus inactivation based on packed beds. New Biotechnol 55:98–107
93. Jungbauer A (2019) Continuous virus inactivation: how to generate a plug flow. Biotechnol J 14(2):e1800278
94. Feidl F et al (2020) Process-wide control and automation of an integrated continuous manufacturing platform for antibodies. Biotechnol Bioeng 117(5):1367–1380
95. Chen R et al (2022) Converting a mAb downstream process from batch to continuous using process modeling and process analytical technology. Biotechnol J 17(11):e2100351
96. Gerstweiler L et al (2022) Control strategy for multi-column continuous periodic counter current chromatography subject to fluctuating inlet stream concentration. J Chromatogr A 1667:462884
97. Sencar J, Hammerschmidt N, Jungbauer A (2020) Modeling the residence time distribution of integrated continuous bioprocesses. Biotechnol J 15(8):e2000008
98. Thakur G, Hebbi V, Rathore AS (2020) An NIR-based PAT approach for real-time control of loading in protein A chromatography in continuous manufacturing of monoclonal antibodies. Biotechnol Bioeng 117(3):673–686
99. Rudt M et al (2017) Real-time monitoring and control of the load phase of a protein A capture step. Biotechnol Bioeng 114(2):368–373

100. Rudt M, Briskot T, Hubbuch J (2017) Advances in downstream processing of biologics – spectroscopy: an emerging process analytical technology. J Chromatogr A 1490:2–9
101. Schwaminger SP, Zimmermann I, Berensmeier S (2022) Current research approaches in downstream processing of pharmaceutically relevant proteins. Curr Opin Biotechnol 77: 102768

Chapter 6
Towards Digital Twin for Biopharmaceutical Processes: Concept and Progress

Chaoying Ding, Ou Yang, and Marianthi Ierapetritou

Abstract The capabilities of digital twins (DT) and the successes they have demonstrated in various industries have attracted significant attention from the biopharmaceutical manufacturing field. The realization of DT could help reduce cost, improve productivity, maintain high and constant product quality, and meet Quality-by-Design (QbD) guidelines. Nonetheless, the development of DT is still at its infancy stage and an integrated DT in biopharma is rarely reported. To assist with further advancement in the development of DT, this book chapter gives an overview of the current state of key components involved in a digital twin for biopharmaceutical manufacturing. Currently available software for DT development, process analytical technologies (PATs) for data acquisition and information communication, various modeling strategies to construct the digital replica, and the integration between physical and virtual plants are comprehensively reviewed.

Keywords Digital twin · Biopharmaceutical manufacturing · Process modeling

Abbreviations

ANN	artificial neural network
CADET	Chromatography Analysis and Design Toolkit
CFD	computational fluid dynamic
CPP	critical process parameter

Chaoying Ding and Ou Yang contributed equally with all other contributors.

C. Ding · M. Ierapetritou (✉)
Department of Chemical and Biomolecular Engineering, University of Delaware, Newark, DE, USA
e-mail: cyding@udel.edu; mgi@udel.edu

O. Yang
Department of Chemical and Biochemical Engineering, Rutgers, The State University of New Jersey, Piscataway, NJ, USA
e-mail: oy21@scarletmail.rutgers.edu

R. Pörtner (ed.), *Biopharmaceutical Manufacturing*, Cell Engineering 11,
https://doi.org/10.1007/978-3-031-45669-5_6

CQA	critical process attributes
DCS	Distributed Control System
DDS	Data Distribution Service
DT	Digital Twins
ELN	Electronic Lab Notebook
FBA	flux balance analysis
FEM	finite element analysis
GRM	general rate model
GRM2D	two dimensional general rate model
HCP	host cell protein
HMI	Human-Machine Interface
HPLC	high-performance liquid chromatography
IF	Isolation Forest
IoT	Internet of things
LIMS	Laboratory Information and Management System
LSTM	Long Short-Term Memory autoencoder
mAb	monoclonal antibody
MES	Manufacturing and Execution System
MQTT	message Queuing Telemetry Transport
NIR	near-infrared
OPC	open Platform Communications
PAT	process analytical technology
PCA	principal component analysis
PID	proportional–integral–derivative
PLC	programmable logic controllers
PLM	Product Lifecycle Management
PLS	partial least square regression
QbD	Quality-by-Design
RCL	regulatory control layer
ROS	Robot Operating System
RPC	remote procedure calls
RTD	residence time distribution
SCADA	Supervisory control and data acquisition
SCL	supervisory control layer
SKIMpy	Symbolic Kinetic Models in Python
SMA	steric mass action
UA	Unified Architecture
UV-vis	UV-visible

Introduction

The concept of the Digital Twin (DT) can be traced back to 2002 when the "Conceptual Ideal for Product Lifecycle Management (PLM)" was first proposed by Dr. Grieves and presented to the industry at the University of Michigan [1–

3]. Since the inception of DT, this terminology has progressively evolved over time, but the basic idea of DT has barely changed. The most widely accepted interpretation comes from Glaessegen and Stargel, that a DT is a combination of multi-physics, multiscale, and probabilistic simulation for a complicated process with the utilization of physical data, sensors, and models to reflect the behavior of its corresponding physical twin [4, 5]. A complete DT is composed of three key elements: a physical component that currently exists or will exist in the physical world ("Physical Twin"), a virtual/digital part that exists in the virtual ("Digital Twin"), and the automated data and information exchange between the two components ("Digital Thread") [6, 7]. The first demonstration of the commercial application for a fully integrated DT was documented by General Electric (GE) in a patent application, in which "Predix" platform is developed for generating DT to supply historical and real-time process information and to run data analytics and optimization [7]. Nonetheless, from the perspective of practical application, simplified or partial DT is more favorable for the industry because it is much easier to realize in the actual production process. In particular, a digital shadow is one of the simplified DT concepts in which there is only a one-way flow from physical to digital component but no channel for reverse data communication [4, 8].

The digital twin is attracting more and more attention recently due to the rapid development of the Internet of things (IoT), Industry 4.0 technologies, advanced data analytics, and artificial intelligence [9, 10]. A complete DT could provide a comprehensive and holistic understanding of the process by constructing a network of dependencies between real-time data and their underlying meta information. Specifically, the establishment of DT provides an efficient way for achieving the cyber-physical integration of manufacturing practices, which could help obtain all accessible process information across different scales and sites and enable real-time information communication with the physical process control system [11, 12]. By virtue of the digital object like process analysis and optimization, *in silico* process model could be developed and updated in sync with data from the physical component to provide guidance for the actual procedures and further improve the overall efficiency and mitigate the risks in operation and maintenance. In general, the applications of the digital twins have demonstrated their capabilities in facilitating remote sensing, real-time monitoring, data acquisition, information exchange, process visualization, and process knowledge extraction [4, 13–16]. Integrated DTs have been applied in various industries including aerospace, semiconductor, energy production, smart product manufacturing, automotive transportation, and healthcare [5, 7, 10].

Tremendous growth has been witnessed in the biopharmaceutical market over the past few years, as well as the biological drug approvals by the EU and the US regulatory agencies [17–20]. Currently, the biopharmaceutical industry is under pressure to speed up the development of biological products, especially due to the COVID-19 pandemic [21]. However, the development of resource-intensive bioprocesses is invariably accompanied by prohibitive manufacturing costs, which makes biotherapeutics inaccessible to a significant part of the population [22]. To increase biologics production while maintaining the requirements for quality management and documentation, processes become increasingly complex and intensified, for example moving from the straightforward fed-batch bioreactor to more complicated perfusion processes [23, 24]. This shift increases the complexity of bioprocess and the number of process variables to be monitored. Moreover, the Quality-by-Design (QbD) initiative introduced by FDA to take product quality into account during the process design, is driving the biopharma industry to acquire deeper insights into the connection between bioprocess and bioproduct [25]. The implementation of digital twin in biopharmaceutical manufacturing could help accelerate the process development and gain thorough understanding of the process while reducing the efforts of performing time and resource intensive experiments [4, 26]. The realization of DT would allow for real-time monitoring through PATs, prediction of crucial quality attributes, identification of design space, and process optimization to provide real-time feedback and guide the operating procedures [17, 19].

Nowadays, with the development of simulation platforms and the fast-growing biological drugs market, biopharmaceutical companies have been working toward applying digital twin concepts during process development and commercial manufacturing. The advanced computational hardware expands the capabilities of modeling which allows faster and more reliable model building. To accelerate process scale-up and reduce experiments required, Pfizer built a two-phase bioreactor computational fluid dynamic (CFD) simulation to predict the transport and fluid dynamics inside of bioreactor from small to large scale. Their GPU workstation allows parallel computations which can complete massive simulations within a limited time and rule out nonideal physical conditions [27]. Biopharmaceutical companies intend to collaborate with digital transformation companies to build their digital twin system. For example, GlaxoSmithKline (GSK) cooperates with Siemens and ATOS to operationalize digital twins by building a real-time vaccine manufacturing simulation platform. The project aims to reduce the development times and optimizes the quality of their vaccine product [28, 29]. In addition, BioNTech also worked with Siemens to digitalize their facility to produce COVID-19 vaccine and reduce the transformation timeline from 1 year to 5 months by applying automation and digitalization technologies [30]. Sanofi announced a partnership with Dassault Systems in order to use their 3DEXPERIENCE platform to visualize and optimize operation activities. The collaboration covers the end-to-end data management system, virtual operation, process development scale-up and production industrialization [31]. From the above examples, it is clear that digital twin platforms contain key components including a robust modeling platform, data acquisition and centralized data management system. Control strategies should also be included for virtual plant and physical plant "communication" [32]. This book chapter reviews the current digital industrialization software available in the market and discusses modeling methodologies and PAT tools for biopharma digital twin development.

Current Software Available for Digital Twin Development

Digital twin system contains a physical plant and a virtual plant where the virtual plant receives data dynamically from the physical plant to achieve real-time monitoring, process analysis, optimization, and sends signals back to the physical plant for operation guidance or execution [33]. The virtual plant receives process data, meta data, customers' feedback, market requirements, investment plans and selects strategies for physical plant to execute. Thus, data acquisition and data storage systems, modeling and analysis software, control, and digital/cloud platforms are designed to achieve these tasks as illustrated in Fig. 6.1.

Data Flow and Management

In biomanufacturing, heterogeneous data are obtained with high dimensions [34]. In general, there are two types of data: time-series data and event-based data. Data including pH, temperature, and agitation speed sensors during the cell culture, can be generated dynamically representing time-series data. Event-based data refers to triggers, batch reports, and alarms which are also important to batch quality and operation activities. Data can be obtained by sensors, actuators, control equipment for example programmable logic controllers (PLC) or proportional–integral–derivative (PID) controllers, or online and offline testing equipment such as spectroscopy, BioProfile® FLEX2, high-performance liquid chromatography (HPLC) [19]. To transfer data from sensors or different measurement equipment to other machines

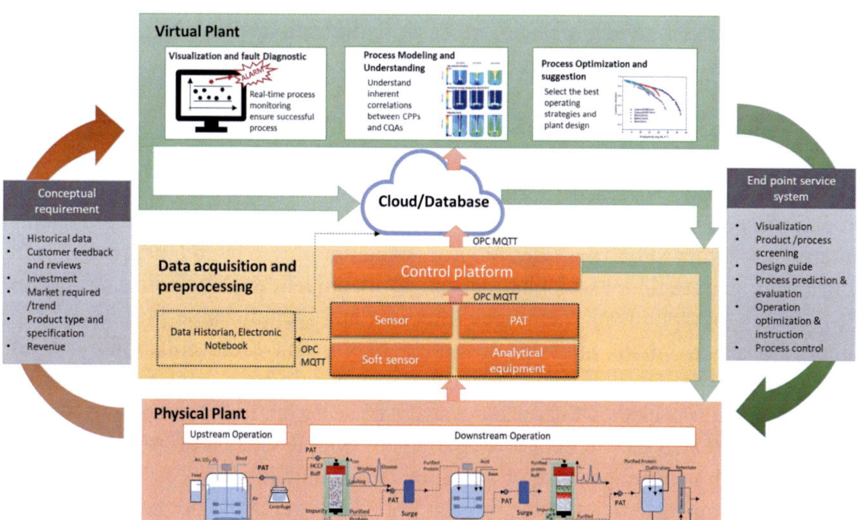

Fig. 6.1 Schematic illustration of different components of DT

or interfaces for visualizations and calculations, Open Platform Communications (OPC) server and Message Queuing Telemetry Transport (MQTT) are two widely used and dominant methods. The most recent version of OPC is OPC Unified Architecture (UA) which expands OPC compatibilities to different types of platforms for users. Other methods such as Robot Operating System (ROS) and Data Distribution Service (DDS) are also available. One benefit of OPC UA is that it supports a wide range of data types, including semantic data. Comparing the MQTT, OPC UA allows remote procedure calls (RPC). On the flip side, MQTT has the quality-of-service function while OPC UA does not include such feature. The comparisons between different servers can be found in [35].

OPC server or MQTT broker can connect to an OPC client or subscriber. This includes data management and storage system such as Electronic Lab Notebook (ELN), Laboratory Information and Management System (LIMS); control systems for example Supervisory control and data acquisition (SCADA), Manufacturing and Execution System (MES), Distributed Control System (DCS), Human-Machine Interface (HMI) [19, 36, 37]. The server can also connect to the data historian platform for real-time monitoring. Data historian, such as OSIsoft PI [19] centralizes the different data in a logical way, especially cloud-based data historian can store a large amount of data and can be integrated with other software for monitoring and analysis. A simplified data flow that integrates physical and virtual plant is shown in Fig. 6.1.

Process Modeling Software

Different types of modeling software bring a broad capability of digital twin in data analysis, operation guidance, and process optimization. Building models and performing data analysis for single unit operation and in-silico experiment saves cost and time resources during the process development stage. In addition, the development of flowsheet models brings opportunities for industries to build an integrated digital twin system.

In recent years, process simulation software tools are available specifically for biopharmaceutical operations. gPROMS FormulatedProducts from Process Systems Enterprise (Siemens PSE) provides bioprocessing libraries to build unit operations in both batch and continuous biopharmaceutical manufacturing, including bioreactor [38], chromatography [39], membrane filtration, and lyophilization [40]. Users can either choose an inbuilt mechanistic model or customize the equations to their specific processes. The software contains parameter estimation, sensitivity analysis, and optimization packages for scenario analysis, scale-up, and operation optimization [41]. gPROMS can integrate with Siemens STAR CCM+ or ANSYS Fluent software which is especially useful for bioreactor simulation in order to capture the fluid dynamic characteristics of the process [42]. Although it has not been applied to biopharmaceutical manufacturing yet, gPROMS Digital application platform and operational excellence solutions provide integrated software frameworks for data exchange, process monitoring, soft-sensing, and optimization [43].

SuperPro designer (Scotch Plains, NJ, USA) and BioSolve (Biopharm) are static modeling simulation tools that can capture mass balance, operation parameters, equipment, and scheduling in biopharmaceutical manufacturing [44]. One of the widely used functions of these two software tools is economic analysis for early-stage decision-making. These software platforms can evaluate different types of operating strategies (process cycles, duration, and production rate) in batch, continuous or hybrid operating modes by estimating their cost of goods, which supports to select the most cost-effective process [45–49]. SuperPro Designer enable integration with SchedulePro with labor, equipment, material, and utility information to reduce operating downtime and provides detailed schedules for target processes [50].

Using computational fluid dynamic (CFD) can help to understand complex multiphases systems, including bioreactor, chromatography, ultrafiltration and diafiltration [26, 51]. This type of model usually focuses on single-unit operation and results in high-fidelity models for process understanding, equipment selection, and scale-up activities. Software tools such as Ansys Fluent (Ansys, Inc., Canonsburg, PA), COMSOL Multiphysics, OpenFoam are used for CFD simulation in biopharmaceutical industry. COMSOL is based on finite element analysis (FMA) and can be used to solve multiphysics phenomena simultaneously. It contains model builder, application builder, and model manager functions that allow users to customize the model building, convert the model to an application and archive models [52]. COMSOL has been used for both upstream and downstream analysis, in order to compare different agitator geometries [53], simulate hydrodynamic regimes of flow in the bioreactor [54], and describe protein adsorption profiles in chromatography [55]. Ansys Fluent is also a user-friendly CFD simulation software that allows building advanced physics fluid dynamic models [56]. It has been used to understand oxygen mass transfer [57] and provide insights into shear [58], and mixing [59] in bioreactors and support scale-up [60]. It has also been used to predict fluid behavior and pressure profile for filtration processes [61]. The most recent version of Ansys Fluent released in July 2022 includes GPU technology and open-source accessibility which significantly reduces the computational time and improve the simulation flexibility. In addition, Ansys also provides Ansys-managed cloud service which allows uploading and visualizing the Ansys Fluent results on web client or apps [62]. OpenFOAM is an open-source, C++ object-oriented library for CFD simulation that runs in Linux, macOS and Windows systems [63, 64]. The software is highly customizable with source code available. The simulation is configured by text files that have high compatibility with other simulation software tools [65]. One drawback of OpenFoam is that there is no graphical user interface which brings a steep learning curve for beginners [66].

Biopharmaceutical development and manufacturing tend to generate large amount of information from materials and processes [67], especially with the development of PATs and high-throughput automated operations [68, 69]. In order to interpret the spectroscopic data and extract information, statistical and data analysis is indispensable to monitor the critical process attributes (CQAs) and critical process parameters (CPPs). Several software are available to help with this task. Many of them, such as JMP, Unscrambler, SIMCA do not require expertise is coding methodologies and offer user-friendly platform. Functions from these software including design of experiments, linear and nonlinear regression, multivariate analysis can be used to select CPPs among large scale of parameters and correlate CPP and CQAs for real-time monitoring and scale up [70]. For example, Kirdar et al. [71] use SIMCA-P+ to build partial least square regression (PLS) model for 2 L scale and 2000 L scale bioreactor to understand the change of cell performance among scales and perform batch control chart for fault diagnosis during manufacturing. Using Unscrambler software, Clavaud et al. [72] applied PLS method and process near infrared spectroscopy data to monitor osmolality, glucose, protein titer and cell density during bioreactor operation. The authors also applied principal component analysis method to understand inter-batch variability. Using JMP can create interactive visualization dashboard for real time process monitoring [73]. Unscrabler and SIMCA can connect to platform such as Aspen Process Pulse and SIMCA online [74] to achieve the same function.

In addition to the software mentioned above, modelers can also use MATLAB (MathWorks, Inc., Cambridge, UK) and Python to process data and build different unit operations in-silico. For example, Ding and Ierapetritou [75] used MATLAB to simulate multi-column periodic counter-current chromatography and applied surrogate-based feasibility analysis to obtain design space of the operation. Whelan et al. [76] used MATLAB together with Eigenvector PLS toolbox as chemometric data analysis tool to process Raman spectroscopy data and monitor glucose, glutamine, glutamate, lactate, ammonia, and cell density in fed-batch bioreactor in real-time. Weilandt et al. [77] established Symbolic Kinetic Models in Python (SKIMpy) toolbox to build large-scale kinetic models for bioreactor operation, which can support the understanding of metabolism and gene expression. MATLAB has packages such as App designer [78] and Simulink [79] which can be used for data visualization and control. On the other hand, Python as an open-source software can link to a python web-client for visualization and Python-based control software for process control [80, 81].

Digital/Cloud Platform Toward Digital Twin

Software companies develop different platforms to integrate data acquisition and analysis within virtual platforms [82]. In biopharmaceutical area, SIEMENS provides a series of products such as COMOS, gPROMS, and SIMIT to support data transfer, modeling and visualization [83]. COMOS allows the data hardware flows into the COMOS platform from hardware equipment [84] and then data can be transferred to gPROMS [85] or Star-CCM+ software [86] for biopharmaceutical upstream and downstream equipment design, process analysis and optimization [87]. Finally, the models can be deployed on the SIMIT platform for process training, virtual commissioning, and validation. ATOS collaborates with SIEMENS and develops an integrated process digital twin system specifically for pharmaceutical companies. The platform includes process sensors, PAT system, data centralization platforms, an automation and control system, predictive and preventive models, and finally 2D/3D monitoring and visualization platform with the support of a web application [88]. Dassault Systems provides "Made to Cure for Biopharma" solution based on 3DEXPERIENCE platform to analyze data and optimize production [89]. 3DEXPERIENCE provides a cloud-based environment that grants users access to data and virtual plants for process monitoring and data sharing. The platform integrates with the MODSIM system which has capabilities for data transfer, geometry clean-up, and model building [90].

Development of Digital Twin in Biopharma

In recent years, efforts have been done toward the development of digital twin in biopharmaceutical operations. To realize a digital twin for existed physical plant, a complete virtual plant can be built based on the software mentioned in the previous section and a physical-virtual plant communication need to be established. This section reviews the progress and research in biopharmaceutical operations from three perspectives, including process analytical technologies, process model building, and physical-virtual plant integration. Figure 6.2 depicts the digital twin platform including these three components, highlights the methodologies, and lists detailed inputs and outputs of each unit operation that can be measured and optimized.

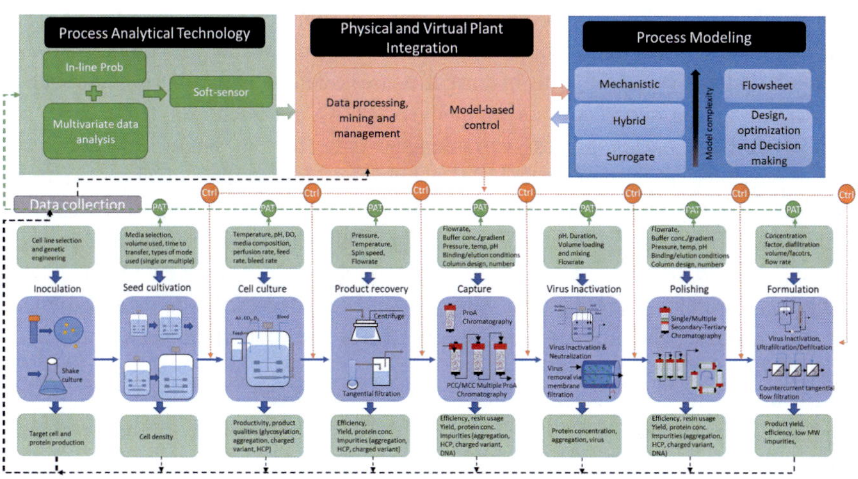

Fig. 6.2 A Fully integrated DT framework for biopharmaceutical manufacturing

PAT Tools

Another critical element in the achievement of DT is to obtain real-time information from the ongoing process and to translate the information into process knowledge and understanding of the process [91]. Accordingly, it is necessary to place well-suited sensors at appropriate locations for on-line monitoring to measure the operating variables and estimate the product attributes [17]. Process analytical technology initiative was introduced by FDA in 2004 [92] to change from off-line testing to real-time on-line testing and holistic process control. The implementation of PAT could help improve the overall process understanding and to ensure the pre-defined quality goal of biological products [93, 94].

In bioprocess monitoring, different types of in-line sensors are used to measure the different types of process parameters (physical, chemical, and biological parameters) [95, 96]. Details regarding sensors, techniques, and PAT installation can be found be in Maruthamuthu et al. [97]. Although in-line or near-line probes and systems are well developed, there are challenges in quickly detecting product-related contamination in subtle structural differences and determining the absence of virus loads [98]. Spectroscopic methods are important because of rapid and automated detection with no sample preparation/destruction, and improvements in measurability and accuracy [99]. The spectroscopic sensors could be combined with other non-spectroscopic measurements to measure multiple attributes and increase the process reliability [100–102].

More importantly, multivariate data analytics can be combined with physical sensors to interpret the spectroscopic data and extract information about the process- and product-related attributes [99]. This combination allows for on-line monitoring and is referred to as soft sensors which can be either model-based or data-driven [103]. The model-driven soft sensors mainly rely on the first-principle knowledge of the process. The application of this type of soft sensor is limited by the requirement of in-depth process knowledge and the increased computational complexity. The data-driven soft sensors are dependent on chemometric models, which correlate the readily measured variables with historical process data. Chemometric techniques, such as PLS and principal component analysis (PCA), are often used to describe the relationship between the process variables and data from measurements collected during previous or current runs [104]. However, due to the advances in data collection, storage, and management [105], machine learning techniques are applied to capture the correlation accurately. The application of artificial neural networks (ANNs), support vector regression, and Gaussian processes have been demonstrated in the literature [17, 106]. To deal with the issues related to the noisy and diluted signal acquisition, the Kalman filter approach has been incorporated into the soft sensors [96, 107].

Soft sensors to monitor critical process parameters have been implemented in the biopharmaceutical industry [95]. The spectroscopic application in the upstream production mainly focuses on the cell growth monitoring and culture component quantifications [4]. UV-visible (UV-vis) spectroscopy has been used for in-line measurements of cell density and amino acids [97, 108]. Near-infrared (NIR) technology is applied to measure biomass, metabolites (glucose, acetate, lactate, glutamate, etc.), cell density and viability [96, 109]. Two-dimensional fluorescence spectroscopy has also been used to analyze process parameters like biomass and product titer and to monitor metabolic activity [110]. Raman spectroscopy has also been employed to monitor the concentrations of glucose, glutamine, ammonia and biomass during the bioprocess [111]. For the downstream operations, the sensors should be able to monitor product and impurity concentrations [100]. It is noted that the applications of spectroscopic sensors in the downstream process are low compared to those in the upstream. Currently, UV-vis spectroscopy is the most commonly used technique in the downstream purification procedures for the measurement of protein concentration and quantification [17]. Moreover, this technology has been recently applied into the ultrafiltration/diafiltration process to monitor product concertation, particle sizes and buffer exchange [112]. Additionally, NIR has been applied to measure the mAb concentration before the inlet of column and the column breakthrough in real time at the column outlet. Those data were used by a Python-based control software to achieve the real-time control of loading into continuous Protein A chromatography [113]. Diffuse transmission fluorescence spectroscopy was also utilized to observe the fouling of Protein A resin [114]. In-line Raman spectroscopy was first applied in the downstream chromatography to monitor the breakthrough of the Protein A column [95].

Process Modeling

To realize DT in biopharmaceutical manufacturing, a complete representation of the physical plant is required, and that's where process modeling is very important. Process modeling, which transforms process knowledge into mathematical representations, could aid in reducing the experiment effort and gaining deeper understanding of the process behavior. Based on the types of unit operations and modeling objectives, different modeling approaches such mechanistic, surrogate and hybrid modeling are commonly employed. After the development of unit operation models, these models can be integrated together to construct the flowsheet model. In this section, current progress on process modeling in the biopharma field will be reviewed based on each type of modeling strategy.

Mechanistic Modeling

In biopharmaceutical manufacturing, protein-based, or mRNA-based products are produced using a series of production steps including a bioreactor followed by separation, purification steps and final formulation steps. For bioreactor modeling, mechanistic models can correlate operating conditions such as dissolved oxygen, pH, temperature, feeding strategy, media compositions to cell activities, metabolite concentrations, system homogeneity, and fluid dynamics. The models can predict and optimize process productivity, and product quality [99]. Inside of bioreactor, there is a complex system with multiphase transport phenomena, dynamically changed cell populations characterized by complicated bioreactions [115]. These phenomena can be captured by the structured or unstructured kinetic model, flux balance analysis model, and CFD simulation. The selective examples are shown in the Table 6.1 below to represent the capabilities of different modeling methods. It shows that the kinetic model can be majorly used to predict and optimize titer and product quality by adjusting the temperature downshift and feeding strategies. Feeding strategies include feeding frequency, amount [116] as well as additional components [117, 118]. Apart from titer and quality, Alhuthali and Kontoravdi [38] used the population balance model to predict the host cell protein (HCP) accumulation during the cell culture. However, kinetic model cannot include all metabolic reactions to avoid computational expensive calculations. In stoichiometric model (such as flux balance analysis (FBA) and metabolic network analysis), the main assumption is considering that the intracellular reaction is under a pseudo-steady state and net flux is zero [119]. Using this model fluxes can be estimated for metabolic cycles inside the cell to understand metabolic bottlenecks under different operating conditions [120–122]. To capture the cell dynamics, a dynamic metabolic flux analysis is developed, which discretizes the time domain to multiple intervals

Table 6.1 Applications of different mechanistic modeling methods

Models	Purposes	Reference
Kinetic model	Predict and optimize titer or quality by adjusting temperature shift and feeding strategy	[116–118, 210, 211]
	Improve process understanding of temperature and pH effect on product titer, quality attributes, and impurities for operation guidance and suggestions	[38, 212–214]
	Improve process understanding of metabolic cycles	[215]
Stoichiometric model	Culture media and feeding optimization for titer improvement.	[216, 217]
	Intracellular process understanding and metabolic bottlenecks under different conditions (temperature, pH, pCO2)	[120–122]
Stoichiometric model with Markov chain modeling	Cell line selection	[218]
Kinetic and Stoichiometric model integration	Accurately predict effect of temperature shift, seed density and nutrient concentration	[125]
CFD simulation	Oxygen transfer and understanding shear and mixing.	[60, 219]
CFD and kinetic model integration	Understand the effect of agitation on titer Couple kinetic and thermodynamic for process scale-up; Optimize feeding strategy, agitation and aeration to maximize titer	[58, 59, 220, 221]

and calculates the flux change during each of these intervals [123, 124]. Another approach is to integrate the stoichiometric model with the kinetic model [125]. As mentioned in the previous section, CFD simulation can be used to capture gas-liquid transfer and mixing heterogeneity of the bioreactor. CFD simulation can be integrated with kinetic modeling methods to predict titer under different operating conditions by using oxygen addition, nutrient heterogeneity, and agitation rate.

Among the downstream operations, column chromatography has played a dominant role in biopharmaceutical protein purification with high selectivity over the past two decades [126, 127]. Generally, Protein A chromatography is applied during the primary capture step to remove process-related impurities [128, 129], while different modes of chromatography (like ion-exchange, mix-mode, and hydrophobic chromatography) are utilized in the polishing step to further remove product-related and other residual impurities [130–132]. Current research on chromatography modeling has focused on the understanding of solute transport in moving phase (inter- and intraparticle) and kinetic adsorption in stationary phase [126]. With different assumptions and simplifications, there are different mechanistic models to describe mass transport of the moving phase inside the column, including equilibrium dispersive model, lumped kinetic model, and general rate model (GRM) [133–135]. Among them, general rate model is the most comprehensive one that accounts for convection and axial dispersion in the interstitial column volume, film mass transfer of solute from the interstitial to the intraparticle pore space, pore and/or surface diffusion within the resin particle, and protein adsorption [136, 137]. The transport model is coupled with a kinetic model to describe the adsorption phenomena that drives the chromatographic separation. There are different mathematical forms of kinetic expressions in the literature [126, 138], and among these, the Langmuir isotherm is the simplest and most widely used model to illustrate the intraparticle binding mechanism. Nonetheless, since the Langmuir model assumes independent binding sites and cannot account for the dependence of salt on the adsorption, the steric mass action (SMA) model, which incorporates the steric hindrance effects, is introduced for ion-exchange chromatography [138, 139]. After the mechanistic model building, parameter estimation and model validation are required to avoid overfitting and ensure the reliability and generality of the model. These developed mechanistic models can further contribute to the prediction of performance metrics like purity, yield and productivity, identification of the design space and process optimization [75, 130, 140–142]. Additionally, in order to capture the heterogeneity of internal flow in the radial direction of the column, CFD modeling is also applied to simulate the dynamic fluid flow [143, 144], but the application of CFD is limited due to the expensive computational cost, high knowledge requirement of the internal column geometry, and difficulty in experimental validation. It is worth noting that two dimensional general rate model (GRM2D) is recently developed by the CADET (Chromatography Analysis and Design Toolkit), which is an open-source package under the supervision of Dr. Eric von Lieres [145]. The newly built GRM2D model is adapted from GRM by introducing a radial coordinate to consider the inhomogeneity transport along the radial dimension [146].

In the biopharmaceutical downstream process, membrane-based unit operations like virus filtration and ultrafiltration are also indispensable [89, 147]. Virus filters are widely employed to ensure the safety of biotherapeutic proteins, and size-exclusion is normally considered the dominant mechanism for virus clearance [148]. The blocking model theory with varying blocking indexes combined with cake filtration model could accurately characterize the virus filtration behaviors such as the membrane fouling and filtration flux under different buffer conditions. In addition to size-exclusion mechanism, there exist other mechanisms like adsorption described by the Langmuir and Freundlich theories [148–153]. Ultrafiltration modules are typically used to adjust the concentration of the target protein component and the buffer media composition [154]. Reliable thermodynamic models have been developed to represent the ultrafiltration process, which is modeled via the combination of Stagnant Film Model for concentration polarization, Osmotic Pressure Model and Boundary Layer Model for the mass transfer through the membrane [26, 155, 156]. Binabaji et al. [157, 158] developed a mechanistic model to describe the behavior of ultrafiltration under highly concentrated antibody solution and validated the model through experimental efforts. This model, capturing the intermolecular interactions and back-filtration phenomenon at high concentration, could help increase the filtrate flux and maximum achievable concentration by evaluating the effect of membrane modules and operating protocols via simulation.

Surrogate Modeling

Mechanistic model suffers from the high computation power, and its accuracy highly relies on the obtained process knowledge. Surrogate model, also known as data-driven or black-box model, is an efficient approach to simulate the process with limited understanding and high complexity. With large amounts of data involved, the computationally cheap surrogate model in a lower dimensional space can be obtained to represent the original complicated process. Surrogate-based model has been applied both in the upstream and downstream operations in biopharmaceutical processes [159, 160].

In upstream, data-driven model has been used for cell-line selection [161], media optimization [162], process prediction, optimization, and scale up [163, 164]. Multivariate analysis has capabilities of handing large amount of data and information. Sokolov et al. [165] used PCA analysis to find correlations between pH, temperature shift, and media composition with quality attributes including aggregate, fragments, charge variants, and glycans. The authors also used PLS regression coupled with genetic algorithm to predict product titer and other quality attributes. Similarly Green and Glassey [166] use PLS and multi-way PLS to predict titer and quality attributes. The results show that online measurements such as pH and DO could not provide good prediction on quality attributes. Combing with amino acids data, the quality attributes could be predicted, which is not limited to lab scale but also manufacturing scale [167]. The predictability of the model can potentially be used as a control chart to identify the deviations during the operation [163]. One limitation of the PLS model is that it can only captures the linear correlations. To deal with this issue, modeler can choose machine learning methods such as ANNs, random forests, extreme gradient boosting, support vector machines, Gaussian processes etc. Alavijeh et al. [163] reviewed the advantages and disadvantages of each of the methods from the requirement of data source, model efficiency and robustness. Bashokouh et al. [162] used ANN with multilayer feed-forward structure to perform a predictive model of titer and then apply the optimization algorithm to find optimum amount of fetal bovine serum (FBS) in media, temperature and incubation time for maximum productivity.

The different applications of data-driven model have also been demonstrated in downstream unit operations like column chromatography [168, 169]. Wang et al. [170] applied ANNs modeling into protein chromatography for the estimation of adsorption and mass transfer parameters by mapping the *in silico* experimental chromatograms to the corresponding model parameters. Li et al. [171] developed two different types of surrogate model for efficient optimization of the simulated moving bed chromatography and compared the performance of original full-order and reduced-order models in terms of the computational time and accuracy. Additionally, a computational-efficient framework of surrogate-based feasibility analysis with adaptive sampling was employed to identify the design space of Protein A chromatography in our previous work [75]. Besides the application in chromatographic processes, ANNs and regression models are trained based on the CFD simulation data of the membrane spacers under a wide range of potential commercial designs and the resulting surrogate model could accurately predict pressure loss and mass transfer coefficients [172]. In order to capture the complicated fouling dynamics, Zhang et al. [173] utilized the data collected from a pilot-scale ultrafiltration membrane system to obtain three different data-driven models (linear regression, ANNs, and random forest), compared the performance of three modeling techniques, and optimized the backwash sequence timing.

Hybrid Modeling

Mechanistic and surrogate models could both be used to simulate the process, but they have different requirements for the process understanding and data. Mechanistic model requires a priori basis of knowledge about the process, and it is challenging to postulate appropriate assumptions and develop a precise model, but the model can be highly generalizable with physical or empirical interpretation of the process. Surrogate model is determined exclusively from data, so the quantity and quality of data significantly affect the model performance. Due to the lack of first-principle knowledge, the computationally efficient surrogate model has poor interpretability and generalizability. Thus, hybrid modeling is introduced to combine the advantages and alleviate the drawbacks of the two models. The missing knowledge in the mechanistic model can be represented by a surrogate model, constructing a hybrid model. It was observed that this strategy has the potential to significantly improve the model predictive performance [11, 174, 175].

The hybrid model of bioreactor modeling shows higher accuracy, extrapolation and interpolation capabilities than both the mechanistic and data-driven models [176]. Narayanan et al. [177] used kinetic models to build mass balance equations. Then the authors applied BlackBox-PLS2 and stepwise-PLS2 to predict rate constant from operating parameters and metabolite concentrations. This method integrates kinetic models and data-driven models. Similarly, Stosch et al. [94] also used hybrid model to analyze the effects of temperature, pH, cell concentration, and feed rate on productivity. Another approach is to have kinetic model and data-driven model connected in series. Kotidis and Kontoravdi [178] developed HyGlycoM model to predict glycan concentrations under different feeding strategies. The model composes two kinetic model for cell metabolism and nucleotide sugar donors (NSDs) and used the output – NSDs concentration as inputs of ANN model to estimate the glycosylation process. Instead of using machine learning methods, model-based design of experiment was applied to optimize media composition and feeding strategies. In this method, mechanistic model built in MATLAB is integrated with DesignExpert (AspenTech) a software for design of experiment and basic statistical analysis [179]. In this case, the DOE can run in silico which will significantly reduce the resource consumption for lab experiments and at the same time accelerate process development. Data driven model can be also integrated with stoichiometric model to understand metabolic shifts during the cell culture [180], and predict amino acid concentration [181], or quality attributes [182].

In addition to its application on the upstream bioreactors, hybrid modeling strategies are also applied in the chromatography and ultrafiltration procedures [183]. Narayanan et al. [184, 185] developed different hybrid models with varying degrees of process knowledge incorporated and evaluated their performance in terms of the interpolation and extrapolation capabilities, prediction accuracy and application into continuous chromatography system. Specifically, lumped kinetic model (LKM) is selected for the mass transport because the fluid dynamics inside the column are well described, while ANN is introduced to represent the missing knowledge in the adsorption dynamics. The developed hybrid model outperformed the mechanistic model in terms of prediction accuracy and robustness. To investigate the Protein A resin lifetime, Gaussian Processes were first applied to correlate the aging parameters with cleaning and loading conditions, followed by integrating the GP to the lumped kinetic model to construct the hybrid model [107]. The developed model could give insights into the aging mechanism, which could further provide guidance about how to extend the resin lifetime and decrease the cost of goods through process analysis and optimization. With respect to the ultrafiltration, Krippl et al. [186] established a hybrid model to predict the flux evolution and duration of ultrafiltration process under different operating conditions like various proteins, membrane types, input parameters and filtration modes. The prediction and robustness of this hybrid model are superior to the mechanistic film theory model, which could be further integrated into the virtual plant of digital twin. Thiess et al. [187] combined a regression model with a physicochemical model to characterize and quantify the effect of module geometries of ultrafiltration cassettes on mass transfer and pressure drop.

Flowsheet Model

Besides individually modeling the unit operations, flowsheet models could be built by integrating these single unit procedures. The resulting flowsheet models could help capture process dynamics under process variation, and further identify design space and optimum operating conditions through system analysis and optimization [188]. Flowsheet model is an essential part of a complete DT, but it is considerably challenging to develop a flowsheet that can fully represent the physical plant due to the high complexity of the integrated process and intense requirement of computational power resources. Consequently, current research of flowsheet development either focuses on the coupling the downstream unit operations, or build the integrated process model in the commercial simulation software [50].

Several attempts have been observed in building the integrated process model for downstream operations. Coupling with the control system, the mechanistic models of different chromatographic systems are integrated to construct a flowsheet model to realize real-time decision making and automatically optimize the operations conditions [189]. In order to save computational time, data-driven and hybrid models are employed to build the flowsheet mode as well. Given the upstream product and antigen-binding fragment manufacturing process, Liu et al. [190] developed an integrated data-driven model for multiscale optimization to find the optimum chromatography decisions for purification like the number of chromatography columns and their sizes, the number of cycles per batch, and the operational flow velocities. Zahel et al. [191] established an integrated process model for three chromatographic steps with regression approaches and then incorporated Monte Carlo simulation into the model, which can be used to estimate process capabilities, examine the product quality under process variations, and provide risk-based decision. Hybrid model is also developed by Pirrung et al. [192] to optimize the chromatography decisions in terms of the chromatography modes, operating sequence and conditions. Moreover, an integrated residence time distribution (RTD) model for an integrated continuous downstream process is developed to describe the inside mass flow, accelerate the start-up phase, and track the propagation of disturbances [193].

On the other hand, flowsheet model for the biopharmaceutical process has been constructed in different process simulators, such as SuperPro Designer and BioSolve [49, 194]. Although these modeling simulations are based on mass balance calculations and cannot provide detailed analyses of interconnections between different units, the simplified flowsheet models can help develop process design, evaluate various processing scenarios and achieve decision-making from the economic and environmental aspects, especially at the early-stage development. Petrides et al. [50] built an integrated flowsheet model for a batch biopharmaceutical process of mAb in SuperPro Designer for process scheduling, debottlenecking and economic analysis. Recently, a transition from batch to continuous biopharmaceutical operations has been noticed [89, 195, 196]. It is significant to assess the feasibility and viability of this transition and make decisions with the help of the flowsheet model. A lot of efforts have been made for the comprehensive comparison between different operating modes (such as batch, continuous and hybrid) and scenarios in terms of economic and ecological assessment [47, 49, 197–199].

Virtual and Physical Integration

The realization of a complete DT in biopharmaceutical manufacturing is rarely observed since it requires a highly automated system, complete dataflow integration, rigorous modeling system with self-learning capabilities, and a real-time automated control system. The previous sections introduced modeling from virtual plants and PAT methods from physical plants. Methods for virtual and physical plant integration will be introduced in this section.

Raw data cannot be directly used for data analysis and process modeling. Data pre-processing and data mining are required. Due to different types of data obtained during the process, data from online or off-line measurements can be generated with different frequencies. Missing data happens and causes analytical issues [145]. Ganadharan et al. [200] reviewed methodologies including conventional methods, imputation-based methods, likelihood-based methods and similarity methods to handle missing data. This review covered conventional methods by deleting or ignoring data based on the bio-expert's knowledge and modern methods such as using statistical, multivariate and machine learning methods to fill the gaps. In addition, reducing the dimension of the data and only maintaining critical information is important. Besides, outlier detection requirements, identifying the "golden batch" for model training is also relative to the model creditability. Tiwari et al. [201] applied PCA followed by Isolation Forest (IF) algorithm and Long Short-Term Memory autoencoder (LSTM) for chromatography data pre-processing and successfully predicted the column integrity.

To establish the digital twin, communications from virtual plant to physical plant need to be established. Gomis-Fons et al. [189] presented a model-based control system to monitor and control the integrated mAb production process on small scale. The work used Orbit, a research software from Lund University to communicate with UNICORN control software. UNICORN connects the pump and valve which is used to control the downstream process. Orbit receives results from the process model and sends commands to UNICORN to control flow rate and loading factors in chromatography. Feidl et al. [37] facilitated a SCADA system to control continuous monoclonal antibody (mAb) production by using SIPAT tool for data collection and centralization. The tool connected to MATLAB to build an integrated GUI and controller. The controller integrated the regulatory control layer (RCL) and supervisory control layer (SCL) to provide manual or automatic adjustment to the process. The researchers built CaptureSMB model and integrated it into the control system for automatic control and to ensure maximum capacity utilization. Most of the advanced control strategies reported in the literature are developed for single-unit operation. For example, Kim et al. [202] showed the application of a two-stage optimal control framework with model-based reinforcement learning in predictive control to optimize feeding strategy and maximize productivity in bioreactor. One highlight of this control strategy is to identify model-plant mismatches and real-time perturbations. For downstream, Thakur et al. [113] demonstrated a python-based monitoring and control system that integrates with spectroscopy PAT technology to control loading during chromatography operation. Instead of using conventional UV-based measurement, this PAT technology provides a faster response for more accurate real-time control. From the examples above, although fully automated integrated control has not been achieved in current biopharmaceutical manufacturing, a basic digital-physical integrated prototype system has been established and different model-based control strategies are under development for faster, dynamic, and more accurate control.

Challenges and Opportunities

The fast development of digital twin capabilities and the successful attempts in engineering applications such as aerospace have drawn significant attention from the biopharmaceutical industry and the associated policymakers. Although building blocks of digital twin (e.g., models and PAT tools) and policy guidelines (e.g., United States' Executive Order on Advancing Biotechnology and Biomanufacturing Innovation [203], China's Five-Year Plan for Development of Smart Manufacturing [204], and European Union's Digitizing European Industry Initiative [205]) are in place, several challenges and gaps still exist, inhibiting the adoption of the technology in biopharmaceutical industry.

In order to have an accurate virtual representation of digital twin in biopharmaceutical industry, having appropriate models to describe the physical biopharmaceutical manufacturing processes is critical for digital twin implementation. In current literature, even though various modeling methods have been developed and libraries of models are being incorporated in simulation software, many of the detailed models only target a single unit operation, rather than an integrated flowsheet. Part of the reason may be attributed to the high development and computational costs of these models, hindering the implementation of systems-based modeling framework and the use of integrated models. However, it must be noted that for a fully integrated digital twin to be functional, the different unit operations models of various degrees of complexity need to be properly connected, enabling efficient information flow from upstream to downstream. To resolve this challenge, there are opportunities in developing an end-to-end modeling platform [26, 37] that include key unit operations and auxiliary equipment (e.g., buffer preparation, cleaning-in-place), establishing efficient computational algorithms and reduced order modeling approaches (e.g., pre-analysis to reduce system dimensionality, hybrid modeling), and utilizing cloud and distributed computing resources.

Another key aspect in model development of biopharmaceutical manufacturing is that the process is subject to a high degree of uncertainty mainly due to the variability introduced based on the cells and their complex biological functions that are difficult to control. These uncertainties need to be accounted for and propagated in the modeling framework to ensure that the digital twin accurately reflects the state of process. Certain uncertainty propagation techniques have been applied in powder-based pharmaceutical manufacturing processes, but the applications in biopharmaceutical manufacturing area are rarely reported.

In addition, the developed models are highly dependent on the appropriate parameterization using data. In early pharmaceutical development stage, the available data may be biased as they are not collected for model development, interfering with the evaluation and validation of the developed models. Such constraint in data quantity and data quality often leads to the development of static models that only reflect a specific system's state. To tackle this challenge, streamlined data collection and model validation workflow needs to be established, along with model maintenance workflows [206].

In terms of process data, challenges are observed in data collection, data processing, and data storage. For data collection, the use of traditional sensors and PATs (e.g., NIR, Raman) have become ubiquitous. However, as data from the manufacturing cycle are generated from different sources, the collected data are oftentimes high dimensional and heterogeneous, so advanced data processing and analytics capabilities are needed to process all the data before usage. An increasing number of equipment and software vendors are adopting standard formats and communication protocols, and there are service providers that can facilitate the integration of data, but in general, the coordination and processing of manufacturing data is more of a business decision. For data storage, as the volume of data would increase dramatically after adopting digital twin in biopharmaceutical manufacturing, the data storage system should have large storage capability, along with fast and stable access of data for system analysis and data communications, and reliable security features.

Information security is an emerging challenge to the biopharmaceutical industry as the data collected and stored need to be compliant with regulatory guidelines [207] for data integrity purposes, and the system should be protected from cybersecurity risks such as digital intrusion and manipulation. This specific challenge has been noted in multiple national level governing policy documents and has led to opportunities in research and development of cyber-physical security systems.

With the virtual platform being developed, there are also challenges associated with the integration with the physical plant. Most of the applications in biopharmaceutical industry highlight the one-way data transfer from sensors in a single unit operation to the virtual platform, whereas the reverse is rarely seen except for few feed-back control cases. It is important to note that to accomplish the fully integrated biopharmaceutical digital twin, such information flow from virtual to physical component is also required to help identify abnormalities, conduct real-time optimization, and ensure process and plant-wide control.

Finally, being a highly regulated industry, perspectives from regulatory agencies towards digital twin implementation are an important consideration. With increasing efforts and successes of using modeling and simulation tools [208], regulatory bodies have demonstrated their evolving capabilities and expertise in this area. Recently, FDA has also published a draft guidance for industry on computer software assurance for production and quality system software, detailing some perspectives in validation and approvals of computer systems, making a key step forward in guiding the industry towards accepting digital twin components [209].

Declaration of Competing Interests The authors declare that they have no known competing financial interests or personal relationships that could have appeared to influence the work reported in this paper.

Acknowledgements The authors would like to acknowledge financial support from the U.S. Food and Drug Administration through grant U01FD007695A.

References

1. Grieves M (2022) Intelligent digital twins and the development and management of complex systems. Digital Twin 2(8):8. https://doi.org/10.12688/digitaltwin.17574.1
2. Grieves M, Vickers J (2017) Digital twin: mitigating unpredictable, undesirable emergent behavior in complex systems. In: Transdisciplinary perspectives on complex systems. Springer, pp 85–113
3. Grieves M (2014) Digital twin: manufacturing excellence through virtual factory replication. White Paper 1(2014):1–7
4. Chen Y, Yang O, Sampat C et al (2020) Digital twins in pharmaceutical and biopharmaceutical manufacturing: a literature review. Processes 8(9). https://doi.org/10.3390/pr8091088
5. Glaessgen E, Stargel D (2012) The digital twin paradigm for future NASA and US Air Force vehicles. In: 53rd AIAA/ASME/ASCE/AHS/ASC structures, structural dynamics and materials conference 20th AIAA/ASME/AHS adaptive structures conference 14th AIAA, 2012; p 1818
6. Kritzinger W, Karner M, Traar G et al (2018) Digital Twin in manufacturing: a categorical literature review and classification. IFAC-PapersOnLine 51(11):1016–1022. https://doi.org/10.1016/j.ifacol.2018.08.474
7. Fuller A, Fan Z, Day C et al (2020) Digital twin: enabling technologies, challenges and open research. IEEE Access 8:108952–108971. https://doi.org/10.1109/ACCESS.2020.2998358
8. Bao J, Guo D, Li J et al (2019) The modelling and operations for the digital twin in the context of manufacturing. Enterp Inf Syst 13(4):534–556. https://doi.org/10.1080/17517575.2018.1526324
9. Tao F, Qi Q, Wang L et al (2019) Digital twins and cyber–physical systems toward smart manufacturing and industry 4.0: correlation and comparison. Engineering 5(4):653–661. https://doi.org/10.1016/j.eng.2019.01.014
10. Tao F, Cheng J, Qi Q et al (2018) Digital twin-driven product design, manufacturing and service with big data. J Adv Manuf Technol 94(9):3563–3576. https://doi.org/10.1007/s00170-017-0233-1
11. Sokolov M, von Stosch M, Narayanan H et al (2021) Hybrid modeling—a key enabler towards realizing digital twins in biopharma? Curr Opin Chem Eng 34:100715. https://doi.org/10.1016/j.coche.2021.100715
12. Qi Q, Tao F, Zuo Y et al (2018) Digital twin service towards smart manufacturing. Procedia Cirp 72:237–242. https://doi.org/10.1016/j.procir.2018.03.103
13. Li X, Li D, Wan J et al (2017) A review of industrial wireless networks in the context of industry 4.0. Wirel Netw 23(1):23–41. https://doi.org/10.1007/s11276-015-1133-7

14. Uhlemann TH-J, Schock C, Lehmann C et al (2017) The digital twin: demonstrating the potential of real time data acquisition in production systems. Procedia Manuf 9:113–120. https://doi.org/10.1016/j.promfg.2017.04.043
15. Damiani L, Demartini M, Guizzi G et al (2018) Augmented and virtual reality applications in industrial systems: a qualitative review towards the industry 4.0 era. IFAC-PapersOnLine 51(11):624–630. https://doi.org/10.1016/j.ifacol.2018.08.388
16. Zhuang C, Liu J, Xiong H (2018) Digital twin-based smart production management and control framework for the complex product assembly shop-floor. J Adv Manuf Technol 96(1):1149–1163. https://doi.org/10.1007/s00170-018-1617-6
17. Narayanan H, Sponchioni M, Morbidelli M (2022) Integration and digitalization in the manufacturing of therapeutic proteins. Chem Eng Sci 248. https://doi.org/10.1016/j.ces.2021.117159
18. Walsh G (2018) Biopharmaceutical benchmarks 2018. Nat Biotechnol 36(12):1136–1145. https://doi.org/10.1038/nbt.4305
19. Gargalo C, Heras S, Jones M et al (2020) Towards the development of digital twins for the bio-manufacturing industry. Digital Twins 1–34. https://doi.org/10.1007/10_2020_142
20. Kaplon H, Muralidharan M, Schneider Z et al (2020) Antibodies to watch in 2020. MAbs 12(1):1703531. https://doi.org/10.1080/19420862.2019.1703531
21. Sokolov M (2020) Decision making and risk management in biopharmaceutical engineering—opportunities in the age of covid-19 and digitalization. Ind Eng Chem Res 59(40): 17587–17592. https://doi.org/10.1021/acs.iecr.0c02994
22. Khanal O, Lenhoff AM (2021) Developments and opportunities in continuous biopharmaceutical manufacturing. MAbs 13(1):1903664. https://doi.org/10.1080/19420862.2021.1903664
23. Bielser JM, Wolf M, Souquet J et al (2018) Perfusion mammalian cell culture for recombinant protein manufacturing – a critical review. Biotechnol Adv 36(4):1328–1340. https://doi.org/10.1016/j.biotechadv.2018.04.011
24. Jones W, Gerogiorgis DI (2022) Dynamic simulation, optimisation and economic analysis of fed-batch vs. perfusion bioreactors for advanced mAb manufacturing. Comput Chem Eng 165. https://doi.org/10.1016/j.compchemeng.2022.107855
25. Del Val IJ, Kontoravdi C, Nagy JM (2010) Towards the implementation of quality by design to the production of therapeutic monoclonal antibodies with desired glycosylation patterns. Biotechnol Prog 26(6):1505–1527. https://doi.org/10.1002/btpr.470
26. Smiatek J, Jung A, Bluhmki E (2020) Towards a digital bioprocess replica: computational approaches in biopharmaceutical development and manufacturing. Trends Biotechnol 38(10): 1141–1153. https://doi.org/10.1016/j.tibtech.2020.05.008
27. Pfizer (2020) M-Star – how Pfizer leveraged digital twins to create a process scale-up roadmap. https://mstarcfd.com/wp-content/uploads/2021/12/M-Star_Pfizer-CaseStudy.pdf. Accessed 10 Dec 2022
28. Siemens (2021). Stepping up the pace in vaccine development and production. https://www.siemens.com/global/en/company/stories/industry/2021/pharma-vaccine-digitalization.html. Accessed 10 Dec 2022
29. GlaxoSmithKline (2022). Digital twin: using advanced technology to accelerate vaccine development. https://www.gsk.com/en-gb/behind-the-science-magazine/digital-twin-using-advanced-technology-to-accelerate-vaccine-development/. Accessed 10 Dec 2022
30. SIEMENS (2021), Siemens accelerated BioNTech Covid-19 vaccine production setup with automation and digitalization solutions. https://new.siemens.com/th/en/company/press-centre/siemens-accelerated-biontech-covid-19-vaccine-production-setup-.html. Accessed 10 Dec 2022
31. DassaultSystèmes (2022), Dassault Systèmes partners with Sanofi to optimize tech transfer and industrialization at its future "EVolutive Facilities". https://investor.3ds.com/news-releases/news-release-details/dassault-systemes-partners-sanofi-optimize-tech-transfer-and. Accessed 10 Dec 2022

32. Jones D, Snider C, Nassehi A et al (2020) Characterising the digital twin: a systematic literature review. CIRP J Manuf Sci Technol 29:36–52. https://doi.org/10.1016/j.cirpj.2020.02.002
33. Cimino C, Negri E, Fumagalli L (2019) Review of digital twin applications in manufacturing. Comput Ind 113:103130. https://doi.org/10.1016/j.compind.2019.103130
34. Hong MS, Severson KA, Jiang M et al (2018) Challenges and opportunities in biopharmaceutical manufacturing control. Comput Chem Eng 110:106–114. https://doi.org/10.1016/j.compchemeng.2017.12.007
35. Profanter S, Tekat A, Dorofeev K, et al. (2019) OPC UA versus ROS, DDS, and MQTT: performance evaluation of industry 4.0 protocols. 2019 IEEE International Conference on Industrial Technology (ICIT), 955–962. https://doi.org/10.1109/ICIT.2019.8755050
36. OPCDataHub (2010) Introduction of OPC. https://www.opcdatahub.com/WhatIsOPC.html#:~:text=The%20OPC%20client%20software%20is,and%20more%20options%20for%20users. Accessed 10 Dec 2022
37. Feidl F, Vogg S, Wolf M et al (2020) Process-wide control and automation of an integrated continuous manufacturing platform for antibodies. Biotechnol Bioeng 117(5):1367–1380. https://doi.org/10.1002/bit.27296
38. Alhuthali S, Kontoravdi C (2022) Population balance modelling captures host cell protein dynamics in CHO cell cultures. PLoS One 17(3):e0265886. https://doi.org/10.1371/journal.pone.0265886
39. Close EJ, Salm JR, Bracewell DG et al (2014) Modelling of industrial biopharmaceutical multicomponent chromatography. Chem Eng Res Des 92(7):1304–1314. https://doi.org/10.1016/j.cherd.2013.10.022
40. De-Luca R, Bano G, Tomba E et al (2020) Accelerating the development and transfer of freeze-drying operations for the manufacturing of biopharmaceuticals by model-based design of experiments. Ind Eng Chem Res 59(45):20071–20085. https://doi.org/10.1021/acs.iecr.0c03115
41. Siemens PSE (2022) Overview of gPROMS formulated products's bioprocessing libraries https://www.psenterprise.com/sectors/biopharmaceuticals/bioprocessing. Accessed 10 Dec 2022
42. Siemens PSE (2022), Hybrid multizonal gPROMS–CFD. https://www.psenterprise.com/products/gproms/hybrid-multizonal. Accessed 10 Dec 2022
43. Siemens PSE (2022) gPROMS digital application platform. https://www.psenterprise.com/products/gproms/digital-applications/platform. Accessed 10 Dec 2022
44. Intelligen, Inc (2022) SuperPro designer overview. https://www.intelligen.com/products/superpro-overview/. Accessed 10 Dec 2022
45. Pleitt K, Somasundaram B, Johnson B et al (2019) Evaluation of process simulation as a decisional tool for biopharmaceutical contract development and manufacturing organizations. Biochem Eng J 150:107252. https://doi.org/10.1016/j.bej.2019.107252
46. Varadaraju H, Schneiderman S, Zhang L et al (2011) Process and economic evaluation for monoclonal antibody purification using a membrane-only process. Biotechnol Prog 27(5):1297–1305. https://doi.org/10.1002/btpr.639
47. Xenopoulos A (2015) A new, integrated, continuous purification process template for monoclonal antibodies: process modeling and cost of goods studies. J Biotechnol 213:42–53. https://doi.org/10.1016/j.jbiotec.2015.04.020
48. Walther J, Godawat R, Hwang C et al (2015) The business impact of an integrated continuous biomanufacturing platform for recombinant protein production. J Biotechnol 213:3–12. https://doi.org/10.1016/j.jbiotec.2015.05.010
49. Yang O, Prabhu S, Ierapetritou M (2019) Comparison between batch and continuous monoclonal antibody production and economic analysis. Ind Eng Chem Res 58(15):5851–5863. https://doi.org/10.1021/acs.iecr.8b04717
50. Petrides D, Carmichael D, Siletti C et al (2014) Biopharmaceutical process optimization with simulation and scheduling tools. Bioengineering (Basel) 1(4):154–187. https://doi.org/10.3390/bioengineering1040154

51. Babi DK, Griesbach J, Hunt S et al (2022) Opportunities and challenges for model utilization in the biopharmaceutical industry: current versus future state. Curr Opin Chem Eng 36: 100813. https://doi.org/10.1016/j.coche.2022.100813
52. COMSOL (2022) COMSOL Multiphysics® simulation software. https://www.comsol.com/comsol-multiphysics. Accessed 10 Dec 2022
53. Buss A, Suleiko A, Rugele K et al. (2017) CFD analysis of a stirred vessel bioreactor with double pitch blade and rushton type impellers. In Proceedings of the 2017 COMSOL conference in Rotterdam, 2017
54. Zarei S, Mousavi SM, Amani T et al (2021) Three-dimensional CFD simulation of anaerobic reactions in a continuous packed-bed bioreactor. Renew Energy 169:461–472. https://doi.org/10.1016/j.renene.2021.01.029
55. Gerontas S, Shapiro MS, Bracewell DG (2013) Chromatography modelling to describe protein adsorption at bead level. J Chromatogr A 1284:44–52. https://doi.org/10.1016/j.chroma.2013.01.102
56. ANSYS (2022) Overview of Ansys fluent fluid simulation software. https://www.ansys.com/products/fluids/ansys-fluent. Accessed 10 Dec 2022
57. Seidel S, Maschke RW, Werner S et al (2021) Oxygen mass transfer in biopharmaceutical processes: numerical and experimental approaches. Chemie Ingenieur Technik 93(1–2): 42–61. https://doi.org/10.1002/cite.202000179
58. Liu Y, Wang Z-J, Xia J-Y et al (2016) Application of Euler–Lagrange CFD for quantitative evaluating the effect of shear force on Carthamus tinctorius L. cell in a stirred tank bioreactor. Biochem Eng J 114:209–217. https://doi.org/10.1016/j.bej.2016.07.006
59. Farzan P, Ierapetritou MG (2017) Integrated modeling to capture the interaction of physiology and fluid dynamics in biopharmaceutical bioreactors. Comput Chem Eng 97:271–282. https://doi.org/10.1016/j.compchemeng.2016.11.037
60. Cappello V, Plais C, Vial C et al (2021) Scale-up of aerated bioreactors: CFD validation and application to the enzyme production by Trichoderma reesei. Chem Eng Sci 229:116033. https://doi.org/10.1016/j.ces.2020.116033
61. Radoniqi F, Zhang H, Bardliving CL et al (2018) Computational fluid dynamic modeling of alternating tangential flow filtration for perfusion cell culture. Biotechnol Bioeng 115(11): 2751–2759. https://doi.org/10.1002/bit.26813
62. ANSYS (2022) Ansys Cloud Direct Cloud-based engineering Simulation. https://www.ansys.com/products/cloud/ansys-cloud. Accessed 10 Dec 2022
63. OpenFOAM (2022), Overview of OpenFOAM capabilities. https://openfoam.org/. Accessed 10 Dec 2022
64. Jasak H (2009) OpenFOAM: open source CFD in research and industry. Int J Naval Arch Ocean Eng 1(2):89–94. https://doi.org/10.2478/IJNAOE-2013-0011
65. Ansoni JL, Seleghim P Jr (2016) Optimal industrial reactor design: development of a multiobjective optimization method based on a posteriori performance parameters calculated from CFD flow solutions. Adv Eng Softw 91:23–35. https://doi.org/10.1016/j.advengsoft.2015.08.008
66. Medina H, Beechook A, Saul J et al (2015) Open source computational fluid dynamics using OpenFOAM. In: Royal Aeronautical Society, general aviation conference, London
67. Banner M, Alosert H, Spencer C et al (2021) A decade in review: use of data analytics within the biopharmaceutical sector. Curr Opin Chem Eng 34:100758. https://doi.org/10.1016/j.coche.2021.100758
68. Manahan M, Nelson M, Cacciatore JJ et al (2019) Scale-down model qualification of ambr® 250 high-throughput mini-bioreactor system for two commercial-scale mAb processes. Biotechnol Prog 35(6):e2870. https://doi.org/10.1002/btpr.2870
69. Velugula-Yellela SR, Williams A, Trunfio N et al (2018) Impact of media and antifoam selection on monoclonal antibody production and quality using a high throughput micro-bioreactor system. Biotechnol Prog 34(1):262–270. https://doi.org/10.1002/btpr.2575

70. Ündey C, Ertunç S, Mistretta T et al (2010) Applied advanced process analytics in biopharmaceutical manufacturing: challenges and prospects in real-time monitoring and control. J Process Control 20(9):1009–1018. https://doi.org/10.1016/j.jprocont.2010.05.008
71. Kirdar AO, Conner JS, Baclaski J et al (2007) Application of multivariate analysis toward biotech processes: case study of a cell-culture unit operation. Biotechnol Prog 23(1):61–67. https://doi.org/10.1021/bp060377u
72. Clavaud M, Roggo Y, Von Daeniken R et al (2013) Chemometrics and in-line near infrared spectroscopic monitoring of a biopharmaceutical Chinese hamster ovary cell culture: prediction of multiple cultivation variables. Talanta 111:28–38. https://doi.org/10.1016/j.talanta.2013.03.044
73. JMP (2022) Introduction of JMP: data analysis software for Mac and Windows. https://www.jmp.com/en_us/software/data-analysis-software.html. Accessed 10 Dec 2022
74. SARTORIUS (2022) Real-time process intelligence to raise your manufacturing IQ. https://www.sartorius.com/en/products/process-analytical-technology/data-analytics-software/real-time-process-monitoring-software/simca-online. Accessed 10 Dec 2022
75. Ding C, Ierapetritou M (2021) A novel framework of surrogate-based feasibility analysis for establishing design space of twin-column continuous chromatography. Int J Pharm 609:121161. https://doi.org/10.1016/j.ijpharm.2021.121161
76. Whelan J, Craven S, Glennon B (2012) In situ Raman spectroscopy for simultaneous monitoring of multiple process parameters in mammalian cell culture bioreactors. Biotechnol Prog 28(5):1355–1362. https://doi.org/10.1002/btpr.1590
77. Weilandt DR, Salvy P, Masid M et al. (2022) Symbolic Kinetic Models in Python (SKiMpy): intuitive modeling of large-scale biological kinetic models. bioRxiv 2022.01.17.476618. https://doi.org/10.1101/2022.01.17.476618
78. MathWorks (2022) Introduction of MATLAB app designer. https://www.mathworks.com/products/matlab/app-designer.html. Accessed 10 Dec 2022
79. Boojari MA, Perra S, Colombo G et al (2022) Dynamic modeling and control of a continuous biopharmaceutical manufacturing plant. In: Optimization of pharmaceutical processes. Springer, pp 323–353
80. Nikita S, Thakur G, Jesubalan NG et al (2022) AI-ML applications in bioprocessing: ML as an enabler of real time quality prediction in continuous manufacturing of mAbs. Comput Chem Eng 164:107896. https://doi.org/10.1016/j.compchemeng.2022.107896
81. Liu Y, Zhang C, Chen J et al (2022) A fully integrated online platform for real time monitoring of multiple product quality attributes in biopharmaceutical processes for monoclonal antibody therapeutics. J Pharm Sci 111(2):358–367. https://doi.org/10.1016/j.xphs.2021.09.011
82. Sultanow E, Chircu A (2019) A review of IoT technologies, standards, tools, frameworks and platforms. In: The Internet of Things in the industrial sector, Springer International Publishing, pp 3–34
83. SIEMENS (2022) Digital twin and simulation. https://new.siemens.com/global/en/markets/pharmaceutical-life-science-industries/pharma-industry/focus-topics/digital-twin.html?gclid=EAIaIQobChMIg9Xi0oLa-wIVB8DICh3MMQYXEAMYAiAAEgI-IPD_BwE&acz=1. Accessed 10 Dec 2022
84. SIEMENS (2022) Efficient immersive trainings in a virtual environment for plant operators. https://new.siemens.com/global/en/products/automation/industry-software/plant-engineering-software-comos/virtual-reality-training.html. Accessed 10 Dec 2022
85. SIEMENS (2022) gPROMS – digital process twin technology https://new.siemens.com/global/en/products/automation/industry-software/gproms-digital-process-design-and-operations.html. Accessed 10 Dec 2022
86. SIEMENS (2022) Engineer innovation with multiphysics computational fluid dynamics (CFD) simulation. https://www.plm.automation.siemens.com/global/en/products/simcenter/STAR-CCM.html. Accessed 10 Dec 2022

87. SIEMENS (2022) Virtual commissioning and operator training with SIMIT. https://new.siemens.com/global/en/products/automation/industry-software/simit.html#Areasofapplication forSIMIT). Accessed 10 Dec 2022
88. Atos (2022) Process digital twin for pharma. https://atos.net/en/industries/healthcare-life-sciences/pharma-digital-twin. Accessed 10 Dec 2022
89. Erickson J, Baker J, Barrett S et al (2021) End-to-end collaboration to transform biopharmaceutical development and manufacturing. Biotechnol Bioeng 118(9):3302–3312. https://doi.org/10.1002/bit.27688
90. Dassault Systèmes (2022) Science-based modeling and simulation. https://www.3ds.com/3dexperience/integrated-modeling-and-simulation. Accessed 10 Dec 2022
91. Schmidt A, Helgers H, Lohmann LJ et al (2022) Process analytical technology as key-enabler for digital twins in continuous biomanufacturing. J Chem Technol Biotechnol 97(9): 2336–2346. https://doi.org/10.1002/jctb.7008
92. Food Drug Administration (2004) Guidance for industry, PAT-A framework for innovative pharmaceutical development, manufacturing and quality assurance. https://www.fda.gov/media/71012/download. Accessed 10 Dec 2022
93. Stosch M, Davy S, Francois K et al (2014) Hybrid modeling for quality by design and PAT-benefits and challenges of applications in biopharmaceutical industry. Biotechnol J 9(6):719–726. https://doi.org/10.1002/biot.201300385
94. Stosch M, Hamelink J-M, Oliveira R (2016) Hybrid modeling as a QbD/PAT tool in process development: an industrial E. coli case study. Bioprocess Biosyst Eng 39(5):773–784. https://doi.org/10.1007/s00449-016-1557-1
95. Claßen J, Aupert F, Reardon KF et al (2017) Spectroscopic sensors for in-line bioprocess monitoring in research and pharmaceutical industrial application. Anal Bioanal Chem 409(3): 651–666. https://doi.org/10.1007/s00216-016-0068-x
96. Biechele P, Busse C, Solle D et al (2015) Sensor systems for bioprocess monitoring. Eng Life Sci 15(5):469–488. https://doi.org/10.1002/elsc.201500014
97. Maruthamuthu MK, Rudge SR, Ardekani AM et al (2020) Process analytical technologies and data analytics for the manufacture of monoclonal antibodies. Trends Biotechnol 38(10): 1169–1186. https://doi.org/10.1016/j.tibtech.2020.07.004
98. Bakeev KA (2010) Process analytical technology: spectroscopic tools and implementation strategies for the chemical and pharmaceutical industries. Wiley
99. Chopda V, Gyorgypal A, Yang O et al (2021) Recent advances in integrated process analytical techniques, modeling, and control strategies to enable continuous biomanufacturing of monoclonal antibodies. J Chem Technol Biotechnol 97(9):2317–2335. https://doi.org/10.1002/jctb.6765
100. Rolinger L, Rudt M, Hubbuch J (2020) A critical review of recent trends, and a future perspective of optical spectroscopy as PAT in biopharmaceutical downstream processing. Anal Bioanal Chem 412(9):2047–2064. https://doi.org/10.1007/s00216-020-02407-z
101. Walch N, Scharl T, Felföldi E et al (2019) Prediction of the quantity and purity of an antibody capture process in real time. Biotechnol J 14(7):1800521. https://doi.org/10.1002/biot.201800521
102. Sauer DG, Melcher M, Mosor M et al (2019) Real-time monitoring and model-based prediction of purity and quantity during a chromatographic capture of fibroblast growth factor 2. Biotechnol Bioeng 116(8):1999–2009. https://doi.org/10.1002/bit.26984
103. Mandenius CF, Gustavsson R (2015) Mini-review: soft sensors as means for PAT in the manufacture of bio-therapeutics. J Chem Technol Biotechnol 90(2):215–227. https://doi.org/10.1002/jctb.4477
104. Pomerantsev AL, Rodionova OY (2012) Process analytical technology: a critical view of the chemometricians. J Chemom 26(6):299–310. https://doi.org/10.1002/cem.2445
105. Narayanan H, Luna MF, von Stosch M et al (2020) Bioprocessing in the digital age: the role of process models. Biotechnol J 15(1):1900172. https://doi.org/10.1002/biot.201900172

106. Gopakumar V, Tiwari S, Rahman I (2018) A deep learning based data driven soft sensor for bioprocesses. Biochem Eng J 136:28–39. https://doi.org/10.1016/j.bej.2018.04.015
107. Feidl F, Luna MF, Podobnik M et al (2020) Model based strategies towards protein A resin lifetime optimization and supervision. J Chromatogr A 1625:461261. https://doi.org/10.1016/j.chroma.2020.461261
108. Ude C, Schmidt-Hager J, Findeis M et al (2014) Application of an online-biomass sensor in an optical multisensory platform prototype for growth monitoring of biotechnical relevant micro-organism and cell lines in single-use shake flasks. Sensors 14(9):17390–17405. https://doi.org/10.3390/s140917390
109. Classen J, Aupert F, Reardon KF et al (2017) Spectroscopic sensors for in-line bioprocess monitoring in research and pharmaceutical industrial application. Anal Bioanal Chem 409(3):651–666. https://doi.org/10.1007/s00216-016-0068-x
110. Bonk S, Sandor M, Rüdinger F et al. (2011) In-situmicroscopy and 2D fluorescence spectroscopy as online methods for monitoring CHO cells during cultivation. In BMC proceedings, Springer, 2011, pp 1–3
111. Berry BN, Dobrowsky TM, Timson RC et al (2016) Quick generation of Raman spectroscopy based in-process glucose control to influence biopharmaceutical protein product quality during mammalian cell culture. Biotechnol Prog 32(1):224–234. https://doi.org/10.1002/btpr.2205
112. Rolinger L, Rüdt M, Diehm J et al (2020) Multi-attribute PAT for UF/DF of proteins—monitoring concentration, particle sizes, and buffer exchange. Anal Bioanal Chem 412(9):2123–2136. https://doi.org/10.1007/s00216-019-02318-8
113. Thakur G, Hebbi V, Rathore AS (2020) An NIR-based PAT approach for real-time control of loading in Protein A chromatography in continuous manufacturing of monoclonal antibodies. Biotechnol Bioeng 117(3):673–686. https://doi.org/10.1002/bit.27236
114. Pathak M, Lintern K, Chopda V et al (2017) Fluorescence based real time monitoring of fouling in process chromatography. Sci Rep 7(1):1–8. https://doi.org/10.1038/srep45640
115. Flevaris K, Chatzidoukas C (2021) Facilitating the industrial transition to microbial and microalgal factories through mechanistic modelling within the Industry 4.0 paradigm. Curr Opin Chem Eng 33. https://doi.org/10.1016/j.coche.2021.100713
116. Wang Z, Wang C, Chen G (2022) Kinetic modeling: a tool for temperature shift and feeding optimization in cell culture process development. Protein Expr Purif 198:106130. https://doi.org/10.1016/j.pep.2022.106130
117. Xu J, Tang P, Yongky A et al (2019) Systematic development of temperature shift strategies for Chinese hamster ovary cells based on short duration cultures and kinetic modeling. MAbs 11(1):191–204. https://doi.org/10.1080/19420862.2018.1525262
118. Kotidis P, Jedrzejewski P, Sou SN et al (2019) Model-based optimization of antibody galactosylation in CHO cell culture. Biotechnol Bioeng 116(7):1612–1626. https://doi.org/10.1002/bit.26960
119. Llaneras F, Picó J (2008) Stoichiometric modelling of cell metabolism. J Biosci Bioeng 105(1):1–11. https://doi.org/10.1263/jbb.105.1
120. Hutter S, Villiger TK, Brühlmann D et al (2017) Glycosylation flux analysis reveals dynamic changes of intracellular glycosylation flux distribution in Chinese hamster ovary fed-batch cultures. Metab Eng 43(Pt A):9–20. https://doi.org/10.1016/j.ymben.2017.07.005
121. Martínez VS, Buchsteiner M, Gray P et al (2015) Dynamic metabolic flux analysis using B-splines to study the effects of temperature shift on CHO cell metabolism. Metab. Eng. Commun 2:46–57. https://doi.org/10.1016/j.meteno.2015.06.001
122. Brunner M, Doppler P, Klein T et al (2018) Elevated pCO2 affects the lactate metabolic shift in CHO cell culture processes. Eng Life Sci 18(3):204–214. https://doi.org/10.1002/elsc.201700131
123. Leighty RW, Antoniewicz MR (2011) Dynamic metabolic flux analysis (DMFA): a framework for determining fluxes at metabolic non-steady state. Metab Eng 13(6):745–755. https://doi.org/10.1016/j.ymben.2011.09.010

124. Ahn WS, Antoniewicz MR (2012) Towards dynamic metabolic flux analysis in CHO cell cultures. Biotechnol J 7(1):61–74. https://doi.org/10.1002/biot.201100052
125. Nolan RP, Lee K (2011) Dynamic model of CHO cell metabolism. Metab Eng 13(1):108–124. https://doi.org/10.1016/j.ymben.2010.09.003
126. Kumar V, Lenhoff AM (2020) Mechanistic modeling of preparative column chromatography for biotherapeutics. Annu Rev Chem Biomol Eng 11:235–255. https://doi.org/10.1146/annurev-chembioeng-102419-125430
127. Gerstweiler L, Bi J, Middelberg APJ (2021) Continuous downstream bioprocessing for intensified manufacture of biopharmaceuticals and antibodies. Chem Eng Sci 231. https://doi.org/10.1016/j.ces.2020.116272
128. Gao Z-Y, Zhang Q-L, Shi C et al (2020) Antibody capture with twin-column continuous chromatography: effects of residence time, protein concentration and resin. Sep Purif Technol 253. https://doi.org/10.1016/j.seppur.2020.117554
129. Shi C, Gao ZY, Zhang QL et al (2020) Model-based process development of continuous chromatography for antibody capture: a case study with twin-column system. J Chromatogr A 1619:460936. https://doi.org/10.1016/j.chroma.2020.460936
130. Kumar V, Leweke S, Heymann W et al (2021) Robust mechanistic modeling of protein ion-exchange chromatography. J Chromatogr A 1660:462669. https://doi.org/10.1016/j.chroma.2021.462669
131. Andris S, Hubbuch J (2020) Modeling of hydrophobic interaction chromatography for the separation of antibody-drug conjugates and its application towards quality by design. J Biotechnol 317:48–58. https://doi.org/10.1016/j.jbiotec.2020.04.018
132. Halan V, Maity S, Bhambure R et al (2019) Multimodal chromatography for purification of biotherapeutics–a review. Curr Protein Pept Sci 20(1):4–13. https://doi.org/10.2174/1389203718666171020103559
133. Baur D, Angarita M, Muller-Spath T et al (2016) Optimal model-based design of the twin-column CaptureSMB process improves capacity utilization and productivity in protein A affinity capture. Biotechnol J 11(1):135–145. https://doi.org/10.1002/biot.201500223
134. Khanal O, Kumar V, Westerberg K et al (2019) Multi-column displacement chromatography for separation of charge variants of monoclonal antibodies. J Chromatogr A 1586:40–51. https://doi.org/10.1016/j.chroma.2018.11.074
135. Piątkowski W, Antos D, Kaczmarski K (2003) Modeling of preparative chromatography processes with slow intraparticle mass transport kinetics. J Chromatogr A 988(2):219–231. https://doi.org/10.1016/s0021-9673(02)02060-5
136. Püttmann A, Schnittert S, Naumann U et al (2013) Fast and accurate parameter sensitivities for the general rate model of column liquid chromatography. Comput Chem Eng 56:46–57. https://doi.org/10.1016/j.compchemeng.2013.04.021
137. Shekhawat LK, Rathore AS (2019) An overview of mechanistic modeling of liquid chromatography. Prep Biochem Biotechnol 49(6):623–638. https://doi.org/10.1080/10826068.2019.1615504
138. Shekhawat KL, Rathore SA (2019) An overview of mechanistic modeling of liquid chromatography. Prep Biochem Biotechnol 49(6):623–638. https://doi.org/10.1080/10826068.2019.1615504
139. Chen Y-C, Yao S-J, Lin D-Q (2022) Parameter-by-parameter method for steric mass action model of ion exchange chromatography: theoretical considerations and experimental verification. J Chromatogr A 1680:463418. https://doi.org/10.1016/j.chroma.2022.463418
140. Dimartino S, Boi C, Sarti GC (2011) Influence of protein adsorption kinetics on breakthrough broadening in membrane affinity chromatography. J Chromatogr A 1218(26):3966–3972. https://doi.org/10.1016/j.chroma.2011.04.062
141. He Q-L, von Lieres E, Sun Z et al (2020) Model-based process design of a ternary protein separation using multi-step gradient ion-exchange SMB chromatography. Comput Chem Eng 138. https://doi.org/10.1016/j.compchemeng.2020.106851

142. Osberghaus A, Hepbildikler S, Nath S et al (2012) Optimizing a chromatographic three component separation: a comparison of mechanistic and empiric modeling approaches. J Chromatogr A 1237:86–95. https://doi.org/10.1016/j.chroma.2012.03.029
143. Umatheva U, Chen G, Ghosh R (2019) Computational fluid dynamic (CFD) simulation of a cuboid packed-bed chromatography device. Chem Eng Res Des 152:393–401. https://doi.org/10.1016/j.cherd.2019.10.005
144. Schmidt I, Lottes F, Minceva M et al (2011) Estimation of chromatographic columns performances using computer tomography and CFD simulations. Chemie Ingenieur Technik 83(1–2):130–142. https://doi.org/10.1002/cite.201000171
145. Leweke S, von Lieres E (2018) Chromatography analysis and design toolkit (CADET). Comput Chem Eng 113:274–294. https://doi.org/10.1016/j.compchemeng.2018.02.025
146. CADET (2022) Two dimensional general rate model (GRM2D)¶. https://cadet.github.io/master/modelling/unit_operations/2d_general_rate_model.html#d-general-rate-model-model. Accessed 10 Dec 2022
147. Coolbaugh MJ, Varner CT, Vetter TA et al (2021) Pilot-scale demonstration of an end-to-end integrated and continuous biomanufacturing process. Biotechnol Bioeng 118(9):3287–3301. https://doi.org/10.1002/bit.27670
148. Namila N (2020) The effects of solution condition on virus filtration performance. University of Arkansas
149. Rathore AS, Kumar V, Arora A et al (2014) Mechanistic modeling of viral filtration. J Membr Sci 458:96–103. https://doi.org/10.1016/j.memsci.2014.01.037
150. Suh D, Kim M, Lee C et al (2022) Virus filtration in biopharmaceutical downstream processes: key factors and current limitations. Sep Purif Rev 1–14. https://doi.org/10.1080/15422119.2022.2143379
151. Shirataki H (2022) Analysis of filtration with virus removal filters using the characteristic form of blocking model. Biochem Eng J 183:108460. https://doi.org/10.1016/j.bej.2022.108460
152. Johnson SA, Chen S, Bolton G et al (2022) Virus filtration: a review of current and future practices in bioprocessing. Biotechnol Bioeng 119(3):743–761. https://doi.org/10.1002/bit.28017
153. Hadpe SR, Sharma AK, Mohite VV et al (2017) ATF for cell culture harvest clarification: mechanistic modelling and comparison with TFF. J Chem Technol Biotechnol 92(4):732–740. https://doi.org/10.1002/jctb.5165
154. Liu HF, Ma J, Winter C et al (2010) Recovery and purification process development for monoclonal antibody production. MAbs 2(5):480–499. https://doi.org/10.4161/mabs.2.5.12645
155. Huter MJ, Strube J (2019) Model-based design and process optimization of continuous single pass tangential flow filtration focusing on continuous bioprocessing. Processes 7(6):317. https://doi.org/10.3390/pr7060317
156. Grote F, Froehlich H, Strube J (2011) Integration of ultrafiltration unit operations in biotechnology process design. Chem Eng Technol 34(5):673–687. https://doi.org/10.1002/ceat.201000484
157. Binabaji E, Ma J, Rao S et al (2016) Ultrafiltration of highly concentrated antibody solutions: experiments and modeling for the effects of module and buffer conditions. Biotechnol Prog 32(3):692–701. https://doi.org/10.1002/btpr.2252
158. Binabaji E, Ma J, Rao S et al (2015) Theoretical analysis of the ultrafiltration behavior of highly concentrated protein solutions. J Membr Sci 494:216–223. https://doi.org/10.1016/j.memsci.2015.07.068
159. Bhosekar A, Ierapetritou M (2018) Advances in surrogate based modeling, feasibility analysis, and optimization: a review. Comput Chem Eng 108:250–267. https://doi.org/10.1016/j.compchemeng.2017.09.017
160. Reis MS, Saraiva PM (2022) Data-driven process system engineering–contributions to its consolidation following the path laid down by George Stephanopoulos. Comput Chem Eng 159:107675. https://doi.org/10.1016/j.compchemeng.2022.107675

161. Xie Y (2019) Data-driven predictive modeling for cell line selection in biopharmaceutical production. Massachusetts Institute of Technology
162. Bashokouh F, Abbasiliasi S, Tan JS (2019) Optimization of cultivation conditions for monoclonal IgM antibody production by M1A2 hybridoma using artificial neural network. Cytotechnology 71(4):849–860. https://doi.org/10.1007/s10616-019-00330-5
163. Alavijeh MK, Baker I, Lee YY et al (2022) Digitally enabled approaches for the scale up of mammalian cell bioreactors. Digital Chemical Engineering 100040. https://doi.org/10.1016/j.dche.2022.100040
164. Rathore AS, Mittal S, Pathak M et al (2014) Chemometrics application in biotech processes: assessing comparability across processes and scales. J Chem Technol Biotechnol 89(9): 1311–1316. https://doi.org/10.1002/jctb.4428
165. Sokolov M, Ritscher J, MacKinnon N et al (2017) Enhanced process understanding and multivariate prediction of the relationship between cell culture process and monoclonal antibody quality. Biotechnol Prog 33(5):1368–1380. https://doi.org/10.1002/btpr.2502
166. Green A, Glassey J (2015) Multivariate analysis of the effect of operating conditions on hybridoma cell metabolism and glycosylation of produced antibody. J Chem Technol Biotechnol 90(2):303–313. https://doi.org/10.1002/jctb.4481
167. Tsang VL, Wang AX, Yusuf-Makagiansar H et al (2014) Development of a scale down cell culture model using multivariate analysis as a qualification tool. Biotechnol Prog 30(1): 152–160. https://doi.org/10.1002/btpr.1819
168. Japel RC, Buyel JF (2022) Bayesian optimization using multiple directional objective functions allows the rapid inverse fitting of parameters for chromatography simulations. J Chromatogr A 1679:463408. https://doi.org/10.1016/j.chroma.2022.463408
169. Nikita S, Tiwari A, Sonawat D et al (2021) Reinforcement learning based optimization of process chromatography for continuous processing of biopharmaceuticals. Chem Eng Sci 230. https://doi.org/10.1016/j.ces.2020.116171
170. Wang G, Briskot T, Hahn T et al (2017) Estimation of adsorption isotherm and mass transfer parameters in protein chromatography using artificial neural networks. J Chromatogr A 1487: 211–217. https://doi.org/10.1016/j.chroma.2017.01.068
171. Li S, Feng L, Benner P et al (2014) Using surrogate models for efficient optimization of simulated moving bed chromatography. Comput Chem Eng 67:121–132. https://doi.org/10.1016/j.compchemeng.2014.03.024
172. Binger ZM, Achilli A (2023) Surrogate modeling of pressure loss & mass transfer in membrane channels via coupling of computational fluid dynamics and machine learning. Desalination 548:116241. https://doi.org/10.1016/j.desal.2022.116241
173. Zhang B, Kotsalis G, Khan J et al (2020) Backwash sequence optimization of a pilot-scale ultrafiltration membrane system using data-driven modeling for parameter forecasting. J Membr Sci 612:118464. https://doi.org/10.1016/j.memsci.2020.118464
174. Von Stosch M, Oliveira R, Peres J et al (2014) Hybrid semi-parametric modeling in process systems engineering: past, present and future. Comput Chem Eng 60:86–101. https://doi.org/10.1016/j.compchemeng.2013.08.008
175. Chen Y, Ierapetritou M (2020) A framework of hybrid model development with identification of plant-model mismatch. AICHE J 66(10):e16996. https://doi.org/10.1002/aic.16996
176. Tsopanoglou A, del Val IJ (2021) Moving towards an era of hybrid modelling: advantages and challenges of coupling mechanistic and data-driven models for upstream pharmaceutical bioprocesses. Curr Opin Chem Eng 32:100691. https://doi.org/10.1016/j.coche.2021.100691
177. Narayanan H, Sokolov M, Morbidelli M et al (2019) A new generation of predictive models: the added value of hybrid models for manufacturing processes of therapeutic proteins. Biotechnol Bioeng 116(10):2540–2549. https://doi.org/10.1002/bit.27097
178. Kotidis P, Kontoravdi C (2020) Harnessing the potential of artificial neural networks for predicting protein glycosylation. Metab Eng Commun 10:e00131. https://doi.org/10.1016/j.mec.2020.e00131

179. Möller J, Kuchemüller KB, Steinmetz T et al (2019) Model-assisted design of experiments as a concept for knowledge-based bioprocess development. Bioprocess Biosyst Eng 42(5): 867–882. https://doi.org/10.1007/s00449-019-02089-7
180. Zalai D, Koczka K, Párta L et al (2015) Combining mechanistic and data-driven approaches to gain process knowledge on the control of the metabolic shift to lactate uptake in a fed-batch CHO process. Biotechnol Prog 31(6):1657–1668. https://doi.org/10.1002/btpr.2179
181. Schinn SM, Morrison C, Wei W et al (2021) A genome-scale metabolic network model and machine learning predict amino acid concentrations in Chinese Hamster Ovary cell cultures. Biotechnol Bioeng 118(5):2118–2123. https://doi.org/10.1002/bit.27714
182. Antonakoudis A, Strain B, Barbosa R et al (2021) Synergising stoichiometric modelling with artificial neural networks to predict antibody glycosylation patterns in Chinese hamster ovary cells. *Comput Chem Eng* 154:107471. https://doi.org/10.1016/j.compchemeng.2021.107471
183. Wang G, Briskot T, Hahn T et al (2017) Root cause investigation of deviations in protein chromatography based on mechanistic models and artificial neural networks. J Chromatogr A 1515:146–153. https://doi.org/10.1016/j.chroma.2017.07.089
184. Narayanan H, Seidler T, Luna MF et al (2021) Hybrid models for the simulation and prediction of chromatographic processes for protein capture. J Chromatogr A 1650:462248. https://doi.org/10.1016/j.chroma.2021.462248
185. Narayanan H, Luna M, Sokolov M et al (2021) Hybrid models based on machine learning and an increasing degree of process knowledge: application to capture chromatographic step. Ind Eng Chem Res 60(29):10466–10478. https://doi.org/10.1021/acs.iecr.1c01317
186. Krippl M, Dürauer A, Duerkop M (2020) Hybrid modeling of cross-flow filtration: predicting the flux evolution and duration of ultrafiltration processes. Sep Purif Technol 248:117064. https://doi.org/10.1016/j.seppur.2020.117064
187. Thiess H, Leuthold M, Grummert U et al (2017) Module design for ultrafiltration in biotechnology: hydraulic analysis and statistical modeling. J Membr Sci 540:440–453. https://doi.org/10.1016/j.memsci.2017.06.038
188. Metta N, Ghijs M, Schäfer E et al (2019) Dynamic flowsheet model development and sensitivity analysis of a continuous pharmaceutical tablet manufacturing process using the wet granulation route. Processes 7(4):234. https://doi.org/10.3390/pr7040234
189. Gomis-Fons J, Schwarz H, Zhang L et al (2020) Model-based design and control of a small-scale integrated continuous end-to-end mAb platform. Biotechnol Prog 36(4):e2995. https://doi.org/10.1002/btpr.2995
190. Liu S, Papageorgiou LG (2019) Optimal antibody purification strategies using data-driven models. Engineering 5(6):1077–1092. https://doi.org/10.1016/j.eng.2019.10.011
191. Zahel T, Hauer S, Mueller EM et al (2017) Integrated process modeling—a process validation life cycle companion. Bioengineering 4(4):86. https://doi.org/10.3390/bioengineering4040086
192. Pirrung SM, van der Wielen LA, van Beckhoven RF et al (2017) Optimization of biopharmaceutical downstream processes supported by mechanistic models and artificial neural networks. Biotechnol Prog 33(3):696–707. https://doi.org/10.1002/btpr.2435
193. Sencar J, Hammerschmidt N, Jungbauer A (2020) Modeling the residence time distribution of integrated continuous bioprocesses. Biotechnol J 15(8):e2000008. https://doi.org/10.1002/biot.202000008
194. Taras S, Woinaroschy A (2011) Simulation and multi-objective optimization of bioprocesses with Matlab and SuperPro Designer using a client–server interface. Chem Eng Trans 25: 207–212
195. Fisher AC, Kamga MH, Agarabi C et al (2019) The current scientific and regulatory landscape in advancing integrated continuous biopharmaceutical manufacturing. Trends Biotechnol 37(3):253–267. https://doi.org/10.1016/j.tibtech.2018.08.008
196. Godawat R, Konstantinov K, Rohani M et al (2015) End-to-end integrated fully continuous production of recombinant monoclonal antibodies. J Biotechnol 213:13–19. https://doi.org/10.1016/j.jbiotec.2015.06.393

197. Ding C, Ardeshna H, Gillespie C et al (2022) Process design of a fully integrated continuous biopharmaceutical process using economic and ecological impact assessment. Biotechnol Bioeng 119(12):3567–3583. https://doi.org/10.1002/bit.28234
198. Gupta P, Kateja N, Mishra S et al (2021) Economic assessment of continuous processing for manufacturing of biotherapeutics. Biotechnol Prog 37(2):e3108. https://doi.org/10.1002/btpr.3108
199. Badr S, Okamura K, Takahashi N et al (2021) Integrated design of biopharmaceutical manufacturing processes: operation modes and process configurations for monoclonal antibody production. Comput Chem Eng 153. https://doi.org/10.1016/j.compchemeng.2021.107422
200. Gangadharan N, Turner R, Field R et al (2019) Metaheuristic approaches in biopharmaceutical process development data analysis. Bioprocess Biosyst Eng 42(9):1399–1408. https://doi.org/10.1007/s00449-019-02147-0
201. Tiwari A, Bansode V, Rathore AS (2022) Application of advanced machine learning algorithms for anomaly detection and quantitative prediction in protein A chromatography. J Chromatogr A 1682:463486. https://doi.org/10.1016/j.chroma.2022.463486
202. Kim JW, Park BJ, Oh TH et al (2021) Model-based reinforcement learning and predictive control for two-stage optimal control of fed-batch bioreactor. Comput Chem Eng 154:107465. https://doi.org/10.1016/j.compchemeng.2021.107465
203. Food Drug Administration (2022). Executive order on advancing biotechnology and biomanufacturing innovation for a sustainable, safe, and secure American bioeconomy. https://www.whitehouse.gov/briefing-room/presidential-actions/2022/09/12/executive-order-on-advancing-biotechnology-and-biomanufacturing-innovation-for-a-sustainable-safe-and-secure-american-bioeconomy/. Accessed 10 Dec 2022
204. Huang Q (2022) Intelligent manufacturing. In: Understanding China's manufacturing industry. Springer, pp 111–127
205. Teixeira JE, Tavares-Lehmann ATC (2022) Industry 4.0 in the European union: policies and national strategies. Technol Forecast Soc Chang 180:121664. https://doi.org/10.1016/j.techfore.2022.121664
206. Ferreira AP, Tobyn M (2015) Multivariate analysis in the pharmaceutical industry: enabling process understanding and improvement in the PAT and QbD era. Pharm Dev Technol 20(5):513–527. https://doi.org/10.3109/10837450.2014.898656
207. Food Drug Administration (2003), Part 11, Electronic records; electronic signatures – scope and application. https://www.fda.gov/regulatory-information/search-fda-guidance-documents/part-11-electronic-records-electronic-signatures-scope-and-application. Accessed 10 Dec 2022
208. Food Drug Administration (2022), Successes and Opportunities in Modeling & Simulation for FDA https://www.fda.gov/media/163156/download. Accessed 10 Dec 2022
209. Food Drug Administration (2022), Computer software assurance for production and quality system software. https://www.fda.gov/regulatory-information/search-fda-guidance-documents/computer-software-assurance-production-and-quality-system-software. Accessed 10 Dec 2022
210. Radhakrishnan D, Robinson AS, Ogunnaike BA (2017) Controlling the glycosylation profile in mAbs using time-dependent media supplementation. Antibodies 7(1):1. https://doi.org/10.3390/antib7010001
211. Kiparissides A, Pistikopoulos E, Mantalaris A (2015) On the model-based optimization of secreting mammalian cell (GS-NS0) cultures. Biotechnol Bioeng 112(3):536–548. https://doi.org/10.1002/bit.25457
212. Sou SN, Jedrzejewski PM, Lee K et al (2017) Model-based investigation of intracellular processes determining antibody Fc-glycosylation under mild hypothermia. Biotechnol Bioeng 114(7):1570–1582. https://doi.org/10.1002/bit.26225

213. Agarwal N, Mason A, Pradhan R et al (2019) Kinetic modeling as a tool to understand the influence of cell culture process parameters on the glycation of monoclonal antibody biotherapeutics. Biotechnol Prog 35(5):e2865. https://doi.org/10.1002/btpr.2865
214. Villiger TK, Scibona E, Stettler M et al (2016) Controlling the time evolution of mAb N-linked glycosylation-Part II: model-based predictions. Biotechnol Prog 32(5):1135–1148. https://doi.org/10.1002/btpr.2315
215. Ghorbaniaghdam A, Henry O, Jolicoeur M (2014) An in-silico study of the regulation of CHO cells glycolysis. J Theor Biol 357:112–122. https://doi.org/10.1016/j.jtbi.2014.04.035
216. Xing Z, Kenty B, Koyrakh I et al (2011) Optimizing amino acid composition of CHO cell culture media for a fusion protein production. Process Biochem 46(7):1423–1429. https://doi.org/10.1016/j.procbio.2011.03.014
217. Fouladiha H, Marashi S-A, Torkashvand F et al (2020) A metabolic network-based approach for developing feeding strategies for CHO cells to increase monoclonal antibody production. Bioprocess Biosyst Eng 43(8):1381–1389. https://doi.org/10.1007/s00449-020-02332-6
218. Spahn PN, Hansen AH, Hansen HG et al (2016) A Markov chain model for N-linked protein glycosylation–towards a low-parameter tool for model-driven glycoengineering. Metab Eng 33:52–66. https://doi.org/10.1016/j.ymben.2015.10.007
219. Thomas JA, Liu X, DeVincentis B et al (2021) A mechanistic approach for predicting mass transfer in bioreactors. Chem Eng Sci 237:116538. https://doi.org/10.1016/j.ces.2021.116538
220. Farzan P, Ierapetritou MG (2018) A framework for the development of integrated and computationally feasible models of large-scale mammalian cell bioreactors. Processes 6(7): 82. https://doi.org/10.3390/pr6070082
221. Janoska A, Buijs J, van Gulik WM (2022) Predicting the influence of combined oxygen and glucose gradients based on scale-down and modelling approaches for the scale-up of penicillin fermentations. Process Biochem. https://doi.org/10.1016/j.procbio.2022.11.006

Chapter 7
Integrated Process and Supply Chain Design and Optimization

Niki Triantafyllou, Miriam Sarkis, Nilay Shah, Cleo Kontoravdi, and Maria M. Papathanasiou

Abstract Advanced Therapy Medicinal Products (ATMPs) are an emerging class of therapeutic the development of which is primarily based on genetically engineering genes, cells, or tissues. Owing to their promising clinical outcomes in the treatment of rare and life-threatening diseases, ATMPs, have been gaining increasing interest from the global scientific and industrial community. The rapid growth in the demand for ATMPs is challenging manufacturers who are required to scale up production under process and demand uncertainties. In this chapter, we discuss the complexities of the global ATMP supply chain, highlighting the underlying interdependencies across raw materials, stakeholders and trade-offs with respect to supply chain costs, product availability, and environmental footprints of the manufacturing process. We present a comprehensive literature review on Process Systems Engineering tools in this space and discuss how those can assist systematic and proactive decision-making throughout stages of process development, design and optimization, and supply chain planning for ATMPs. We propose an integrated computational framework for process design and optimization and conclude with its demonstration on an industrially relevant case study, focusing on the quantification of uncertainty in process performance, sensitivity analysis, and process optimization for plasmid DNA production, a key ATMP raw material.

Keywords Advanced therapeutics · Process design · Process optimization · Flowsheet optimization · Supply chain · Sensitivity analysis

Abbreviations

AAV	Adeno-associated Virus
ATMPs	Advanced Therapy Medicinal Products

N. Triantafyllou · M. Sarkis · N. Shah · C. Kontoravdi · M. M. Papathanasiou (✉)
The Sargent Centre for Process Systems Engineering, Imperial College London, London, UK

Department of Chemical Engineering, Imperial College London, London, UK
e-mail: maria.papathanasiou11@imperial.ac.uk

R. Pörtner (ed.), *Biopharmaceutical Manufacturing*, Cell Engineering 11,
https://doi.org/10.1007/978-3-031-45669-5_7

AV	Adenovirus
BO	Bayesian Optimization
C&GT	Cell and Gene Therapies
CapEX	Capital Expenditures
CAR	Chimeric Antigen Receptor
CDMOs	Contract Manufacturing and Development Organizations
cGMP	Current Good Manufacturing Practice
CMOs	Contract Manufacturing Organizations
COGS	Cost of Goods Sold
COM	Component Object Model
CPU	Central Processing Unit
DFO	Derivative-free Optimization
DSP	Downstream Process
EI	Expected Improvement
GP	Gaussian Process
GSA	Global Sensitivity Analysis
GWP	Global Warming Potential
HSV	Herpes Simplex Virus
IQR	Interquartile Range
KPIs	Key Performance Indicators
LCA	Life Cycle Assessment
LCB	Lower Confidence Bound
LHS	Latin Hypercube Sampling
LSA	Local Sensitivity Analysis
LV	Lentivirus
M	Median
MES	Max-value Entropy Search
NLP	Non-linear Programming
OpEx	Operating Expenditures
pDNA	Plasmid DNA
PI	Probability of Improvement
PSE	Process Systems Engineering
QA	Quality Assurance
QC	Quality Control
RBF	Radial Basis Function
RS-HDMR	Random Sampling High Dimensional Model Representation
RV	Retrovirus
SA	Sensitivity Analysis
TCR	T-Cell Receptor
UCB	Upper Confidence Bound
UF/DF	Ultrafiltration/Diafiltration
USP	Upstream Process
VV	Viral Vector

Introduction

Advanced Therapy Medicinal Products (ATMPs) are a novel class of therapeutics that have demonstrated promising results in the cure of life-threatening diseases, including cancer and neurodegenerative disorders. Various advanced therapies classify as ATMPs, including cell [1], gene [2] and tissue-engineering therapies [3]. Owing to their promising clinical results, ATMPs have attracted global interest that translates into 27 approved products, 1454 ATMP developers worldwide, and 2220 active clinical trials, as of January 2023 [4, 5]. Since 2017, there has been a remarkable number of market authorizations in the cell and gene therapy (C&GT) space [6]. The approvals of new gene therapy products to treat rare diseases have hit a record in 2022 [7]. Of particular interest are also the latest approvals of Kite Pharma's *Yescarta* and Bristol-Myers Squibb's *Breyanzi* as second-line treatments, which up until 2022 were considered last-line treatments [4, 7].

The ATMP sector has attracted record-breaking investments, particularly in the cell and gene therapy (C&GT) space, reaching $23.1 billion in 2021 [8]. C&GT are at the forefront of ATMPs development and account for 98% of ongoing clinical trials [7]. Currently, innovation in C&GT is led by small- and medium-sized companies, accounting for 89% of the products and preclinical assets. Big pharmaceutical industries are involved in later clinical trial stages and commercialization of the products [9]. C&GT currently target ultra-rare diseases or late-stage cancer patients. This translates into high development and manufacturing costs that are reflected in a few patients, and therefore relatively high prices (US$110,920-US $814,780 for cell therapies and US$357,309–US$1,206,751 for gene therapies), possibly hindering wider adoption [10]. The patient-centric manufacture of C&GT that does not benefit from economies of scale and the, often, cold chain logistics can further increase the capital and operating costs. This can further complicate reimbursement procedures and lead to product discontinuations and withdrawals [10]. An example of this is Bluebird which withdrew two gene therapies (Zynteglo and Skysona) from the European market [11]. As more products reach commercialization, the sector is experiencing a need for innovative solutions for process development, scale-up, cost reduction, and infrastructure planning to ensure the availability and affordability of these therapies.

In this chapter, we discuss advancements in Process Systems Engineering (PSE) that support decision-making in the design and optimization of biopharmaceuticals and ATMPs, both in manufacturing and supply chain. We also present a methodological framework we have designed for the development of decision-support tools for process design and optimization for advanced therapeutics within the C&GT space. The capabilities of the framework are demonstrated through a case study on the identification of efficient timelines and scalability of the plasmid DNA (pDNA) and viral vector (VV) production processes.

The Global ATMP Supply Chain Network: Challenges and Opportunities

Manufacturers are faced with key decisions that are tightly tied to the specifications of the therapy under development and which can impact their global supply. ATMP development and manufacturing are highly reliant on the delivery of genetic information to modify the target cell, gene, and/or tissue. This can be facilitated via viral or non-viral delivery methods, with the former being the current state-of-the-art. Owing to their natural ability to transduce their genomes into cells, viral vectors are ideal for delivering genetic payloads to targeted cells [12]. Retroviruses, adenoviruses (AV), lentiviruses (LV), adeno-associated viruses (AAVs), and herpes simplex viruses (HSVs) are some of the ones used for genetic modifications [13–15]. AAV finds an application at *in vivo* gene therapy, with successes through the commercialization of Luxturna and Zongelsma [16, 17]. In the space of cell therapies, LV vectors are widely established. The recently approved Chimeric Antigen Receptor (CAR) T-cell therapy Kymriah uses LV vectors to genetically engineer patient cells [18, 19]. The supply of the vectors can be either via in-house manufacturing or purchasing from a third party. In both cases, as well as in the case of using naked pDNA, manufacturers are called to make key decisions on how to source the pDNA and/or viral vectors, ensuring adequate supply. This is a challenging task as both the clinical trial outcomes and the global demand are typically uncertain in this space. At the same time, viral vectors and pDNA are coupled with lengthy and complex manufacturing processes, as well as stringent regulatory and quality criteria.

Despite technological advancements in unit operation manufacturing, viral vector production yields remain low compared to other biological products [20, 21]. Process standardization and intensification across scales remain a challenging task in both upstream process (USP) and downstream process (DSP) trains. Established USP platforms for LV and AAV manufacturing feature suspension-based mammalian cell cultures, often relying on pDNA transfection for viral propagation which limits the scalability of the process [21–23]. In addition, different DSP options are available for primary recovery, clarification, and purification of viral vectors, including a range of chromatography and flow-through processes. Ongoing efforts are focusing on minimizing viral losses in DSP steps through an increased understanding of underlying biological processes and differences across VV serotypes [24–30].

Irrespective of the delivery method, pDNA is used as the critical starting material for the C&GT supply chain as it contains the gene of interest [31]. pDNA products required by the C&GT space can be diverse and manufacturers need to produce both gene-specific and off-the-shelf standard plasmids [32]. pDNA production costs and manufacturing timelines must be reduced to decrease the Cost of Goods Sold (COGS) and maximize the production of C&GT products. The USP in pDNA manufacturing is based on *E. coli* fermentation techniques in large stainless-steel bioreactors and is an inherently slow and costly process, prone to batch failure [33]. DSP purification is similarly expensive, lengthy, and challenging to scale [33]. Overall, the pDNA manufacturing process is characterized by long lead times, high capital investment, operating, and labor costs. It is evident that pDNA and VVs come with their own complex supply chains and the integration of all the separate elements under one holistic end-to-end C&GT supply chain is an ongoing challenge [31]. Manufacturers need to plan well ahead for the acquisition of raw materials to avoid shortages, long wait times, lot-to-lot variability, and high COGS [34].

COVID-19 highlighted shortcomings in the C&GT supply chain, as one-third of the cell therapy manufacturing companies reported delays or complete halt in manufacturing, while one-fifth reported disruptions in the procurement of supplies [35]. This can possibly be attributed partially to social distancing rules but also, critically, to repurposing of existing facilities toward vaccine manufacturing ramp-up [21]. Generally, supply chain issues can occur, as specifications for pDNA release from a plasmid manufacturer may not be aligned with the VV manufacturer's pDNA entry specifications [31]. Such disruptions could jeopardize the viability of the C&GT industry and critically impact patient lives [35]. Once an order for a C&GT product has been made, developers need to consider the different timelines regarding the acquisition of pDNA for VV transfection, the acquisition of VVs, and the acquisition of other raw materials and components, as well as the availability of manufacturing capacity and quality control [22]. It is essential to build a robust and responsive supply chain between pDNA manufacturers, VV manufacturers, and C&GT manufacturers to ensure the efficient operation of the supply chain. Manufacturers need to make decisions such as in-house or outsourced manufacturing of critical raw materials, centralized or decentralized manufacturing, batch sizes, and inventory management decisions. Figure 7.1 illustrates the possible alternative pathways for raw material acquisition to produce *ex vivo* autologous CAR or TCR T-cell therapies.

Contract manufacturing organizations (CMOs) and contract manufacturing and development organizations (CDMOs) are widely used in the C&GT industry, offering outsourcing services for the development and manufacturing of intermediate products, active ingredients (AI), or final products [36]. C&GT manufacturers may outsource the production of cell and gene therapies or raw materials (i.e pDNA and VVs), to increase their capacity, accelerate their processes by securing advanced production technology and decrease investment-related risks [37, 38]. There are limited CDMOs worldwide with the required capabilities that can manufacture pDNA and VVs at a large scale, leading to long waiting lists and substantial backlogs [33]. Lack of key materials in a timely manner may be a threat that can ultimately disrupt smooth operation in C&GT. To overcome this bottleneck, both CDMOs and biopharmaceutical manufacturers have invested in new manufacturing facilities to increase the current capacity and fulfill future demands [33, 39].

The adoption of a scalable and cost-effective manufacturing process, as well as the decision of in-house manufacturing or outsourcing to one (centralized manufacturing) or multiple (decentralized manufacturing) CDMOs, is critical to meeting the demand. Centralized manufacturing has been the norm for large-scale, efficient, and standardized production of pharmaceuticals. This manufacturing and distribution paradigm has demonstrated its benefits in more efficient resource planning, easier monitoring, and a decrease in cost per dose by dividing regulatory,

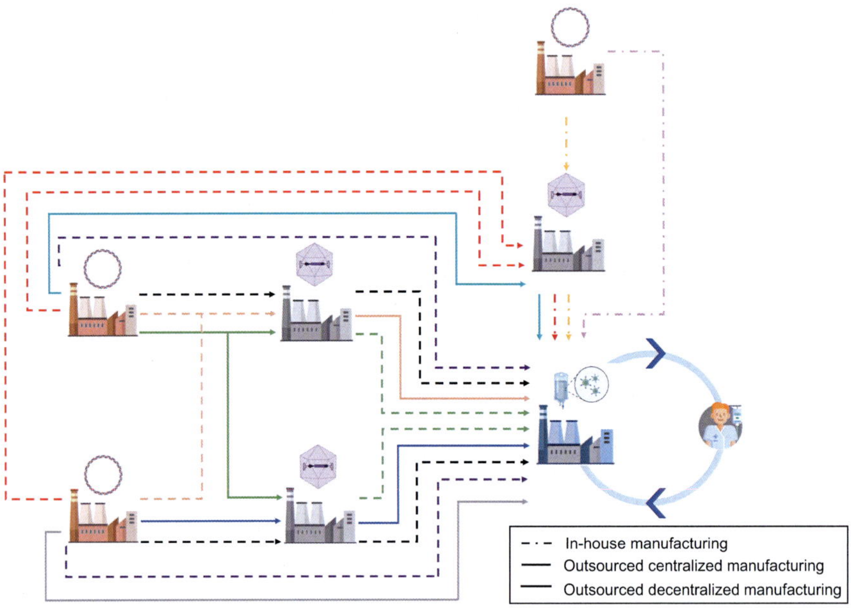

Fig. 7.1 Alternative pathways of raw material acquisition in autologous ex vivo cell therapy manufacturing, where each color represents a different possible route. (—): in-house non-viral manufacturing, (—): in-house viral manufacturing, (—): outsourced centralized pDNA manufacturing and in-house VV manufacturing, (—): outsourced decentralized pDNA manufacturing and in-house VV manufacturing, (—): outsourced centralized non-viral manufacturing, (—): outsourced fully decentralized non-viral manufacturing, (—): outsourced fully centralized viral manufacturing, (—): outsourced fully decentralized viral manufacturing, (—): outsourced decentralized pDNA and centralized VV manufacturing, (—): outsourced centralized pDNA and decentralized VV manufacturing. [Image created using BioRender.com]

operating, and capital costs over large product scales [40]. Autologous cell therapies, however, given their bespoke nature do not benefit from economies of scale [41, 42]. Decentralized or distributed smaller-scale manufacturing sites that distribute production into separate geographical regions and are nearby to the end users can provide greater operational flexibility [42, 43]. Decentralized raw material production can be a more appealing alternative as it can decrease lead times, batch failure rates, and transportation-based costs and risks arising from the raw materials' sensitive nature (e.g. specific handling and cryopreservation/thawing cycles). Nevertheless, a fully decentralized network may lead to higher production costs, which are mainly attributed to economies of scale and skills shortages (QPs) [40, 43, 44]. Furthermore, by using multiple suppliers for raw material acquisition, the risk for short-term supply chain failure due to raw material shortages deriving from failure to comply with cGMP manufacturing practices, voluntary supplier withdrawal, or the 'just-in-time' inventory practice of many healthcare suppliers, is minimized [44]. The curative nature of these treatments also requires a new tailored commercial approach, changing the way pharma and biotech companies operate.

One of the potential changes might be a shift to a more supply chain- and manufacturing-driven operating model.

The Role of Decision-Support Tools in ATMP Manufacturing and Supply Chain

The overarching goal of making ground-breaking therapies available to those in need comes with several product- and process-related challenges that impact the performance of the supply chain [45]. Informed decision-making in the ATMP landscape requires an assessment of trade-offs and alternatives to fulfill a set of key objectives, which can be grouped into the following broad categories:

Social Ensuring that the right therapy is returned to the right patient at the right time comes with several challenges, including careful planning of shipments from hospitals to manufacturing sites and back to the starting patient [42, 46]. Furthermore, making sure enough capacity is in place to manufacture the therapy comes with a need for informed investment planning, scale-up, and accurate demand forecasting. Timely delivery of the therapy also requires raw materials to be available and stored *on-site* to avoid delays in the manufacturing of the C&GT therapy. Decentralized manufacturing facilities for C&GT therapies further reduce transport delays.

Economic The industry is facing a need to reduce C&GT therapy costs to make them more affordable from clinical to commercial stages. In this context, patient-nonspecific products (e.g. *in vivo* gene therapy) and raw materials for patient-specific ATMPs benefit from economies of scale and costs per dose can be lowered by selecting suitable scales of manufacturing, which favors centralized manufacturing networks. C&GT therapeutics (e.g. *ex vivo* autologous gene therapy) require patient-specific manufacturing steps, which hinder scale-up and lead to significantly higher manufacturing costs compared to other biologics. This opens opportunities to opt for decentralized manufacturing and minimize therapy return times where possible. The emergence of allogeneic *ex vivo* therapies does promise higher flexibility in scaling up the manufacturing process as cells from a single donor may be compatible with more than one patient; this would allow manufacturing in small batches rather than individual ones and benefit from economies of scale. Overall, in the current C&GT landscape, maximizing asset utilization is a key driver in lowering therapy costs. Manufacturers are pressured to identify the maximum scalability of available resources, assess the requirement of capacity expansions, and quantify the associated risks of investment.

Environmental Awareness of environmental sustainability of supply chain operations in the life science sector is becoming increasingly relevant in recent years [47, 48]. Biopharmaceutical process steps are water intensive and energy intensive while pollutants are emitted during distribution. In addition, manufacturing platforms may rely on stainless steel or single-use equipment, which respectively result in the use of high-pressure steam for sterilization or disposal of plastic waste at the end of each campaign or batch. ATMP manufacturers are going to face pressure from governmental agencies to improve sustainability and demonstrate green credentials of raw material suppliers, which highlights a need to quantify the environmental footprints of different manufacturing setups and minimize them where possible.

Resilience Cost-effective and timely delivery of ATMP products is challenged by several uncertainties related to demand and process capabilities, which render manufacturing and investment planning intrinsically at *risk*. On-going development and intensification efforts in upstream productivity and recovery of purification steps in VV manufacturing impact the attainable batch sizes, costs and process times. This manufacturing uncertainty challenges the estimation of the required scale-up investments and may create delays in supplying raw materials for C&GT therapy manufacturing. Furthermore, the demand for autologous C&GT end-therapies is not seasonal or stable over the demand horizon, and this pressures manufacturers to ensure enough manufacturing lines are available at the cell therapy manufacturing site to cope with unforeseen demand fluctuations. In addition to this, raw material manufacturers supplying other markets (e.g. vaccine manufacturing) should account for potential demand increases and prevent backlogs in manufacturing slots for gene therapy applications. An important performance metric becomes the resilience of ATMPs supply chains to these uncertainties, which ultimately impacts the successful delivery of a cost-effective therapy to the patient in need.

Designing supply chains that perform in each of these aspects and ensure successful manufacturing and delivery of ATMPs requires a systematic analysis of alternatives and trade-offs [49]. Embedding resilience in these manufacturing strategies starts from an in-depth understanding of how underlying manufacturing uncertainties in emerging platforms may propagate to key performance indicators for (i) cost-effectiveness, (ii) manufacturing capacity and throughputs, and (iii) environmental process footprints.

Research in PSE focuses on developing *in silico* platforms to aid the analysis and optimization of process performance. In the space of pharmaceutical manufacturing, research has focused on flowsheet synthesis and optimization to assess and optimize the processes of interest. Optimization approaches have been proposed in the literature for several processes. We review the below contributions that have focused on ATMP manufacturing and other transferable methodologies for flowsheet optimization.

Overview of Decision-Support Tools

Computer-aided techno-economic analysis is a widely established methodology for the quantification of process performance and relies on an integrated system of mass and heat balance equations, scheduling, equipment sizing, and costing calculations.

Output metrics of interest may include batch size, capital, and operating costs. As model inputs are often uncertain, uncertainty and sensitivity analyses are performed to quantify and characterize the uncertainty of process performance. In the context of ATMPs, contributions have focused on assessing techno-economic performance and feasibility of different manufacturing setups. Uncertainty and sensitivity analyses are performed as a second step to quantify the impact of underlying manufacturing uncertainties (e.g. raw material costs, process yields) on therapy costs and investment requirements.

Pereira Chilima et al. [50] presented a cost assessment comparing scale-out (autologous) and scale-up (allogeneic) approaches for C&GT therapy manufacturing. The framework relied on a bioprocess and costing model for the estimation of facility footprint and fixed capital investment of each cell therapy application. The scale-up capabilities of allogeneic therapies have received further attention in the literature, with further work looking into integrating techno-economic analysis with a decisional tool [51]. Pereira Chilima et al. [52] proposed a framework assessing the impact of manufacturing decisions on COGS, process robustness, and reimbursement for allogeneic C&GT products. The propagation of uncertainty was assessed via Monte Carlo sampling from pre-defined probability distributions assigned to key process parameters. This allowed the quantification of uncertainty around KPIs of interest. Uncertainty characterization was performed via local sensitivity analysis (LSA), which highlighted the impact of viral transduction and electroporation efficiencies, viral transduction cost per lot, and DSP capacity as single-input contributions on COGS and other KPIs of interest.

Lower therapy costs for ATMPs can also be achieved via cost-effective raw material manufacturing platforms. Comisel et al. [53, 54] developed a bioprocess techno-economic model for lentivirus (LV) and adeno-associated vector (AAV) manufacturing with the aim of identifying a cost-effective scale-up strategy from clinical to commercial stages. The authors assessed how raw material cost and process inputs may propagate to COGS/dose for LV-based and AAV-based gene therapy products manufactured via different technologies. An LSA approach quantified the highest-ranking single-input contributions, namely DSP recovery yields, viral titers as well as dose size.

Recent work in techno-economic modelling and sensitivity analysis has also focused on emerging classes of next-generation vaccine platforms, which are largely based on ATMPs. Bioprocess simulations of VV-based vaccines were presented in Ferreira et al. [55] to conduct assessments between alternative batch and perfusion USP setups and assess the impact of USP titer uncertainty on manufacturing KPIs. Kis et al. [56] developed SuperPro models for emerging vaccine platforms and integrated a supply chain design model. LSA was performed to quantify the impact of USP uncertainty on manufacturing and supply chain KPIs, including total supply chain costs and product delivery time. An integrated techno-economic and supply chain optimization framework is also presented in Sarkis *et al* [57]. In this work, uncertainty in techno-economic outputs was characterized via systematic variation of all inputs simultaneously through Sobol sampling within a predefined range and iterative flowsheet simulations. The work showcased how variations within the characterized uncertainty range for the duration of manufacturing bottlenecks propagated to optimal supply chain capacity allocation decisions. In a study focusing on COVID-19 vaccine manufacturing, Kis et al. [58] also considered the variation of a set of manufacturing uncertainties simultaneously, including batch failure rate, process scale, production titers, and resource costs. Uncertain inputs were sampled via Sobol sequence, while the uncertainty space of outputs generated via the iterative techno-economic simulation helps quantify the total production capacities and investments required to meet global demands. Global Sensitivity Analysis (GSA) was then performed to identify the most-impactful input parameters in terms of single-input contribution to output variability. GSA enables quantification of whether second and higher-order interactions for each input combination are significant and represents a powerful tool for a systematic understanding of how input uncertainty propagates to the output uncertainty space.

Flowsheet optimization is the process of designing and optimizing the configuration of a biochemical process flowsheet with the scope of meeting specific objectives such as maximizing productivity, minimizing costs, or decreasing environmental impact. This typically entails the use of mathematical models and optimization algorithms to identify the optimal operating conditions and equipment configurations for the process at hand. Flowsheet optimization can be implemented at various levels, such as at the unit operation level, where the focus is on individual process units like bioreactors or chromatography columns, or at the flowsheet level, where the focus is on upstream or downstream or the end-to-end process. The optimization can be performed using various techniques, including mathematical programming, heuristics, or metaheuristics.

Bio-chemical process simulators (e.g., SuperPro Designer, ASPEN, and BioSolve) are widely used owing to their extensive model libraries, chemicals, consumables, equipment, and other databases and their ease of use. Most of the commercially available simulators rely on steady-state process and economic models, heuristics, and expert knowledge to find feasible process configurations. Typically, the underlying algebraic equations used by most process simulators are unavailable to the user. This can challenge further optimization studies as the derivatives and convexity properties are unknown. However, it is possible to obtain an estimate of the objective function (chosen KPI) by simulating scenarios through quasi-random sampling (e.g., Sobol sequence, Halton sequence, Latin Hypercube Sampling (LHS)) or adaptive sampling. An effective approach to simulation-based flowsheet optimization is derivative-free optimization (DFO) methods, such as Nelder-Mead, pattern search, and Bayesian Optimization. Specifically, data-driven optimization where the flowsheet model is treated as an inexpensive black-box function is a very promising solution.

Caballero and Grossmann replaced the process flowsheet with Kriging metamodels and developed a heuristics-based optimization algorithm that entails refining and bound contraction stages [59]. Boukouvala and Ierapetritou proposed an optimization methodology in order to optimize expensive continuous pharmaceutical manufacturing flowsheets, in which Kriging surrogate modeling and black-box feasibility analysis were combined to solve constrained and noisy optimization problems [60]. Cozad et al. developed the ALAMO methodology for simulation-based optimization where a low-complexity surrogate model was built using a best subset technique that takes advantage of integer programming formulations and then the model was iteratively improved through the use of derivative-free optimization solvers by adaptively sampling new simulation points [61]. Quirante and Caballero substituted individual units that caused significant numerical noise or large CPU times in a sour water stripping process flowsheet with surrogates based on Kriging interpolation and solved the resulting hybrid simulation-optimization model [62]. Lin et al. formulated optimization problems that included Kriging surrogate models, which substituted each unit of the flowsheet, simplified unit models, and solved them with commercial NLP solvers [63]. Wang et al. proposed a framework for process analysis and optimization of pharmaceutical flowsheets, in which global sensitivity analysis was performed to identify influential inputs, feasibility analysis with an RBF surrogate-based adaptive sampling approach was conducted to characterize the design space, and process optimization with commercial solvers was performed to find the optimal operating conditions [64]. Kim and Boukouvala investigated the efficiency of five different subset selection regression techniques for developing surrogate models for the optimization of simulation-based systems [65]. Khezri et al. replaced large-scale process simulation models with various types of surrogate models, namely multilayer perceptron neural networks, radial basis function neural networks, support vector machine, and adaptive neuro-fuzzy inference system models, and implemented them on the self-optimizing control procedure to obtain the optimal operating points [66]. Janus and Engell developed an iterative surrogate-based flowsheet optimization methodology, in which different strategies were tested, namely an evolution strategy, candidate generation, candidate rejections, and a combination of the two latter, to reduce the number of simulations [67]. Ludl et al. proposed an adaptive sampling technique based on machine learning models to explore the solution space of flowsheet-based models, where the trained classification model could predict whether each point in the design space is feasible [68]. Van de Berg et al. investigated how model-based (i.e. Bayesian optimization, CUATRO, Py-BOBYQA, SQSnobfit) and direct search (i.e. Newton's method, BFGS, Adam, Nelder-Mead, DIRECT) DFO algorithms perform for simulation-based flowsheet optimization [69].

In this chapter, we showcase and utilize the basic principles of Bayesian optimization in an integrated framework in order to improve the process performance of the mostly empirically based next-generation therapeutics. We do this by identifying optimal operating conditions that minimize/maximize KPIs such as batch size, production costs, and lead times.

Integrated Process Design and Optimization Framework

Here we present a methodological framework for integrated process design and flowsheet optimization for therapeutics within the C&GT space. The proposed approach is based on the development of an *in silico* platform to quantify process performance and optimize it leveraging on knowledge obtained via Global Sensitivity Analysis (GSA) and uncertainty analysis. The framework consists of three main steps: (i) techno-economic analysis, (ii) global sensitivity analysis, and (iii) Bayesian optimization (Fig. 7.2). A process simulation model is used to calculate process throughputs, equipment sizing, schedule of operations, and capital and operating expenditures. Techno-economic output metrics of interest (KPIs) can be obtained from the process simulation. Next, GSA is performed to assess the impact of input uncertainty on output metrics variability. A set of critical inputs contributing to output uncertainty above a pre-defined threshold is obtained via this exercise. Finally, Bayesian optimization is performed in the critical input design space, to obtain optimal manufacturing recipes for improved process performance.

Flowsheet Simulation and Techno-Economic Analysis

Techno-economic analysis is an established methodology that combines principles of process modeling, process design, and capital cost and operating cost estimation to analyze the economic performance of a process [70–73]. As a first step, process modeling calculations solve systems of algebraic and differential equations to compute mass and energy balances. Inputs required are process data, including reaction stoichiometries, kinetic rates, and recovery yields in the separation processes. The output of the process model can be used to estimate equipment sizes and utility requirements. The equipment purchase cost can be estimated based on equipment size and factored in the capital cost. Operating costs can be estimated based on input resource costs and resources consumed by the process, which were computed via process modeling.

Several KPIs can be selected for the process according to the required level of detail of the analysis and can be broadly grouped into *throughput*-related quantities and *cost*-related quantities. In batch pharmaceutical manufacturing, the throughput, or intrinsic rate of the process, can be quantified by *batch size* and *batch time*. Batch times correspond to the time to process a batch from the start to the end of the process. An important KPI is the *cycle time,* which is defined as the minimum required time before another batch can be started (i.e. bottleneck step). The number of batches per year can therefore be quantified accordingly. The *capital cost* (CapEx) includes direct fixed capital (DFC) costs, start-up and validation, and working capital costs. DFC includes equipment purchase costs and estimates costs of installation, piping, and building. Start-up and validation cost is calculated as a percentage of DFC. Working capital accounts for the upfront expenditure required to cover 30 days of operation in terms of variable costs. The *operating cost* (OpEx) (per year or batch) includes raw material costs (media, buffers, transfecting agents, cell banks), consumables (chromatography resins, filter membranes), utilities, labor costs, QC/QA costs, and facility-dependent costs. Facility-dependent costs account for equipment depreciation and maintenance and quantify the expense required to pay back the capital investment.

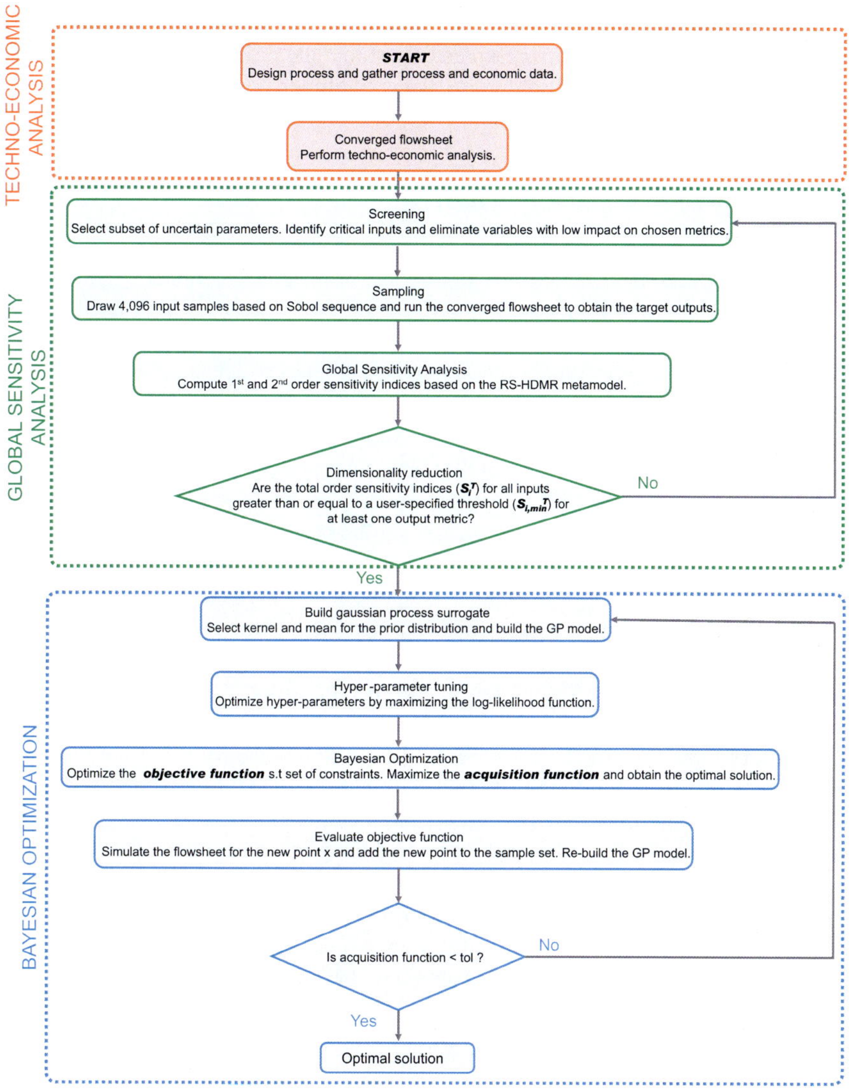

Fig. 7.2 Main steps of the proposed computational framework for integrated process design and optimization

A key metric to evaluate the cost-effectiveness of the manufacturing platform is the *unit production cost*, which can be calculated based on the operating cost and the throughput of the process. *Environmental* KPIs for the production process can also be computed through techno-economic analysis. Quantities such as water and energy consumption, and waste production (solvents vs single-use equipment) can, in fact, be integrated with life cycle-assessment methods (LCA) [74] to estimate process footprint on impact categories of interest, for instance, human health, climate change, and water usage. In the presented case study, the process flowsheet is created using SuperPro Designer ® User Guide. www.intelligen.com.

Uncertainty and Sensitivity Analysis

Sensitivity analysis (SA) is the study of how the uncertainty in model output can be apportioned, *quantitively* or *qualitatively*, to the uncertainty of the model inputs [75, 76]. The techno-economic models that describe the manufacturing processes for ATMPs consist of a high number of parameters, some of which are considered well-known, while others are highly uncertain. Uncertain parameters can be identified based on literature, cGMP manufacturers, suppliers, and expert feedback on the design of specific processes considered. Uncertain parameters in the manufacturing of ATMPs include process parameters, resource costs, and failure rates among others. Uncertainty and sensitivity analysis can be performed for the user-identified process and economic parameters and quantify their contribution to key output metrics. The latter can be critical KPIs, important to the smooth operation of the overall ATMP supply chain such as productivity (i.e batch sizes), throughput, lead times, CapEx, OpEx, production costs, and/or environmental sustainability metrics.

Once the techno-economic model for the specific process at hand is developed, the next step is to perform uncertainty analysis. The input of the uncertainty analysis is the input-output data of the techno-economic simulations. By using the Bayesian analysis, we can incorporate prior knowledge into the model by assigning different probability distributions to each uncertain parameter [77]. This is done by propagating the uncertainty by sampling the model and the parameters according to specific probability distributions, to produce model output samples. A uniform distribution can be used when there is equal probability for every value in the specified range for each parameter, while a triangular distribution is preferable when the nominal/peak value has the highest probability. Information on the selection of the right probability distribution can be obtained from literature and expert feedback. Once the distribution probabilities are defined, we perform uncertainty analysis by drawing an input vector sample from the respective distributions of all the uncertain inputs and we compute ('run') the flowsheet model to obtain the sample output vector. Once we have produced a significant number of input-output samples in order to avoid convergence issues, we can calculate the median output, the standard deviation, the quartiles of the output distribution, and the confidence bounds and plot the variability distributions [77]. By doing this, we are able to quantify and visualize the uncertainty for specific KPIs. After the uncertainty analysis is performed, we can continue with sensitivity analysis.

Sensitivity analysis can be either local or global. Typically, local SA methods are performed by evaluating the derivative of the model output with respect to a single input parameter while the remaining model inputs are fixed at a nominal value

[76, 78, 79]. On the other hand, global sensitivity analysis (GSA) estimates the effect of varying a specific input by varying all other inputs simultaneously [76, 80]. By doing so, GSA accounts for the interactions between variables and thus explores the whole input space [80, 81]. GSA is used to examine which model inputs are most influential in defining the uncertainty of the model output and obtain information about the input-output mapping [78, 81]. This information can be used to rank variables and decrease model dimensionality by fixing the nonessential variables [80, 82]. One of the most popular GSA methods is the variance-based global sensitivity method based on Sobol sensitivity indices [81]. Sobol sensitivity indices can be divided into first-order indices (S_i), which estimate the individual contribution of an input parameter to the output variance, second-order indices (S_{ij}), which estimate the contribution based on the interactions of two input parameters to the output variance, and total-order indices (S_T), which measure the contribution of a single input including the interactions with all other inputs [75, 80]. Uncertainty quantification of the key metrics and their apportionment to specific parameters provides confidence about the trends and comparisons required to steer decision-making in the biopharmaceutical industry.

Sobol indices can be computed either with variance-based or metamodeling methods [80]. Variance-based methods, despite their popularity and efficiency, can be impractical for large models, as they require a large number of function evaluations in order to achieve convergence. Metamodeling techniques only map inputs and outputs and are usually a better alternative in cases of computationally intractable models. In the presented case study, we use the Random Sampling High Dimensional Model Representation (RS-HDMR) metamodeling method implemented in the SobolGSA software [82].

Bayesian Optimization

Flowsheet optimization is used to improve the performance of a process in terms of efficiency, cost, and environmental impact by identifying optimal operating conditions that minimize/maximize the chosen key performance indicators. Flowsheet optimization is also useful in cases where the process is not performing as expected, or where new regulations or market settings have changed the process target requirements. In this chapter, we present Bayesian Optimization (BO) based on Gaussian Process (GP) surrogates combined with GSA as an effective approach to biopharmaceutical flowsheet optimization.

Bayesian optimization is a data-driven optimization method for finding the optimum of an expensive-to-evaluate, possibly noisy, black-box objective function [83, 84]. The main idea behind Bayesian optimization is to estimate the objective function with a surrogate function, such as the posterior mean function of a GP [85]. The optimization process is guided by the acquisition function, allowing for informed decisions to be made about the next point to sample, resulting in a reduction in the number of required function evaluations compared to other optimization methods [83, 84]. The acquisition function is maximized/minimized at each iteration and determines where to evaluate next the objective function [85]. The objective function is then sampled at the argmax/argmin of the acquisition function and the GP is updated [83]. GPs are chosen because they can reliably quantify uncertainty and point to unexplored regions of the search space and due to their ability to incorporate prior knowledge into the model [85, 86]. BO requires that all inputs (i.e dimensions) have user-defined bounds that identify the search space [83]. It should be noted that BO performs sufficiently in problems of moderate dimensions [84], hence the use of screening methods to discard non-essential variables or dimensionality reduction techniques is crucial.

Mathematically, given a set of N uncertain parameters $\boldsymbol{X} = \{x_1, x_2, \ldots, x_N\}$ and their acceptable ranges $[\boldsymbol{X}_{min}, \boldsymbol{X}_{max}]$, the GP approximates the objective function $f(\boldsymbol{x})$ (chosen KPI from the flowsheet) as a Gaussian distribution with mean $m(\boldsymbol{x})$ and covariance $k(\boldsymbol{x}, \boldsymbol{x}')$ [86, 87]:

$$f(\boldsymbol{x}) \sim \mathcal{GP}(m(\boldsymbol{x}), k(\boldsymbol{x}, \boldsymbol{x}'))$$

The mean function $m(\boldsymbol{x})$ can be represented as a linear combination of basis functions and parameters but usually, the prior mean is assumed zero, while the covariance function $k(\boldsymbol{x}, \boldsymbol{x}')$ can be defined using a kernel function, such as the radial basis function, the squared exponential kernel, or the Matérn kernel [83, 84, 86]. Incorporating prior knowledge into Bayesian optimization by choosing the mean and covariance functions of the GP can improve the efficiency of the optimization process and lead to more accurate solutions [87].

The GP posterior (i.e predictive) distribution for the objective function given the set of uncertain inputs is then given by [86, 87]:

$$\mathrm{p}(f(\boldsymbol{x})|\boldsymbol{X}, \boldsymbol{y}) = N(m(\boldsymbol{x}), k(\boldsymbol{x}, \boldsymbol{x}'))$$

where $\boldsymbol{X}$ is the set of observed inputs and $\boldsymbol{y}$ is the set of observed outputs.

Subsequently, the acquisition function is defined, to balance exploration (sampling in uncertain regions) and exploitation (sampling near the current optimal point) [84, 87]. Popular choices for the acquisition function include the probability of improvement (PI) [88, 89], expected improvement (EI) [90, 91], the upper/lower confidence bound (UCB or LCB) [91, 92], and max-value entropy search (MES) [93]. At each iteration, the acquisition function is used to select the next point to sample, and the process is repeated until a stopping criterion is met.

The proposed methodology is presented in Fig. 7.2. By quantifying the uncertainty and sensitivity of the parameters, GSA can identify critical input parameters that affect the variability of key output metrics (objective function). This allows for preliminary dimensionality reduction by reducing the number of parameters that need to be optimized in BO and therefore reducing the computational cost while improving the efficiency. A probabilistic model (GP surrogate model) of the objective function using is built, using an acquisition function to guide the search process, and updating the model based on new function evaluations (simulation runs). The final results are the optimal values for the critical input parameters and the corresponding maximum/minimum of the chosen output metric, along with the uncertainty estimates of the model.

Integrated Process Design and Optimization for pDNA Manufacturing

The capabilities of the framework presented in the previous section are showcased through a case study on the production of pharmaceutical plasmid DNA. We investigate the process uncertainties in the production of pDNA used as a critical raw material in the manufacture of lentiviral-based CAR T-cell therapies (i.e *Kymriah, Breyanzi, Abecma*, *Carvykti*). We examine a future demand scenario of 40,000 therapies, for which 60 g of pDNA (or 250 μg pDNA per dose) are required.

Flowsheet Simulation and Techno-Economic Analysis

First, a fed-batch process is modeled to produce 60 g of pharmaceutical pDNA in a batch time of 130 h, using SuperPro Designer. The pDNA production process consists of cell cultivation, pDNA recovery, purification, formulation, and fill & finish as shown in Fig. 7.3. Information on the process and economic parameters and their accepted ranges are based on literature, cGMP manufacturers, suppliers, and expert feedback [94–100]. Once the flowsheet is developed, techno-economic analysis is conducted with SuperPro Designer.

Global Sensitivity and Uncertainty Analysis

Global Sensitivity Analysis (GSA) is performed to quantify manufacturing uncertainty based on the contribution of uncertain input parameters to the output variance of specific KPIs, namely batch size and OpEx. 4096 samples are generated for the uncertain input parameter vector based on Sobol quasi-random sequence within pre-defined ranges.

GSA is performed using a SobolGSA-Matlab-Component Object Model (COM)-SuperPro interface. Critical parameters have, for at least one of the KPIs, total sensitivity indices greater than a user-specified threshold ($S_{T,\ min} = 1\%$). A noticeably low value of $S_{T,\ min}$ is chosen for the purpose of this work, to create a high-dimensional design space and test the capabilities of the proposed algorithm. The screening for critical parameters considers an initial set of more than 25 parameters

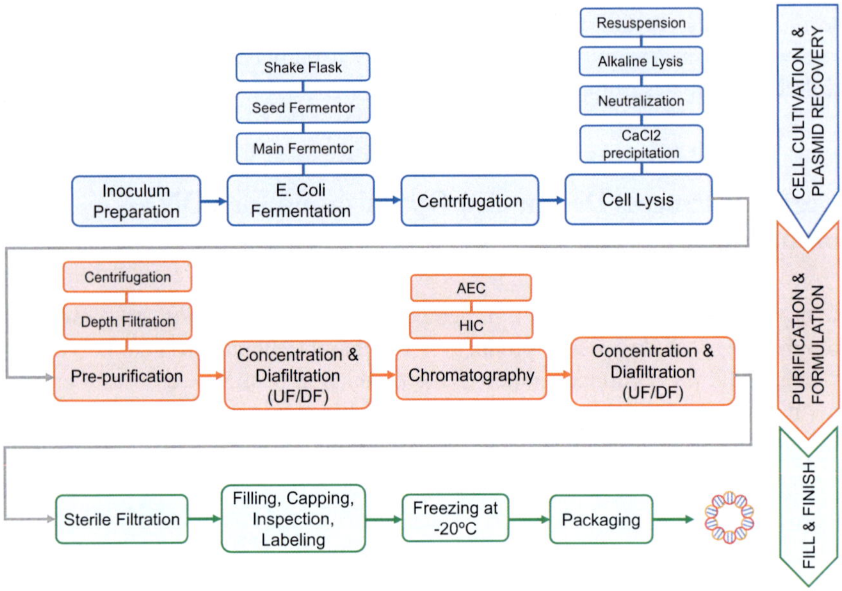

Fig. 7.3 Process flowsheet for the production of pharmaceutical pDNA

including task lengths, conversions, buffer volumes, working volumes, rejection and concentration coefficients, failure rates, etc. Parameters that had a minor effect on the output metrics were fixed to their nominal values without adding extra uncertainty to the model. GSA was repeated until all remaining parameters were critical. The final set of 7 critical parameters that account for at least 1% of the output variance for at least one output metric and their corresponding ranges are shown in Table 7.1.

Manufacturing uncertainty is illustrated as a violin plot, that represents the median value of the output samples (M) and the interquartile range (IQR), which quantifies the spread of the middle 50% of the output samples. The quantification of the uncertainty distribution improves the confidence about the trends and comparisons required for decision-making across scales and production routes. Violin plots in Fig. 7.4 illustrate the total variability of the two KPIs when all critical input parameters are simultaneously varied based on specific ranges and distributions (i.e triangular for the diafiltration volumes and uniform for the remaining inputs). Figure 7.5 depicts the explained variance to specific input parameters and their interactions as defined by the GSA by summarizing the first (S_i) and second (S_{ij}) order sensitivity indices for the two KPIs presented in this chapter. It should be noted that the critical parameters are ranked and contribute differently to each output metric.

The median value of the batch size (M = 64.1 g) reported in the violin plot in Fig. 7.4a is very close to the target batch size (nominal value) of 60 g. The batch size variability is IQR= 52.4–78.1 g. Major contributors to the batch size variance, as

Table 7.1 Critical parameters in pharmaceutical pDNA production, their acceptable ranges, nominal values, and optimal values for the maximization of the batch size and the minimization of OpEx

Input	Process step	Nominal	Range	Optimal (batch-size)	Optimal (OpEx)	Unit
Fermentation duration (main)	Main fermentor	40	18–40	20.75	18	*h*
Fed-batch medium volume		38.2	19.1–57.3	33.4	19.1	*L*
Starting working volume		204.5	102.2–306.8	178.9	102.2	*L*
Resuspension buffer volume	Resuspension	578	289.0–867.1	361.3	289.0	*L*
Lysis buffer volume	Lysis	578	289.0–867.1	361.3	289.0	*L*
Neutralization buffer volume	Neutralization	578	289.0–867.1	361.3	289.0	*L*
Diafiltration volumes 1	1st UF/DF	50	10–50	42	10	–
Diafiltration volumes 2	2nd UF/DF	40	10–50	43	10	–
Flush volume per filter area 2		1.3	0.67–2	1.17	0.67	L/m^2

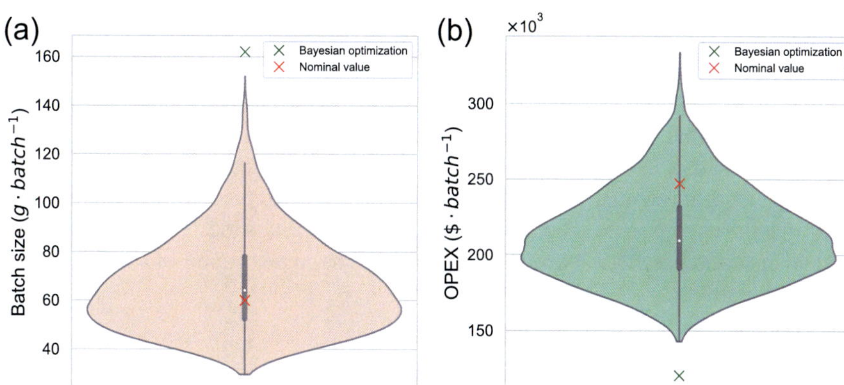

Fig. 7.4 Violin plots for batch size and OpEx. The black bar in the center is the interquartile range, the white dot inside the black bar is the median value, and the black lines stretched from the bar are the lower/upper adjacent values. Bayesian optimization is used to (**a**) maximize the batch size and (**b**) minimize the OpEx

shown in Fig. 7.5a, are the fed-batch medium volume (36.7%), and the initial medium working volume (16.9%), with a significant combined second-order interaction for the former two input volumes (6.2%). This is expected as variations in the volume of the cell culture through the fed-batch operating mode and the working

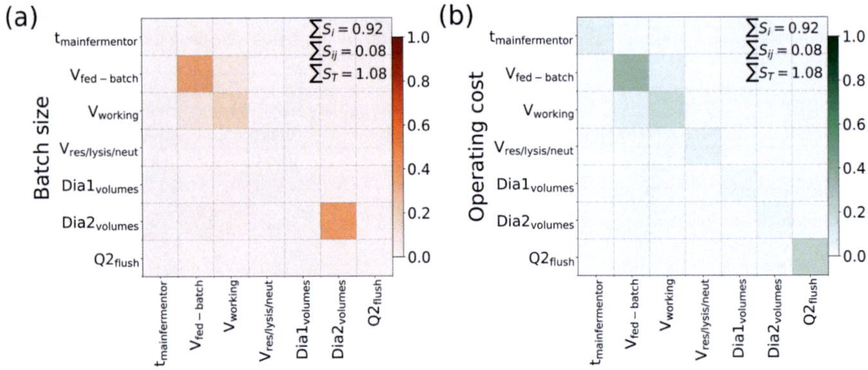

Fig. 7.5 Global sensitivity analysis for the nine critical input parameters on the batch size and OpEx for the production of pDNA. First-order effects (S_i) are displayed in the diagonal and second-order effects (S_{ij}) are presented in the upper and lower triangular

volume of the culture directly impact USP productivity. Furthermore, the most impactful DSP recovery step becomes UF/DF 2, which is responsible for buffer exchange to the final formulation buffer, as it impacts the pDNA losses during polishing. In this filtration step, the target stream is stripped of most impurities after chromatography and is more concentrated in pDNA. Thus, any losses in pDNA at UF/DF 2 result in a significantly lower batch size compared to UF/DF 1. Specifically, 38.3% of the batch size variability is attributed to the number of diafiltration volumes in UF/DF 2.

The Operating expenditure per batch variability (IQR = 191,256–231,640 $\$\cdot$batch^{-1}) (Fig. 7.4b) is distributed across inputs with the fed-batch medium volume (34.7%), the initial medium working volume (16.6%) and the volume of buffer flushed in UF/DF 2 (21.7%) being the most sensitive inputs, as shown in Fig. 7.5b. This suggests that raw material costs prevail over labor and facility-dependent costs at this manufacturing scale. The nominal point of the OpEx is located far up from the median since the base case flowsheet operates on the upper bound of the fermentation task duration.

Bayesian Optimization

After all the critical parameters are identified and their contributions to the output variability are quantified, Bayesian optimization is performed to obtain the optimal operating conditions to (a) maximize the batch size, and (b) minimize the operating cost per batch. We have created an interface between Python and SuperPro Designer to conduct Bayesian optimization. This interface utilizes Bayesian optimization to optimize (bio)pharmaceutical flowsheets and enables real-time adjustment of the process parameters, leading to improved efficiency and enhanced performance. The BO fragment of the algorithm is developed using the Python packages PyTorch and

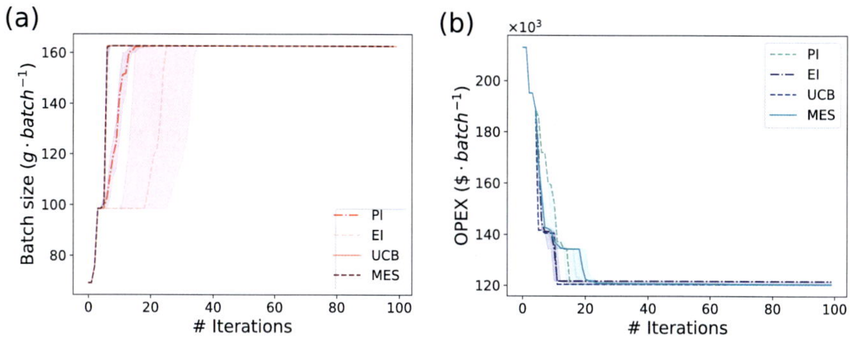

Fig. 7.6 Bayesian optimization progress for different acquisition functions for (**a**) batch size maximization and (**b**) OpEx minimization. The plot presents the median and confidence intervals (first and third quartiles) from 15 random seeds

BoTorch [101]. GPs are used as the surrogate with zero mean and the Matérn kernel. Firstly, the GP surrogate is trained with 5 samples based on Sobol quasi-random sequence and then the Bayesian optimization algorithm iteratively updates the GP. We test four of the most popular acquisition functions, namely probability of improvement (PI), expected improvement (EI), upper (or lower) confidence bound (UCB or LCB), and max-value entropy search (MES). Before the GP is trained, inputs are normalized and outputs are standardized.

The BO algorithm is performed in a 7-dimensional design space and through adaptive sampling it can identify optimal values outside of the violin plots by discovering optimal combinations of the input parameters that were not generated with quasi-random sampling as seen in Fig. 7.4. The BO algorithm leads to (a) an increase of 171% in the batch size and (b) a reduction of 50.5% in the OpEx per batch from the nominal points while operating within the acceptable ranges for all critical inputs (Fig. 7.4). The optimal manufacturing recipes are displayed in Table 7.1. Figure 7.6 presents the progress of the BO algorithm for four different acquisition functions for 100 iterations. It should be noted that different acquisition functions perform differently for different objective functions and the right acquisition function for each KPI is key to the best algorithm performance.

Concluding Remarks and Future Perspectives

As the Advanced Therapy Medicinal Product sector expands, manufacturers, regulators, and suppliers need to develop strategies to ensure adequate and timely supply. The need for innovative approaches to process development, design, and optimization, and supply chain planning becomes eminent, aiming to lower therapy costs and the environmental footprints of processes.

In this chapter, state-of-the-art methodologies in techno-economic, uncertainty, and sensitivity analysis, as well as flowsheet optimization have been discussed. In recent years, transferable methodologies from the area of Process Systems Engineering (PSE) have been applied in (bio-)pharmaceutical processes and tackling challenges related to process design, scale-up, and robust investment planning and scheduling. Harnessing established PSE methodologies, we developed and presented a computational framework for integrated process design and optimization to produce advanced therapeutics. We demonstrated how uncertainty analysis and Global Sensitivity Analysis (GSA) can accelerate the development of effective manufacturing strategies. Lastly, by utilizing the principles of Bayesian optimization, we identified the optimal set of process parameters that would improve the scalability, cost-effectiveness, and sustainability of each manufacturing node.

The works discussed in this chapter, as well as the presented case study, demonstrate how research in the interface of Life Sciences and PSE can advance standard practice in (bio-)pharmaceutical development and manufacturing, especially when it comes to strategic planning. Quantifying process and economic uncertainties, early on, can support informed decision-making. This can accelerate and de-risk the end-to-end supply chain by acquiring key raw materials on time and ensuring the timely return of the therapies to the patient population (*social* sustainability). Simultaneously, quantifying uncertainty and apportioning it to specific economic and design parameters can assist in the identification of the most cost-competitive supply chain network (*economic* sustainability). This can further enable the assessment of cost-effective solutions, such as large-scale centralized manufacturing for raw materials that benefit from economies-of-scale and decentralized manufacturing for patient-specific therapeutics. Model-based tools can also allow the integration of environmental KPIs, such as Global Warming Potential (GWP), and the identification of process inputs that are the highest contributors to improving the environmental footprint (*environmental* sustainability) of the process and the distribution network.

Acknowledgements MMP and NS would like to acknowledge funding from the UK Engineering & Physical Sciences Research Council (EPSRC) for the Future Targeted Healthcare Manufacturing Hub hosted at University College London with UK university part-.

ners (Grant Reference: EP/P006485/1). Financial and in-kind support from the consortium of industrial users and sector organizations is also acknowledged. NT is thankful for the Marit Mohn Scholarship awarded by the Department of Chemical Engineering, Imperial College London.

Competing Interests The authors declare no competing interests.

References

1. Schumacher TNM (2002) T-cell-receptor gene therapy. Nat Rev Immunol 2:512–519. https://doi.org/10.1038/nri841
2. Naldini L (2015) Gene therapy returns to centre stage. Nature 526:351–360. https://doi.org/10.1038/nature15818

3. Neves LS, Rodrigues MT, Reis RL, Gomes ME (2016) Current approaches and future perspectives on strategies for the development of personalized tissue engineering therapies. Expert Rev Precis Med Drug Dev 1:93–108. https://doi.org/10.1080/23808993.2016.1140004
4. Hunt TD (2023) Cell & gene: state of the industry briefing
5. The TransMission: Current FDA Approved Gene and Cell Therapies. https://www.mirusbio.com/blog/fda-approved-gene-cell-therapies. Accessed 27 Feb 2023
6. Approved Cellular and Gene Therapy Products | FDA. https://www.fda.gov/vaccines-blood-biologics/cellular-gene-therapy-products/approved-cellular-and-gene-therapy-products. Accessed 28 Feb 2023
7. The Alliance for Regenerative Medicine (2022) Regenerative medicine: the pipeline momentum builds H1
8. Regenerative TA for (2021) Regenerative medicine: disrupting the status quo. 1–40
9. Burnell S, Siciliano P Cell and gene therapy in 2040
10. Seoane-Vazquez E, Shukla V, Rodriguez-Monguio R (2019) Innovation and competition in advanced therapy medicinal products. EMBO Mol Med 11:1–5. https://doi.org/10.15252/emmm.201809992
11. Bluebird Bio has serious cash flow concerns; inability to strike reimbursement deals in Europe partly to blame
12. Bulcha JT, Wang Y, Ma H, Tai PWL, Gao G (2021) Viral vector platforms within the gene therapy landscape. Signal Transduct Target Ther 6:1–23. https://doi.org/10.1038/s41392-021-00487-6
13. Lukashev AN, Zamyatnin AA (2016) Viral vectors for gene therapy: current state and clinical perspectives. Biochem Mosc 81:700–708. https://doi.org/10.1134/S0006297916070063
14. Lundstrom K (2018) Viral vectors in gene therapy. Diseases 6:42. https://doi.org/10.3390/diseases6020042
15. Fisher PB, Savino R, Betapudi V, Goswami R, Subramanian G, Silayeva L, Newkirk I, Doctor D, Chawla K, Chattopadhyay S, Chandra D, Chilukuri N (2019) Gene therapy leaves a vicious cycle. Front Oncol 1:297. https://doi.org/10.3389/fonc.2019.00297
16. CHMP (2018) Package insert – LUXTURNA (voretigene neparvovec-rzyl)
17. CHMP (2019) Package insert: ZONGELSMA (onasemnogene abeparvovec-xioi)
18. CHMP (2019) Package insert: KYMRIAH (tisagenlecleucel)
19. U.S. Food & Drug Administration (FDA) (2017) FDA approves CAR-T cell therapy to trat adults with certain types of large B-cell lymphoma. https://www.fda.gov/news-events/press-announcements/fda-approves-car-t-cell-therapy-treat-adults-certain-types-large-b-cell-lymphoma. Accessed 17 Mar 2021
20. Capra E, Gennari A, Loche A, Temps C (2022) Viral-vector therapies at scale: today's challenges and future opportunities
21. Lambrix E, Kent-Egan H (2022) Viral vectors: shortages and innovation. In: Biotech review of the year
22. Petitt S, Glover C, Madsen J, Pratik J (2018) Key considerations in gene therapy manufacturing for commercialization
23. Cameau E, Pedregal A, Glover C (2019) Cost modelling comparison of adherent multi-trays with suspension and fixed-bed bioreactors for the manufacturing of gene therapy products. Cell Gene Ther Insights 5:1663–1674. https://doi.org/10.18609/cgti.2019.175
24. Valkama AJ, Oruetxebarria I, Lipponen EM, Leinonen HM, Käyhty P, Hynynen H, Turkki V, Malinen J, Miinalainen T, Heikura T, Parker NR, Ylä-Herttuala S, Lesch HP (2020) Development of large-scale downstream processing for lentiviral vectors. Mol Ther Methods Clin Dev 17:717–730. https://doi.org/10.1016/j.omtm.2020.03.025
25. Merten O-W, Schweizer M, Chahal P, Kamen A (2014) Manufacturing of viral vectors: part II. Downstream processing and safety aspects. Pharm Bioprocess 2:237–251. https://doi.org/10.4155/pbp.14.15

26. de Rooij J, DeConto J, Schaenzler G, Bauer D, Barre K, Duskin M, Kohli A, Watanabe K (2019) Upstream and downstream solutions for AAV manufacturing. Cell Gene Ther Insights 5:1017–1029. https://doi.org/10.18609/cgti.2019.110
27. Terova O, Soltys S, Hermans P, de Rooij J, Detmers F (2018) Overcoming downstream purification challenges for viral vector manufacturing: enabling advancement of gene therapies in the clinic. Cell Gene Ther Insights 4:101–111. https://doi.org/10.18609/cgti.2018.017
28. Qu W, Wang M, Wu Y, Xu R (2015) Scalable downstream strategies for purification of recombinant adeno-associated virus vectors in light of the properties. Curr Pharm Biotechnol 16:684–695
29. Perry C, Rayat ACME (2021) Lentiviral vector bioprocessing. Viruses 13:268. https://doi.org/10.3390/v13020268
30. Vellinga J, Smith JP, Lipiec A, Majhen D, Lemckert A, van Ooij M, Ives P, Yallop C, Custers J, Havenga M (2014) Challenges in manufacturing adenoviral vectors for global vaccine product deployment. Hum Gene Ther 25:318–327. https://doi.org/10.1089/hum.2014.007
31. Gosse M, Jones C, Jesus D, D'Costa S (2022) Regulatory & supply chain implications for plasmids as critical starting materials in the manufacture of viral vector gene therapy products. Cell Gene Ther Insights 8:279–286. https://doi.org/10.18609/cgti.2022.044
32. Labbé RP, Vessillier S, Rafiq QA (2021) Lentiviral vectors for T cell engineering: clinical applications, bioprocessing and future perspectives. Viruses 13:1528. https://doi.org/10.3390/v13081528
33. Ohlson J (2020) Plasmid manufacture is the bottleneck of the genetic medicine revolution. Drug Discov Today 25:1891–1893. https://doi.org/10.1016/J.DRUDIS.2020.09.040
34. Capra E, Gennari A, Loche A (2022) Viral-vector therapies at scale: today's challenges and future opportunities. McKinsey. https://www.mckinsey.com/industries/life-sciences/our-insights/viral-vector-therapies-at-scale-todays-challengesand-future-opportunities
35. Loche A, Mossmann W, van der Veken L, Yang G COVID-19 and cell and gene therapy: how to keep innovation on track. https://www.mckinsey.com/industries/life-sciences/our-insights/covid-19-and-cell-and-gene-therapy-how-to-keepinnovation-on-track
36. Shah N (2004) Pharmaceutical supply chains: key issues and strategies for optimisation. Comput Chem Eng 28:929–941
37. Kurata H, Ishino T, Ohshima Y, Yohda M CDMOs play a critical role in the biopharmaceutical ecosystem. https://doi.org/10.3389/fbioe.2022.841420
38. Elekidis AP, Georgiadis MC (2022) Optimal contract selection for contract manufacturing organizations in the secondary pharmaceutical industry. Comput Chem Eng 165:107957. https://doi.org/10.1016/J.COMPCHEMENG.2022.107957
39. Gene Therapy Manufacturing Fails to Meet Demand: Implications for Biopharma. L.E.K. Consulting. https://www.lek.com/insights/ei/gene-therapy-manufacturing-fails-meet-demand-implications-biopharma. Accessed 23 Aug 2022
40. Lam C, Meinert E, Yang A, Cui Z (2021) Comparison between centralized and decentralized supply chains of autologous chimeric antigen receptor T-cell therapies: a UK case study based on discrete event simulation. Cytotherapy 23:433–451. https://doi.org/10.1016/j.jcyt.2020.08.007
41. Papathanasiou MM, Stamatis C, Lakelin M, Farid S, Titchener-Hooker N, Shah N (2020) Autologous CAR T-cell therapies supply chain: challenges and opportunities? Cancer Gene Ther. https://doi.org/10.1038/s41417-019-0157-z
42. Triantafyllou N, Bernardi A, Lakelin M, Shah N, Papathanasiou MM (2022) A digital platform for the design of patient-centric supply chains. Sci Rep 12:17365. https://doi.org/10.1038/s41598-022-21290-5
43. Harrison RP, Rafiq QA, Medcalf N (2018) Centralised versus decentralised manufacturing and the delivery of healthcare products: a United Kingdom exemplar. Cytotherapy 20:873–890. https://doi.org/10.1016/j.jcyt.2018.05.003

44. Harrison RP, Ruck S, Rafiq QA, Medcalf N (2018) Decentralised manufacturing of cell and gene therapy products: learning from other healthcare sectors. Biotechnol Adv 36:345–357
45. Sarkis M, Bernardi A, Shah N, Papathanasiou MM (2021) Emerging challenges and opportunities in pharmaceutical manufacturing and distribution. Processes 9:1–16. https://doi.org/10.3390/pr9030457
46. Bernardi A, Papathanasiou MM, Lakelin M, Shah N (2021) Assessment of intermediate storage and distribution nodes in personalised medicine. Comput Aided Chem Eng 50: 1997–2002. https://doi.org/10.1016/B978-0-323-88506-5.50309-0
47. NHS (2022) Delivering a "net zero" national health service classification: official
48. HM Government (2021) Life sciences vision
49. Sarkis M, Bernardi A, Shah N, Papathanasiou MM (2021) Decision support tools for next-generation vaccines and advanced therapy medicinal products: present and future. Curr Opin Chem Eng 32
50. Pereira Chilima TD, Moncaubeig F, Farid SS (2020) Estimating capital investment and facility footprint in cell therapy facilities. Biochem Eng J 155. https://doi.org/10.1016/j.bej.2019.107439
51. Jenkins MJ, Farid SS (2018) Cost-effective bioprocess design for the manufacture of allogeneic CAR-T cell therapies using a decisional tool with multi-attribute decision-making analysis. Biochem Eng J 137:192–204. https://doi.org/10.1016/j.bej.2018.05.014
52. Pereira Chilima TD, Moncaubeig F, Farid SS (2018) Impact of allogeneic stem cell manufacturing decisions on cost of goods, process robustness and reimbursement. Biochem Eng J 137:132–151. https://doi.org/10.1016/J.BEJ.2018.04.017
53. Comisel RM, Kara B, Fiesser FH, Farid SS (2021) Gene therapy process change evaluation framework: transient transfection and stable producer cell line comparison. Biochem Eng J 176:108202. https://doi.org/10.1016/J.BEJ.2021.108202
54. Comisel RM, Kara B, Fiesser FH, Farid SS (2021) Lentiviral vector bioprocess economics for cell and gene therapy commercialization. Biochem Eng J 167:107868. https://doi.org/10.1016/j.bej.2020.107868
55. Ferreira RG, Gordon NF, Stock R, Petrides D (2021) Adenoviral vector covid-19 vaccines: process and cost analysis. Processes 9. https://doi.org/10.3390/pr9081430
56. Kis Z, Papathanasiou M, Calvo-Serrano R, Kontoravdi C, Shah N (2019) A model-based quantification of the impact of new manufacturing technologies on developing country vaccine supply chain performance: a Kenyan case study. J Adv Manuf Process. https://doi.org/10.1002/amp2.10025
57. Sarkis M, Tak K, Chachuat B, Shah N, Papathanasiou MM (2022) Towards resilience in next-generation vaccines and therapeutics supply chains. In: Symposium on computer aided process engineering (ESCAPE 32)
58. Kis Z, Tak K, Ibrahim D, Papathanasiou MM, Chachuat B, Shah N, Kontoravdi C (2022) Pandemic-response adenoviral vector and RNA vaccine manufacturing. NPJ Vaccines 7. https://doi.org/10.1038/s41541-022-00447-3
59. Caballero JA, Grossmann IE (2008) Rigorous flowsheet optimization using process simulators and surrogate models. Comput Aided Chem Eng 25:551–556. https://doi.org/10.1016/S1570-7946(08)80097-1
60. Boukouvala F, Ierapetritou MG (2013) Surrogate-based optimization of expensive flowsheet modeling for continuous pharmaceutical manufacturing. J Pharm Innov 8:131–145. https://doi.org/10.1007/s12247-013-9154-1
61. Cozad A, Sahinidis NV, Miller DC (2014) Learning surrogate models for simulation-based optimization. AICHE J 60:2211–2227. https://doi.org/10.1002/aic.14418
62. Quirante N, Caballero JA (2016) Large scale optimization of a sour water stripping plant using surrogate models. Comput Chem Eng 92:143–162. https://doi.org/10.1016/j.compchemeng.2016.04.039

63. Lin Z, Wang J, Nikolakis V, Ierapetritou M (2017) Process flowsheet optimization of chemicals production from biomass derived glucose solutions. Comput Chem Eng 102:258–267. https://doi.org/10.1016/j.compchemeng.2016.09.012
64. Wang Z, Escotet-Espinoza MS, Ierapetritou M (2017) Process analysis and optimization of continuous pharmaceutical manufacturing using flowsheet models. Comput Chem Eng 107: 77–91. https://doi.org/10.1016/j.compchemeng.2017.02.030
65. Kim SH, Boukouvala F (2020) Machine learning-based surrogate modeling for data-driven optimization: a comparison of subset selection for regression techniques. Optim Lett 14:989–1010. https://doi.org/10.1007/s11590-019-01428-7
66. Panahi M, Khezri V, Yasari E, Skogestad S (2021) Application of surrogate models as an alternative to process simulation for implementation of the self-optimizing control procedure on large-scale process plants-a natural gas-to-liquids (gtl) case study. Ind Eng Chem Res 60: 4919–4929. https://doi.org/10.1021/acs.iecr.0c05715
67. Janus T, Engell S (2021) Iterative process design with surrogate-assisted global flowsheet optimization. Chem Ing Tech 93:2019–2028. https://doi.org/10.1002/cite.202100095
68. Ludl PO, Heese R, Höller J, Asprion N, Bortz M (2022) Using machine learning models to explore the solution space of large nonlinear systems underlying flowsheet simulations with constraints. Front Chem Sci Eng 16:183–197. https://doi.org/10.1007/s11705-021-2073-7
69. van de Berg D, Savage T, Petsagkourakis P, Zhang D, Shah N, del Rio-Chanona EA (2022) Data-driven optimization for process systems engineering applications. Chem Eng Sci 248: 117135. https://doi.org/10.1016/J.CES.2021.117135
70. Seider WD, Lewin D, Seader JD, Widgado RG, Gani R, Ng KM (2016) Product and process design principles: synthesis, analysis and evaluation, 4th edn. Wiley, New York
71. Heinzle E, Biwer AP, Cooney C (2006) Development of sustainable bioprocesses: modeling and assessment, 1st edn. Wiley, Chichester
72. Towler G, Sinnott R (2008) Chemical engineering design: principles, practice and economics of plant and process design. Elsevier, Oxford
73. Jagschies G, Lindskog E, Lacki K, Galliher P (2018) Biopharmaceutical processing: development, design and implementation of manufacturing processes, 1st edn. Elsevier, Amsterdam
74. Hauschild MZ, Rosenbaum RK, Irving Olsen S (2018) Life cycle assessment: theory and practice, 1st edn. Springer, Cham
75. Saltelli A, Annoni P, Azzini I, Campolongo F, Ratto M, Tarantola S (2010) Variance based sensitivity analysis of model output. Design and estimator for the total sensitivity index. Comput Phys Commun 181:259–270. https://doi.org/10.1016/j.cpc.2009.09.018
76. Kucherenko S, Rodriguez-Fernandez M, Pantelides C, Shah N (2009) Monte Carlo evaluation of derivative-based global sensitivity measures. Reliab Eng Syst Saf 94:1135–1148. https://doi.org/10.1016/j.ress.2008.05.006
77. Saltelli A, Ratto M, Andres T, Campolongo F, Cariboni J, Gatelli D, Saisana M, Tarantola S (2008) Global sensitivity analysis. The primer. Wiley, Chichester
78. Cucurachi S, Borgonovo E, Heijungs R (2016) A protocol for the global sensitivity analysis of impact assessment models in life cycle assessment. Risk Anal 36:357–377. https://doi.org/10.1111/risa.12443
79. Groen EA, Bokkers EAM, Heijungs R, de Boer IJM (2017) Methods for global sensitivity analysis in life cycle assessment. Int J Life Cycle Assess 22:1125–1137. https://doi.org/10.1007/s11367-016-1217-3
80. Kucherenko S, Song S (2017) Different numerical estimators for main effect global sensitivity indices. Reliab Eng Syst Saf 165:222–238. https://doi.org/10.1016/j.ress.2017.04.003
81. Sobol IM (2001) Global sensitivity indices for nonlinear mathematical models and their Monte Carlo estimates. Math Comput Simul 55:271
82. Kucherenko S (2013) SOBOLHDMR: a general-purpose modeling software. Methods Mol Biol 1073:191–224. https://doi.org/10.1007/978-1-62703-625-2_16

83. Brochu E, Cora V, Freitas N (2010) A tutorial on Bayesian optimization of expensive cost functions, with application to active user modeling and hierarchical reinforcement learning. CoRR abs/1012.2599
84. Shahriari B, Swersky K, Wang Z, Adams RP, de Freitas N (2016) Taking the human out of the loop: a review of Bayesian optimization. Proc IEEE 104:148–175. https://doi.org/10.1109/JPROC.2015.2494218
85. Snoek J, Larochelle H, Adams RP (2012) Practical Bayesian optimization of machine learning algorithms. https://doi.org/10.48550/arxiv.1206.2944
86. Rasmussen CE, Williams CKI (2005) Gaussian processes for machine learning. MIT Press, Cambridge
87. Schulz E, Speekenbrink M, Krause A (2018) A tutorial on Gaussian process regression: modelling, exploring, and exploiting functions. J Math Psychol 85:1–16. https://doi.org/10.1016/j.jmp.2018.03.001
88. Kushner HJ, Yin G (1997) Stochastic approximation algorithms and applications, Applied mathematics. Springer, New York
89. Kushner HJ (1964) A new method of locating the maximum point of an arbitrary multipeak curve in the presence of noise. J Basic Eng 86:97–106. https://doi.org/10.1115/1.3653121
90. Jones DR, Schonlau M, Welch WJ (1998) Efficient global optimization of expensive black-box functions. J Glob Optim 13:455–492. https://doi.org/10.1023/A:1008306431147
91. Mockus J, Tiesis V, Zilinskas A (2014) The application of Bayesian methods for seeking the extremum. In: Towards global optimization. North Holland, pp 117–129
92. Cox DD, John S (1992) A statistical method for global optimization. In: Proceedings 1992 IEEE international conference on systems, man, and cybernetics, vol. 2, pp 1241–1246
93. Wang Z, Jegelka S (2017) Max-value entropy search for efficient bayesian optimization. https://doi.org/10.48550/arxiv.1703.01968
94. Przybylowski M, Bartido S, Borquez-Ojeda O, Sadelain M, Rivière I (2007) Production of clinical-grade plasmid DNA for human phase I clinical trials and large animal clinical studies. Vaccine 25:5013–5024. https://doi.org/10.1016/j.vaccine.2007.04.077
95. Urthaler J, Ascher C, Wöhrer H, Necina R (2007) Automated alkaline lysis for industrial scale cGMP production of pharmaceutical grade plasmid-DNA. J Biotechnol 128:132–149. https://doi.org/10.1016/j.jbiotec.2006.08.018
96. Prather KJ, Sagar S, Murphy J, Chartrain M (2003) Industrial scale production of plasmid DNA for vaccine and gene therapy: plasmid design, production, and purification. Enzym Microb Technol 33:865–883. https://doi.org/10.1016/S0141-0229(03)00205-9
97. Carnes A, Williams J (2008) Plasmid DNA manufacturing technology. Recent Pat Biotechnol 1:151–166. https://doi.org/10.2174/187220807780809436
98. da Gama R, Petrides D (2021) Manufacturing of pharmaceutical grade plasmid DNA (pDNA) process modeling and cost analysis for the process manufacturing industries. https://doi.org/10.13140/RG.2.2.12780.28800
99. Sun B, Yu XH, Yin Y, Liu X, Wu Y, Chen Y, Zhang X, Jiang C, Kong W (2013) Large-scale purification of pharmaceutical-grade plasmid DNA using tangential flow filtration and multi-step chromatography. J Biosci Bioeng 116:281–286
100. Eon-Duval A, Burke G (2004) Purification of pharmaceutical-grade plasmid DNA by anion-exchange chromatography in an RNase-free process. J Chromatogr B Analyt Technol Biomed Life Sci 804:327–335. https://doi.org/10.1016/j.jchromb.2004.01.033
102. Balandat M, Karrer B, Jiang DR, Daulton S, Letham B, Wilson AG, Bakshy E (2019) BoTorch: a framework for efficient Monte-Carlo Bayesian optimization. https://doi.org/10.48550/arxiv.1910.06403

Chapter 8
Production of Complex Proteins in Plants: From Farming to Manufacturing

Henrik Nausch, Matthias Knödler, and Johannes F. Buyel

Abstract Recombinant biopharmaceutical proteins such as growth factors and antibodies are typically produced in microbial or mammalian cells. However, microbial host cells cannot produce complex biopharmaceuticals that require defined oligomerization or specific glycosylation profiles, and mammalian cells are unsuitable for the production of toxic proteins. These issues are readily overcome by plant cell cultures, which have the machinery required for complex protein modifications and can sequester toxic proteins within subcellular compartments. In this chapter, we first compare the features of plant cell cultures and conventional expression systems, including the ability of plant cells to produce complex and/or toxic proteins. We discuss the main benefits of plant cells and whole plants, including high-throughput transient expression for rapid, low-cost process development reflecting the simplicity of upstream production. We then set out current challenges and trends in the upstream production and downstream processing phases of plant molecular farming. Finally, we discuss how these different elements can be

---

Authors Henrik Nausch and Matthias Knödler have contributed equally to this work.

H. Nausch
Department Bioprocess Engineeering, Fraunhofer Institute for Molecular Biology and Applied Ecology IME, Aachen, Germany
e-mail: henrik.nausch@ime.fraunhofer.de

M. Knödler
Department Bioprocess Engineeering, Fraunhofer Institute for Molecular Biology and Applied Ecology IME, Aachen, Germany

Institute for Molecular Biotechnology, Worringerweg 1, RWTH Aachen University, Aachen, Germany
e-mail: matthias.knoedler@rwth-aachen.de

J. F. Buyel (✉)
Institute for Molecular Biotechnology, Worringerweg 1, RWTH Aachen University, Aachen, Germany

Department of Biotechnology (DBT), Institute of Bioprocess Science and Engineering (IBSE), University of Natural Resources and Life Sciences (BOKU), Vienna, Austria
e-mail: johannes.buyel@rwth-aachen.de

R. Pörtner (ed.), *Biopharmaceutical Manufacturing*, Cell Engineering 11,
https://doi.org/10.1007/978-3-031-45669-5_8

combined into an integrated manufacturing facility based on plants and how this might differ from conventional production platforms.

Keywords Bioprocess engineering · Biopharmaceutical manufacturing · Plant molecular farming · Plant cell culture · Recombinant protein production

Abbreviations

ADCs	Antibody–drug conjugates
API	Active pharmaceutical ingredient
ATF	Alternating tangential flow filtration
ATPS	Aqueous two-phase systems
BChE	Butyrylcholinesterase
CFD	Computational fluid dynamics
CHO	Chinese hamster ovary
DSP	Downstream processing
EMA	European Medicines Agency
EPO	Erythropoietin
FDA	U.S. Food and Drug Agency
FM	Fresh mass
GMOs	Genetically modified organisms
GMP	Good manufacturing practice
GOI	Gene of interest
HCPs	Host cell proteins
HTS	High-throughput screening
LBM	Lattice Boltzmann method
LES	Large eddy simulation
mAbs	Monoclonal antibodies
MPC	Model-predictive control
NMR	Nuclear magnetic resonance
NTU	Nephelometric turbidity units
OTR	Oxygen transfer rate
OUR	Oxygen uptake rate
PAT	Process analytical technology
PCC	Plant cell culture
PCPs	Plant cell packs
PMF	Plant molecular farming
PMPs	Plant-made pharmaceuticals
PTM	Post-translational modification
QSAR	Quantitative structure–activity relationship
RANS	Reynolds averaged Navier–Stokes model
RITs	Recombinant immunotoxins
RNAi	RNA interference

scFv	Single-chain fragment variable
STRs	Stirred-tank (bio)reactors
T-DNA	Transfer DNA
TF	Transcription factor
TFF	Tangential flow filtration
USP	Upstream production
UTRs	Untranslated regions
VFUs	Vertical farming units
VLPs	Virus-like particles
VNPs	Viral nanoparticles
vvm	Vessel volumes per minute

Introduction

Recombinant proteins remain a major focus of research due to their importance in the biotechnology and biopharmaceutical industries. The latter has seen annual growth rates of 10–15% since 1982, when insulin was approved as the first recombinant biopharmaceutical produced in *Escherichia coli* (*E. coli*) [1]. This consistent growth has been possible because bottlenecks in manufacturing have repeatedly been overcome. For example, the advent of monoclonal antibodies (mAbs) as a novel product class in 1986 resulted in the development of Chinese hamster ovary (CHO) cells as one of the dominant expression systems in the biopharmaceutical industry [2]. CHO cells address the inability of *E. coli* and other bacteria to fold and modify complex proteins, allowing the formation of multiple disulfide bonds and the addition of glycans, as well as providing an efficient secretion system [3].

More recently, the 2020 SARS-CoV-2 pandemic has dramatically accelerated the development of nucleotide-based medicines, especially for vaccination [4–6]. These products can be altered more easily than their protein-based counterparts, allowing the rapid development and manufacturing of vaccines against new pathogen variants. This has raised questions about the benefits of protein products over nucleotide-based medicines in the future [7]. However, such comparisons focus on speed and adaptability without full consideration of functionality. For example, mRNA vaccines can target certain cells types and trigger the synthesis of non-toxic proteins [4], whereas cancer therapy may require the production of toxic drugs in a time-bound, targeted and localized manner, with minimal systemic off-target effects, and this is substantially more challenging or even impossible with nucleic acid-based therapeutics. Antibody–drug conjugates (ADCs) are antibodies covalently joined to highly toxic small-molecule drugs for cancer therapy, which cannot be synthesized by human cells [8]. Therefore, nucleotide-based medicines are unsuitable for such applications as well. ADCs can be replaced with recombinant immunotoxins (RITs), where the toxic component is also a protein and is translationally fused to the antibody [8]. Such constructs could theoretically be delivered as nucleic acids, but it is unclear how these toxic proteins could be produced in an autologous manner inside a patient's body without causing severe damage to the producer cells and surrounding cell populations. Hence, recombinant proteins remain essential for such applications.

Importantly, RITs and other toxic proteins cannot be produced directly in CHO cells, because their deleterious effects on the host cells would severely limit the resulting titers [9, 10]. Therefore, alternative expression systems are required to manufacture these classes of recombinant proteins. Just as CHO cells have replaced bacteria for the production of novel mAbs, plants and plant cell cultures (PCCs) are more suitable hosts for the manufacture of toxic products, reflecting the additional compartmentalization of plant cells, different signaling pathways, and the absence of targets for many proteins that are toxic in mammals [8, 11–13]. Safety, speed and sustainability are additional benefits of plant cells [13–16]. Importantly, biopharmaceuticals can be produced in plants and PCCs in compliance with good manufacturing practice (GMP), as shown for mAbs [17, 18], vaccines [19] and enzyme replacement therapies [20].

In this chapter, we briefly set out the benefits of plant-made pharmaceuticals (PMPs) before discussing the current challenges of industrial production. Given that most industry production capacity is based on bioreactors, we will focus on PCCs because they are compatible with this equipment, but whole plant-based production is covered elsewhere [21, 22]. We will initially discuss the ability of plant cells to produce complex proteins before considering the challenges during upstream production (USP) in conventional bioreactors and developments that may be required to resolve current shortcomings. Next, we consider downstream processing (DSP) options, how they match with established platform processes, and the benefits of mechanistic and hybrid modeling. Finally, we merge these aspects to identify the gaps between the current state of plant molecular farming and what may be required to convert this into "plant molecular manufacturing".

Advantages and Opportunities of Plants and PCCs for Molecular Farming

The advantages of recombinant protein production in plants have been widely discussed in the context of safety, scalability, speed and sustainability [21–25]. Here, we focus on the advantages from a product perspective, specifically the opportunities that plant cells offer to manufacturers of particular classes of proteins.

Expression of Toxins and Other Complex Biopharmaceutical Proteins in Plants

Plants can express proteins that are toxic when expressed in mammalian or bacterial cells because the molecular target of the toxic protein may not be present in plants, and even if it is, the toxic protein can be sequestered within a subcellular compartment such as the vacuole, which separates the product from its target [8, 26]. For example, ribosome inactivating proteins (RIPs) kill mammalian cells [8] but can be produced safely in plants. This was demonstrated for the lectin viscumin [27, 28], which was produced in Australian tobacco (*Nicotiana benthamiana*) with a yield of ~7 mg kg^{-1} fresh mass (FM) after purification. The plant-derived protein was approximately three-fold more active than the same protein expressed in *E. coli*, presumably because the latter did not add glycan structures necessary for the efficient internalization and intracellular trafficking of type-2 RIPs [27, 29].

RIPs have also been fused to antibodies as candidates for cancer therapy, which have been successfully produced in plants [8, 30]. The first RIT produced in plant cells was a single-chain fragment variable (scFv) antibody fused to the type-I RIP bryodin 1 from bryony (*Bryonia dioica*) [31]. Further examples include a truncated version of the *Pseudomonas aeruginosa* exotoxin A expressed in lettuce (*Lactuca sativa*) [30] and the scFv-RC-RNase fusion protein produced in transgenic tobacco (*Nicotiana tabacum*) with a yield of ~2 mg kg^{-1} FM, which was designed for the treatment of hepatocellular carcinoma [32]. Although no plant-made RITs have yet completed clinical development, there is growing industrial interest. For example, the Belgium start-up ATB Therapeutics[1] develops antibody–toxin fusions in plants for the treatment of recalcitrant and refractory cancers.

Another promising class of products for plant molecular farming is derived from plant viruses, including plant virus-like particles (VLPs), which lack nucleic acid [33], and plant viral nanoparticles (VNPs), which retain nucleic acid [34]. These products are based on the capsids of plant viruses, which spontaneously assemble from multiple copies of the same coat protein or different coat protein subunits. Because plant viruses have evolved to replicate and assemble in plants, heterologous VLPs and VNPs are efficiently assembled in tobacco cells. VLPs and VNPs can be used as imaging reagents, immunomodulators, vaccines and for anti-cancer therapy [35].

Another complex protein that can be manufactured in plants is spider silk, which is formed when one or more polypeptides known as spidroins assemble into helical structures. These products are valuable for nanotechnology and biomedicine because of their extraordinary toughness and elasticity [36, 37]. Spidroins can be used for nanoparticle-based drug delivery and for the preparation of hydrogels or 3D tissue implants for wound healing and tissue regeneration [38]. Such applications have already been validated in animal models [39]. Recombinant spidroins can be produced in transgenic plants such as rice (*Oryza sativa*) and alfalfa (*Medicago sativa*), which are more suitable for industrial-scale production than *E. coli* or mammalian cells [40].

Similarly, collagen triple helices are readily produced in plants [41–48]. Plant-derived collagen is currently marketed by CollPlant Biotechnologies[2] for

---

[1] https://www.atbtherapeutics.com/media/press-releases

[2] https://collplant.com/

regenerative and aesthetic medicine applications, and the recombinant protein fibers are highly biocompatible.

Oligomeric butyrylcholinesterase (BChE), used as bioscavenger for neurotoxic organophosphorus compounds present in pesticides and chemical warfare agents, is another example of a protein that is more suited to production in plants than mammalian cells [49–51]. Whereas mammalian cells produced mostly inactive mono- and dimeric BChE, active tetrameric BChE was formed in plant cells [49].

Production of Biopharmaceuticals with Engineered Glycans

Plants also offer distinct manufacturing advantages for therapeutic glycoproteins. One of the most common post-translational modifications (PTMs) in eukaryotes is *N*-linked glycosylation, but this is uncommon in bacteria [52]. The glycans produced in eukaryotes and bacteria are structurally distinct [53] and it is challenging to achieve humanized glycosylation profiles in bacteria [54]. In contrast, the glycosylation profiles of proteins produced in CHO cells feature both *N*-linked and *O*-linked glycans that are similar (albeit not identical) to human glycans, and CHO cells are therefore favored for the manufacturing of glycoproteins, currently accounting for more than 60% of approved biopharmaceuticals [55–57]. Even so, the non-human glycan structures formed in CHO cells sometimes have a more pertinent impact on manufacturing, and cannot always be fine-tuned by altering cell culture conditions [55]. For example, up to 80% of human erythropoietin (EPO) produced in CHO cells is discarded because the non-human glycosylation patterns interfere with its bioactivity during the treatment of anemia [58].

Plant glycans also differ from those found on human glycoproteins, but this can be advantageous. For example, recombinant human glucocerebrosidase, which is used for the treatment of Gaucher's disease (a lysosomal storage disorder), is active when expressed in plants and directed to accumulate in the vacuole because this results in the addition of mannose-tipped glycans that are recognized by receptors on human macrophages in contrast, CHO cells extend the glycans and render the protein inactive unless the glycans are trimmed *in vitro* [59]. Similarly, some plant-derived antibodies have proven to be less immunogenic and more efficacious compared to the same proteins expressed in CHO cells [60–65].

Where glycan structures are critical for recombinant protein activity, the production host can be genetically modified to favor the production of suitable glycosylation profiles. This can include the overexpression or inhibition of individual enzymes, such as sialyltransferases or sialidases [66]. This precise strategy has been used to improve the glycosylation profile of EPO produced in plants [61]. Genetic engineering can even achieve the complete humanization of plant-derived glycoproteins, which may support the approval of biopharmaceuticals with glycans that strongly affect pharmacokinetics (e.g., serum half-life) or immunogenicity [67, 68]. Plants do not typically add terminal sialic acid to *N*-acetylglucosamine (GlcNAc) residues in *N*-linked glycans, a modification typically found in humans

[12]. However, they do add core α-1,3-fucose and β-1,2-xylose, which are not found in mammals, and several modifications are therefore required to "humanize" the plant glycosylation machinery [69, 70]. Such modifications have been achieved by strategies such as the overexpression of missing enzymes and the suppression of unwanted reactions using tools such as RNA interference (RNAi) and CRISPR/Cas9 [11, 61, 71–73].

Building an Engineered Plant Line as a Platform for Biopharmaceutical Production

Although production hosts such as tobacco plants do not compete with food or feed crops, they contain toxic metabolites such as the alkaloid nicotine, which must be removed during DSP in order to comply with GMP regulations [11]. This has been addressed by the use of CRISPR/Cas9 to abolish the synthesis of nicotine, allowing the purification of recombinant proteins without additional steps to remove this metabolite [74]. The pathway was blocked by selectively knocking out six tobacco genes encoding berberine bridge enzyme-like (BBL) proteins and thereby preventing the final oxidation step in the nicotine synthesis pathway, leading to a >99% reduction in nicotine levels [74].

Further modifications to host plants can be introduced to increase product yields, including tRNA pool modifications, the suppression of gene silencing or protein degradation, and the optimization of stress tolerance [11]. Additional options include the reduction of endogenous oxidase [75] and protease [76] activities, or the co-expression of chaperones for improved protein folding [77].

Bottlenecks in the Production of Complex Biopharmaceuticals Using Plant Cells

Although plant-based expression platforms can produce complex biopharmaceutical proteins, the competitiveness of plant cells compared to traditional microbial and mammalian systems is currently limited by challenges that include (1) the USP yield, (2) DSP recovery, and (3) technological differences with conventional systems designed for microbial and mammalian cells.

A major challenge is the relatively low product yields of PCCs, which typically fall within the range 10–100 mg L^{-1}, equivalent to a productivity of 1–15 mg L^{-1} d^{-1} [1, 16, 78, 79]. The highest yields reported in PCCs are ~1 g L^{-1} or ~0.2 g L^{-1} d^{-1} [80]. This is at the low end of the yields achieved in microbial and mammalian cells, which generally fall within the range 5–20 g L^{-1} or 0.3–1.7 g L^{-1} d^{-1} [81–83] and can reach 50 g L^{-1} or 4.2 g L^{-1} d^{-1} [84].

In contrast to whole plants, PCCs typically do not require a complex clarification process because the products can be secreted into and recovered from the medium, similar to perfusion processes using mammalian cells, for example based on tangential flow filtration (TFF) or alternating tangential flow filtration (ATF) [84]. Chromatographic purification can be challenging if affinity-based separation steps are not available, as is often the case for non-antibody products.

Finally, there is limited industry acceptance and willingness to invest in novel plant-based expression systems. The barriers preventing broad market penetration as perceived from within the molecular farming community may include low productivity, high DSP costs and regulatory hurdles, but the outside perspective may simply be a lack of knowledge about the technology as well as the lack of large-scale manufacturing infrastructure [13, 85].

Plant-Made Pharmaceuticals: From Small-Scale Screening to Large-Scale Manufacturing

Transient Expression Systems to Assess the Manufacturability of Biopharmaceutical Candidates

Biopharmaceutical development can be accelerated if rapid, small-scale, high-throughput screening (HTS) tools are available to assess different product candidates [86, 87]. Initially, candidate screening solely relied on transient expression in whole plants. This system is based on *Agrobacterium tumefaciens* (recently renamed *Rhizobium radiobacter*) and its natural ability to mobilize transfer DNA (T-DNA) from a resident plasmid and translocate it into plant cells. The T-DNA can be modified to carry any gene of interest [88], which will be expressed in plant cells 3–5 days after the syringe injection or infiltration of bacteria, typically into the leaves. Several different vector systems are available for transient expression, including viral replicons based on geminivirus hyper-transcription [89], the magnICON platform [90, 91] and TRBO [92], as well as the pEAQ hyper-translation system [93]. The yield of recombinant protein can reach up to 6 g kg^{-1} FM [94]. Transient expression avoids the lengthy selection of transgenic plants (~6 months), therefore providing useful amounts of the target protein within 1 week. However, transient expression is often sensitive to changes in environmental conditions, resulting in substantial batch-to-batch variations [95].

The screening of the recombinant protein production in leaf discs achieved the first step toward more consistent expression [96], but the development of plant cell packs (PCPs) in a 96-well plate format has substantially improved the throughput and comparability of transient expression data in plant cells [86, 87]. PCPs are porous masses of plant cells molded to the shape of different containers, facilitating transient expression and recombinant protein production in a scalable manner. By casting the PCPs in 96-well plates, at least 2000 construct variants, including those with different promoters, untranslated regions (UTRs) and targeting signals, can be screened per day at costs of ~0.5 € per sample. Protein expression is often achieved within 3–5 days and yields can reach 2.5 g kg^{-1} [97]. Importantly, PCPs can be cast from diverse cells lines at scales ranging from 50–100 mg to 200–300 g FM, allowing their use for screening and production of protein for initial activity/functionality assessment. In this context, PCPs can also be used for the rapid and cost-efficient isotope labeling of proteins for nuclear magnetic resonance spectroscopy and subsequent structure elucidation, which is especially relevant for complex proteins that cannot be produced in mammalian or microbial systems [98].

Expression levels in PCPs often correlate well with transient expression in intact plants, facilitating scale-up as the products and processes mature [12, 17, 18, 21, 22, 85, 99, 100]. PCPs are also advantageous over transient expression in intact plants because PCPs offer a higher level of containment, which is an important safety measure when handling toxic products https://doi.org/10.1080/21655979.2023.2244235 and genetically modified organisms (GMOs) [101]. The consistent environment of PCCs also overcomes seasonal variations that affect whole plants, such as peaks of illumination and leaf temperature that reduce product yields in the summer [100]. These effects can be protein-specific, because some reports show that product yields are lower in the winter [17, 18].

PCCs and compatible screening technologies therefore increase the number of candidates that can be tested and optimized during early-stage biopharmaceutical development [26, 102] and can facilitate scale-up to expression in intact plants should this scale be required. However, expression levels in PCCs must be improved for the technology to be used not just for screening but also for manufacturing.

Plant Cell Cultures for the Production of Biopharmaceuticals

The industrial use of PCCs currently appears advantageous compared to intact plants because the latter require unique facilities such as greenhouses or vertical farming units (VFUs) [22]. Carrot (*Daucus carota*), rice (*Oryza sativa*) and tobacco (*Nicotiana tabacum*) cell lines have been used for the production of biopharmaceuticals because these cells (i) achieve a growth rate of (μ_{max}) of 0.03–0.04 h^{-1}, which is comparable to mammalian cells with a μ_{max} of 0.02–0.05 h^{-1}, (ii) can be grown to high cell densities of 600 g L^{-1} FM, and (iii) contain few, non-toxic secondary metabolites [14, 15, 103, 104]. However, the major current bottleneck is low protein accumulation and we consider some strategies to overcome this below.

Tailoring Bioreactor Design and Operation to Plant Cell Cultivation

Most strategies to increase protein accumulation in plant cells focus on the optimization of cultivation conditions and protein expression. Stirred-tank (bio)reactors (STRs) are typically used for the cultivation of plant cells because they enable batch, fed-batch, semi-continuous and continuous fermentation regimes [14, 15, 105–108] and the cultivation volume typically ranges between 50 and 100,000 L [109–117]. However, STRs are often designed for microbial or mammalian cells, so the design and cultivation conditions may not be suitable for plant cells [106]. For example, plant cells are substantially larger (50–100 μm diameter) than bacteria (~1 μm), yeast (5–10 μm) or mammalian cells (10–20 μm) and the use of optimal impeller types and configurations (e.g., Rushton, pitch-blade or marine) as well as optimal aeration strategies may not be sufficient to account for these differences [106, 118]. Approaches that overcame similar problems with the cultivation of filamentous fungi in STRs might be suitable for PCCs [119]. This includes options to handle the high viscosity of PCCs and the non-Newtonian rheology at high cell densities $>100\ g\ L^{-1}$, which can cause uneven cell distributions in the reactor and impair oxygen as well as nutrient supply [120].

Process optimization can also be achieved using computational fluid dynamic (CFD) models, which are applied in many microbial and mammalian cell systems. However, current CFD models often consider the cultivation medium as a two-phase-system, composed of a gas and liquid phase with the cells homogenously distributed in the liquid phase. Due to the larger size and volume of plant cells, this assumption might not be appropriate. Plant cells behave rather like solid particles and therefore need to be regarded as a solid phase in a three-phase-system as have recently been modeled.[3] Such three-phase CFD models can be created by numerical simulation, large eddy simulation (LES) or the Reynolds averaged Navier–Stokes (RANS) and Lattice Boltzmann method (LBM) models, but these currently require supercomputers [121, 122]. Lower-order models are therefore needed for the implementation of three-phase CFD models.

Cell distribution is also affected by foaming in the bioreactor, with up to 75% of the cell biomass accumulating on the foam [123–125]. Foam arises from extracellular polysaccharides, fatty acids and secreted plant proteins, and the severity of foaming depends on the bioreactor settings, the aeration rate, and the composition of the medium. To reduce foaming, high-speed vortexing marine impellers are typically used for PCCs because the downward stream submerges and dispenses the foam in the liquid phase. However, this may cause lethal shear stress [126]. For example, increasing the stirrer tip speed from 2.4 to 7.7 $m\ s^{-1}$ reduced the growth of tobacco cells by more than two-fold [113].

Notably, typical aeration rates for PCCs are 0.1–0.4 vessel volumes per minute (vvm) [109, 110, 113], but the oxygen demand of PCCs is less than that of microbial and mammalian cells [127], suggesting the gassing rate can be reduced below 0.1 vvm and down to 0.01 vvm, which would also limit the formation of foam. Similarly, bubble-free cultivation might prevent foaming, and can be achieved by

[3]Efficient Modeling of Bioreactors – From Lab Scale to Industrial Scale J. Khinast[1]; C. Witz[2]; P. Eibl[1]; [1]Graz University of Technology, Graz/A; [2]Simvantage GmbH, Graz/A; ProcessNet and DECHEMA-BioTechNet Jahrestagung 2022 "ProcessNet and DECHEMA-BioTechNet Jahrestagung 2022", 12–15 September 2022, Eurogress Aachen, Germany.

integrating gas-permeable membranes in the sparger. The feasibility of this approach was demonstrated in several studies in which cell growth was unaffected, even at high cell densities >500 g L^{-1} FM [124, 128–130].

Foaming can also be mitigated by using antifoams, although such additives reduce the oxygen transfer rate (OTR) from the gas to the liquid phase and the oxygen uptake rate (OUR) into cells from the liquid phase by up to 80% [127]. Therefore, antifoams are only added when foaming affects bioreactor sterility in microbial and mammalian cell systems [131]. For the high-density fermentation of PCCs, where the gassing rate cannot be reduced (or reducing it may be ineffective), antifoams can be added directly to the medium. Many types of antifoam agents are available for cell cultures [132], among which polypropylene glycol-based Pluronics such as L-61, F-68 and F-127 have already been used at concentrations of 0.01–0.05% for tobacco, carrot and rice cells [133–141].

Finally, tailored lighting conditions can be used adjust intermediate metabolite levels and plant cell growth [97, 142, 143]. Because both depend on protein and thus enzyme activity and abundance, illumination may also be exploited to improve PMP expression levels in the future. However, costs for photobioreactors are likely to be higher than for conventional counterparts, so this extra effort has to be balanced against the potentially increased product yields.

Improving Cell Growth and Product Yields by Medium Optimization

Several minimal media are available for plant cells, such as Murashige and Skoog [144], Linsmaier and Skoog [145], Gamborg's B5 [146], Schenk and Hildebrandt SH [147], and Chu N6 medium [148] (Table 8.1). These media contain a carbon source (often sucrose), ammonium nitrate and potassium nitrate as nitrogen sources, as well as potassium dihydrogen phosphate and dipotassium hydrogen phosphate as phosphate sources. Several mineral salts, vitamins and phytohormones/growth regulators are also included.

The carbon source is required because carbon dioxide assimilation is inhibited by the phytohormone 2,4-D [150], and sucrose is typically included for this purpose at a concentration of 87 mM (30 g L^{-1}), which appears optimal for cell growth. For example, replacing 87 mM sucrose with 197 mM glucose or mannitol reduced the growth and FM accumulation of tobacco cells by ~20% and ~80%, respectively [151, 152]. This may reflect the higher osmolality, causing osmotic stress that inhibits cell growth [97]. In rice cell cultures, 87 mM sucrose can be replaced with 80 mM pyruvic acid, 80 mM glyoxylic acid and a combination of amino acids (10 mM glycine, 60 mM glutamine, 20 mM aspartic acid and 13 mM arginine) but this did not improve the cell growth, whereas alternative carbon sources such as 80 mM mannitol, 80 mM glycerol and 80 mM lactic acid reduced cell growth [153]. Similarly, carrot cell suspension cultures were able to grow with glycerol as the sole carbon source [154]. These effects may reflect the different osmolarities of the medium or the different amounts of carbon available (e.g., the same concentration of sucrose and glucose yield different amounts of carbon because sucrose has

Table 8.1 Composition of basal media for plant cell suspension cultures, including vitamins

Component	Ingredient	Chemical formula	Unit	MS	B5	N6	SH	AA
Macro elements	Sucrose	$C_{12}H_{22}O_{11}$	mM	87.64	87.64	87.64	87.64	87.64
	Potassium nitrate	KNO_3	mM	18.79	24.73	27.99	24.73	[–]
	Ammonium nitrate	NH_4NO_3	mM	20.61	[–]	[–]	[–]	[–]
	Diammonium sulfate	$(NH_4)_2SO_4$	mM	[–]	1.01	3.50	[–]	[–]
	Ammonium dihydrogen phosphate	$(NH_4)H_2PO_4$	mM	[–]	[–]	[–]	2.61	[–]
	Potassium dihydrogen phosphate	KH_2PO_4	mM	1.25	[–]	2.92	[–]	1.31
	Sodium dihydrogen phosphate	NaH_2PO_4	mM	[–]	1.09	[–]	[–]	[–]
	Glycine	$C_2H_5NO_2$	mM	0.27	[–]	0.27	[–]	1.0
	Glutamine	$C_5H_{10}N_2O_3$	mM	[–]	[–]	[–]	[–]	6.0
	Aspartic acid	$C_4H_7NO_4$	mM	[–]	[–]	[–]	[–]	2.0
	Arginine	$C_6H_{14}N_4O_2$	mM	[–]	[–]	[–]	[–]	1.5
Micro elements	Cobalt(II) chloride	$CoCl_2 \times 6H_2O$	nM	110.00	110.00	[–]	440.00	100.00
	Copper(II) sulfate	$CuSO_4 \times 5H_2O$	nM	100.00	100.00	[–]	200.00	100.00
	Sodium ferric ethylenediaminetetraacetate	FeNaEDTA	μM	100.00	100.00	100.00	53.95	[–]
	Ferric sulfate	$FeSO_4 \times 7H_2O$	μM	[–]	[–]	[–]	[–]	100.17
	Disodium ethylenediaminetetraacetate	Na_2EDTA	μM	[–]	[–]	[–]	[–]	100.79
	Boric acid	H_3BO_3	μM	100.27	48.52	25.88	80.86	100.27
	Potassium iodide	KI	μM	5.00	4.52	4.81	6.02	5.00
	Manganese(II) sulfate	$MnSO_4 \times H_2O$	μM	100.00	59.16	19.70	59.17	131.95
	Sodium molybdate	$Na_2MoO_4 \times 2H_2O$	μM	1.03	1.03	[–]	0.41	1.03
	Zinc sulfate	$ZnSO_4 \times 7H_2O$	μM	29.91	6.96	5.22	3.48	29.91
	Calcium chloride	$CaCl_2$	mM	2.99	1.02	1.13	1.36	3.96
	Potassium chloride	KCl	mM	[–]	[–]	[–]	[–]	39.44
	Magnesium sulfate	$MgSO_4$	mM	1.50	1.01	0.75	1.62	3.07

Vitamins	Myo-inositol	$C_6H_{12}O_6$	μM	554.94	554.94	[–]	5549.40	554.94
	Nicotinic acid	$C_6H_5NO^2$	μM	4.06	8.12	4.06	40.60	4.06
	HCl-pyridoxine	$C_8H_{11}NO_3 \times HCl$	μM	2.43	4.86	2.96	2.43	0.49
	HCl-thiamine	$C_{12}H_{17}ClN_4OS \times HCl$	μM	0.30	29.65	2.43	1.50	1.50
Phytohormone	2,4-dichlorophenoxyacetic acid	$C_8H_6Cl_2O_3$	μM	0.91	0.91	0.91	0.91	0.91

AA Amino acid medium [149], *B5* Gamborg's B5 medium including vitamins (G0210) [146], *MS* Murashige and Skoog medium including vitamins (M0222) [144], *N6* Chu N6 medium including vitamins (0204) [148], *SH* Schenk and Hildebrandt medium including vitamins (S0225) [147]

six carbon atoms and glucose only three). Both aspects must be taken into account when changing the carbon source. A moderate osmolality and sufficient carbon source is best achieved by using fed-batch cultivation as the default strategy for PCCs, although concentrated fed-batch, perfusion or continuous cultures may perform even better.

Whereas the carbon source has a moderate effect on cell growth, the nitrogen source has a much more profound impact (Table 8.1). Standard MS medium contains 18.8 mM potassium nitrate and 20.6 mM ammonium nitrate, corresponding to 20.6 mM ammonium and 39.4 mM nitrate and thus a ratio of 1.00:0.52 [144]. The same is true for Linsmaier and Skoog medium, which is optimized for the cultivation of tobacco cells [145]. However, other media have much lower levels of ammonium, including Gamborg's B5 medium (24.7 mM potassium nitrate, 1.0 mM diammonium sulfate = ammonium/nitrate ratio of 1.00:0.08) [146], SH medium (24.7 mM potassium nitrate, 2.6 mM ammonium dihydrogenphosphate = ammonium/nitrate ratio of 1.00:0.11) [147], and Chu N6 medium (28.0 mM potassium nitrate, 3.5 mM diammonium sulfate = ammonium/nitrate ratio of 1.00:0.25) [148]. This is because higher ammonium concentrations can be toxic and also inhibit nitrate uptake [155, 156]. In systematic studies of the ammonium/nitrate ratio, the standard MS medium was found to promote the optimal growth of BY-2 cells [80], although others have reported that higher ammonium levels increase the growth of these cells even more (18.8 mM potassium nitrate, 20.6 mM ammonium nitrate, 20 mM ammonium chloride = ammonium/nitrate ratio of 1.00:1.03) [152]. The growth rate of tobacco cells fell by 30% when the nitrate level was increased (18.8 mM potassium nitrate, 20.6 mM ammonium nitrate, 100 mM potassium nitrate = ammonium/nitrate ratio of 1.00:0.15), but this resulted in a 17-fold increase in product accumulation (recombinant antibody 2G12) from 0.7 mg to 12.0 mg L^{-1} [133]. Although ammonium seems to have a beneficial effect on the growth of BY-2 cells in suspension, using ammonium as the sole nitrogen source (20 mM ammonium chloride) resulted in no growth [133]. Moreover it has been shown that the addition of all proteinogenic amino acids to the medium can increase recombinant protein yields, as shown for a mAb, by threefold from 15 to 45 mg kg^{-1} FM [117]. Notably, the AA medium commonly used for rice cell suspension cultures lacks a dedicated source of ammonium and nitrate, relying on amino acids as the nitrogen source instead: namely 1.0 mM glycine, 6.0 mM glutamine, 2.0 mM aspartic acid and 1.54 mM arginine [149]. The amino acids are assumed to reduce tendency of rice cells to aggregate [157]. The rational optimization of ammonium, nitrate and amino acid components in minimal media therefore has the potential to improve both cell growth and product formation. Because nitrate promotes cell growth and biomass formation and ammonium promotes the expression of recombinant proteins, PCCs can be grown in two different media, the first being a growth medium and the second an induction medium. The latter can be added as a pulse feed before harvesting. However, an alternative strategy is the semi-continuous fermentation of rice cells followed by the induction of protein expression by replacing the sucrose-containing medium with a sucrose-free induction medium [49–51].

Dihydrogenphosphate is used as the phosphate source in all media (Table 8.1). Standard MS medium contains 2.7 mM potassium dihydrogenphosphate compared to 1.1 mM sodium dihydrogenphosphate in Gamborg's B5 medium, 2.6 mM sodium dihydrogenphosphate in SH medium, and 2.9 mM potassium dihydrogenphosphate in Chu N6 medium. Increasing the phosphate concentration in MS medium from 2.7 to 10.0 mM boosted biomass accumulation fivefold from 10% to 50% packed cell volume (PCV) (30 to 240 g L^{-1}) as well as the yield of the recombinant antibody 2G12 ~10-fold from 2 to 19 mg L^{-1} [134]. Accordingly, elevating the phosphate concentration in the medium or as an individual feed can improve growth and product formation. In this context, a separate phosphate feed could be used to adjust the phosphate concentration in the medium and thereby control cell growth and biomass formation and accordingly the harvest time.

Mineral salts and vitamins were found to be essential for the growth of tobacco BY-2 cells, but increasing their concentration did not have any beneficial effect [134]. In the case of phytohormones/growth regulators, different types of auxins such as 2,4-D, 1-naphthaleneacetic acid (NAA), indoleacetic acid (IAA) and indole-3-butyric acid (IBA) can be used, sometimes combined with cytokinins such as kinetin or 6-benzylaminopurine (BAP), and gibberellic acid (GA) in equal concentrations in the medium. Auxins were found to be essential for the growth of BY-2 cells but the type of auxin was not important, and the addition of cytokinins and gibberellic acid had no pronounced effect [158].

Improving Biomass and Protein Yield by Decoupling Growth and Product Formation

The decoupling of plant cell growth and recombinant protein expression ensures that all resources are initially directed to support cell growth and biomass formation before switching to product formation when a high cell density is achieved. As discussed above, transient expression decouples growth and product formation in intact plants [13] and the PCP format achieves effectively the same strategy in PCCs [87, 159]. However, the scalability of PCPs is currently limited, and a substantial amount of cell handling is required for PCP casting and infiltration. These limitations do not apply to stably-transformed PCCs, and other means for the temporal decoupling of growth and product formation are therefore applied during large-scale manufacturing.

One decoupling strategy is the replacement of constitutive promoters with inducible counterparts. This is already widely used in microbial systems, such as IPTG-inducible expression in *E. coli* [160] and methanol-inducible expression in yeast [161]. Various inducible promoters are available in plants (Table 8.2), among which the rice α-amylase 3D (RAmy3D) [172], dexamethasone (DEX) [175], tetracycline (TetA) [159], and oxidative stress-inducible peroxidase (POD) [171] promoters have already been tested in PCCs, as well as a red light-inducible promoter in protoplasts [169]. However, only the RAmy3D system has achieved significant improvements in yields, boosting them from 10–100 to 100–250 mg L^{-1} [184]. The RAmy3D

Table 8.2 Inducible expression cassettes used for the production of recombinant proteins in plants

Promoter	Induction type	Inductor	Plant species	References
Os.hb2	Physical	CO_2 increase (5%)	Rice	[162]
Adh	Physical	O_2 depletion (−)	Thale cress	[163]
HvDhn4s	Physical	Water depletion (leaf water content of 75%)	Bread wheat	[164].
HSP18.1	Physical	Heat shock (37–40 °C, 1–2 h)	Thale cress	[165]
HvHSP17	Physical	Heat shock (37–40 °C, 1–2 h)	Barley	[166]
TdCor39	Physical	Cold shock (4 °C, 4 h)	Durum wheat	[167]
OsWRKY71	Physical	Cold shock (4 °C, 4 h)	Rice	[167]
rd29	Physical	Cold shock (4 °C, 4 h)	Thale cress	[168]
PhyB	Physical	Active: red light, 660 nm (8 μmol $m^{-2} s^{-1}$) Inactive: far-red light 740 nm (20 μmol $m^{-2} s^{-1}$)	Tobacco	[169]
CarH	Physical	Active: green light, 525 nm (5 μmol $m^{-2} s^{-1}$)	Thale cress	[170]
SWPA2	Physical and chemical	UV-C (1 μmol $m^{-2} s^{-1}$) Hydrogen peroxide (1 mM = 34 mg L^{-1})	Tobacco	[171]
RAmy3D	Chemical	Sucrose starvation (−)	Rice	[172]
PR-1α	Chemical	Salicylic acid analog BTH (600 μg mL^{-1})	Tobacco	[173]
P35S-tetO_3	Chemical	Tetracycline (10 μM = 4 mg L^{-1})	Tobacco	[159]
PhCMV-etr$_8$	Chemical	Clarithromycin (100 μg mL^{-1})	Tobacco	[169]
P35S-PIR3$_9$	Chemical	Pristinamycin (200 μg mL^{-1})	Tobacco	[174]
P35S-4UAS$_4$	Chemical	Dexamethasone (10 mg L^{-1})	Tobacco, rice, cotton, slash pine	[175]
P35S-OLEXA$_8$	Chemical	17-β-estradiol (5 μM = 1 mg L^{-1})	Thale cress, maize, tobacco	[176, 177]
P35-GRE$_6$	Chemical	Muristerone (100 μM = 49 mg L^{-1})	Tobacco	[178]
CsV-EcR UBI-EcR	Chemical	Methoxyfenozide (10 μM = 4 mg L^{-1})	Thale cress, tobacco, maize	[179, 180]

(continued)

Table 8.2 (continued)

Promoter	Induction type	Inductor	Plant species	References
In2–2	Chemical	Safener Sa2 (50 mg L^{-1})	Thale cress	[181]
P35S-ACE1	Chemical	Copper sulfate (50 μM = 8 mg L^{-1})	Tobacco	[182]
P35S-AlcA	Chemical	Ethanol (0.1% v v^{-1})	Tobacco	[183]

promoter is induced by sucrose starvation, which is typically achieved by exchanging a sucrose-containing growth medium with a sucrose-free induction medium so that the gene expression starts at a defined time point [49, 51]. However, this medium exchange is likely to increase process costs. In addition to the Ramy3D promoter, various inducible promoter systems have been tested in whole plants and may be suitable in PCCs. These are divided into two groups according to whether the inductive stimulus is physical or chemical [185]. Members of the first group are usually (but not always) derived from plants, whereas those in the second can originate from plants, microbes or animals.

Physically inducible promoters are activated by abiotic factors, such as oxygen availability, temperature, drought and light. The rice Os.hb2 promoter [162] and the Arabidopsis (*Arabidopsis thaliana*) Adh promoter [163] are upregulated by elevated CO_2 levels (30%) and O_2 depletion, respectively. The Arabidopsis HSP18.1 promoter [165] and the barley (*Hordeum vulgare*) ortholog HvHSP17 [166] are induced by heat shock (37–40 °C for 1–2 h), whereas the cold-inducible TdCor39 and OsWRKY71 promoters from durum wheat (*Triticum durum*) and rice [167] as well as the rd29 promoter from Arabidopsis [168] are induced by exposing whole plants to 4 °C for 4 h. The barley HvDhn4s promoter is induced by drought [164]. Finally, the Arabidopsis PhyB promoter is active under red light (660 nm), but inactive under far-red light (740 nm) in tobacco protoplasts and the moss *Physcomitrella patens* [169]. The CarH/CarO promoter from the bacterium *Thermus thermophilus* is induced by green light (525 nm), and was shown to be functional in Arabidopsis protoplasts [170].

Chemically induced promoters are regulated by the presence or absence of a chemical agent or substrate. As stated above, a pertinent example originating from a plant is the rice RAmy3D promoter induced by sucrose starvation [49–51, 153, 175, 184, 186–196]. Others are directly activated by stress hormones such as abscisic acid (ABA), which affects the cold-inducible promoters mentioned above, or salicylic acid (SA) and its analog benzo-(1,2,3)-thiadiazole-7-carbothioic acid S-methyl ester (BTH), which stimulate the tobacco PR-1α promoter [173, 197]. Other chemically inducible systems consist of a chimeric transcription factor and a cognate promoter from a heterologous source. The inducer binds to the transcription factor, causing it to bind or release its target promoter segment, which may be fused to a canonical plant promoter such as CaMV 35 s [198–200]. Such promoters can be induced by antibiotics, steroids, pesticides, ethanol or copper, and their activity has been demonstrated in differentiated whole plants. The most common antibiotic-regulated promoters are derived from *E. coli* and are based on the binding of tetracycline analogs (TetR/TetA system) [169, 201–203] or clarithromycin (E/ETR system) [169], whereas the pristinamycin-stimulated PIP/pPIRON promoter system is from *Streptomyces* coelicolor [169, 174]. Steroid-inducible promoters include various systems based on dexamethasone (DEX) cassettes from rodents (GR/GVG, GR/LhG4 and GAL4/UAS) [175, 204, 205] as well as the human XVE/LexA system induced by estradiol, the activity of which has been demonstrated in tobacco, Arabidopsis [176] and maize (*Zea mays*) [177] However, the protein yields were lower than achieved using a constitutive promoter. Pesticide-regulated promoters were developed for ecdysone-agonist insecticides (muristerone, methoxyfenozide or tebufenozide/EcR), which originate from the tobacco budworm (*Heliothis virescens*), the spruce budworm (*Choristoneura fumiferana*) and the European corn borer (*Ostrinia nubilalis*), and were tested in tobacco and Arabidopsis [178, 179] or maize [180]. A herbicide-inducible promoter from maize was based on safener Sa5/In2-2 [181], but requires sulfonylurea-resistant plants. A copper-inducible promoter (ACE1/MT) from yeast (*Saccharomyces cerevisiae*) was used in tobacco (*Nicotiana tabacum*) [182] but is compromised by the toxicity of copper. Another prevalent chemically inducible promoter is the ethanol-inducible AlcR/AlcA system from *Aspergillus nidulans*, which was first established in tobacco [183] and then transferred to other plant species. The ethanol system has been used in transgenic plants and transient expression systems [206, 207]. It remains to be seen which inducible systems can be transferred to PCCs, but they should all work in theory and would be particularly useful for the expression of toxic proteins such as RITs, as discussed above.

Trends in Downstream Processing

Harvesting and Product Recovery from the Culture Supernatant

The DSP of pharmaceutical products has often been identified as a major cost driver in plant molecular farming [208–210]. However, DSP is a cost driver for mAb production in mammalian cells too [211, 212]. In addition, many previous bottlenecks, such as filter capacity and the high concentration of host cell proteins (HCPs), have recently been overcome [213, 214]. Importantly, several of these DSP challenges are more pronounced or even exclusive to intact plants rather than PCCs. For example, plant cells facilitate the secretion of recombinant proteins into the medium, from where the proteins can readily be isolated without the need for immediate cell harvesting or disruption, much like perfusion technologies such as ATF/TFF used in mammalian cell cultures [215, 216] and as already demonstrated for rice cells [50]. If the direct isolation of a secreted product from within the STR is impractical, or if the product is retained in the plant cells to improve stability [217, 218], a combination of TFF and flocculation as used for the dewatering of microalgae is suitable for the harvesting of cells from the culture supernatant [219]. The use of flocculants can improve the clarification of plant extracts [220], but care must be taken when designing the subsequent purification steps because highly charged flocculants can irreversibly bind and block chromatography resins with the opposite charge [221].

An (accelerated) settling step based on inclined plate settlers or centrifuges can also be used to recover plant cells or separate them from the medium. Due to the large size of plant cells (~100 μm diameter), sedimentation is achieved under gentler conditions such as 5 min at 4000 g [110] compared to 35 min at 7500 g for microbes [27] and the technical requirements for the process equipment are therefore less demanding and expensive. Centrifuges may also be used for subsequent clarification.

Extraction for Intracellular Product Recovery and Clarification

Plant cell lysis is required to recover intracellular products. Several mechanical, chemical and enzymatic methods are available for this purpose [222, 223], and they can be combined to improve effectiveness. Although many of these process options were developed for microbial cells, they have been adapted to the requirements of PCCs. For example, high-pressure homogenizers or bead mills can be used for the lysis of rice cells [51]. Plant cells typically have thicker walls than bacteria, but the large size of plant cells (~100 μm diameter) renders them more susceptible to mechanical lysis [224]. If mechanical lysis is inefficient, an optional cellulase pretreatment step may be included to reduce the rigidity of the plant cell wall [223]. However, such enzymes must comply with quality standards (e.g., food grade or pharma grade), thus increasing costs, and the enzymes must be removed during subsequent DSP steps.

Regardless of whether the product is intracellular or secreted, one or several clarification steps are typically necessary for solid–liquid separation because cell debris will carry over from TFF-based harvesting. As discussed above, centrifugation can be used as an initial clarification step, for example using disc-stack centrifuges [51]. This is especially attractive if the same centrifuge can be used at different stages of the process, such as cell recovery and solid–liquid separation. Disc-stack centrifuges can be operated semi-continuously or continuously and can achieve the force necessary for plant cell sedimentation [110]. However, they are expensive to operate and manual cleaning is labor-intensive, especially under GMP conditions. Furthermore, centrifugation alone is often insufficient to achieve a turbidity of ~50 nephelometric turbidity units (NTU), which is required for subsequent purification steps involving chromatography [225]. Therefore, centrifugation is often combined with depth filtration [110]. This increases process complexity (including stock keeping) but can be simplified if solid–liquid separation is solely based on filters that all use the same auxiliary pumps and balances. In contrast to centrifugation, where there is little interaction between the target protein and equipment, common filter components such as diatomaceous earth may bind to the target protein and reduce the percentage recovery [226, 227]. Altering process conditions such as pH and conductivity may help to reduce these interactions, or filters without diatomaceous earth can be used, but the latter can reduce filter performance, for example increasing filtrate turbidity to ~1500 NTU compared to 3 NTU for filters with diatomaceous earth [226]. Whereas conventional depth filters do not permit cell recovery and would therefore be limited to process streams containing secreted products, positive-pressure or vacuum drum filters facilitate the further processing of collected biomass [228, 229]. Importantly, integrated filters have recently been developed that allow cell capture, lysis and product elution, thereby greatly simplifying primary product recovery [230]. Ultimately, the choice of clarification strategy will depend on the separation problem (e.g., difference in size and density between product and colloids to be removed), product properties (e.g., binding to filter material) and process scale (e.g., availability of equipment). Accordingly, these parameters should be determined for each individual feed stream and then used as input for rational process development.

Non-chromatographic Purification

The purity of biopharmaceutical proteins must be high enough to ensure patient safety. Chromatography is the key technique for the purification and polishing of biopharmaceuticals both in the laboratory and in industrial processes because purities >95% can readily be achieved [231]. Unfortunately, the high material costs and the time-intensive optimization of chromatography process conditions are major cost drivers during the development and manufacturing of biopharmaceutical products, and the large proportion of HCPs can overcomplicate this issue for PMPs [232, 233]. Therefore, simple, scalable, non-chromatographic operations are attractive as initial steps to achieve purification, especially when these steps can be "platformized" by applying them to multiple products with minimal process adjustment [234].

The precipitation of HCPs can be induced by heat or low pH and this has been developed as such a platform approach for intact plants [235, 236]. The same approach can be applied during the processing of PCCs, which contain a subset of the HCPs found in whole plants of the same species. Typically, more than 60% of HCPs can be removed in a single step, such as heating to 60 °C or reducing the pH to 4.0, using scalable equipment such as heat exchangers [213]. Heat and low-pH treatment can even be combined to accommodate target proteins sensitives to extremes in either parameter [237].

If heat or pH cannot be used to remove HCPs due to product sensitivity, size-based separation by ultrafiltration/diafiltration has proven a successful alternative.

Because many plant HCPs form oligomers, their effective size is often >100 kDa [209]. Accordingly, membranes with a cut-off of 30, 70 or 100 kDa, sometimes even 300 kDa, may be used to separate HCPs from the product, as shown for cyanovirin N, ferritin and butyrylcholinesterase [50, 227, 238]. Continuous operation is possible [239], but concentration polarization or even membrane fouling can reduce the performance of such membrane-based separation operations [240, 241] and therefore adequate filter materials [242], module design [243] and control strategies need to be implemented, for example using Raman microspectroscopy [244]. Membrane-based purification comes with the additional benefit that membrane cascades can be designed to facilitate both purification and concentration and therefore remove the most abundant impurity in bioprocesses first, namely water [245]. This has been implemented for plant-derived recombinant proteins [238] and can be rapidly adapted to PCC-based processes.

Aqueous two-phase systems (ATPS) are promising chromatography-free purification methods that integrate solid–liquid separation, concentration and purification. This approach has been applied to plant-derived recombinant proteins such as mAbs [246], but extensive screening is needed to identify suitable conditions under which two phases will form with desirable distribution coefficients (typically >10), so the technology may be difficult to platformize [247, 248]. Currently, non-chromatographic purification steps often fail to achieve product purities >95% and chromatography therefore remains the major approach for purification and polishing.

Chromatographic Purification

Chromatography exploits the selective interaction of colloids (e.g., proteins) in a liquid phase with a stationary resin [249]. The resin is often functionalized with ligands to improve and/or alter this selectivity, which also depends on the product properties (e.g., size and surface charge distribution) as well as process conditions such as pH, conductivity and contact time. Therefore, optimizing chromatographic purification is a multi-parameter problem that is highly product dependent [250, 251]. It is difficult to provide generic recommendations for purification strategies [232] and a detailed discussion of chromatography is beyond the scope of this chapter.

A general approach is to use high-throughput methods combined with statistical experimentation [252] to identify suitable process conditions and purification sequences [253, 254]. One route to simplify purification is the use of affinity chromatography with selective ligands, such as Protein A for the capture of antibodies [255]. However, a suitable protein domain for affinity purification must be available for the target. The absence of a natural ligand can be overcome by expressing the target with added affinity tags, such as an N-terminal or C-terminal

His_6 sequence [256]. However, such tags persist in the final product and give rise to regulatory concerns including unintended immunogenicity [257]. As a consequence, cleavable tags have been developed that can be removed after an initial affinity capture step. The first generation of such tags often left one or more amino acids attached to the target protein because the cleavage site was within the tag [258]. More recently, a second generation of cleavable tags has been designed that is cleaved precisely, leaving the target protein in its original state [259]. However, these new technologies are often protected by intellectual property rights so that their use will be accompanied by in-licensing costs.

Chromatography Models Can Reduce Downstream Process Development and Regulatory Approval Times

Although some knowledge transfer is possible between different recombinant protein products, most optimization steps are product-specific. One trend in DSP development that is gaining momentum in academia and industry is chromatography modeling. Mechanistic models describe chromatography based on physicochemical principles which allow *a priori* predictions about the separation of proteins within a chromatography column [260]. Mechanistic models consist of equations describing mass transport and protein sorption to a given chromatography solid phase, such as an ion exchange resin [261]. They must be calibrated with empirical data, and the size of the dataset depends on the complexity of the model. If the dataset is limited, models may be simplified (e.g., by lumping mass transport parameters) to reduce the risk of overfitting [251]. A recent study indicated that mechanistic models can reduce the experimental effort by ~75% compared to empirical process characterization, as shown for a mAb purification step [262]. Furthermore, the models can be directly applied as digital twins to facilitate model-predictive control. This is well in line with the continuous optimal performance concept as proposed by the European Medicines Agency (EMA) and U.S. Food and Drug Administration (FDA) [251, 262–265].

Mechanistic models can be complemented or augmented by data-driven approaches yielding so-called hybrid models that are particularly useful for the early-stage development of chromatographic purification processes, when feed stocks are often too complex for a mechanistic description. For example, a recent study described the use of a quantitative structure–activity relationship (QSAR) model for the *in silico* prediction of antibody Fab fragments on different chromatography resins after generating a training data set using a library of different product variants [266]. Such approaches can reduce process development times substantially and could become an essential tool for the development of chromatography steps in the future. Detailed guidance about the establishment of such a modeling approach has recently been published [251].

Transforming Plant Molecular Farming in Plant Molecular Manufacturing

In this chapter, we have shown that PCCs can be advantageous for the production of complex and potentially toxic recombinant proteins that require post-translational modifications such as glycosylation and disulfide bonds. Whereas substantial improvements have been achieved for the accumulation of recombinant proteins in intact plants, the system is scalable in theory but currently lacks global production capacity, especially in a GMP context. In contrast, PCCs are largely compatible with existing fermentation infrastructure and rapid screening tools are available, but the performance of plant cells in STRs is currently inferior to microbial systems and mammalian cell cultures due to low product yields. Accordingly, we think that one route to advance plant molecular farming (the production of recombinant proteins and metabolites in plants) to "plant molecular manufacturing", which we define here as the cost-efficient, economically viable and scalable production of said products, is to first adapt PCCs to the existing global manufacturing infrastructure (STRs) and once industry and regulators have built understanding and trust in the platform explore the additional benefits and opportunities of production in intact plants.

To do so, it will be important to establish genetically customized plant cell lines that can be used as platforms for transient and stable expression. The rational development of defined, high-performance media along with corresponding cultivation strategies (e.g., reactor operation, feeding regimes and inducible product formation), should go hand in hand with this activity. Finally, DSP operations should be further tailored to the requirements of PCCs, including cell separation by drum filters together with corresponding small-scale models and devices. Once these improvements are in place, plants, plant cells and plant molecular manufacturing can complement existing host systems for the production of novel biopharmaceuticals and can help to facilitate global, democratic and sustainable access to high-quality medicines in the future.

Acknowledgements We wish to thank Dr. Richard M Twyman for editorial assistance. This work was supported by the Deutsche Forschungsgemeinschaft (DFG) in the framework of the Research Training Group "Tumor-targeted Drug Delivery" under the grant no. 331065168. The authors have no conflict of interest to declare.

Conflict of Interest Statement The authors have no conflict of interest to declare.

References

1. Karki U, Fang H, Guo W et al (2021) Cellular engineering of plant cells for improved therapeutic protein production. Plant Cell Rep. https://doi.org/10.1007/s00299-021-02693-6
2. Lu R-M, Hwang Y-C, Liu I-J et al (2020) Development of therapeutic antibodies for the treatment of diseases. J Biomed Sci 27:1. https://doi.org/10.1186/s12929-019-0592-z

3. O'Flaherty R, Bergin A, Flampouri E et al (2020) Mammalian cell culture for production of recombinant proteins: a review of the critical steps in their biomanufacturing. Biotechnol Adv 43:107552. https://doi.org/10.1016/j.biotechadv.2020.107552
4. Kyriakidis NC, López-Cortés A, González EV et al (2021) SARS-CoV-2 vaccines strategies: a comprehensive review of phase 3 candidates. npj Vaccines 6:28. https://doi.org/10.1038/s41541-021-00292-w
5. Barbier AJ, Jiang AY, Zhang P et al (2022) The clinical progress of mRNA vaccines and immunotherapies. Bio/Technology 40:840–854. https://doi.org/10.1038/s41587-022-01294-2
6. Tenchov R, Bird R, Curtze AE et al (2021) Lipid nanoparticles─from liposomes to mRNA vaccine delivery, a landscape of research diversity and advancement. ACS Nano 15:16982–17015. https://doi.org/10.1021/acsnano.1c04996
7. Damase TR, Sukhovershin R, Boada C et al (2021) The limitless future of RNA therapeutics. Front Bioeng Biotechnol 9. https://doi.org/10.3389/fbioe.2021.628137
8. Knödler M, Buyel JF (2021) Plant-made immunotoxin building blocks: a roadmap for producing therapeutic antibody-toxin fusions. Biotechnol Adv 47:107683. https://doi.org/10.1016/j.biotechadv.2020.107683
9. Tawfiq Z, Caiazza NC, Kambourakis S et al (2020) Synthesis and biological evaluation of antibody drug conjugates based on an antibody expression system: Conamax. ACS Omega 5:7193–7200. https://doi.org/10.1021/acsomega.9b03628
10. Ou J, Si Y, Goh K et al (2018) Bioprocess development of antibody-drug conjugate production for cancer treatment. PLoS One 13:e0206246. https://doi.org/10.1371/journal.pone.0206246
11. Buyel JF, Stöger E, Bortesi L (2021) Targeted genome editing of plants and plant cells for biomanufacturing. Transgenic Res. https://doi.org/10.1007/s11248-021-00236-z
12. Fischer R, Holland T, Sack M et al (2021) Glyco-engineering of plant-based expression systems. In: Rapp E, Reichl U (eds) Advances in glycobiotechnology. Springer, Cham, pp 137–166
13. Fischer R, Buyel JF (2020) Molecular farming – the slope of enlightenment. Biotechnol Adv 40:107519. https://doi.org/10.1016/j.biotechadv.2020.107519
14. Santos RB, Abranches R, Fischer R et al (2016) Putting the spotlight back on plant suspension cultures. Front Plant Sci 7:297. https://doi.org/10.3389/fpls.2016.00297
15. Xu J, Zhang N (2014) On the way to commercializing plant cell culture platform for biopharmaceuticals: present status and prospect. Pharm Bioprocess 2:499–518. https://doi.org/10.4155/pbp.14.32
16. Xu J, Ge X, Dolan MC (2011) Towards high-yield production of pharmaceutical proteins with plant cell suspension cultures. Biotechnol Adv 29:278–299. https://doi.org/10.1016/j.biotechadv.2011.01.002
17. Ma JK-C, Drossard J, Lewis D et al (2015) Regulatory approval and a first-in-human phase I clinical trial of a monoclonal antibody produced in transgenic tobacco plants. Plant Biotechnol J 13:1106–1120. https://doi.org/10.1111/pbi.12416
18. Sack M, Rademacher T, Spiegel H et al (2015) From gene to harvest: insights into upstream process development for the GMP production of a monoclonal antibody in transgenic tobacco plants. Plant Biotechnol J 13:1094–1105. https://doi.org/10.1111/pbi.12438
19. Ward BJ, Séguin A, Couillard J et al (2021) Phase III: randomized observer-blind trial to evaluate lot-to-lot consistency of a new plant-derived quadrivalent virus like particle influenza vaccine in adults 18-49 years of age. Vaccine 39:1528–1533. https://doi.org/10.1016/j.vaccine.2021.01.004
20. Grabowski GA, Golembo M, Shaaltiel Y (2014) Taliglucerase alfa: an enzyme replacement therapy using plant cell expression technology. Mol Genet Metab 112:1–8. https://doi.org/10.1016/j.ymgme.2014.02.011
21. Buyel JF, Twyman RM, Fischer R (2017) Very-large-scale production of antibodies in plants: the biologization of manufacturing. Biotechnol Adv 35:458–465. https://doi.org/10.1016/j.biotechadv.2017.03.011

22. Huebbers JW, Buyel JF (2021) On the verge of the market – plant factories for the automated and standardized production of biopharmaceuticals. Biotechnol Adv 46:107681. https://doi.org/10.1016/j.biotechadv.2020.107681
23. Commandeur U, Twyman RM (2004) Biosafety aspects of molecular farming in plants. In: Molecular farming. Wiley, Weinheim, pp 251–266
24. Buyel JF (2019) Plant molecular farming – integration and exploitation of side streams to achieve sustainable biomanufacturing. Front Plant Sci 9:1893. https://doi.org/10.3389/fpls.2018.01893
25. Tusé D, Nandi S, McDonald KA et al (2020) The emergency response capacity of plant-based biopharmaceutical manufacturing – what it is and what it could be. Front Plant Sci 11:1573. https://doi.org/10.3389/fpls.2020.594019
26. Opdensteinen P, Dietz SJ, Gengenbach BB et al (2021) Expression of biofilm-degrading enzymes in plants and automated high-throughput activity screening using experimental Bacillus subtilis biofilms. Front Bioeng Biotechnol 9:714. https://doi.org/10.3389/fbioe.2021.708150
27. Gengenbach BB, Keil LL, Opdensteinen P et al (2019) Comparison of microbial and transient expression (tobacco plants and plant-cell packs) for the production and purification of the anticancer mistletoe lectin viscumin. Biotechnol Bioeng 116:2236–2249. https://doi.org/10.1002/bit.27076
28. Buyel JF (2018) Plants as sources of natural and recombinant anti-cancer agents. Biotechnol Adv 36:506–520. https://doi.org/10.1016/j.biotechadv.2018.02.002
29. Taubenschmid J, Stadlmann J, Jost M et al (2017) A vital sugar code for ricin toxicity. Cell Res 27:1351–1364. https://doi.org/10.1038/cr.2017.116
30. Mirzaee M, Jalali-Javaran M, Moieni A et al (2018) Expression of VGRNb-PE immunotoxin in transplastomic lettuce (Lactuca sativa L.). Plant Mol Biol 97:103–112. https://doi.org/10.1007/s11103-018-0726-9
31. Francisco JA, Gawlak SL, Miller M et al (1997) Expression and characterization of bryodin 1 and a bryodin 1-based single-chain immunotoxin from tobacco cell culture. Bioconjug Chem 8:708–713. https://doi.org/10.1021/bc970107k
32. Cui L, Peng H, Zhang R et al (2012) Recombinant hHscFv-RC-RNase protein derived from transgenic tobacco acts as a bifunctional molecular complex against hepatocellular carcinoma. Biotechnol Appl Biochem 59:323–329. https://doi.org/10.1002/bab.1039
33. Peyret H, Steele JFC, Jung J-W et al (2021) Producing vaccines against enveloped viruses in plants: making the impossible, difficult. Vaccines 9:780
34. Rybicki EP (2020) Plant molecular farming of virus-like nanoparticles as vaccines and reagents. WIREs Nanomed Nanobiotechnol 12:e1587. https://doi.org/10.1002/wnan.1587
35. Dickmeis C, Kauth L, Commandeur UH (2020) From infection to healing: the use of plant viruses in bioactive hydrogels. Wiley interdisciplinary reviews: WIREs (2020). https://doi.org/10.1002/wnan.1662. doi:https://doi.org/10.18154/RWTH-2020-08445
36. Hauptmann V, Weichert N, Rakhimova M et al (2013) Spider silks from plants – a challenge to create native-sized spidroins. Biotechnol J 8:1183–1192. https://doi.org/10.1002/biot.201300204
37. Scheller J, Gührs KH, Grosse F et al (2001) Production of spider silk proteins in tobacco and potato. Nat Biotechnol 19:573–577. https://doi.org/10.1038/89335
38. Hardy JG, Torres-Rendon JG, Leal-Egaña A et al (2016) Biomineralization of engineered spider silk protein-based composite materials for bone tissue engineering. Materials (Basel):9. https://doi.org/10.3390/ma9070560
39. Debabov VG, Bogush VG (2020) Recombinant spidroins as the basis for new materials. ACS Biomater Sci Eng 6:3745–3761. https://doi.org/10.1021/acsbiomaterials.0c00109
40. Whittall DR, Baker KV, Breitling R et al (2021) Host systems for the production of recombinant spider silk. Trends Biotechnol 39:560–573. https://doi.org/10.1016/j.tibtech.2020.09.007

41. Brodsky B, Kaplan DL (2013) Shining light on collagen: expressing collagen in plants. Tissue Eng A 19:1499–1501. https://doi.org/10.1089/ten.TEA.2013.0137
42. Stein H, Wilensky M, Tsafrir Y et al (2009) Production of bioactive, post-translationally modified, heterotrimeric, human recombinant type-I collagen in transgenic tobacco. Biomacromolecules 10:2640–2645. https://doi.org/10.1021/bm900571b
43. Shilo S, Roth S, Amzel T et al (2013) Cutaneous wound healing after treatment with plant-derived human recombinant collagen flowable gel. Tissue Eng A 19:1519–1526. https://doi.org/10.1089/ten.TEA.2012.0345
44. Willard JJ, Drexler JW, Das A et al (2013) Plant-derived human collagen scaffolds for skin tissue engineering. Tissue Eng A 19:1507–1518. https://doi.org/10.1089/ten.TEA.2012.0338
45. Wang T, Lew J, Premkumar J et al (2017) Production of recombinant collagen: state of the art and challenges. Eng Biol 1:18–23. https://doi.org/10.1049/enb.2017.0003
46. Seror J, Stern M, Zarka R et al (2021) The potential use of novel plant-derived recombinant human collagen in aesthetic medicine. Plastic Reconstr Surg:148
47. Ruggiero F, Exposito J-Y, Bournat P et al (2000) Triple helix assembly and processing of human collagen produced in transgenic tobacco plants. FEBS Lett 469:132–136. https://doi.org/10.1016/S0014-5793(00)01259-X
48. Merle C, Perret S, Lacour T et al (2002) Hydroxylated human homotrimeric collagen I in agrobacterium tumefaciens-mediated transient expression and in transgenic tobacco plant. FEBS Lett 515:114–118. https://doi.org/10.1016/s0014-5793(02)02452-3
49. Corbin JM, Hashimoto BI, Karuppanan K et al (2016) Semicontinuous bioreactor production of recombinant butyrylcholinesterase in transgenic rice cell suspension cultures. Front Plant Sci 7:412. https://doi.org/10.3389/fpls.2016.00412
50. Corbin JM, Kailemia MJ, Cadieux CL et al (2018) Purification, characterization, and N-glycosylation of recombinant butyrylcholinesterase from transgenic rice cell suspension cultures. Biotechnol Bioeng 115:1301–1310. https://doi.org/10.1002/bit.26557
51. Corbin JM, McNulty MJ, Macharoen K et al (2020) Technoeconomic analysis of semicontinuous bioreactor production of biopharmaceuticals in transgenic rice cell suspension cultures. Biotechnol Bioeng 117:3053–3065. https://doi.org/10.1002/bit.27475
52. Helenius A, Aebi M (2001) Intracellular functions of N-linked glycans. Science 291:2364–2369. https://doi.org/10.1126/science.291.5512.2364
53. Weerapana E, Imperiali B (2006) Asparagine-linked protein glycosylation: from eukaryotic to prokaryotic systems. Glycobiology 16:91R–101R. https://doi.org/10.1093/glycob/cwj099
54. Valderrama-Rincon JD, Fisher AC, Merritt JH et al (2012) An engineered eukaryotic protein glycosylation pathway in Escherichia coli. Nat Chem Biol 8:434–436. https://doi.org/10.1038/nchembio.921
55. Hossler P, Khattak SF, Li ZJ (2009) Optimal and consistent protein glycosylation in mammalian cell culture. Glycobiology 19:936–949. https://doi.org/10.1093/glycob/cwp079
56. Kim JY, Kim Y-G, Lee GM (2012) CHO cells in biotechnology for production of recombinant proteins: current state and further potential. Appl Microbiol Biotechnol 93:917–930. https://doi.org/10.1007/s00253-011-3758-5
57. Goh JB, Ng SK (2018) Impact of host cell line choice on glycan profile. Crit Rev Biotechnol 38:851–867. https://doi.org/10.1080/07388551.2017.1416577
58. Jacobs P, Callewaert N (2009) N-glycosylation engineering of biopharmaceutical expression systems. Curr Mol Med 9:774–800. https://doi.org/10.2174/156652409789105552
59. Shaaltiel Y, Bartfeld D, Hashmueli S et al (2007) Production of glucocerebrosidase with terminal mannose glycans for enzyme replacement therapy of Gaucher's disease using a plant cell system. Plant Biotechnol J 5:579–590. https://doi.org/10.1111/j.1467-7652.2007.00263.x
60. Shanmugaraj B, I Bulaon CJ, Phoolcharoen W (2020) Plant molecular farming: a viable platform for recombinant biopharmaceutical production. Plants (Basel) 9:842. https://doi.org/10.3390/plants9070842
61. Kallolimath S, Castilho A, Strasser R et al (2016) Engineering of complex protein sialylation in plants. Proc Natl Acad Sci 113:9498. https://doi.org/10.1073/pnas.1604371113

62. Rapp E, Reichl U (2021) Advances in glycobiotechnology, Advances in biochemical engineering/biotechnology, 0724-6145, vol 175. Springer, Cham
63. Chen Q, Davis KR (2016) The potential of plants as a system for the development and production of human biologics. F1000Res 5. https://doi.org/10.12688/f1000research.8010.1
64. Christou P, Dhingra A, Slamet-Loedin IH et al (2021) Preface: genome editing in plants. Transgenic Res 30:317–320. https://doi.org/10.1007/s11248-021-00268-5
65. Navarre C, Smargiasso N, Duvivier L et al (2017) N-glycosylation of an IgG antibody secreted by Nicotiana tabacum BY-2 cells can be modulated through co-expression of human β-1,4-galactosyltransferase. Transgenic Res 26:375–384. https://doi.org/10.1007/s11248-017-0013-6
66. Wang Q, Yin B, Chung C-Y et al (2017) Glycoengineering of CHO cells to improve product quality. Methods Mol Biol 1603:25–44. https://doi.org/10.1007/978-1-4939-6972-2_2
67. Cosgrave E, Kattla JJ, Campbell MP et al (2017) Glycomics. In: Comprehensive biotechnology. Elsevier, Amsterdam, pp 413–434
68. Croset A, Delafosse L, Gaudry J-P et al (2012) Differences in the glycosylation of recombinant proteins expressed in HEK and CHO cells. J Biotechnol 161:336–348. https://doi.org/10.1016/j.jbiotec.2012.06.038
69. Margolin EA, Strasser R, Chapman R et al (2020) Engineering the plant secretory pathway for the production of next-generation pharmaceuticals. Trends Biotechnol 38:1034–1044. https://doi.org/10.1016/j.tibtech.2020.03.004
70. Jansing J, Bortesi L (2022) Knockout of glycosyltransferases in Nicotiana benthamiana by genome editing to improve glycosylation of plant-produced proteins. Methods Mol Biol 2480: 241–284. https://doi.org/10.1007/978-1-0716-2241-4_14
71. Strasser R, Stadlmann J, Schähs M et al (2008) Generation of glyco-engineered Nicotiana benthamiana for the production of monoclonal antibodies with a homogeneous human-like N-glycan structure. Plant Biotechnol J 6:392–402. https://doi.org/10.1111/j.1467-7652.2008.00330.x
72. Son S, Park SR (2022) Challenges facing CRISPR/Cas9-based genome editing in plants. Front Plant Sci 13:902413. https://doi.org/10.3389/fpls.2022.902413
73. Strasser R, Altmann F, Steinkellner H (2014) Controlled glycosylation of plant-produced recombinant proteins. Curr Opin Biotechnol 30:95–100. https://doi.org/10.1016/j.copbio.2014.06.008
74. Schachtsiek J, Stehle F (2019) Nicotine-free, nontransgenic tobacco (Nicotiana tabacum l.) edited by CRISPR-Cas9. Plant Biotechnol J 17:2228–2230. https://doi.org/10.1111/pbi.13193
75. González MN, Massa GA, Andersson M et al (2019) Reduced enzymatic browning in potato tubers by specific editing of a polyphenol oxidase gene via ribonucleoprotein complexes delivery of the CRISPR/Cas9 system. Front Plant Sci 10:1649. https://doi.org/10.3389/fpls.2019.01649
76. Hoernstein SNW, Fode B, Wiedemann G et al (2018) Host cell proteome of Physcomitrella patens harbors proteases and protease inhibitors under bioproduction conditions. J Proteome Res 17:3749–3760. https://doi.org/10.1021/acs.jproteome.8b00423
77. Wakasa Y, Yasuda H, Oono Y et al (2011) Expression of ER quality control-related genes in response to changes in BiP1 levels in developing rice endosperm. Plant J 65:675–689. https://doi.org/10.1111/j.1365-313X.2010.04453.x
78. Owczarek B, Gerszberg A, Hnatuszko-Konka K (2019) A brief reminder of systems of production and chromatography-based recovery of recombinant protein biopharmaceuticals. Biomed Res Int 2019:4216060. https://doi.org/10.1155/2019/4216060
79. Tripathi NK, Shrivastava A (2019) Recent developments in bioprocessing of recombinant proteins: expression hosts and process development. Front Bioeng Biotechnol 7:420. https://doi.org/10.3389/fbioe.2019.00420
80. Häkkinen ST, Reuter L, Nuorti N et al (2018) Tobacco BY-2 media component optimization for a cost-efficient recombinant protein production. Front Plant Sci 9:45. https://doi.org/10.3389/fpls.2018.00045

81. Yang WC, Minkler DF, Kshirsagar R et al (2016) Concentrated fed-batch cell culture increases manufacturing capacity without additional volumetric capacity. J Biotechnol 217:1–11. https://doi.org/10.1016/j.jbiotec.2015.10.009
82. Zhu W, Xu R, Gong G et al (2021) Medium optimization for high yield production of human serum albumin in Pichia pastoris and its efficient purification. Protein Expr Purif 181:105831. https://doi.org/10.1016/j.pep.2021.105831
83. Jiang H, Horwitz AA, Wright C et al (2019) Challenging the workhorse: comparative analysis of eukaryotic micro-organisms for expressing monoclonal antibodies. Biotechnol Bioeng 116: 1449–1462. https://doi.org/10.1002/bit.26951
84. Zhou H, Fang M, Zheng X et al (2021) Improving an intensified and integrated continuous bioprocess platform for biologics manufacturing. Biotechnol Bioeng 118:3618–3623. https://doi.org/10.1002/bit.27768
85. Schillberg S, Finnern R (2021) Plant molecular farming for the production of valuable proteins – critical evaluation of achievements and future challenges. J Plant Physiol 258–259:153359. https://doi.org/10.1016/j.jplph.2020.153359
86. Rademacher T, Sack M, Blessing D et al (2019) Plant cell packs: a scalable platform for recombinant protein production and metabolic engineering. Plant Biotechnol J 17:1560–1566. https://doi.org/10.1111/pbi.13081
87. Gengenbach BB, Opdensteinen P, Buyel JF (2020) Robot cookies – plant cell packs as an automated high-throughput screening platform based on transient expression. Front Bioeng Biotechnol 8:393. https://doi.org/10.3389/fbioe.2020.00393
88. Komarova TV, Baschieri S, Donini M et al (2010) Transient expression systems for plant-derived biopharmaceuticals. Expert Rev Vaccines 9:859–876. https://doi.org/10.1586/erv.10.85
89. Yamamoto T, Hoshikawa K, Ezura K et al (2018) Improvement of the transient expression system for production of recombinant proteins in plants. Sci Rep 8:4755. https://doi.org/10.1038/s41598-018-23024-y
90. Marillonnet S, Thoeringer C, Kandzia R et al (2005) Systemic agrobacterium tumefaciens–mediated transfection of viral replicons for efficient transient expression in plants. Nat Biotechnol 23:718–723. https://doi.org/10.1038/nbt1094
91. Marillonnet S, Giritch A, Gils M et al (2004) In planta engineering of viral RNA replicons: efficient assembly by recombination of DNA modules delivered agrobacterium. Proc Natl Acad Sci U S A 101:6852. https://doi.org/10.1073/pnas.0400149101
92. Lindbo JA (2007) TRBO: a high-efficiency tobacco mosaic virus RNA-based overexpression vector. Plant Physiol 145:1232. https://doi.org/10.1104/pp.107.106377
93. Lomonossoff GP, D'Aoust M-A (2016) Plant-produced biopharmaceuticals: a case of technical developments driving clinical deployment. Science 353:1237. https://doi.org/10.1126/science.aaf6638
94. Castilho A, Windwarder M, Gattinger P et al (2014) Proteolytic and N-glycan processing of human α1-antitrypsin expressed in Nicotiana benthamiana. Plant Physiol 166:1839–1851. https://doi.org/10.1104/pp.114.250720
95. Buyel JF, Fischer R (2012) Predictive models for transient protein expression in tobacco (Nicotiana tabacum L.) can optimize process time, yield, and downstream costs. Biotechnol Bioeng 109:2575–2588. https://doi.org/10.1002/bit.24523
96. Piotrzkowski N, Schillberg S, Rasche S (2012) Tackling heterogeneity: a leaf disc-based assay for the high-throughput screening of transient gene expression in tobacco. PLoS One 7: e45803. https://doi.org/10.1371/journal.pone.0045803
97. Opdensteinen P, Buyel JF (2022) Reducing water uptake into BY-2 cells by systematically optimizing the cultivation parameters increases product yields achieved by transient expression in plant cell packs. Biotechnol J 17:2200134. https://doi.org/10.1002/biot.202200134
98. Opdensteinen P, Sperl LE, Mohamadi M et al (2022) The transient expression of recombinant proteins in plant cell packs facilitates stable isotope labelling for NMR spectroscopy. Plant Biotechnol J 20:1928–1939. https://doi.org/10.1111/pbi.13873

99. Schillberg S, Spiegel H (2022) Recombinant protein production in plants: a brief overview of strengths and challenges. In: Schillberg S, Spiegel H (eds) Recombinant proteins in plants: methods and protocols. Springer, New York, pp 1–13
100. Knödler M, Rühl C, Emonts J et al (2019) Seasonal weather changes affect the yield and quality of recombinant proteins produced in transgenic tobacco plants in a greenhouse setting. Front Plant Sci 10:1245. https://doi.org/10.3389/fpls.2019.01245
101. Hundleby PAC, Harwood WA (2019) Impacts of the EU GMO regulatory framework for plant genome editing. Food Energy Secur 8:e00161. https://doi.org/10.1002/fes3.161
102. Poborilova Z, Plchova H, Cerovska N et al (2020) Transient protein expression in tobacco BY-2 plant cell packs using single and multi-cassette replicating vectors. Plant Cell Rep 39: 1115–1127. https://doi.org/10.1007/s00299-020-02544-w
103. Zagorskaya AA, Deineko EV (2021) Plant-expression systems: a new stage in production of biopharmaceutical preparations. Russ J Plant Physiol 68:17–30. https://doi.org/10.1134/S1021443721010210
104. Zagorskaya AA, Deineko EV (2017) Suspension-cultured plant cells as a platform for obtaining recombinant proteins. Russ J Plant Physiol 64:795–807. https://doi.org/10.1134/S102144371705017X
105. Moon K-B, Park J-S, Park Y-I et al (2020) Development of systems for the production of plant-derived biopharmaceuticals. Plants 9:30
106. Doran PM (1999) Design of mixing systems for plant cell suspensions in stirred reactors. Biotechnol Prog 15:319–335. https://doi.org/10.1021/bp990042v
107. Huang T-K, McDonald KA (2012) Bioreactor systems for in vitro production of foreign proteins using plant cell cultures. Biotechnol Adv 30:398–409. https://doi.org/10.1016/j.biotechadv.2011.07.016
108. Huang T-K, McDonald KA (2009) Bioreactor engineering for recombinant protein production in plant cell suspension cultures. Biochem Eng J 45:168–184. https://doi.org/10.1016/j.bej.2009.02.008
109. Reuter LJ, Bailey MJ, Joensuu JJ et al (2014) Scale-up of hydrophobin-assisted recombinant protein production in tobacco BY-2 suspension cells. Plant Biotechnol J 12:402–410. https://doi.org/10.1111/pbi.12147
110. Raven N, Rasche S, Kuehn C et al (2015) Scaled-up manufacturing of recombinant antibodies produced by plant cells in a 200-L orbitally-shaken disposable bioreactor. Biotechnol Bioeng 112:308–321. https://doi.org/10.1002/bit.25352
111. Kato K, Shiozawa Y, Yamada A et al (1972) A Jar Fermentor culture of Nicotiana tabacum L. cell suspensions. Agric Biol Chem 36:899–902. https://doi.org/10.1080/00021369.1972.10860355
112. Cho J-M, Kwon J-Y, Lim J-A et al (2007) Increased hGM-CSF production and secretion with Pluronic F-68 in transgenicNicotiana tabacum suspension cell cultures. Biotechnol Bioprocess Eng 12:594. https://doi.org/10.1007/BF02931074
113. Ho C-H, Henderson KA, Rorrer GL (1995) Cell damage and oxygen mass transfer during cultivation of Nicotiana tabacum in a stirred-tank bioreactor. Biotechnol Prog 11:140–145. https://doi.org/10.1021/bp00032a004
114. Castillo J, Alvarez M (2011) Effect of different agitation systems and protocols on the growth of Nicotiana tabacum cells: comparison of growth kinetics in Erlenmeyer flasks, conventional stirred tanks, and eccentrically agitated tank bioreactors. AIChE annual meeting, conference proceedings
115. Hogue RS, Lee JM, An G (1990) Production of a foreign protein product with genetically modified plant cells. Enzym Microb Technol 12:533–538. https://doi.org/10.1016/0141-0229(90)90071-W
116. Gao J, Lee JM (1992) Effect of oxygen supply on the suspension culture of genetically modified tobacco cells. Biotechnol Prog 8:285–290. https://doi.org/10.1021/bp00016a004

117. Fischer R, Liao YC, Drossard J (1999) Affinity-purification of a TMV-specific recombinant full-size antibody from a transgenic tobacco suspension culture. J Immunol Methods 226:1–10. https://doi.org/10.1016/s0022-1759(99)00058-7
118. Kieran P, MacLoughlin P, Malone D (1997) Plant cell suspension cultures: some engineering considerations. J Biotechnol 59:39–52. https://doi.org/10.1016/S0168-1656(97)00163-6
119. Meyer V, Cairns T, Barthel L et al (2021) Understanding and controlling filamentous growth of fungal cell factories: novel tools and opportunities for targeted morphology engineering. Fungal Biol Biotechnol 8:8. https://doi.org/10.1186/s40694-021-00115-6
120. Curtis WR, Emery AH (1993) Plant cell suspension culture rheology. Biotechnol Bioeng 42: 520–526. https://doi.org/10.1002/bit.260420416
121. Hartmann FS, Udugama IA, Seibold GM et al (2022) Digital models in biotechnology: towards multi-scale integration and implementation. Biotechnol Adv 60:108015. https://doi.org/10.1016/j.biotechadv.2022.108015
122. Helleckes LM, Hemmerich J, Wiechert W et al (2022) Machine learning in bioprocess development: from promise to practice. Trends Biotechnol. https://doi.org/10.1016/j.tibtech.2022.10.010
123. Wongsamuth R, Doran PM (1994) Foaming and cell flotation in suspended plant cell cultures and the effect of chemical antifoams. Biotechnol Bioeng 44:481–488. https://doi.org/10.1002/bit.260440411
124. Abdullah MA, Ariff AB, Marziah M et al (2000) Strategies to overcome foaming and wall-growth during the cultivation of Morinda elliptica cell suspension culture in a stirred-tank bioreactor. Plant Cell Tissue Organ Cult 60:205–212. https://doi.org/10.1023/A:1006495107778
125. Cheung CK-L, Leksawasdi N, Doran PM (2018) Bioreactor scale-down studies of suspended plant cell cultures. AICHE J 64:4281–4288. https://doi.org/10.1002/aic.16415
126. Namdev PK, Dunlop EH (1995) Shear sensitivity of plant cells in suspensions present and future. Appl Biochem Biotechnol 54:109–131. https://doi.org/10.1007/BF02787914
127. Smart NJ, Fowler MW (1981) Effect of aeration on large-scale cultures of plant cells. Biotechnol Lett 3:171–176. https://doi.org/10.1007/BF00239657
128. Piehl G-W, Berlin J, Mollenschott C et al (1988) Growth and alkaloid production of a cell suspension culture of Thalictrum rugosum in shake flasks and membrane-stirrer reactors with bubble free aeration. Appl Microbiol Biotechnol 29:456–461. https://doi.org/10.1007/BF00269068
129. Su WW, Humphrey AE (1991) Production of rosmarinic acid from perfusion culture of Anchusa officinalis in a membrane-aerated bioreactor. Biotechnol Lett 13:889–892. https://doi.org/10.1007/BF01022093
130. Bongartz P, Bator I, Baitalow K et al (2021) A scalable bubble-free membrane aerator for biosurfactant production. Biotechnol Bioeng 118:3545–3558. https://doi.org/10.1002/bit.27822
131. Taticek et al. (1991) The scale-up of plant cell culture: engineering considerations. Plant Cell Tiss Org Cult 24:139–158. https://doi.org/10.1007/BF00039742
132. Routledge SJ (2012) Beyond de-foaming: the effects of antifoams on bioprocess productivity. Comput Struct Biotechnol J 3:e201210014. https://doi.org/10.5936/csbj.201210014
133. Holland T, Sack M, Rademacher T et al (2010) Optimal nitrogen supply as a key to increased and sustained production of a monoclonal full-size antibody in BY-2 suspension culture. Biotechnol Bioeng 107:278–289. https://doi.org/10.1002/bit.22800
134. Holland T (2013) Entwicklung von Pflanzenzellkulturen im Hinblick auf die industrielle Produktion biopharmazeutischer Wirkstoffe [Development of plant cell cultures with respect to the industrial production of biopharmaceutical ingredients]. PhD thesis. RWTH Aachen University, Aachen
135. Schmale K (2007) Einsatz von Pflanzenzellkulturen als industriell nutzbare Expressionssysteme für therapeutische Proteine [Plant cell cultures as industrial expression systems for therapeutic proteins]. PhD thesis

136. King AT, Davey MR, Mulligan BJ et al (1990) Effects of Pluronic F-68 on plant cells in suspension culture. Biotechnol Lett 12:29–32. https://doi.org/10.1007/BF01028488
137. Sowana D, Williams D, O'Neill B et al (2002) Studies of the shear protective effects of Pluronic F-68 on wild carrot cell cultures. Biochem Eng J 12:165–173. https://doi.org/10.1016/S1369-703X(02)00038-4
138. Ogawa Y, Sakurai N, Oikawa A et al (2012) High-throughput cryopreservation of plant cell cultures for functional genomics. Plant Cell Physiol 53:943–952. https://doi.org/10.1093/pcp/pcs038
139. Lowe K, Anthony P, Davey M et al (2001) Beneficial effects of Pluronic F-68 and artificial oxygen carriers on the post-thaw recovery of cryopreserved plant cells. Artif Cells Blood Substit Immobil Biotechnol 29:297–316. https://doi.org/10.1081/BIO-100104232
140. Anthony P, Lowe KC, Davey MR et al (1998) Enhanced post-thaw viability of cryopreserved cells by oxygenated perfluorocarbon or Pluronic® F-68. Artif Cells Blood Substit Immobil Biotechnol 26:27–33. https://doi.org/10.3109/10731199809118944
141. Doran PM (2006) Loss of secreted antibody from transgenic plant tissue cultures due to surface adsorption. J Biotechnol 122:39–54. https://doi.org/10.1016/j.jbiotec.2005.08.029
142. Beuel A-K, Jablonka N, Heesel J et al (2021) LEDitSHAKE: a lighting system to optimize the secondary metabolite content of plant cell suspension cultures. Sci Rep 11:23353. https://doi.org/10.1038/s41598-021-02762-6
143. Buyel JF, Buyel JJ, Haase C et al (2015) The impact of pseudomonas syringae type III effectors on transient protein expression in tobacco. Plant Biol (Stuttg) 17:484–492. https://doi.org/10.1111/plb.12264
144. Murashige T, Skoog F (1962) A revised medium for rapid growth and bio assays with tobacco tissue cultures. Physiol Plant 15:473–497. https://doi.org/10.1111/j.1399-3054.1962.tb08052.x
145. Linsmaier EM, Skoog F (1965) Organic growth factor requirements of tobacco tissue cultures. Physiol Plant 18:100–127. https://doi.org/10.1111/j.1399-3054.1965.tb06874.x
146. Gamborg OL, Miller RA, Ojima K (1968) Nutrient requirements of suspension cultures of soybean root cells. Exp Cell Res 50:151–158. https://doi.org/10.1016/0014-4827(68)90403-5
147. Schenk RU, Hildebrandt AC (1972) Medium and techniques for induction and growth of monocotyledonous and dicotyledonous plant cell cultures. Can J Bot 50:199–204. https://doi.org/10.1139/b72-026
148. Chu C-C, Wang C-C, Sun C-S et al (1975) Establishment of an efficient medium for anther culture of rice through comparative experiments on the nitrogen sources. Sci Sinica 18:659–668. https://doi.org/10.1360/ya1975-18-5-659
149. Thompson JA, Abdullah R, Cocking EC (1986) Protoplast culture of rice (Oryza sativa L.) using media solidified with agarose. Plant Sci 47:123–133. https://doi.org/10.1016/0168-9452(86)90059-2
150. Bergmann L (1967) Wachstum grüner Suspensionskulturen von Nicotiana tabacum Var. "Samsun" mit CO2 als Kohlenstoffquelle [Growth of green suspension cultures of nicotiana tabacum var. "Samsun" with CO_2 as carbon source]. Planta 74:243–249. https://doi.org/10.1007/BF00384845
151. Vasilev N, Schmitz C, Grömping U et al (2014) Assessment of cultivation factors that affect biomass and geraniol production in transgenic tobacco cell suspension cultures. PLoS One 9:e104620. https://doi.org/10.1371/journal.pone.0104620
152. Ullisch DA, Müller CA, Maibaum S et al (2012) Comprehensive characterization of two different Nicotiana tabacum cell lines leads to doubled GFP and HA protein production by media optimization. J Biosci Bioeng 113:242–248. https://doi.org/10.1016/j.jbiosc.2011.09.022
153. Terashima M, Ejiri Y, Hashikawa N et al (2001) Utilization of an alternative carbon source for efficient production of human alpha(1)-antitrypsin by genetically engineered rice cell culture. Biotechnol Prog 17:403–406. https://doi.org/10.1021/bp010024p

154. Jones A, Veliky IA (1980) Growth of plant cell suspension cultures on glycerol as a sole source of carbon and energy. Can J Bot 58:648–657. https://doi.org/10.1139/b80-082
155. Behrend J, Mateles RI (1975) Nitrogen metabolism in plant cell suspension cultures: I. Effect of amino acids on growth. Plant Physiol 56:584–589. https://doi.org/10.1104/pp.56.5.584
156. Behrend J, Mateles RI (1976) Nitrogen metabolism in plant cell suspension cultures: II. Role of organic acids during growth on ammonia. Plant Physiol 58:510–512. https://doi.org/10.1104/pp.58.4.510
157. Hayashi T, Ohsumi C, Kato Y et al (1994) Effects of amino acid medium on cell aggregation in suspension-cultured Rice cells. Biosci Biotechnol Biochem 58:256–260. https://doi.org/10.1271/bbb.58.256
158. Vasilev N, Grömping U, Lipperts A et al (2013) Optimization of BY-2 cell suspension culture medium for the production of a human antibody using a combination of fractional factorial designs and the response surface method. Plant Biotechnol J 11:867–874. https://doi.org/10.1111/pbi.12079
159. Bortesi L, Rademacher T, Schiermeyer A et al (2012) Development of an optimized tetracycline-inducible expression system to increase the accumulation of interleukin-10 in tobacco BY-2 suspension cells. BMC Biotechnol 12:40. https://doi.org/10.1186/1472-6750-12-40
160. Rosano GL, Morales ES, Ceccarelli EA (2019) New tools for recombinant protein production in Escherichia coli: a 5-year update. Protein Sci 28:1412–1422. https://doi.org/10.1002/pro.3668
161. Xu N, Zhu J, Zhu Q et al (2018) Identification and characterization of novel promoters for recombinant protein production in yeast Pichia pastoris. Yeast 35:379–385. https://doi.org/10.1002/yea.3301
162. Ruiz C, Pla M, Company N et al (2016) High CO_2 concentration as an inductor agent to drive production of recombinant phytotoxic antimicrobial peptides in plant biofactories. Plant Mol Biol 90:329–343. https://doi.org/10.1007/s11103-015-0419-6
163. Ashraf M, Biswas S et al (2014) Cloning and characterization of alcohol dehydrogenase (Adh) promoter region for expression under submergence and salinity stress. PTCB 24. https://doi.org/10.3329/ptcb.v24i1.19252
164. Xue G-P, Way HM, Richardson T et al (2011) Overexpression of TaNAC69 leads to enhanced transcript levels of stress up-regulated genes and dehydration tolerance in bread wheat. Mol Plant 4:697–712. https://doi.org/10.1093/mp/ssr013
165. Yoshida K, Kasai T, Garcia MR et al (1995) Heat-inducible expression system for a foreign gene in cultured tobacco cells using the HSP18.2 promoter of Arabidopsis thaliana. Appl Microbiol Biotechnol 44:466–472. https://doi.org/10.1007/BF00169945
166. Freeman J, Sparks CA, West J et al (2011) Temporal and spatial control of transgene expression using a heat-inducible promoter in transgenic wheat. Plant Biotechnol J 9:788–796. https://doi.org/10.1111/j.1467-7652.2011.00588.x
167. Kovalchuk N, Jia W, Eini O et al (2013) Optimization of TaDREB3 gene expression in transgenic barley using cold-inducible promoters. Plant Biotechnol J 11:659–670. https://doi.org/10.1111/pbi.12056
168. Yamaguchi-Shinozaki K, Shinozaki K (1993) Characterization of the expression of a desiccation-responsive rd29 gene of Arabidopsis thaliana and analysis of its promoter in transgenic plants. Mol Gen Genet 236:331–340. https://doi.org/10.1007/BF00277130
169. Müller K, Siegel D, Rodriguez Jahnke F et al (2014) A red light-controlled synthetic gene expression switch for plant systems. Mol BioSyst 10:1679–1688. https://doi.org/10.1039/c3mb70579j
170. Chatelle C, Ochoa-Fernandez R, Engesser R et al (2018) A green-light-responsive system for the control of transgene expression in mammalian and plant cells. ACS Synth Biol 7:1349–1358. https://doi.org/10.1021/acssynbio.7b00450

171. Kim K-Y, Kwon S-Y, Lee H-S et al (2003) A novel oxidative stress-inducible peroxidase promoter from sweetpotato: molecular cloning and characterization in transgenic tobacco plants and cultured cells. Plant Mol Biol 51:831–838. https://doi.org/10.1023/A:1023045218815
172. Chen M-H, Liu L-F, Chen Y-R et al (1994) Expression of α-amylases, carbohydrate metabolism, and autophagy in cultured rice cells is coordinately regulated by sugar nutrient. Plant J 6:625–636. https://doi.org/10.1046/j.1365-313X.1994.6050625.x
173. Williams S, Friedrich L, Dincher S et al (1992) Chemical regulation of bacillus thuringiensis ∂-endotoxin expression in transgenic plants. Bio/Technology 10:540–543. https://doi.org/10.1038/nbt0592-540
174. Frey AD, Rimann M, Bailey JE et al (2001) Novel pristinamycin-responsive expression systems for plant cells. Biotechnol Bioeng 74:154–163. https://doi.org/10.1002/bit.1105
175. Tang W, Page M (2013) Inducible expression of Norwalk virus capsid protein gene in plant cell suspension cultures. In Vitro Cell Dev Biol Plant 49:129–136. https://doi.org/10.1007/s11627-012-9487-3
176. Zuo J, Niu QW, Chua NH (2000) Technical advance: an estrogen receptor-based transactivator XVE mediates highly inducible gene expression in transgenic plants. Plant J 24:265–273. https://doi.org/10.1046/j.1365-313x.2000.00868.x
177. Bruce W, Folkerts O, Garnaat C et al (2000) Expression profiling of the maize flavonoid pathway genes controlled by estradiol-inducible transcription factors CRC and P. Plant Cell 12:65. https://doi.org/10.1105/tpc.12.1.65
178. Martinez A, Sparks C, Hart CA et al (1999) Ecdysone agonist inducible transcription in transgenic tobacco plants. Plant J 19:97–106. https://doi.org/10.1046/j.1365-313X.1999.00504.x
179. Padidam M, Gore M, Lily Lu D et al (2003) Chemical-inducible, ecdysone receptor-based gene expression system for plants. Transgenic Res 12:101–109. https://doi.org/10.1023/A:1022113817892
180. Unger E, Cigan AM, Trimnell M et al (2002) A chimeric ecdysone receptor facilitates methoxyfenozide-dependent restoration of male fertility in ms45 maize. Transgenic Res 11: 455–465. https://doi.org/10.1023/A:1020350208095
181. de Veylder L, van Montagu M, Inzé D (1997) Herbicide safener-inducible gene expression in Arabidopsis thaliana. Plant Cell Physiol 38:568–577. https://doi.org/10.1093/oxfordjournals.pcp.a029206
182. Mett VL, Lochhead LP, Reynolds PH (1993) Copper-controllable gene expression system for whole plants. Proc Natl Acad Sci U S A 90:4567–4571. https://doi.org/10.1073/pnas.90.10.4567
183. Caddick MX, Greenland AJ, Jepson L et al (1998) An ethanol inducible gene switch for plants used to manipulate carbon metabolism. Bio/Technology 16:177–180. https://doi.org/10.1038/nbt0298-177
184. McDonald KA, Lo Hong M, Trombly DM et al (2005) Production of human alpha-1-antitrypsin from transgenic rice cell culture in a membrane bioreactor. Biotechnol Prog 21: 728–734. https://doi.org/10.1021/bp0496676
185. Kulshreshtha A, Sharma S, Padilla CS et al (2022) Plant-based expression platforms to produce high-value metabolites and proteins. Front Plant Sci:13, 1043478
186. Liu Y-K, Li Y-T, Lu C-F et al (2015) Enhancement of recombinant human serum albumin in transgenic rice cell culture system by cultivation strategy. New Biotechnol 32:328–334. https://doi.org/10.1016/j.nbt.2015.03.001
187. Liu Y-K, Lu C-W, Chang J-Y et al (2018) Optimization of the culture medium for recombinant protein production under the control of the αAmy3 promoter in a rice suspension-cultured cell expression system. Plant Cell Tissue Organ Cult 132. https://doi.org/10.1007/s11240-017-1337-x

188. Huang L-F, Liu Y-K, Lu C-A et al (2005) Production of human serum albumin by sugar starvation induced promoter and rice cell culture. Transgenic Res 14:569–581. https://doi.org/10.1007/s11248-004-6481-5
189. Trexler MM, McDonald KA, Jackman AP (2002) Bioreactor production of human alpha(1)-antitrypsin using metabolically regulated plant cell cultures. Biotechnol Prog 18:501–508. https://doi.org/10.1021/bp020299k
190. Hong S-Y, Kwon T-H, Jang Y-S et al (2006) Production of bioactive human granulocyte-colony stimulating factor in transgenic rice cell suspension cultures. Protein Expr Purif 47:68–73. https://doi.org/10.1016/j.pep.2005.09.028
191. Shin Y-J, Hong S-Y, Kwon T-H et al (2003) High level of expression of recombinant human granulocyte-macrophage colony stimulating factor in transgenic rice cell suspension culture. Biotechnol Bioeng 82:778–783. https://doi.org/10.1002/bit.10635
192. Kim T-G, Baek M-Y, Lee E-K et al (2008) Expression of human growth hormone in transgenic rice cell suspension culture. Plant Cell Rep 27:885–891. https://doi.org/10.1007/s00299-008-0514-0
193. Chan MT, Chao YC, Yu SM (1994) Novel gene expression system for plant cells based on induction of alpha-amylase promoter by carbohydrate starvation. J Biol Chem 269:17635–17641
194. Kim T-G, Tien N-Q-D, Yang M-S et al (2016) Production of monoclonal antibodies against 53-kDa protein of porphyromonas gingivalis in transgenic rice cell suspension culture. Plant Cell Tiss Organ Cult 126:387–397. https://doi.org/10.1007/s11240-016-1005-6
195. Lee S-J, Park C-I, Park M-Y et al (2007) Production and characterization of human CTLA4Ig expressed in transgenic rice cell suspension cultures. Protein Expr Purif 51:293–302. https://doi.org/10.1016/j.pep.2006.08.019
196. Toyofuku K, Umemura T, Yamaguchi J (1998) Promoter elements required for sugar-repression of the RAmy3D gene for α-amylase in rice. FEBS Lett 428:275–280. https://doi.org/10.1016/S0014-5793(98)00518-3
197. Cao J, Shelton AM, Earle ED (2001) Gene expression and insect resistance in transgenic broccoli containing a bacillus thuringiensis cry1Ab gene with the chemically inducible PR-1a promoter. Mol Breed 8:207–216. https://doi.org/10.1023/A:1013734923291
198. Wang R, Zhou X, Wang X (2003) Chemically regulated expression systems and their applications in transgenic plants. Transgenic Res 12:529–540. https://doi.org/10.1023/A:1025852307127
199. Moore I, Samalova M, Kurup S (2006) Transactivated and chemically inducible gene expression in plants. Plant J 45:651–683. https://doi.org/10.1111/j.1365-313X.2006.02660.x
200. Corrado G, Karali M (2009) Inducible gene expression systems and plant biotechnology. Biotechnol Adv 27:733–743. https://doi.org/10.1016/j.biotechadv.2009.05.006
201. Gatz C, Kaiser A, Wendenburg R (1991) Regulation of a modified CaMV 35S promoter by the Tn10-encoded Tet repressor in transgenic tobacco. Mol Gen Genet 227:229–237. https://doi.org/10.1007/BF00259675
202. Gatz C, Frohberg C, Wendenburg R (1992) Stringent repression and homogeneous de-repression by tetracycline of a modified CaMV 35S promoter in intact transgenic tobacco plants. Plant J 2:397–404. https://doi.org/10.1111/j.1365-313x.1992.00397.x
203. Weinmann P, Gossen M, Hillen W et al (1994) A chimeric transactivator allows tetracycline-responsive gene expression in whole plants. Plant J 5:559–569. https://doi.org/10.1046/j.1365-313x.1994.5040559.x
204. Aoyama T, Chua N-H (1997) A glucocorticoid-mediated transcriptional induction system in transgenic plants. Plant J 11:605–612. https://doi.org/10.1046/j.1365-313X.1997.11030605.x
205. Moore I, Gälweiler L, Grosskopf D et al (1998) A transcription activation system for regulated gene expression in transgenic plants. Proc Natl Acad Sci 95:376. https://doi.org/10.1073/pnas.95.1.376

206. Werner S, Breus O, Symonenko Y et al (2011) High-level recombinant protein expression in transgenic plants by using a double-inducible viral vector. Proc Natl Acad Sci U S A 108: 14061–14066. https://doi.org/10.1073/pnas.1102928108
207. Dugdale B, Mortimer CL, Kato M et al (2013) In plant activation: an inducible, hyperexpression platform for recombinant protein production in plants. Plant Cell 25:2429. https://doi.org/10.1105/tpc.113.113944
208. Wilken LR, Nikolov ZL (2012) Recovery and purification of plant-made recombinant proteins. Biotechnol Adv 30:419–433. https://doi.org/10.1016/j.biotechadv.2011.07.020
209. Buyel JF, Twyman RM, Fischer R (2015) Extraction and downstream processing of plant-derived recombinant proteins. Biotechnol Adv 33:902–913. https://doi.org/10.1016/j.biotechadv.2015.04.010
210. Buyel JF (2015) Process development strategies in plant molecular farming. Curr Pharm Biotechnol 16:966–982. https://doi.org/10.2174/138920101611150902115413
211. Bansode V, Gupta P, Kateja N et al (2022) Contribution of protein a step towards cost of goods for continuous production of monoclonal antibody therapeutics. J Chem Technol Biotechnol 97:2420–2433. https://doi.org/10.1002/jctb.6686
212. Yang O, Prabhu S, Ierapetritou M (2019) Comparison between batch and continuous monoclonal antibody production and economic analysis. Ind Eng Chem Res 58:5851–5863. https://doi.org/10.1021/acs.iecr.8b04717
213. Buyel JF, Hubbuch J, Fischer R (2016) Comparison of tobacco host cell protein removal methods by blanching intact plants or by heat treatment of extracts. J Vis Exp. https://doi.org/10.3791/54343
214. Buyel JF, Opdensteinen P, Fischer R (2015) Cellulose-based filter aids increase the capacity of depth filters during the downstream processing of plant-derived biopharmaceutical proteins. Biotechnol J 10:584–591. https://doi.org/10.1002/biot.201400611
215. Bielser J-M, Wolf M, Souquet J et al (2018) Perfusion mammalian cell culture for recombinant protein manufacturing – a critical review. Biotechnol Adv 36:1328–1340. https://doi.org/10.1016/j.biotechadv.2018.04.011
216. Weinberger ME, Schoch L, Kulozik U (2022) New technical concept for alternating tangential flow filtration in biotechnological cell separation processes. Biotechnol Prog 39:e3309. https://doi.org/10.1002/btpr.3309
217. Goulet C, Khalf M, Sainsbury F et al (2012) A protease activity–depleted environment for heterologous proteins migrating towards the leaf cell apoplast. Plant Biotechnol J 10:83–94. https://doi.org/10.1111/j.1467-7652.2011.00643.x
218. Benchabane M, Goulet C, Rivard D et al (2008) Preventing unintended proteolysis in plant protein biofactories. Plant Biotechnol J 6:633–648. https://doi.org/10.1111/j.1467-7652.2008.00344.x
219. Soomro RR, Ndikubwimana T, Zeng X et al (2016) Development of a two-stage microalgae dewatering process – a life cycle assessment approach. Front Plant Sci 7:113
220. Buyel JF, Fischer R (2014) Flocculation increases the efficacy of depth filtration during the downstream processing of recombinant pharmaceutical proteins produced in tobacco. Plant Biotechnol J 12:240–252. https://doi.org/10.1111/pbi.12132
221. Buyel JF, Fischer R (2014) Downstream processing of biopharmaceutical proteins produced in plants. Bioengineered 5:138–142. https://doi.org/10.4161/bioe.28061
222. Islam M, Aryasomayajula A, Selvaganapathy P (2017) A review on macroscale and microscale cell lysis methods. Micromachines 8:83. https://doi.org/10.3390/mi8030083
223. Giovannoni M, Gramegna G, Benedetti M et al (2020) Industrial use of cell wall degrading enzymes: the fine line between production strategy and economic feasibility. Front Bioeng Biotechnol 8:356
224. Carpita NC (1985) Tensile strength of cell walls of living cells 1. Plant Physiol 79:485–488. https://doi.org/10.1104/pp.79.2.485

225. Buyel JF, Fischer R (2014) Scale-down models to optimize a filter train for the downstream purification of recombinant pharmaceutical proteins produced in tobacco leaves. Biotechnol J 9:415–425. https://doi.org/10.1002/biot.201300369
226. Buyel JF, Gruchow HM, Fischer R (2015) Depth filters containing diatomite achieve more efficient particle retention than filters solely containing cellulose fibers. Front Plant Sci 6:1134
227. Knödler M, Opdensteinen P, Sankaranarayanan RA et al (2022) Simple plant-based production and purification of the assembled human ferritin heavy chain as a nanocarrier for tumor-targeted drug delivery and bioimaging in cancer therapy. Biotechnol Bioeng. https://doi.org/10.1002/bit.28312
228. Sim T-S, Goh A, Becker EW (1988) Comparison of centrifugation, dissolved air flotation and drum filtration techniques for harvesting sewage-grown algae. Biomass 16:51–62. https://doi.org/10.1016/0144-4565(88)90015-7
229. Molina Grima E, Belarbi E-H, Acién Fernández FG et al (2003) Recovery of microalgal biomass and metabolites: process options and economics. Biotechnol Adv 20:491–515. https://doi.org/10.1016/S0734-9750(02)00050-2
230. Almeida A, Chau D, Coolidge T et al (2022) Chromatographic capture of cells to achieve single stage clarification in recombinant protein purification. Biotechnol Prog 38:e3227. https://doi.org/10.1002/btpr.3227
231. Geigert J (2013) The challenge of CMC regulatory compliance for biopharmaceuticals and other biologics. Springer, New York
232. Buyel JF, Fischer R (2014) Generic chromatography-based purification strategies accelerate the development of downstream processes for biopharmaceutical proteins produced in plants. Biotechnol J 9:566–577. https://doi.org/10.1002/biot.201300548
233. Madabhushi SR, Gavin J, Xu S et al (2018) Quantitative assessment of environmental impact of biologics manufacturing using process mass intensity analysis. Biotechnol Prog 34:1566–1573. https://doi.org/10.1002/btpr.2702
234. Shukla AA, Hubbard B, Tressel T et al (2007) Downstream processing of monoclonal antibodies – application of platform approaches. J Chromatogr B Analyt Technol Biomed Life Sci 848:28–39. https://doi.org/10.1016/j.jchromb.2006.09.026
235. Hassan S, van Dolleweerd CJ, Ioakeimidis F et al (2008) Considerations for extraction of monoclonal antibodies targeted to different subcellular compartments in transgenic tobacco plants. Plant Biotechnol J 6:733–748. https://doi.org/10.1111/j.1467-7652.2008.00354.x
236. Buyel JF, Gruchow HM, Boes A et al (2014) Rational design of a host cell protein heat precipitation step simplifies the subsequent purification of recombinant proteins from tobacco. Biochem Eng J 88:162–170. https://doi.org/10.1016/j.bej.2014.04.015
237. Opdensteinen P, Lobanov A, Buyel JF (2021) A combined pH and temperature precipitation step facilitates the purification of tobacco-derived recombinant proteins that are sensitive to extremes of either parameter. Biotechnol J 16:2000340. https://doi.org/10.1002/biot.202000340
238. Opdensteinen P, Clodt JI, Müschen CR et al (2019) A combined ultrafiltration/diafiltration step facilitates the purification of Cyanovirin-N from transgenic tobacco extracts. Front Bioeng Biotechnol 6:206. https://doi.org/10.3389/fbioe.2018.00206
239. Tan R, Franzreb M (2020) Continuous ultrafiltration/diafiltration using a 3D-printed two membrane single pass module. Biotechnol Bioeng 117:654–661. https://doi.org/10.1002/bit.27233
240. Giacobbo A, Moura Bernardes A, Filipe Rosa MJ et al (2018) Concentration polarization in ultrafiltration/nanofiltration for the recovery of polyphenols from winery wastewaters. Membranes (Basel) 8. https://doi.org/10.3390/membranes8030046
241. Le TTH, Vu LTK, Le NL (2021) Effects of membrane pore size and transmembrane pressure on ultrafiltration of red-fleshed dragon fruit (Hylocereus polyrhizus) juice. J Chem Technol Biotechnol 96:1561–1572. https://doi.org/10.1002/jctb.6672

242. Zhang D, Karkooti A, Liu L et al (2018) Fabrication of antifouling and antibacterial polyethersulfone (PES)/cellulose nanocrystals (CNC) nanocomposite membranes. J Membr Sci 549:350–356. https://doi.org/10.1016/j.memsci.2017.12.034
243. Sreedhar N, Thomas N, Al-Ketan O et al (2018) Mass transfer analysis of ultrafiltration using spacers based on triply periodic minimal surfaces: effects of spacer design, directionality and voidage. J Membr Sci 561:89–98. https://doi.org/10.1016/j.memsci.2018.05.028
244. Jung O, Saravia F, Wagner M et al (2019) Quantifying concentration polarization – Raman microspectroscopy for in-situ measurement in a flat sheet cross-flow Nanofiltration membrane unit. Sci Rep 9:15885. https://doi.org/10.1038/s41598-019-52369-1
245. Lightfoot EN, Root TW, O'Dell JL (2008) Emergence of ideal membrane cascades for downstream processing. Biotechnol Prog 24:599–605. https://doi.org/10.1021/bp070335l
246. Platis D, Labrou NE (2006) Development of an aqueous two-phase partitioning system for fractionating therapeutic proteins from tobacco extract. J Chromatogr A 1128:114–124. https://doi.org/10.1016/j.chroma.2006.06.047
247. Wiendahl M, Oelmeier SA, Dismer F et al (2012) High-throughput screening-based selection and scale-up of aqueous two-phase systems for pDNA purification. J Sep Sci 35:3197–3207. https://doi.org/10.1002/jssc.201200310
248. Zimmermann S, Scheeder C, Zimmermann PK et al (2017) High-throughput downstream process development for cell-based products using aqueous two-phase systems (ATPS) – a case study. Biotechnol J 12:1600587. https://doi.org/10.1002/biot.201600587
249. Wang Y (2018) Theory and principles of interaction chromatography, ACS symposium series. ACS Publication, Washington, DC, pp 19–30
250. Schmidt-Traub H, Schulte M, Seidel-Morgenstern A (eds) (2020) Preparative chromatography. Wiley, Weinheim
251. Bernau CR, Knödler M, Emonts J et al (2022) The use of predictive models to develop chromatography-based purification processes. Front Bioeng Biotechnol 10:1009102. https://doi.org/10.3389/fbioe.2022.1009102
252. Buyel JF (2022) Statistical designs to improve downstream processing. Methods Mol Biol 2480:295–310. https://doi.org/10.1007/978-1-0716-2241-4_16
253. Bernau CR, Jäpel RC, Hübbers JW et al (1652) Precision analysis for the determination of steric mass action parameters using eight tobacco host cell proteins. J Chromatogr A 2021: 462379. https://doi.org/10.1016/j.chroma.2021.462379
254. Osberghaus A, Hepbildikler S, Nath S et al (2012) Determination of parameters for the steric mass action model—a comparison between two approaches. J Chromatogr A 1233:54–65. https://doi.org/10.1016/j.chroma.2012.02.004
255. Hober S, Nord K, Linhult M (2007) Protein a chromatography for antibody purification. J Chromatogr B Analyt Technol Biomed Life Sci 848:40–47. https://doi.org/10.1016/j.jchromb.2006.09.030
256. Bornhorst JA, Falke JJ (2000) Purification of proteins using polyhistidine affinity tags. Methods Enzymol 326:245–254. https://doi.org/10.1016/s0076-6879(00)26058-8
257. Young CL, Britton ZT, Robinson AS (2012) Recombinant protein expression and purification: a comprehensive review of affinity tags and microbial applications. Biotechnol J 7:620–634. https://doi.org/10.1002/biot.201100155
258. Zhang B, Rapolu M, Kumar S et al (2017) Coordinated protein co-expression in plants by harnessing the synergy between an intein and a viral 2A peptide. Plant Biotechnol J 15:718–728. https://doi.org/10.1111/pbi.12670
259. Lingg N, Kröß C, Engele P et al (2022) CASPON platform technology: ultrafast circularly permuted caspase-2 cleaves tagged fusion proteins before all 20 natural amino acids at the N-terminus. New Biotechnol 71:37–46. https://doi.org/10.1016/j.nbt.2022.07.002
260. Shekhawat LK, Rathore AS (2019) An overview of mechanistic modeling of liquid chromatography. Prep Biochem Biotechnol 49:623–638. https://doi.org/10.1080/10826068.2019.1615504

261. Schmidt-Traub H (ed) (2006) Preparative chromatography: of fine chemicals and pharmaceutical agents, 1st edn. Wiley-VCH, Weinheim
262. Saleh D, Wang G, Rischawy F et al (2021) In silico process characterization for biopharmaceutical development following the quality by design concept. Biotechnol Prog 37:e3196. https://doi.org/10.1002/btpr.3196
263. Andris S, Hubbuch J (2020) Modeling of hydrophobic interaction chromatography for the separation of antibodydrug conjugates and its application towards quality by design. J Biotechnol 317:48–58. https://doi.org/10.1016/j.jbiotec.2020.04.018
264. Moser A, et al. (2021) Mechanistic mathematical models as a basis for digital twins. Adv Biochem Eng Biot 176:133–180. https://doi.org/10.1007/10_2020_152
265. Mollerup J, et al. (2008) Quality by design–thermodynamic modelling of chromatographic separation of proteins. J Chromatogr A 1177:200–206. https://doi.org/10.1016/j.chroma.2007.08.059
266. Robinson JR, Karkov HS, Woo JA et al (2017) QSAR models for prediction of chromatographic behavior of homologous Fab variants. Biotechnol Bioeng 114:1231–1240. https://doi.org/10.1002/bit.26236

Chapter 9
Manufacturing of Recombinant Proteins Using Quality by Design (QbD) Methodology: Current Trend and Challenges

Younok Dumortier Shin and Karlheinz Landauer

Abstract Since the introduction in 2009, the Quality by Design (QbD) is a methodology widely used in the development of therapeutic proteins. The fundamental concept of QbD is that the development of a therapeutic protein starts with the comprehensive understanding of the target product profile in terms of identity, strength, quality, purity, and potency. The manufacturing is then designed to ensure the product produced from the designed process meet the target product profile consistently and reproducible matter. Presently, all marketing application for therapeutic proteins submitted to FDA or EU regulatory agencies require the QbD methodology implemented during their product development and manufacturing stages. Despite of its effectiveness and regulatory requirements, the implementation of QbD concept in its entirety has been limited due to (1) insufficient time to fully characterize the target protein profile in the beginning of development process, (2) complexity of input and output parameters prohibiting the full design of experiments, and (3) challenges in collecting and analyzing the end-to-end product and manufacturing data. In order to overcome those challenges, many efforts are on-going with the most noticeable progress made in the advanced data analytics and machine learning perspective.

Keywords Quality by Design (QbD) · Therapeutic proteins · Target product profile · Design of experiments (DoE) · Advanced data analytics · Machine learning

Abbreviation

ADC	Antibody Drug Conjugate
AI	Artificial Intelligence
API	Active Pharmaceutical Ingredient

Y. D. Shin · K. Landauer (✉)
QBDC GmbH, Nusshof, Switzerland
e-mail: yshin@alaired.com; karlheinz.landauer@qbdc.ch

R. Pörtner (ed.), *Biopharmaceutical Manufacturing*, Cell Engineering 11,
https://doi.org/10.1007/978-3-031-45669-5_9

ATMP	Advanced Therapy Medical Product
BPOG	Biophorum Operations Group
CDC	Complement Dependent Cytotoxicity
CHO	Chinese Ovary Hamster
cIEF	Capillary Isoelectric Focusing
CPPs	Critical Process Parameters
CPV	Continued Process Verification
CQAs	Critical Quality Attributes
DoE	Design of Experiment
DP	Drug Product
DS	Drug Substance
ELIZA	Enzyme-Linked Immunosorbent Assay
GMP	Good Manufacturing Practice
HMW	High Molecular Weight
ISPE	International Society for Pharmaceutical Engineering
LMW	Low Molecular Weight
ML	Machine Learning
NOR	Normal Operating Range
OPLS	Orthogonal Partial Least Squares
PAR	Proven Acceptable Range
PAT	Process Analytical Technology
PC	Process Characterization
pCPPs	Potential Critical Process Parameters
pCQAs	Potential Critical Quality Attributes
PDA	Parenteral Drug Association
PLS	Projections to Latent Structures
PPQ	Process Performance Qualification
QbD	Quality by Design
QTPP	Quality Target Product Profile

Introduction: Quality by Design (QbD)

The development of a human medicinal product is a highly regulated venture, in which applicants are asked to understand the molecule, its functionality and how the manufacturing process influences the manufactured product. There are two guidelines tackling this point: ICH Q11 Development and manufacture of drug substances [1] and ICH Q8 Pharmaceutical Development [2]. Additionally, the relatively new EU GMP Annex 15: Qualification and Validation [3] as well as the US FDA's Guidance for the Industry Process Validation: General Principles and Practices [4] need to be considered.

In former days, the regular approach to cover pharmaceutical development was essentially not really described by guidelines and companies were essentially free to develop their molecules as wished. Today ICH Q8 states that it is important to recognize that quality cannot be tested into products meaning that quality should be built in by design [2]. Changes in formulation and manufacturing processes during development and lifecycle management should be looked upon as opportunities to gain additional knowledge and further support establishment of the design space. Based on the advances in Project Management in all Industries, these Project Management concepts were adopted into the pharmaceutical world with the intention to design a quality product and its manufacturing process to consistently deliver the intended performance of the product.

Project Management and the Quality Target Product Profile (QTPP) is the process of organizing and controlling the resources and activities necessary to produce a product. It involves the coordination of people, finances, materials, and other resources in order to achieve the desired objectives. Project management is essential for successful product development and delivery. The QTPP is a key component of the quality and project management process. The QTPP is a detailed description of the desired characteristics of a product or service that must meet the project objectives. It includes such elements as:

- Desired performance characteristics
- Design specifications
- Cost constraints
- Quality requirements
- Schedule requirements
- Risk assessment
- Regulatory considerations

QTPP is used to help ensure that the project meets the stakeholders' expectations and that the end product is of the highest and regulatory needed quality. It is also used to identify potential problems and risks that could arise during the project and to develop solutions for them. The development of the QTPP is an important part of the project planning process. This involves gathering and analyzing information from stakeholders and other sources to identify the features and capabilities that the product must have in order to be successful and meet the project objectives. This information is then used to create a detailed description of the desired product. Once the QTPP is developed, it serves as a blueprint for the project. It is used to define the scope of the project, set the timeline, assign tasks, and manage resources. The QTPP is also used to evaluate progress and ensure that the project is on track to meet its objectives. Project management and the target product profile are essential components of any successful product development and delivery project. By properly defining the desired characteristics of the product, and then using the QTPP to guide the project, organizations can ensure that their projects are successful and meet the needs of their stakeholders.

Following Parameter/topics need to be included and considered during the development of a new drug, starting with the QTPP as the "master document" followed by the needed risk assessments and the to be developed control strategy:

- QTPP – Target Quality Product Profile. It essentially describes what features the product should fulfil. Next to the efficacy for a certain indication also other attributes such as doses, container, stability, storage conditions, etc. are described in this document.
- Active Pharmaceutical Ingredient (API)'s critical quality attributes (CQAs); and thus, the basis for the whole control strategy of the manufacturing process needs to be described. It focuses on which of the attributes of an API are critical, why those are critical and how to control these attributes.
- API's critical manufacturing process steps – as a reflection of the CQAs the process needs to be analysed were and which steps may and do have an impact on a CQA.
- Development and description of a control strategy for the manufacturing to ensure the safety, efficacy, and quality of the product. The control strategy is verified during process validation and needs to be continuously verified via the continued/continuous process verification.

The above described implementation of this Project Management tool leads to the development of the QbD strategy. QbD is a systematic approach to product development that focuses on quality from the very beginning, as described in the QTPP. It is a proactive approach to process development that focuses on understanding the molecule, the manufacturing process and the analytical control rather than simply responding to a linear development. QbD involves understanding the product and process needs, designing the process to meet those needs, and then continuously monitoring and improving the product to ensure that it is meeting the regulatory expectations and specifically the patient needs.

QbD involves an iterative process of design, verification, and validation, with the goal of creating a manufacturing process that meets the needs and is free of errors and it is used to span the design space. Within this design space the manufacturing process can be realized, changing set-points within the design space is not considered to be a regulatory change based on the dossier, while movement out of the design space is a change and would normally initiate a regulatory post approval change process. Thus, the design space is an important tool in the product manufacturing process, as it allows for a more efficient and effective adaptation of the process based on better understanding over time. By using the design space, Good Manufacturing Practice (GMP) can effectively identify the best design solution for a particular application if all relevant data is stored and made accessible via data historian.

QTPP and subsequently the QbD approach is usable and applicable for all products in development, yet a special emphasis should be considered for biosimilar as well as Advanced Therapy Medical Product (ATMP) or personalized medicine development projects. Regarding the biosimilar development the basis for the QTPP is coming from the originator product, which needs to be thoroughly analyzed and the manufacturing process needs to hit the originator target. The biosimilar might be differ from the originator, but only in the clinically not-relevant features. For ATMPs and personalized medicines the main issue is the low number of batches or the fact that every batch might be different (personalized). For those projects it is of outmost importance to understand the process as well as the molecule thoroughly to consistently produce the same safety, efficacy, and quality profile.

Practical Examples of QbD

QbD is a systematic approach to the development, manufacture, and control of biopharmaceutical products that aims to ensure quality and consistency in the final product. This chapter will explore the latest trends and practical examples of QbD in the biopharmaceutical industry, with a specific focus on the development of monoclonal antibodies. The chapter will also provide examples of successful QbD implementation in monoclonal antibody development and explore the future of QbD in this field.

In the last 15 years, the QbD approach has evolved significantly within the biopharmaceutical industry. Initially, QbD was primarily applied to small molecule drug development, but it has since been adopted for the development of biologics, including monoclonal antibodies. The FDA's guidance for Industry Q8 (R2) Pharmaceutical Development [2] and Q9 Quality Risk Management [5] have provided guidance on the implementation of QbD in the biopharmaceutical industry. These guidelines outline the principles of QbD and provide a framework for the integration of science and risk management in the development and control of biopharmaceutical products. Additionally, industry forums such as the International Society for Pharmaceutical Engineering (ISPE), Biophorum Operations Group (BPOG) and the Parenteral Drug Association (PDA) have also focused on the implementation of QbD in the biopharmaceutical industry. These organizations provide resources, training, and networking opportunities for professionals in the industry to learn about and implement QbD.

Over the years, the industry has seen a growing awareness and acceptance of QbD as an important approach to ensure quality and consistency in biopharmaceutical products. This has led to an increase in the number of companies implementing QbD, and to a greater focus on the use of quality risk management in the development and control of biopharmaceutical products. In 2008, representatives from several pharmaceutical companies came together to advance the principles of Quality by Design as outlined in ICH Q8(R2), Q9, and Q10 [2, 5, 6]. The group focused on the unique challenges of manufacturing and developing biotechnology products and aimed to create a comprehensive case study to stimulate discussion and advance the application of these guidelines in real-world scenarios. The group's goals included creating a case study for teaching and learning, exemplifying advanced principles, demonstrating the concept of 'prior knowledge', enabling effective techniques for continual improvement, and examining the potential for enhancing science and risk-based regulatory approaches. This case study, "A-Mab: A Case Study in Bioprocess Development" [7] with the objective to demonstrate a QbD approach to product development demonstrates an enhanced approach including a systematic evaluation, understanding and refining of the formulation and manufacturing process, including identifying material attributes and process parameters that can have an effect on product quality, determining the relationships between material attributes and process parameters to product quality, and using the enhanced product and process understanding in combination with quality risk management to establish an appropriate control strategy, such as a proposal for a design space and/or real-time release testing. The approach is illustrated in a sequence of activities that starts with the design of the molecule and spans the development process ultimately resulting in the final process and control strategy used for commercial manufacturing. This case study has been the starting point for most biopharmaceutical companies which have adopted versions of describe tools in place.

In recent years different cross industry working groups have been formed to further align different tools and to share best practices for biopharmaceutical products. BPDG is an industry-wide consortium for networking and sharing best practices for the development of biopharmaceuticals. The BPDG-Formulation Point Share group conducted an inter-company collaboration exercise to gain a better understanding of current industry approaches for establishing biopharmaceutical drug product robustness [8]. Their results revealed that overall drug product robustness is defined by both the formulation and the manufacturing process robustness, and it integrates the principles of QbD. Robustness is an important factor in setting critical quality attribute control strategies and commercial specifications. The BPDG-Formulation Point Share group identified industry challenges with regard to biopharmaceutical drug product robustness and presented some recommendations for best practices.

The identification of CQAs and subsequent characterization in process development studies are key elements of QbD for biopharmaceutical products. A survey was conducted by multiple companies participating in an International Consortium of Innovation and Quality working group on best practices for identifying CQAs with linkages to process characterization (PC) studies. Alt et al. [9] reported the survey results indicating that most companies are using similar steps: (1) determination of QTPP, (2) CQA identification, (3) Determining of the acceptable ranges from process development activities, (4) testing and confirmation of the acceptable process ranges via process characterization activities, and (5) life cycle management to continuously monitor and update the product and process understanding and control.

QTPP outlines the quality attributes that are evaluated using a risk ranking tool. A final list of CQAs is refined by considering product understanding, which is achieved through structure-function relationship studies and pre-clinical results, such as pharmacokinetic studies. Platform knowledge and relevant literature is used to identify initial or potential CQAs (pCQAs) when limited information on the molecule is available. As more understanding of the product is gained during product characterization studies, pCQAs are either redefined to CQAs or considered to be non-critical. An initial risk assessment may be performed at different stages of product development. A common approach is an assessment prior to first in-human studies. Listing CQAs is usually performed in categories comprising broad categories such as product-related impurities and process-related impurities and further detailed categories such as identity and purity. The most frequently used risk assessment approach is the "Impact x Uncertainty" approach similar to the A-Mab Case Study "Quality Attribute Assessment Tool #1" [7]. The "Impact" is scored based on biological activity and safety and most companies are also taking PK and immunogenicity into account.

Process characterization studies are performed to understand how (critical) quality attributes are linked to certain process parameters and thus to establish a process control strategy and defining critical process parameters (CPPs). However, knowledge gained throughout development activities are used to define potential CPPs and used as basis for process characterization studies. Process characterization studies are conducted by a multi- and cross-functional team covering unit operations in both Drug Substance (DS) and Drug Product (DP) manufacturing prior to process performance qualification (PPQ). Prior to conducting process characterization studies, it is mandatory to select unit operations that must be included. A common approach is based on linkage of unit operations to product attributes and the definition of input and output parameters and corresponding in-process testing and process monitoring. Those parameters are further categorized based on the impact on CQAs and allow a risk-based approach to select parameters to be included in process characterization studies. It is worth noting that CPPs are the only parameter defined within ICH Q8 and other categories company specific. As the biopharmaceutical industry has gained extensive prior knowledge during the last decades most companies have implemented an integrated control strategy for similar modalities and are continuously updating their approach. Such control strategies based on acceptable ranges defined within the ICH Q8 R2. The majority of companies provide an acceptable range for process parameters in regulatory license submissions, known as the "Proven Acceptable Range" (PAR), which implies that variation within the range while maintaining other parameters constant within a normal operating range (NOR) will result in a material meeting relevant quality criterion. There is alignment that movement within the NOR can be managed through internal change management. Exceeding the PAR for a CPP does not result automatically in batch rejection but triggers an investigation into its impact on CQAs. The final decision on batch disposition is typically made by a quality review board.

Since the manufacturing science and technology has been significantly improved within last decades there might be the need for re-development to meet the market demand, reduce manufacturing cost, or other reasons. Implementing new production methods (i.e. a next-generation process) in a global regulatory environment can be challenging for companies that lack prior experience. The decision to make such changes requires a careful consideration of the feasibility of regulatory approval and the ability to maintain a steady supply of the original product, taking into account patient safety and efficacy. On the other hand, the trend towards patient-specific treatments and ambition to speed up time to achieve proof of clinical principle leads also to accelerated and novel validation processes. As health authorities do not lower their standards for process validation, the submission content remains unchanged. Companies tend to use prior and platform knowledge and a risk-based approach to shorten the timeline and focus on the most important pCQAs and potential critical process parameters (pCPPs). Some process validation work may also be deferred to post-marketing if it does not affect product safety or efficacy, but this can lead to less manufacturing flexibility and an increased analytical testing for controlling the product.

Even though each company may have individual QbD principles, the A-Mab Study remains the corner stone of QbD approach within the biopharmaceutical industry [10]. However, the A-Mab Study is providing general guidance and therefore the IQ Consortium Biologics Working Group described 2021 the use of enhanced understanding of product attributes to link acceptance criteria to clinical performance, building on previous guidance for product control strategies and detailed assessments of product attribute criticality to fill the gap in information about how to establish suitable acceptance criteria for biotherapeutics. Commercial specifications are a crucial aspect of a biotherapeutic product's control strategy. Specifications ensure product quality and robust supply chain based on the potential impact of an attribute range on patient safety and efficacy. Testing is one part of the control strategy, but it is essential to explore alternatives like gaining deeper process and cross-product understanding, use of innovative at-line process and product assessments, and real-time release efforts to improve process performance, reduce time and cost, and deliver innovative medicines. Ruesch at al [10] summarize specification setting approaches, including three mAb case studies and one antibody drug conjugate (ADC) case study of science and risk-based decision-making. Solely relying on statistical ranges for commercial acceptance criteria may lead to either rejecting acceptable product or releasing batches of unacceptable quality. Therefore, specification setting should be based on risk-based assessments of product quality attributes with impact on safety and efficacy, using prior knowledge and product-specific assessments.

Current Challenges in the Full Implementation of QbD Methodology

Since the first introduction in 2009, as described in the previous section, the QbD methodology has been incorporated in the development of many therapeutic proteins including monoclonal antibodies and fusion proteins. Many of those proteins utilize IgG1 or IgG4 platform produced by Chinese Ovary Hamster (CHO) cell line. To this date, the proteins produced by CHO cell line still require extensive characterization for the product variants such as aggregations, charge variants and glycosylation profiles. The full implementation of the QbD methodology for the development of the new therapeutic protein modalities is still limited due to the significant financial and resources constraints. In this section, three challenges to the QbD full implementation are explained in detail.

Challenges with QTTP Assessment

Table 9.1 summarizes different types of potential protein modifications and their impact to safety and efficacy of the molecule in the body. Glycosylation is a posttranslational modification in eukaryotic cells that affects biologically relevant functions such as pharmacokinetics, pharmacodynamics, and immunogenicity [11]. Zhang et al. [12] identified more than seven different N-glycan species in an IgG1 monoclonal antibody highlighting the heterogeneous nature of protein population including N-linked glycosylation, O-linked glycosylation, α-2,6-sialylation, α-1,3/4-fucosylation and different number of galactose residues in the outer arms of biantennary glycans (G0, G1, G2). The amount of high mannose content affects how quickly proteins are cleared from the blood stream (efficacy concern), but it may also cause adverse events (safety concern). Terminal galactose is a key element in complement-dependent cytotoxicity (CDC) activity [11]. In addition to the glycosylation, the hydrophobic and hydrophilic nature of the protein molecule impacts the protein solubility, accessibility and stability in the body [13]. High molecular weight (HMW) species may trigger immunogenicity reaction. Truncated specifies may get cleared from the body quickly. Oxidation or deamidation of methionine or asparagine site may have negative impact on the protein binding to the target.

Ideally, for a proper QbD implementation, the acceptable ranges of those protein modification in Table 9.1 are defined by the safety and efficacy data obtained from non-human and human *in vivo* studies. This requires first obtaining protein solutions

Table 9.1 Examples of different biochemical and biophysical modifications and its effect

Type of modification	Cause for variation	Effect
Glycosylation	Cell culture media composition and feeding strategy Bioreactor operating conditions (cell density, agitation, pH, temperature, etc.)	Effector functions; efficacy
De-glycosylation		Effector functions; efficacy
C-terminal cleavage		Charge variants change (surrogate for manufacturing repeatability)
C-terminal amidation		Charge variants change
Deamidation (asparagin) Isomerization (aspartate)		Changed affinities – safety and efficacy, charge variants
Cyclization of N-terminal Gln or Glc		Possibly none; charge variants and mass (surrogate for manufacturing)
Oxydation (Met, Cys, His and Trp)		Changed affinities – safety and efficacy; charge variants
Glycation (N-terminus, Lys)		Changed affinities – safety and efficacy; charge variants
Citrate or succinate adduct		Changed affinities – safety and efficacy; charge variants

with various ranges of the product variants and then test them for safety and efficacy. Unfortunately, this step is extremely expensive and time consuming due to the complexity in generating sufficient amount of each product impurity, the cost associated in conducting in vivo studies to measure the safety and efficacy of the drug. As a result, the implementation of QbD methodology starts with the 'genetic QTPPs' for the therapeutic proteins, which makes the subsequent steps challenging.

Challenges with Process Design and Scale-Up

In addition to the limitation in defining meaningful CQAs, the manufacturing process design and control also faces significant financial and resources constraints. Table 9.2 summarizes the list of unit operations and important process parameters that are known to influence CQAs. It is widely known that different elements in the cell culture media (i.e., copper, zinc), the upstream operating conditions (i.e., temperature, pH), and purification conditions (i.e., buffer pH, peak collection criteria), have direct impact to the CQAs. As such, all those parameters must be examined as input parameters to find meaningful acceptable ranges. Unfortunately, the design of experiment (DoE) to examine the impact of different input parameters listed in Table 9.2 can be in the order of thousands of small-scale experiments which presents substantial cost and time, a roadblock for any drug development program.

Table 9.2 List of cell culture media components, process parameters and target CQA

Cell culture media components (input)	Process parameters (input)	Product characterization assays (output)
Cell culture media	Upstream steps:	Potency
Inorganic elements	Cell viability at thaw	Safety aggregation
Trace elements	Seeding density	α-1,3-gal
Vitamins	Temperature	Cell based binding assay
Lipids	pH	Binding characteristics
Carbon sources	pCO_2	Quality
Amino acids	Culture duration	Oxidation
Other reagents added during cell culture process	Feeding strategy	Deamidation
	Agitation rate	Truncation
Glucose solution	Harvest steps:	Glycation
Yeastolate	Viability	C-terminal Lysin/lysine clipping
Insuline	Cell density	
Chelating reagent	Downstream steps:	Glycosylation profile
	Protein loading ratio	Mannose structures (Man5, Man6, Man8)
	Buffer conductivity	
	Buffer pH	
	Buffer volume	
	Chromatography peak collection criteria	
	Flow rates	
	Pressure	

In addition, those small-scale experiments must be conducted not only for each unit operation separately, but also in combination of upstream and downstream together (i.e., upstream material produced under different conditions tested under various purification conditions). However, due to the financial and timeline limitation, such end-to-end ranging finding studies are rarely conducted during the process development. Subsequently, similar to the CQAs, the process parameters to control the manufacturing parameters begin with generic ranges applied to most therapeutic proteins in the development stages, rather than tailored ranges for a given protein for a given indication. For example, for protein aggregate, HMW of less than 5% is defined as an acceptable range for all proteins regardless of whether the HMW for that specific protein is proven to be immunogenic or not. The DoE experiments are then utilized to *verify* the ranges generically defined for therapeutic proteins (i.e., bioreactor temperature of 36–38 °C, glucose feeding addition, day 14 harvest, etc.) successfully produce the protein of interest within the generically defined QTPP (i.e., HMW of less than 5%).

In addition to the sheer number of DoE experiments needed to conduct a full process understanding, the difficulty in interpretating the product quality data obtained from those DoE studies also creates a challenge in implementing a meaningful QbD methodology. Tables 9.3 and 9.4 provide an example of the product quality data obtained from three different experimental conditions. Three small scale batches tested six different input parameters: types of media (two types), seeding density (high versus standard), feeding strategy (fixed versus variable), bioreactor temperature (fixed versus temperature shift), agitation rate (high versus standard). The proteins from these three runs were characterized using various analytical methods. As shown in Table 9.4, the results obtained from the three runs showed that different operating conditions indeed produced the protein material that showed quantifiable differences in terms of glycosylation profile, HMW, charge variants, titer, and functional binding efficiency. However, the interpretation of those numbers, for example the difference in the HMW between 1.9% versus 2.8%, is challenging due to the lack of its effect on the safety and efficacy. In order words, we need the immunogenicity data on the HMW species to be able to select the suitable processing condition, which is often impossible to obtain without an expensive human *in vivo* study. Furthermore, it is unknown how stable (or unstable) this protein is upon long term storage. If the protein is stable, the HMW of 2.8% is acceptable. However, if the protein degrades fast, even the HMW

Table 9.3 Conditions utilized in small scale runs

	Conditions
Run 1	Medium type 1, standard seeding density, bolus feeding at a fixed volume, fixed bioreactor temperature during 14 day cell culture, standard agitation rate
Run 2	Medium type 2, high seeding density, variable feeding to maintain a target glucose concentration, temperature shift after day 5 cell culture, standard agitation rate
Run 3	Medium type 1, high seeding density, variable feeding to maintain a target glucose concentration, temperature shift after day 5 cell culture, high agitation rate

Table 9.4 Peptide mapping data for protein A: product variants

Product characterization	Group description	Run 1	Run 2	Run 3	Comments
Deamidation	Segment 1	7.0%	10.0%	6.8%	Is the difference between 10% and 6.8% meaningful?
Oxidation	Segment 2	10.9%	17.2%	16.9%	Is the difference between 10.9% and 17.2% meaningful?
Glycosylation	A2G0F	85.9%	78.2%	69.0%	Is the difference between 85.9% and 69.0% meaningful?
	A2G2F	7.7%	11.1%	21.3%	Is the difference between 7.7% and 21.3% meaningful?
	M5	0.5%	1.8%	0.7%	Is 1.8% considered significantly higher than the other two?
	M8	0.4%	0.8%	0.3%	Can these values be considered similar or different?
Size variants	HMW[a] (%)	2.2%	2.8%	1.9%	Between the titer and HMW, which run to choose as 'suitable' process?
	LMW[b] (%)	1.2%	1.2%	1.0%	Considered similar?
Charge variants (cIEF)	Acidic peak (%)	19%	14%	28%	What can be concluded in these values? How would the difference in these values translate to efficacy or safety?
	Basic peak (%)	22%	30%	19%	
Productivity (titer)		2.0 g/L	2.5 g/L	1.4 g/L	Were there other parameters (i.e., seeding density) that could be the cause for these differences?
In vitro potency measurement (ELIZA)		102%	98%	110%	Considered similar?

[a]*HMW* high molecular weight
[b]*LMW* low molecular weight

of 1.9% may not be acceptable as such needing to redesign the upstream and downstream process control. Similarly, the effect of different glycosylation and charge variant forms on the safety and efficacy of the drug is remains unknown.

Challenges with Integrating End-to-End Process and Product Data

The challenges in the implementation of QbD methodology can addressed through the Continued Process Verification (CPV) concept, which were introduced by both FDA and EU agencies more than a decade ago [3, 4]. The CVP program allows the pharmaceutical companies to continuously gather the process and process knowledge so that the process control strategies for manufacturing of proteins are updated real time. The continued monitoring requires a centralized data historian tool where all process related data from lab scale, pilot scale and full manufacturing scale runs are consolidated along with the preclinical and clinical results from the clinics. However, the integration of all related data into a centralized data historian is a challenging task due to the (1) lack of necessary testing of incoming raw materials and cell culture materials creating a gap in the dataset, (2) shear number of data set difficult to analyze without advanced analytical tools, and (3) the safety and efficacy signals from the clinics confounded by a limited number of batches with a narrow product profile.

The cell culture media components such as inorganic and trace metals are widely known to have a direct effect on the glycosylation characteristics of therapeutic proteins. For example, the amount of zinc, copper and manganese in a chemically defined cell culture medium significantly impacted critical quality attributes highlighting the degree to which trace metal availability can affect CHO cell culture performance [14]. Not only the trace metals, but also the amino acid as well as the vitamins concentrations in the incoming cell culture materials can influence the product quality profile. Despite of their importance, the actual concentration of those key elements in the cell culture media are hardly tested prior use, due to the lack of sufficient analytical tool available in the manufacturing process development environment. The only information available for the incoming raw material and prepared media are pH, conductivity, and limited number of ions measurements, creating a gap in the central historian data set required for the robust CPV analysis. Not only the incoming raw material data, but also more comprehensive bioreactor data need to be monitored in real time. For example, the lactate and ammonia concentration in the bioreactor can have a significant effect on the cell growth and cell death, as such the real time monitoring of those parameters are necessary to link back to the change in the product quality profile. Currently, those upstream bioreactor parameters are measured using an off-line instrument which are not truly representative of what is happening in the production bioreactor environment.

In the Continued Process Verification (CPV) program context, grappling with the intricate data analysis web poses a primary challenge. This difficulty is amplified by the diverse input and output parameters of bioreactor processes, as Table 9.2 elucidates. Prevailing data analysis methods are predominantly visualization-based, employing techniques such as overlays of multi-batch data over cell growth timelines. These provide some retrospective analysis for single-parameter behavior, yet they fall short when predicting future outcomes or deciphering interactions among different parameters.

Advanced data analytics technologies emerge as vital tools, furnishing multidimensional interpretations of complex datasets and offering actionable insights. Researchers, including Papathansiou et [15]. underscored the necessity for such enhanced analytic tools. This necessity extends beyond analyzing upstream process parameters, such as pH and temperature, to include the actual concentrations of assorted components and understanding the impact of different bioreactor conditions on protein profiles [16].

Notably, Process Analytical Technology (PAT) could significantly augment the role of data analytics in Quality by Design (QbD). Described by the FDA in 2004 [17], PAT emphasizes design, analysis, and control of manufacturing processes by measuring critical quality and performance attributes of raw and in-process materials, as well as final products, in a timely manner. Integrating PAT with advanced analytics enhances real-time monitoring of process parameters and product quality attributes, fostering a dynamic, adaptive approach to process control and optimization ("Guidance for Industry PAT – A Framework for Innovative Pharmaceutical Development, Manufacturing, and Quality Assurance" 2004).

Further, by assimilating omics data, such as genomics, transcriptomics, proteomics, and metabolomics, advanced data analytics capabilities in QbD can be considerably amplified. Omics data offer insights into cellular system interactions with the bioprocess at the molecular level, promoting a more profound understanding of cellular mechanisms affecting product quality and performance [ref]. When used in tandem with multiscale data analysis, advanced analytics techniques can provide a holistic understanding of process variability and process-product relationships.

However, as promising as they are, incorporating advanced data analytics tools in the pharmaceutical industry is not without challenges. One significant issue pertains to data quality, availability, and integration at different stages of development and production. Incomplete or inconsistent data can cripple the effectiveness of advanced analytics tools in revealing meaningful patterns and relationships [18]. Hence, robust data management strategies and establishing common data standards are essential to ensure data quality and consistency for advanced analytics applications.

Another issue arises from the expectations of regulatory authorities and industry stakeholders for explanations and justifications for model predictions and decisions derived from advanced analytics methods. These expectations can be challenging, especially with black-box models [19] Furthermore, the complex mathematical and computational methods often employed by advanced data analytics tools may raise concerns about the interpretability and transparency of models and outcomes [20]. Developing more interpretable, explainable, and transparent AI and Machine Learning (ML) models is crucial to assuage these concerns and facilitating regulatory acceptance.

In vitro binding assays can provide a limited efficacy measure but do not indicate how the drug will behave within the human body. Safety-related functions like immunogenicity are also inadequately addressed by current assays. Beyond these technological challenges, achieving true QbD and CPV requires an in-depth understanding of different protein variants, such as various glycosylation forms, aggregates, and fragments, and their effects on the overall safety and efficacy of the product.

The most reliable safety and effectiveness data are typically derived from clinical trials. However, the number of batches used in these trials is often constrained, usually to one or two, which leaves a considerable gap in the Quality Target Product Profile (QTPP) definition essential for the QbD program. The reasons for this limitation are to minimize product variation, increase the probability of achieving statistically significant clinical outcomes, and manage the high costs and time involved in producing multiple product batches with diverse QTPP profiles.

Addressing these challenges and capitalizing on the full potential of QbD and CPV programs will necessitate significant investment from the pharmaceutical industry involving developing and implementing innovative data analytics technologies and robust, transparent AI and ML models. Moreover, collaboration is key among researchers, industry stakeholders, and regulatory authorities in establishing data management strategies and universal data standards. These measures will ensure that data across various development and production stages are high quality, consistent, and accessible.

Additionally, expanding the number of product batches used in clinical trials and employing advanced analytics tools to monitor and optimize the bioprocess can provide a more nuanced understanding of the QTPP, improving product safety and effectiveness. Such an approach can also enhance manufacturing processes, enabling more dynamic and adaptive control in bioprocessing, ultimately contributing to the overall success of QbD and CPV programs.

Efforts to Overcome Current Challenges in QbD Implementation

The full implementation of QbD methodology for the manufacturing of therapeutic proteins will be possible when right tools and systems can be established in the beginning of the drug development process. The lack of raw material data can be address by implementing a proactive raw material sampling program where incoming media batches are analyzed for trace metals, vitamins and amino acids during process development, scale up and full-scale manufacturing steps. In addition to the raw material data, a proactive sampling program for the cell culture media during production is critical in further enhancing the data set that can provide a meaningful information on the protein characterization. In fact, there has been a significant progress in the implementation of real time bioreactor data using advanced Process Analytical Technology (PAT). The cell metabolites and media components are continuously monitored using an at-line liquid chromatography-mass spectrometry (LC-MS) connected to the production bioreactor [21]. Timely detection of apoptosis, where cell deaths result in undesired enzymatic digestion of the target proteins, was accomplished using capacitance dielectric spectroscopy [22]. Another approach could be the characterization of cultured cells via epigenetic characterization where the environmental effects on the cell DNA regulation are monitored [23]. The data obtained from those additional sample testing and PAT technologies are then analyzed using advanced analytics tools.

Some utilized an orthogonal experimental design (OED) and response surface methodology (RSM) to determine the best parameters to achieve a satisfactory performance of bioprocesses [24]. Multivariate analytic tool such as orthogonal partial least squares (OPLS) or Partial Least Squares or Projections to Latent Structures (PLS) and principal component analysis was successfully deployed for the process monitoring and early detection of batch deviations [25]. Some software companies now offer off-the-shelf statistical programs, such as Minitab and JMP, that can perform multivariate analyses in addition to the data visualization. Discoverant® Software by Aegis Analytical Corporation offers fully integrated package where the bioreactor and QC testing results are automatically updated and tracked onto their analytical tool. QBDC GmbH, with its METIS data historian software allows to track and trace all input and output parameters as well as product quality attributes to verify the success and to optimize the implemented control strategies for manufacturing. Some companies have their in-house computational scientist groups to overcome the broad array of expertise required to generate high quality data with these instruments, pull the data together, and model them in a way that can be predictive and useful [23, 28]. Whether purchasing a built-in software or having in-house computational group, the comprehensive data analysis remains as a key enabling factor in the true concept closer to the full QbD implementation.

In more recent years, the advanced data analytics is used during cell cloning step to select a cell line, as a machine learning tool based on the *in silico* predictive simulation, the field also referred to as bioinformatics and artificial intelligence (AI) across the biotech communities. A high titer cell line was successfully selected based on the cell metabolism and process productivity information can be elation between biological information, such as cell metabolism, and process productivity. Barberi et al. [26] successfully estimated the end-point titer using the cell metabolism information to select a high-performing cell line. There are many companies now offer custom-designed bioinformatics tools used during the gene sequencing and cell line design step as it allows the modification of amino acid sequence to avoid potential aggregation and truncation spots. The AI based predictive modeling is also used to control protein glycosylation profile. Kontoravdi et al. [27] developed a model to design a desired feeding schedules that provide good agreement on the cell growth kinetics, cell metabolism, and protein glycosylation profile.

Even with such progress in the advanced analytics, the true concept of QbD can't be realized until more investment becomes available to gather wide range of product profiles and its effect on the safety and efficacy of the drug. A more advancement is needed in the development of lab scale *in vivo* assays that can measure immunogenicity of various glycosylation profiles and product variants. The process development should include additional funding to isolate the various product variants and tested for the *in vivo* binding assays.

Key for the development and implementation of the QbD process is using and implementing the future potential control strategies already during the cell line development process and all the way down to the GMP process performance qualification program and beyond. Due to the sheer numbers of parameters and attributes which need to take into account, the current approach of the step-wise single experiment assessments will not lead to success. Similar to the QTPP we need to implement a target process profile and a forward looking risk assessment, which anticipates in the early beginning of the development the final manufacturing process. Potential obstacles, failures, effects and dependences need to be implemented into the process development design and analyzed throughout the development phases. The usage of PAT systems and sophisticated spend media analysis are key for the development of process models. Yet, this systems should not only focus on upstream and cell culture parameters, but need to include downstream steps and even the formulation of the drug product. The data amount is and will be enormous and regular statistical tools will not be able to accommodate meaningful data interpretation, but machine learning and AI tools are inevitable. Further all data needs to be in one data base, in one system which allows the data extraction throughout the whole development, as mentioned above starting with the cell line development.

Conclusion

The QbD mythology is such a powerful tool that helps the development of therapeutic proteins be effective and robust. While this methodology is widely adapted across the pharmaceutical industry, the implementation of the QbD concept in its entirety is still limited due to (1) insufficient time to fully characterize the target protein profile in the beginning of development process, (2) complexity of input and output parameters prohibiting the full design of experiments, and (3) challenges in collecting and analyzing the end-to-end product and manufacturing data. Substantial progress has been made in the advanced data analytics and machine learning perspective. However, more comprehensive data gathering in the incoming raw materials and safety and efficacy information are still missing. Due to the unmet medical needs, many of new therapeutic proteins are developed based on a rapid timeline which prohibits time required to conduct such safety and efficacy finding *in vivo* studies. More advancement in the in vivo studies combined with the advanced data analytics will help in the full implementation of QbD methodology.

Conflict of Interests The authors declare no conflict of interest.

References

1. ICH Q11: development and manufacture of drug substances (chemical entities and biotechnological/biological entities) – scientific guideline
2. ICH Q8 (R2) pharmaceutical development – scientific guideline
3. EU guidelines for good manufacturing practice for medicinal products for human and veterinary use, Volume 4, Annex 15: qualification and validation
4. US FDA guidance for industry process validation: general principles and practices
5. ICH Q9 (R1) quality risk management-guidance for industry
6. ICH Q10 pharmaceutical quality system – scientific guideline
7. CMC Biotech Working Group (2009) A-Mab: a case study in bioprocess development. https://qbdworks.com/storage/2014/06/A-MabCaseStudyVersion.pdf
8. Morar-Mitrica S et al (2018) An intercompany perspective on biopharmaceutical drug product robustness studies. J Pharm Sci 107(2):529–542

9. Alt N et al (2016) Determination of critical quality attributes for monoclonal antibodies using quality by design principles. Biologicals 44(5):291–305
10. Ruesch MN et al (2021) Strategies for setting patient-centric commercial specifications for biotherapeutic products. J Pharm Sci 110(2):771–784
11. Schmidt SR (2017) Controlling glycosylation in fusion protein manufacturing to generate potent biobetters. BioProcess Int, September 19
12. Zhang P et al (2016) Challenges of glycosylation analysis and control: an integrated approach to producing optimal and consistent therapeutic drugs. Drug Discov Today 21(5):740–765
13. Raju TS (2008) Terminal sugars of Fc glycans influence antibody effector functions of IgGs. Curr Opin Immunol 20(4):471–478
14. Graham RJ et al (2019) Consequences of trace metal variability and supplementation on Chinese hamster ovary (CHO) cell culture performance: a review of key mechanisms and considerations. Biotechnol Bioeng 116(12):3446–3456
15. Papathanasiou M, Kontoravdi C (2020) Engineering challenges in therapeutic protein product and process design. Curr Opin Chem Eng 27:81–88
16. Khanal SK, Tarafdar A, You S (2023) Artificial intelligence and machine learning for smart bioprocesses. Bioresour Technol 375:128826–128826
17. U.S. Department of Health and Human Services (ed) (2004) Guidance for industry PAT—a framework for innovative pharmaceutical development, manufacturing, and quality assurance. U.S. Department of Health and Human Services, Washington, DC
18. Li B-H, Hou B-C, Wen-Tao Y, Lu X-B, Yang C-W (2017) Applications of artificial intelligence in intelligent manufacturing: a review. Front Inf Technol Electron Eng 18:86–96
19. Pedreschi D, Giannotti F, Guidotti R, Monreale A, Ruggieri S, Turini F, Aaai (2019) Meaningful explanations of black box AI decision systems. In: 33rd AAAI conference on artificial intelligence/31st innovative applications of artificial intelligence conference/9th AAAI symposium on educational advances in artificial intelligence, 9780-84. Honolulu, HI
20. Seong MA, Schiele B, Fritz M (2018) Towards reverse-engineering black-box neural networks. arXiv pre-print server
21. Maruthamuthu MK et al (2020) Process analytical technologies and data analytics for the manufacture of monoclonal antibodies. Trends Biotechnol 38(10):1169–1186
22. Wu S et al (2021) Rapid at-line early cell death quantification using capacitance spectroscopy. Biotechnol Bioeng 119(3):857–567
23. Hernandez I et al (2019) Epigenetic regulation of gene expression in Chinese Hamster Ovary cells in response to the changing environment of a batch culture. Biotechnol Bioeng 116(3): 677–692
24. Cheng Y et al (2023) Artificial intelligence technologies in bioprocess: opportunities and challenges. Bioresour Technol 369. https://doi.org/10.1016/j.biortech.2022.128451
25. Gao J, Adamec E (2020) Multivariate analysis of cell culture bioprocess data. Am Pharm Review, April 27
26. Xu J et al (2022) Upstream cell culture process characterization and in-process control strategy development at pandemic speed. MAbs 14(1). https://doi.org/10.1080/19420862.2022.2060724
27. Barberi G et al (2021) Anticipated cell lines selection in bioprocess scale-up through machine learning on metabolomics dynamics. IFAC-PapersOnLine 54(3):85–90
28. Kontoravdi C et al (2007) Development of a dynamic model of monoclonal antibody production and glycosylation for product quality monitoring. Comput Chem Eng 31(5–6):392–400

Chapter 10
Biopharma 4.0 for Biologics Manufacturing Under Pandemic Constraints

Axel Schmidt, Alina Hengelbrock, and Jochen Strube

Abstract Biologics are the fastest growing molecular group in pharmaceutical therapies as supplying large most human-body like molecules with high efficacy at low doses and less side effects. Biopharma 4.0 relates to the fourth industrial revolution based on digitalization. Digitalization has reached biopharmaceutical industry along the whole development and value chain. Biologics manufacturing of such complex three-dimensional molecules in correct folding and glycosylation pattern for best bioavailability is still a challenge for world scale amount supply. As the variety of new entities vary recently quite broad from peptides, virus like particles, pDNA and mRNA to cell therapies new process design and development methods are needed to speed up robust product supply for patients. A sound theoretical basis is needed to utilize process modeling tools as digital twins for autonomous operation of manufacturing processes. Qualified operators are the rare resource bottleneck to guarantee product quality. Therefore, continuous biomanufacturing as well as the consequent application of the quality by design approachare key-enablers towards biopharma 4.0.

Keywords Biopharma · Digitalization · Quality-by-Design · Automation · Digital Twin

Abbreviations

APC	advanced process control
AMWHV	Arzneimittel- und Wirkstoffherstellungsverordnung
CBM	continuous biomanufacturing
cGMP	current Good-Manufacturing-Practice
CHO	chinese hamster ovary

A. Schmidt · A. Hengelbrock · J. Strube (✉)
Institute for Separation and Process Technology, Clausthal University of Technology, Clausthal-Zellerfeld, Germany
e-mail: schmidt@itv.tu-clausthal.de; hengelbrock@itv.tu-clausthal.de; strube@itv.tu-clausthal.de

R. Pörtner (ed.), *Biopharmaceutical Manufacturing*, Cell Engineering 11,
https://doi.org/10.1007/978-3-031-45669-5_10

COG	cost of goods
CQA	critical quality attributes
DARPA	Defense Advanced Research Projects Agency
DoE	Desing-of-Experiments
FDA	Food and Drug Administration
FMEA	failure-mode-error-analysis
GMP	Good-Manufacturing-Practice
GWP	global warming potential
HIV	human immunodeficiency viruses
ICH	International Council for Harmonisation of Technical Requirements for Pharmaceuticals for Human Use
LNP	lipid nano particle
mAbs	monoclonal antibodies
mRNA	messenger RNA
NN	neural network
OLS	ordinary least squares
PAT	Process-Analytical-Technology
pDNA	plasmid DNA
PLS	partial least squares
QbD	Quality-by-Design
QTPP	quality-target-product-profile
RTRT	Real-time-release-testing
scFv	single chain variable fragment
SU	single use
VLP	Virus-like particle

Introduction

Biopharma 4.0 relates to the fourth industrial revolution based on digitalization. Digitalization has reached biopharmaceutical industry along the whole development and value chain, starting with drug discovery [1], in-silico patient studies [2–5], toxicology studies [6], big data mining technologies for clinical patient data evaluation [7], product design [8, 9] up to recently biomanufacturing [10, 11] based on process engineering tools [12–15]. As all other steps and methods have been broadly reviewed still biomanufacturing is a traditional bottleneck in sufficient supply of drugs under economic constraints of ageing society and limited health care systems.

Biologics are the fastest growing group in pharmaceutical therapies [16, 17]. Biologics manufacturing enables with high efficacy and less side effects the supply of active components which are of large molecular size [18, 19].

The last decades monoclonal antibodies dominated [20], but nowadays again a manyfold of different molecular types like peptides, fragments, virus like particles, pDNA and mRNA as well as allogenic cell therapies [21] are under clinical research

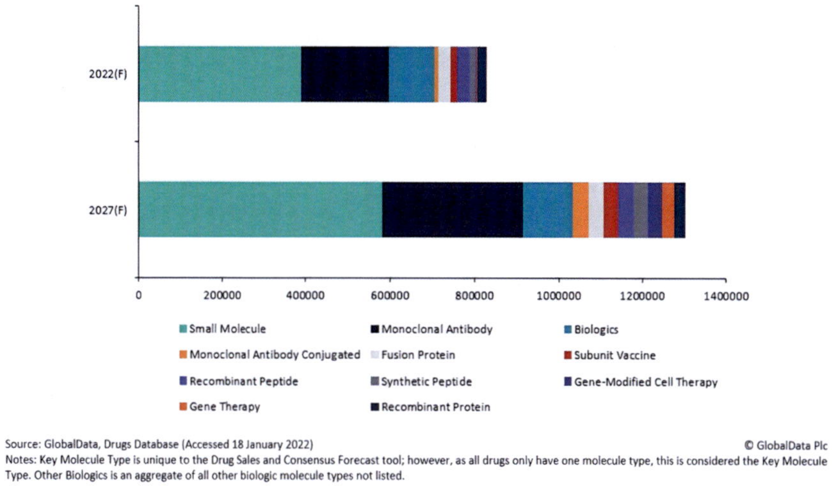

Fig. 10.1 Biologics versus small molecules by 2027 forecast sales [16]

[22, 23], see Fig. 10.1 [24, 25]. This molecular variety causes increasing complexity for process development and design teams. The class of antibodies had been successfully efficiently supplied by defining so-called platform technologies which is the definition of a process sequence, the unit operations inside and ranges for operation parameters. Each new antibody entity has only to be fitted by reduced experiments within this given range. Benchmark is a magnitude of 2 antibody candidates in process design and test amount supply per year for each laboratory unit. But if there is now a broad molecular type variety again this approach is not feasible. The efforts for any full experimental study from scratch are too high and more importantly too long. And the design of a platform technology for each molecular type would take decades each [26, 27].

Therefore, innovative methods are needed. A pure experimental design by statistical experimental plans in down scaled laboratory equipment or with aid of screening robotics would be too slow and of too exceeding efforts. Additional theory support is needed. As much theory as feasible causes therefore the application of scalable predictive physical-chemical process modeling. Such modeling need few a priori predicted laboratory scale experiments for model parameter determination and final model validation. Only validated models are of use in regulatory environment to file decisions. In addition, recent machine learning approaches would support the data evaluation and optimization to reduce further efforts. Any calculations are of much less efforts than any experimental setups in general. E.g., well known process chromatography design with few thousand simulations to exemplify design and control space definitions would not even be possible by experiments. Whereas experimental model parameter determination and model validation takes only about 2 laboratory team weeks, see Fig. 10.2.

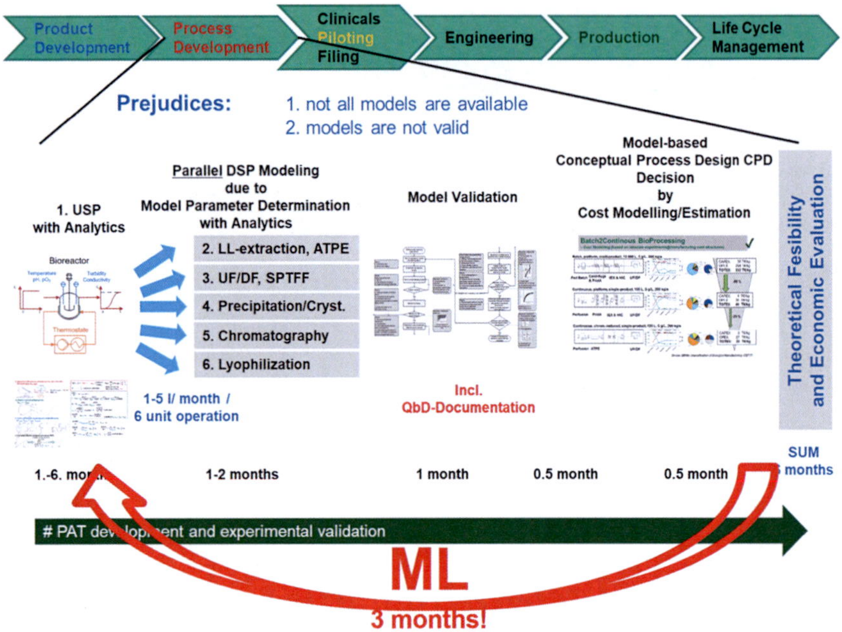

Fig. 10.2 Efficient process development and design with the aid of process modeling [28]

Due to process modeling of each unit after first fermentation broth is available all downstream units feasible could be designed in parallel and the best sequence calculated. This is finally piloted and statistically validated accompanied with the supply of test amounts and data for filling decisions. In the meantime upstream optimization has been accomplished as well and final process integration is gained [28]. This approach is feasible for any molecular entity, efforts is about 5–6 employees for about 6 months. Of course, there will be efforts reduction and speed gain if the molecular group has been under development a few times already, down to about 4–5 months. In addition, over the time the consequent knowledge-based process comprehension with sound theoretical fundamentals summarized in process modeling would lead to a data base which is large enough to be worth analyzing. Preferably by machine learning methods which will again speed up and reduce efforts to about 3 months in total.

Accelerated process development and design for supply of test amount is therefore a crucial task for pharmaceutical industry management. Methods to be learned and transferred are already well established in chemical engineering as chemical industry is already under economic pressure since decades and has survived by industrialization of innovations like:

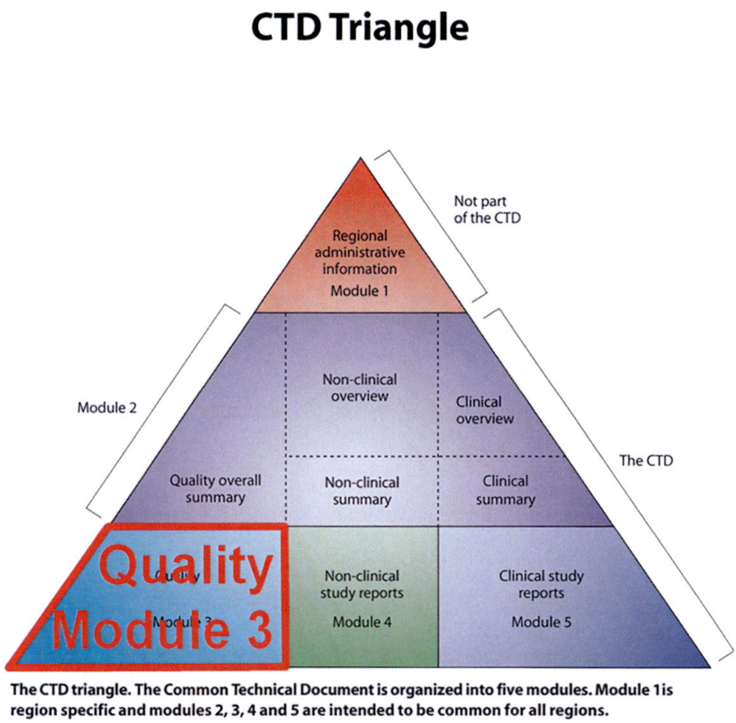

Fig. 10.3 The common technical document triangle

- Continuous instead of batch processing
- Process automation with integration of process analytical technology esp. inline analytics
- Process development and design with aid of mathematical process modeling
- Physical properties determination by experiments and molecular modeling
- Process intensification i.e., process integration of upstream and downstream processing
- as well as hybrid intensified processes
- Fast, flexible and modular plant engineering design

Of course, regulatory of biologics manufacturing is different as the process defines the product quality as well as a setup of orthogonal analytical quality assurance methods. But, since about more than 10 years the regulatory authorities themselves demand continuous manufacturing [29] and the application of the Quality-by-Design approach [30] which enables defined changes in manufacturing operation during filing approval defined process design and control space which guarantees product quality as well. The consequent application prevents, that speedy developed lab- and piloting scale processes, by which the test amounts are supplied, are needed to

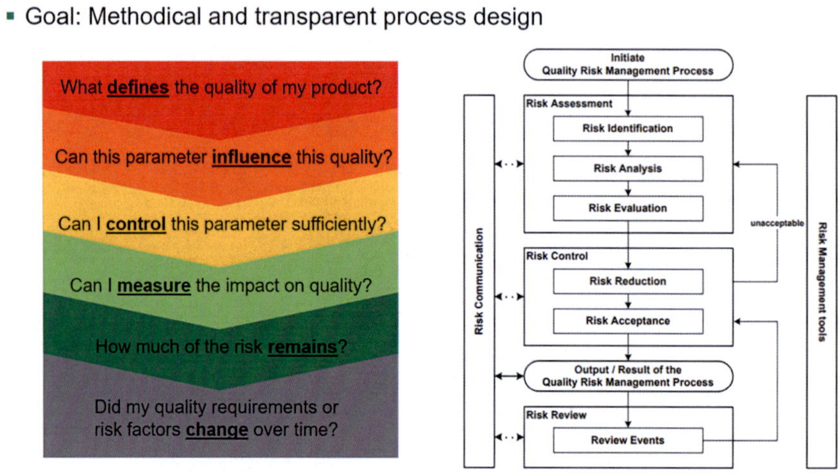

Fig. 10.4 Risk-based process development approach

be scaled up and be operated later in manufacturing – which causes an unnecessary amount of batch variability and batch failures – related to product quality issues. FDA states product quality is much more robust in continuous manufacturing.

The regulatory consideration of the manufacture of biologicals was already considered in the drafting of the EU GMP Guide Part II, which considers the manufacture of active pharmaceutical ingredients. The specifications cover, among other things, requirements for operation with cell banks or cell cultures/fermentation processes, including the recovery and purification processes [31]. Thus, the recommendations of the ICH guidelines of group Q5 were adopted [32–36]. These guidelines have been given additional legal force by national pharmaceutical regulations such as the Arzneimittel- und Wirkstoffherstellungsverordnung (AMWHV) in Germany [37].

The common technical document, see Fig. 10.3 is organized in five modules for regulatory approval which are distributed internally to the specialized departments. Module 3 quality is related to the process manufacturing operation whereas module 4 to non-clinical and module 5 to the clinical study report. Process engineering activities in process development and design as well as operational excellence are therefore focused in module 3. Details on the quality requirements are also provided by the European Pharmacopoeia, which is mandatory for manufacturing and quality control [38].

This includes the QTPP (quality-target-product-profile) and CQA (critical quality attributes) definition as well as formal risk assessment by any tool like FMEA (failure-mode-error-analysis), see Fig. 10.4.

The outcome is a design space within which the control space is included to be operated with aid of PAT (see Fig. 10.5). Further benefit of such approach would be

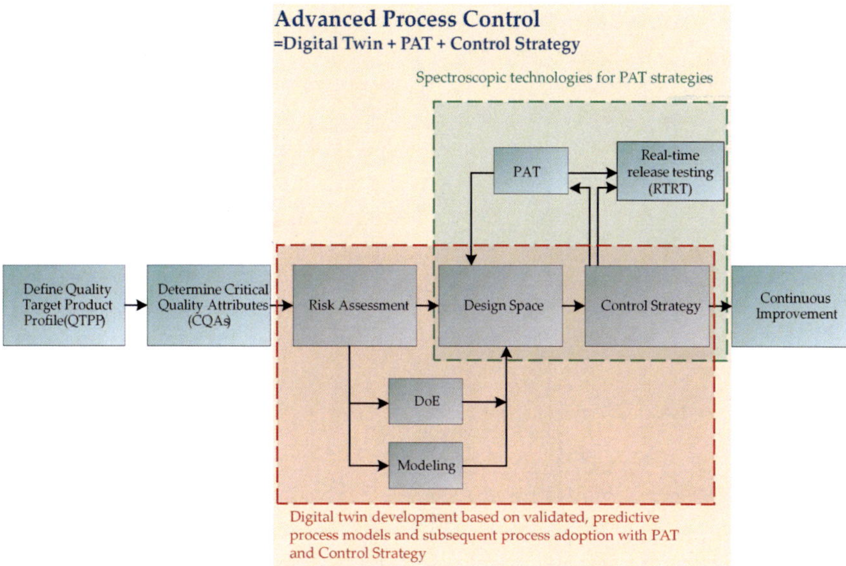

Fig. 10.5 Quality-by-Design based process development [40]

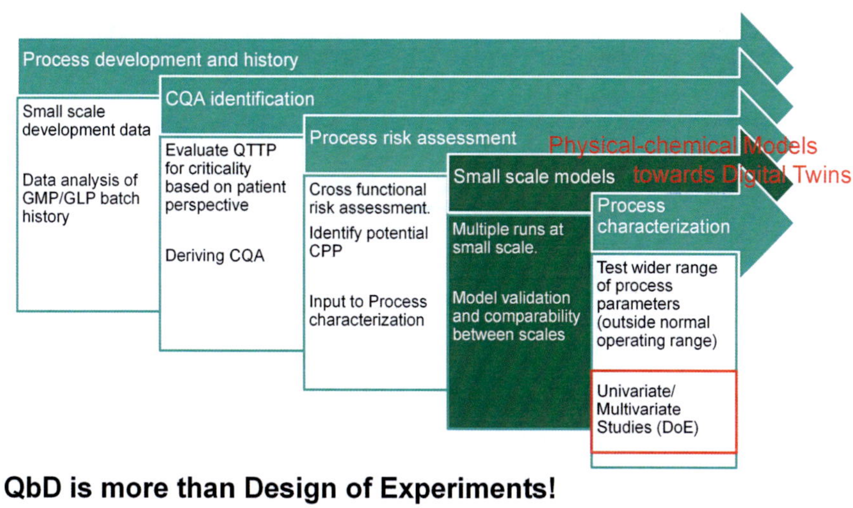

Fig. 10.6 Physico-chemical process models in process development [42]

the opportunity for real time release testing inline methods to reduce quality assurance capacities by factor of 2/3 – 3/4 as well as a speedier batch release to patients. In addition, automatized cleaning and cleaning validation routines would reduce manufacturing plant down time and operator efforts. Skilled operators are

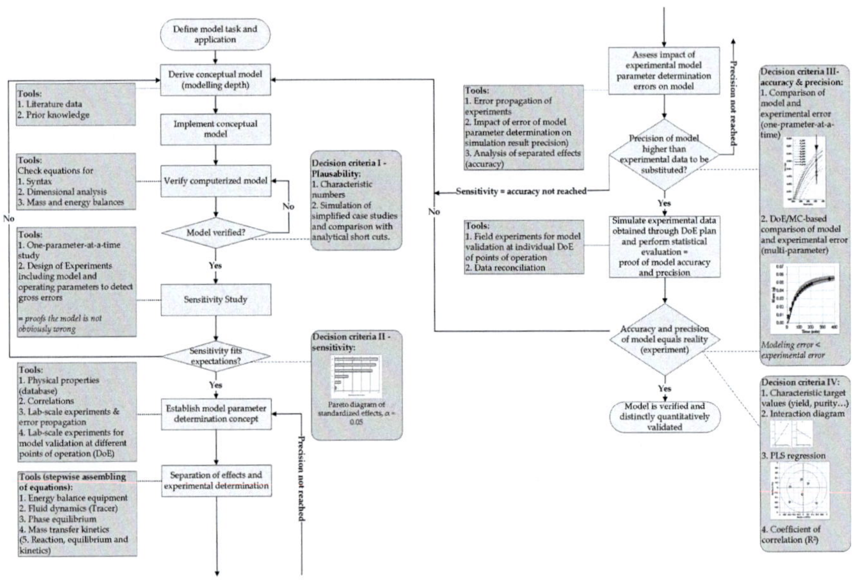

Fig. 10.7 Is any approval without digital twins feasible [11]

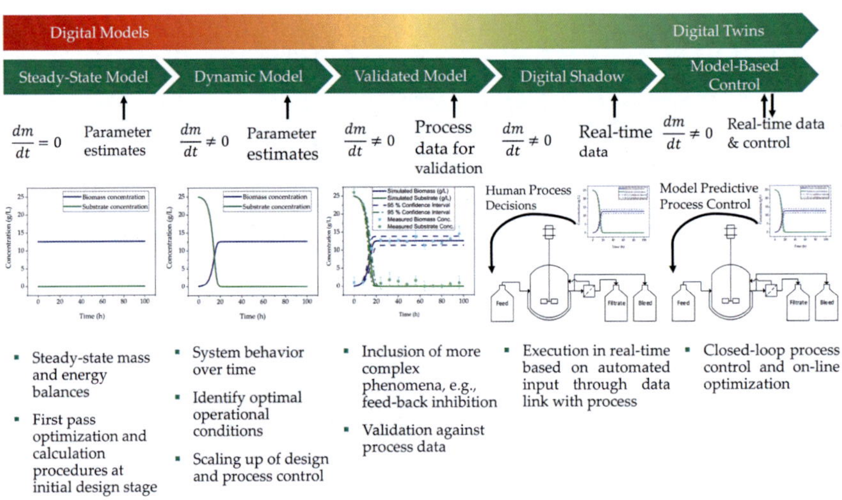

Fig. 10.8 Levels of Digital Twins [43]

nowadays a rare resource and may be identified as the main bottleneck in biologics manufacturing [39].

Single-use disposable technology versus stainless-steel plant configurations will be such under consideration for differing decisions based on speed, capacity and bottlenecks [39, 41]. If design space including control space ranges are defined any

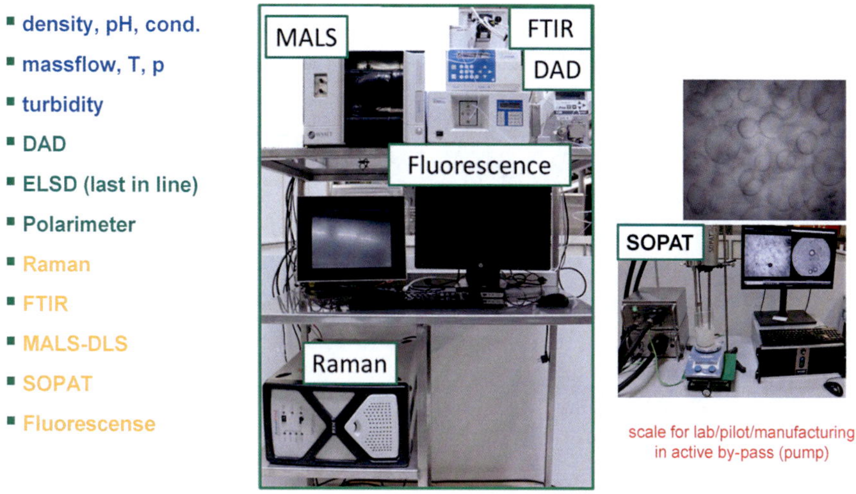

Fig. 10.9 Process-Analytical-Technology skid at the Institute for Separation and Process Technology

experimental study with scaled-down laboratory plants or with aid of experimentally validated process models is feasible (Fig. 10.6). Process models reduce the effort and are therefore to be preferred.

To validate this theory experimentally for regulatory filing decisions, a distinct workplan has been proposed and applied successfully as shown in Fig. 10.7.

The additional benefit of such models is that they are directly applicable as digital twins for real-time process control and optimization (Fig. 10.8).

Digital twins are defined as digital representations of physical processes with set point control automation. Input are real-time information enabled by inline analytics [44, 45].

Inline analytics has in the meantime been manyfold proven as a powerful tool. Of course, any sound statistic data analysis and evaluation is needed like PLS (partial least squares) [46] or NN (neural network) [47] [48–52].

Figure 10.9 exemplifies such a multipurpose PAT tool which is capable to move from unit operation to unit operation through the process during process development and define the best detector or detector combination at each step. The found solution is directly transferable to piloting or even manufacturing scale when operated in an active bypass.

Final model validation and test amount supply is done in a typical multi-purpose piloting plant facility like exemplified in Fig. 10.10, considering the biosafety level needed and additionally cGMP (current Good-Manufacturing-Practice) standards and covering all unit operations to be combined flexible to any process sequence needed. Batch and continuous operation should be feasible. The facility exemplified is equipped with Siemens S7 standard and will allow autonomous operation from an external control room.

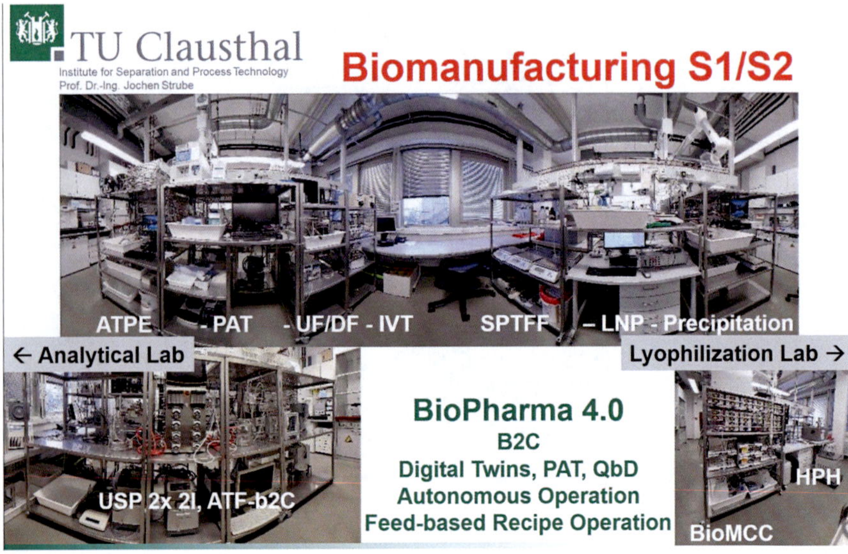

Fig. 10.10 Biomanufacturing in S1/S2 facilties at the Institute for Separation and Process Technology [53]

Case Study on Monoclonal Antibodies IgG1

Historically, mAbs (monoclonal antibodies) have been at first demonstrated in piloting studies to be of benefit and feasible in continuous manufacturing solutions aided by PAT (see Fig. 10.11) [54–57].

As explained as mAbs do allow a platform technology based on protein A capture step after CHO (chinese hamster ovary) mammalian cell cultivation for most companies, due to the quite huge few 100 kg to even tons amount products needed, any transfer and implementation of the effort reduction concepts presented above, is an obvious task [58–60].

Even to achieve the the climate neutrality demand for 2045 in Germany is feasible with CBM (continuous biomanufacturing) resulting in significant COG (cost of goods) and GWP (global warming potential) reduction as depicted in Fig. 10.12 [61–63].

As fragments manufacturing is dominantly based on *E.coli* fermentations, research deals with scFv as an example for a new molecular entitiy.

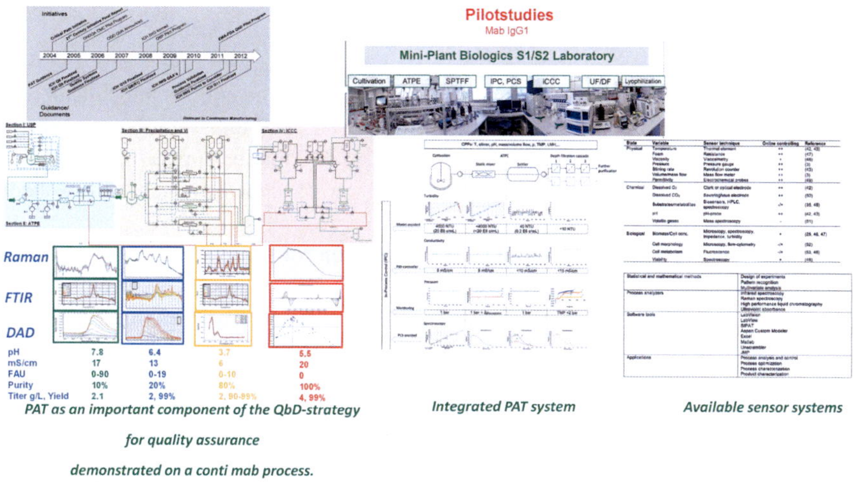

Fig. 10.11 Process-Analytical-Technology for biologics manufacturing

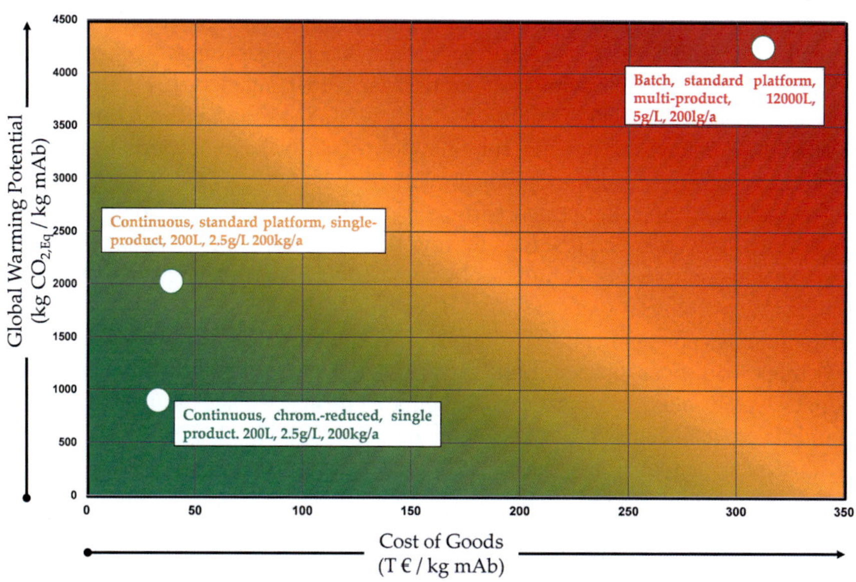

Fig. 10.12 GWP vs COG

Case Study on VLPs: HEK- HIV-Gag

Virus like particles as a totally different group of therapeutic entities are considered to proof generalizability. On the example of HEK (human embryonic kidney) based production of HIV-gag VLPs the methods could be applied successfully to demonstrate APC (advanced process control) by PAT and DT, which is illustrated in Fig. 10.13 [43, 64, 65].

Process development with sound design and control space definition based on digital twins and application of the PAT approach allow a fast and effort reduced data generation. Process control towards autonomous operation is shown as well [43, 66].

Case Study on pDNA2mRNA Vaccines

Recent SARS-CoV-2 pandemics exemplify some methodological lessons learned. E.g., plant design and engineering are considered now to be at best fast, flexible and modular. Different process approaches are shown in Fig. 10.14. DARPA (Defense Advanced Research Projects Agency) project falcon has already exemplified national resilient vaccine supply solutions based on containers [67]. BioNTech announced container based distribution worldwide [68, 69] as well as Evonik [70].

Literature research allows the configuration of an orthogonal purification and polishing step design needed for product safety [39]. Enzymatic *in vitro* transcription with capping and final lipid nano particle formulation are key-technologies to be implemented and approved first time for mRNA [71]. Risk analysis for all units lead to design and control space definitions by about 1000 data points generated with aid of digital twins for each unit (see Figs. 10.15 and 10.16) [72, 73].

Continuous biomanufacturing, digital twins and PAT applications within the QbD framework open up new horizons. First studies (see Fig. 10.17 for process overview) point out a factor 80 capacity improvement. Demands in personnel and chemicals are defined as bottlenecks at the moment [39, 41, 72, 73].

Consequent automation reduces personnel needed further to a minimum and process optimization studies show a potential of further 30% capacity improvement by advanced process control applications [72].

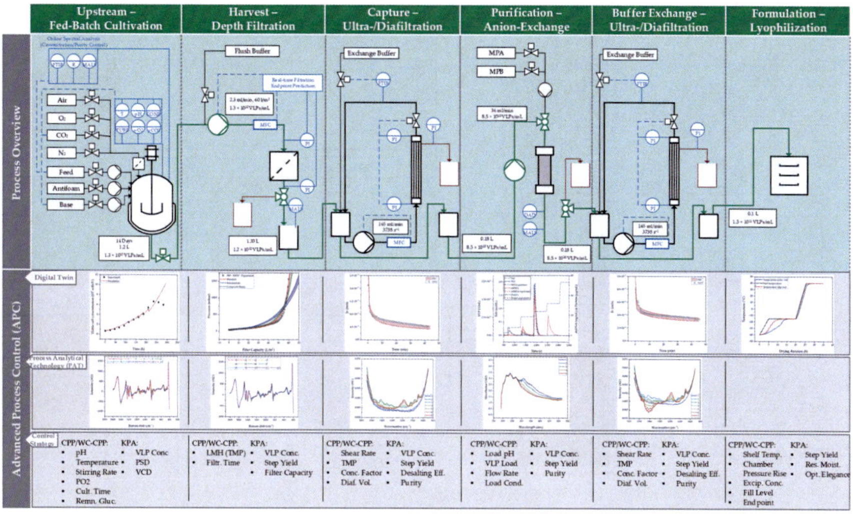

Fig. 10.13 Process flowsheet including Advanced Process Control by Digital Twins and optimized Process-Analytical-Technology

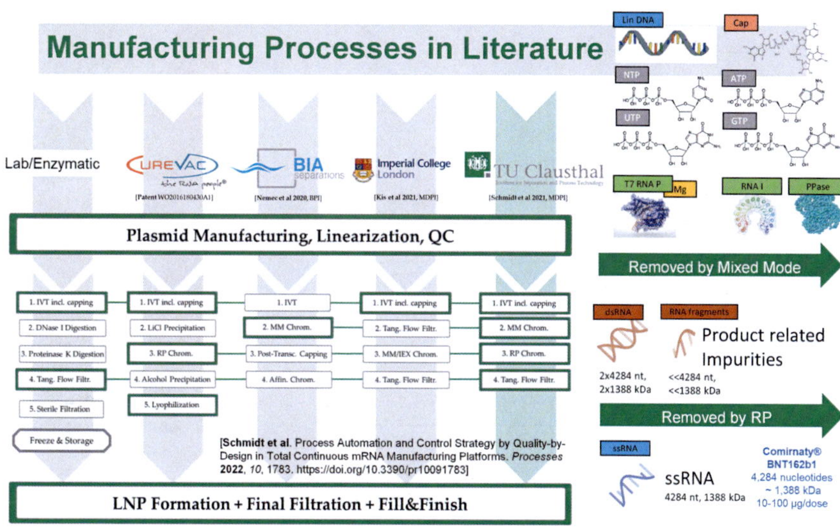

Fig. 10.14 Main process phases in mRNA vaccine manufacturing. Plasmid manufacturing by *E. coli* fermentation followed by purification and linearization, followed by in vitro transcription, capping, and formulation including encapsulation in LNP (lipid nano particle)

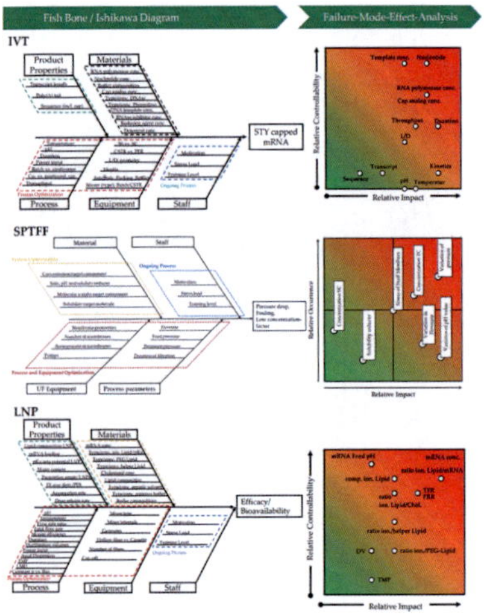

	Factor	Main Effect CQA (purity)	Main Effect PA (STY)	Highest Main Effect Score	Interaction CQA (purity)	Interaction PA (STY)	Highest Interaction Score	Severity
In Vitro Transcription	T7 conc	1	1	1	4	4	4	4
	NTP conc	8	4	8	8	4	8	64
	Cap-analog conc	8	1	8	8	2	8	64
	Template conc	1	1	1	1	2	2	2
	Temp	8	4	8	4	1	4	32
	pH	8	4	8	4	1	4	32
	VF tot	1	4	4	8	4	8	32
	L/D	1	1	1	1	2	2	2
	Factor	**Main Effect CQA (BE)**	**Main Effect PA (VCF)**	**Highest Main Effect Score**	**Interaction CQA (BE)**	**Interaction PA (VCF)**	**Highest Interaction Score**	**Severity**
Inline Diafiltration SPTFF	VF Feed	8	2	8	4	2	4	32
	mRNA conc	2	2	1	2	2	2	2
	SC conc	2	2	1	2	2	2	2
	Viscosity	8	4	8	2	2	2	16
	Fiber Diameter	8	4	8	4	4	4	32
	Fiber Length	8	4	8	4	4	4	32
	Fiber Count	8	4	8	8	8	8	64
	R_m	8	4	8	2	2	2	16
	TMP	8	8	8	8	8	8	64
	VF EB	4	8	8	8	8	8	32
	Factor	**Main Effect CQA (EE)**	**Main Effect PA (c_{LNP})**	**Highest Main Effect Score**	**Interaction CQA (EE)**	**Interaction PA (c_{LNP})**	**Highest Interaction Score**	**Severity**
Lipid-Nanoparticle Formation	mRNA conc	8	2	8	4	2	4	32
	mRNA Feed pH	4	2	4	8	4	8	32
	WR Lipids:mRNA	8	4	8	4	4	4	32
	w% ionizable Lipid	8	4	8	4	4	4	32
	w/w ion./PEG-Lipid	8	4	8	2	2	2	16
	w/w ion./help Lipid	8	4	8	2	2	2	16
	w/w ion./Chol.	8	4	8	4	2	4	32
	VF tot	1	4	4	8	4	8	32
	v/v aq/EtOH	8	4	8	4	4	4	32
	DV	1	4	4	1	2	2	8
	TMP	1	4	4	1	1	1	4

Fig. 10.15 Ishikawa and graphical impact ranking analysis

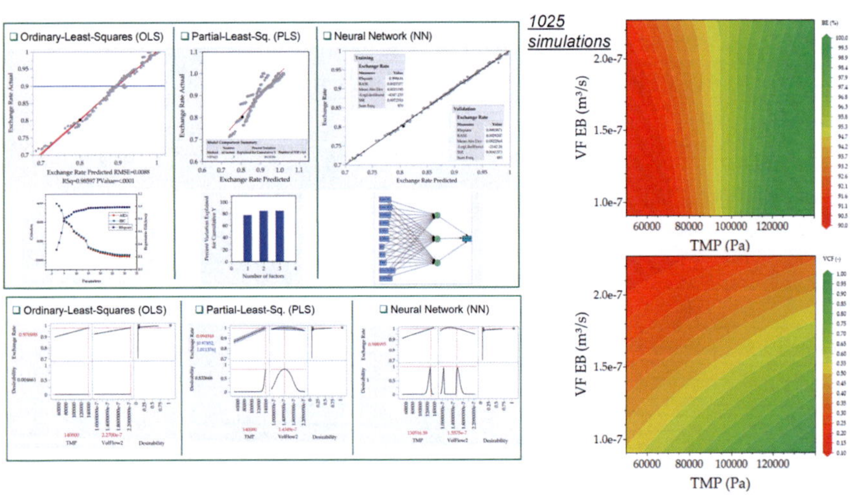

Fig. 10.16 Comparison of regression efficiency, probability and possible effect analysis of OLS (ordinary least squares), PLS, and NN

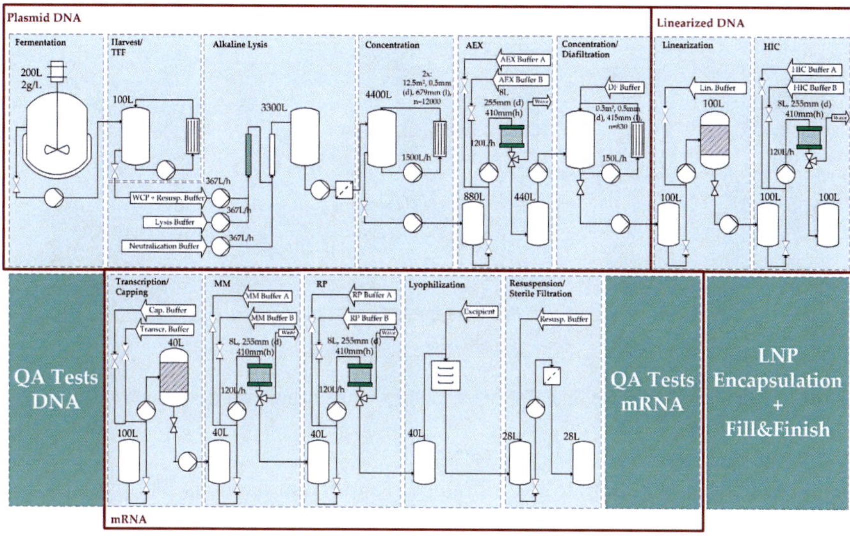

Fig. 10.17 Overview of the plasmid production, quality testing and mRNA manufacturing process

Conclusions

Pandemics did push mRNA as a new entity class and other vaccines under warp-speed regulatory approval. This gives focus on technology demands like

- Accelerate process design and regulatory approval by process modeling within the QbD framework towards digital twins combined with PAT
- establish RTRT (real time release testing) and cleaning automation and validation
- build decentral manufacturing plants, e.g. in containers
- which is enabled by small scale plants due to low μg doses
- bottleneck is on chemicals and personnel, therefore
- reduce chemicals consumption by recycling and continuous operation
- understand chemical consumption and titer-inhibitions based on theoretical molecular level process comprehension
- minimize personnel by autonomous operation even advanced process control methods
- implementation of supply chain hybrid models, i.e., stainless steelpiping, and sensor skids with SU (single use) storage and buffer supply and main internals for unit operations to be changed already pre-validated like chromatography columns, filters and membranes.

The general approach is summarized in Fig. 10.18 how to generate a scalable digital twin. Experimentally validated process models based on laboratory scale model data

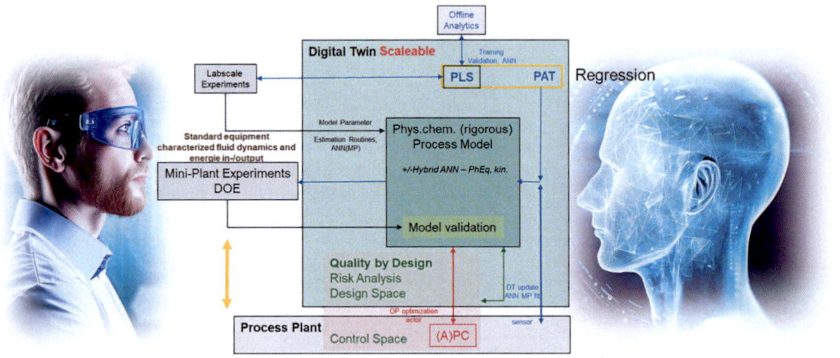

Fig. 10.18 Summary scaleable digital twin technology with potential methods involved

combined with an appropriate PAT strategy. Regression methods like partial least squares or neural networks are applied to fit parameter from data and adapt the digital twin to reality. Advanced process control spaces based on DoE studies predict optimal set-points to keep autonomous operation running even at variations or disturbances.

The dream of Biopharma 4.0 i.e., the application of existing methods like digital twins from process development over design and filing to manufacturing for planning, design, maintenance, and autonomous operation is feasible, of huge benefits and ready-to-project.

Direct scalability on one plant for about 1000 doses clinical trials up to 10 mil. doses in 3 days manufacturing scale have been proven feasible based on IP protected innovations. Speed is gained by the fact that only one qualification and validation plant size is needed.

A significant factor of 80 in productivity throughput improvement is possible to gain with identical personnel and chemical amount – as being the main bottlenecks.

Those concepts of continuous operation in the scale of a plant capable of producing 10 mil. dose per week, operated at the institute, have been compared to state-of-the-art batch operation.

Acknowledgments The authors would like to thank the whole institutes team like Heribert Helgers for cell line development and upstream processing, Florian Vetter and Dr.-Ing. Steffen Zobel-Roos for continuous chromatography and Alex Juckers for lyophilization; Mourad Mouellef and Thomas Knebel for their endless automation efforts; the mechanical and electrical workshop with Nils Hoffmann and Volker Strohmeyer is thanked for their fast-track assembly of IVT and LNP units – under budgeting supervision of Claudia Lacheta; laboratory-head Frank Steinhäuser for his full-time trouble-shooting and total support. Prof. Dr. Martin Tegtmeier as qualified person for his regulatory guidance and supervision.

Conflicts of Interest The authors declare no conflict of interest.

Ethical Approval Statement (If Required) Not applicable/required.

References

1. Rao V, Srinivas K (2011) Modern drug discovery process: an in silico approach. J Bioinf Seq Anal 2:89–94
2. Venkatesh KP, Raza MM, Kvedar JC (2022) Health digital twins as tools for precision medicine: considerations for computation, implementation, and regulation. NPJ Digit Med 5: 150. https://doi.org/10.1038/s41746-022-00694-7
3. Fraunhofer-Gesellschaft med2icin. https://websites.fraunhofer.de/med2icin/. Accessed 05 Jan 2023
4. Erol T, Mendi AF, Dogan D (2020) The Digital Twin revolution in healthcare. In: 2020 4th international symposium on multidisciplinary studies and innovative technologies (ISMSIT). IEEE, pp 1–7
5. Coorey G, Figtree GA, Fletcher DF et al (2021) The health digital twin: advancing precision cardiovascular medicine. Nat Rev Cardiol 18:803–804. https://doi.org/10.1038/s41569-021-00630-4
6. Valerio LG (2012) Application of advanced in silico methods for predictive modeling and information integration. Expert Opin Drug Metab Toxicol 8:395–398. https://doi.org/10.1517/17425255.2012.664636
7. Rønning-andersson A, Daniel Ingemann, Kroier Pedersen (2022) Novo Nordisk and Novartis will participate in large Danish project on decentralized clinical trials. JP/Politikens Hus A/S
8. Boetker J, Raijada D, Aho J et al (2016) In silico product design of pharmaceuticals. Asian J Pharmaceut Sci 11:492–499. https://doi.org/10.1016/j.ajps.2016.02.010
9. Lorenz S, Amsel A-K, Puhlmann N et al (2021) Toward application and implementation of in silico tools and workflows within benign by design approaches. ACS Sustain Chem Eng 9: 12461–12475. https://doi.org/10.1021/acssuschemeng.1c03070
10. Roush D, Asthagiri D, Babi DK et al (2020) Toward in silico CMC: an industrial collaborative approach to model-based process development. Biotechnol Bioeng 117:3986–4000. https://doi.org/10.1002/bit.27520
11. Zobel-Roos S, Schmidt A, Uhlenbrock L et al (2021) Digital Twins in biomanufacturing. Adv Biochem Eng Biotechnol 176:181–262. https://doi.org/10.1007/10_2020_146
12. SKF (2021) Measuring unscheduled downtime. Paper Advance
13. Mowbray M, Vallerio M, Perez-Galvan C et al (2022) Industrial data science – a review of machine learning applications for chemical and process industries. React Chem Eng 7:1471–1509. https://doi.org/10.1039/d1re00541c
14. Thon C, Finke B, Kwade A et al (2021) Artificial intelligence in process engineering. Adv Intell Syst 3:2000261. https://doi.org/10.1002/aisy.202000261
15. Klatt K-U, Marquardt W (2007) Perspectives for process systems engineering – a personal view from academia and industry. In: 17th European symposium on computer aided process engineering, vol 24. Elsevier, pp 19–32
16. GlobalData Healthcare Biologic sales forecast to pass innovative small molecule sales in next five years: GlobalData has forecast biologic sales to significantly overtake sales of small molecules by $120bn by 2027. https://www.pharmaceutical-technology.com/comment/biologic-sales-small-molecule-sales/. Accessed 08 Nov 2022
17. Niazi SK (2022) The coming of age of biosimilars: a personal perspective. Biologics 2:107–127. https://doi.org/10.3390/biologics2020009

18. Nathan JJ, Ramchandani M, Kaur P (2018) Manufacturing of biologics. In: Yamauchi PS (ed) Biologic and systemic agents in dermatology. Springer International Publishing, Cham, pp 101–110
19. Brown DG, Boström J (2018) Where do recent small molecule clinical development candidates come from? J Med Chem 61:9442–9468. https://doi.org/10.1021/acs.jmedchem.8b00675
20. Drug Hunter Inc. (2021) 2020 large molecule drug approvals. https://drughunter.com/resource/2020-large-molecule-drug-approvals/. Accessed 08 Nov 2022
21. de Wilde S, Guchelaar H-J, Zandvliet ML et al (2016) Clinical development of gene- and cell-based therapies: overview of the European landscape. Mol Ther Methods Clin Dev 3:16073. https://doi.org/10.1038/mtm.2016.73
22. US Food and Drug Administration (2022) Novel drug approvals for 2022. https://www.fda.gov/drugs/new-drugs-fda-cders-new-molecular-entities-and-new-therapeutic-biological-products/novel-drug-approvals-2022. Accessed 08 Nov 2022
23. Mullard A (2021) 2020 FDA drug approvals. Nat Rev Drug Discov 20:85–90. https://doi.org/10.1038/d41573-021-00002-0
24. Laffly E, Sodoyer R (2005) Monoclonal and recombinant antibodies, 30 years after. Hum Antibodies 14:33–55
25. Steinitz M (2009) Three decades of human monoclonal antibodies: past, present and future developments. Hum Antibodies 18:1–10. https://doi.org/10.3233/HAB-2009-0196
26. Zurdo J (2013) Developability assessment as an early de-risking tool for biopharmaceutical development. Pharm Bioprocess 1:29–50. https://doi.org/10.4155/pbp.13.3
27. Seyhan AA (2019) Lost in translation: the valley of death across preclinical and clinical divide – identification of problems and overcoming obstacles. Transl Med Commun 4. https://doi.org/10.1186/s41231-019-0050-7
28. Zobel-Roos S, Schmidt A, Mestmäcker F et al (2019) Accelerating biologics manufacturing by modeling or: is approval under the QbD and PAT approaches demanded by authorities acceptable without a Digital-Twin? Processes 7:94. https://doi.org/10.3390/pr7020094
29. Fisher AC, Lee SL, Harris DP et al (2016) Advancing pharmaceutical quality: an overview of science and research in the U.S. FDA's Office of Pharmaceutical Quality. Int J Pharm 515:390–402. https://doi.org/10.1016/j.ijpharm.2016.10.038
30. US Food and Drug Administration (2004) PAT — a framework for innovative pharmaceutical development, manufacturing, and quality assurance: guidance for industry. https://www.fda.gov/regulatory-information/search-fda-guidance-documents/pat-framework-innovative-pharmaceutical-development-manufacturing-and-quality-assurance. Accessed 08 Nov 2022
31. International Council for Harmonisation of Technical Requirements for Pharmaceuticals for Human Use good manufacturing practice guide for active pharmaceutical ingredients Q7: ICH harmonised tripartite guideline. https://database.ich.org/sites/default/files/Q7%20Guideline.pdf. Accessed 08 Nov 2022
32. International Council for Harmonisation of Technical Requirements for Pharmaceuticals for Human Use derivation and characterisation of cell substrates used for production of biotechnological/biological productS Q5D. ICH harmonised tripartite guideline. https://database.ich.org/sites/default/files/Q5D%20Guideline.pdf. Accessed 08 Nov 2022
33. International Council for Harmonisation of Technical Requirements for Pharmaceuticals for Human Use quality of biotechnological products: stability testing of biotechnological/biological products Q5C: ICH harmonised tripartite guideline. https://database.ich.org/sites/default/files/Q5C%20Guideline.pdf. Accessed 08 Nov 2022
34. International Council For Harmonisation of Technical Requirements For Pharmaceuticals For Human Use quality of biotechnological products: analysis of the expression construct in cells used for production of R-DNA derived protein products Q5B: ICH harmonised tripartite guideline. https://database.ich.org/sites/default/files/Q5B%20Guideline.pdf. Accessed 08 Nov 2022

35. International Council for Harmonisation of Technical Requirements for Pharmaceuticals For Human Use viral safety evaluation of biotechnology products derived from cell lines of human or animal origin Q5A(R2). ICH harmonised guideline. https://www.ema.europa.eu/en/documents/scientific-guideline/ich-q-5-r2-viral-safety-evaluation-biotechnology-products-derived-cell-lines-human-animal-origin_en.pdf. Accessed 08 Nov 2022
36. International Council for Harmonisation of Technical Requirements For Pharmaceuticals for Human Use comparability of biotechnological/biological products subject to changes in their manufacturing process Q5E: ICH harmonised tripartite guideline. https://database.ich.org/sites/default/files/Q5E%20Guideline.pdf. Accessed 08 Nov 2022
37. Bundesamt für Justiz Verordnung über die Anwendung der Guten Herstellungspraxis bei der Herstellung von Arzneimitteln und Wirkstoffen und über die Anwendung der Guten fachlichen Praxis bei der Herstellung von Produkten menschlicher Herkunft. https://www.gesetze-im-internet.de/amwhv/. Accessed 08 Nov 2022
38. (2022) European Pharmacopoeia 11th edition print subscription 2023 (11.0, 11.1 and 11.2), 11th edition. European Directorate for the Quality of Medicines & Healthcare
39. Schmidt A, Helgers H, Vetter FL et al (2021) Digital Twin of mRNA-based SARS-COVID-19 vaccine manufacturing towards autonomous operation for improvements in speed, scale, robustness, flexibility and real-time release testing. Processes 9:748. https://doi.org/10.3390/pr9050748
40. Schmidt A, Zobel-Roos S, Helgers H et al (2022) Digital Twins for continuous biologics manufacturing. In: Subramanian G (ed) Process control, intensification, and digitalisation in continuous biomanufacturing, vol 7. Wiley, pp 265–350
41. Schmidt A, Helgers H, Vetter FL et al (2021) Fast and flexible mRNA vaccine manufacturing as a solution to pandemic situations by adopting chemical engineering good practice—continuous autonomous operation in stainless steel equipment concepts. Processes 9:1874. https://doi.org/10.3390/pr9111874
42. Juckers A, Knerr P, Harms F et al (2021) Advanced process analytical technology in combination with process modeling for endpoint and model parameter determination in lyophilization process design and optimization. Processes 9:1600. https://doi.org/10.3390/pr9091600
43. Hengelbrock A, Helgers H, Schmidt A et al (2022) Digital Twin for HIV-gag VLP production in HEK293 cells. Processes 10:866. https://doi.org/10.3390/pr10050866
44. Udugama IA, Bayer C, Baroutian S et al (2022) Digitalisation in chemical engineering: industrial needs, academic best practice, and curriculum limitations. Educ Chem Eng 39:94–107. https://doi.org/10.1016/j.ece.2022.03.003
45. Udugama IA, Lopez PC, Gargalo CL et al (2021) Digital Twin in biomanufacturing: challenges and opportunities towards its implementation. Syst Microbiol Biomanuf 1:257–274. https://doi.org/10.1007/s43393-021-00024-0
46. Clavaud M, Roggo Y, von Daeniken R et al (2013) Chemometrics and in-line near infrared spectroscopic monitoring of a biopharmaceutical Chinese hamster ovary cell culture: prediction of multiple cultivation variables. Talanta 111:28–38. https://doi.org/10.1016/j.talanta.2013.03.044
47. Sagmeister P, Hierzegger R, Williams JD et al (2022) Artificial neural networks and data fusion enable concentration predictions for inline process analytics. Digit Discov 1:405–412. https://doi.org/10.1039/D2DD00006G
48. Schmidt A, Helgers H, Lohmann LJ et al (2022) Process analytical technology as key-enabler for digital twins in continuous biomanufacturing. J Chem Tech Biotech 97:2336–2346. https://doi.org/10.1002/jctb.7008
49. Sixt M, Gudi G, Schulz H et al (2018) In-line Raman spectroscopy and advanced process control for the extraction of anethole and fenchone from fennel (Foeniculum vulgare L. MILL.). Comptes Rendus Chimie 21:97–103. https://doi.org/10.1016/j.crci.2017.12.004
50. Kornecki M, Schmidt A, Strube J (2018) PAT as key-enabling technology for QbD in pharmaceutical manufacturing: a conceptual review on upstream and downstream processing. Chim Oggi Chem Today 36:44–48

51. Kornecki M, Strube J (2018) Process analytical technology for advanced process control in biologics manufacturing with the aid of macroscopic kinetic modeling. Bioengineering (Basel) 5. https://doi.org/10.3390/bioengineering5010025
52. Jensch C, Knierim L, Tegtmeier M et al (2021) Development of a general PAT strategy for online monitoring of complex mixtures—on the example of natural product extracts from bearberry leaf (Arctostaphylos uva-ursi). Processes 9:2129. https://doi.org/10.3390/pr9122129
53. Institute for separation and process technology homepage. https://www.itv.tu-clausthal.de/. Accessed 08 Nov 2022
54. Godawat R, Konstantinov K, Rohani M et al (2015) End-to-end integrated fully continuous production of recombinant monoclonal antibodies. J Biotechnol 213:13–19. https://doi.org/10.1016/j.jbiotec.2015.06.393
55. David L, Schwan P, Lobedann M et al (2020) Side-by-side comparability of batch and continuous downstream for the production of monoclonal antibodies. Biotechnol Bioeng 117:1024–1036. https://doi.org/10.1002/bit.27267
56. Karst DJ, Steinebach F, Soos M et al (2017) Process performance and product quality in an integrated continuous antibody production process. Biotechnol Bioeng 114:298–307. https://doi.org/10.1002/bit.26069
57. Kornecki M, Schmidt A, Lohmann L et al (2019) Accelerating biomanufacturing by modeling of continuous bioprocessing—piloting case study of monoclonal antibody manufacturing. Processes 7:495. https://doi.org/10.3390/pr7080495
58. Thiess H, Zobel-Roos S, Gronemeyer P et al (2017) Engineering challenges of continuous biomanufacturing processes (CBP). In: Subramanian G (ed) Continuous biomanufacturing – innovative technologies and methods. Wiley-VCH Verlag GmbH & Co. KGaA, Weinheim, pp 69–106
59. Zobel-Roos S, Thiess H, Gronemeyer P et al (2017) Continuous chromatography as a fully integrated process in continuous biomanufacturing. In: Subramanian G (ed) Continuous biomanufacturing – innovative technologies and methods. Wiley-VCH Verlag GmbH & Co. KGaA, Weinheim, pp 369–392
60. Gronemeyer P, Thiess H, Zobel-Roos S et al (2017) Integration of upstream and downstream in continuous biomanufacturing. In: Subramanian G (ed) Continuous biomanufacturing – innovative technologies and methods. Wiley-VCH Verlag GmbH & Co. KGaA, Weinheim, pp 481–510
61. Schmidt A, Uhlenbrock L, Strube J (2020) Technical potential for energy and GWP reduction in chemical–pharmaceutical industry in Germany and EU—focused on biologics and botanicals manufacturing. Processes 8:818. https://doi.org/10.3390/pr8070818
62. Schmidt A, Köster D, Strube J (2022) Climate neutrality concepts for the German chemical–pharmaceutical industry. Processes 10:467. https://doi.org/10.3390/pr10030467
63. Uhl A, Schmidt A, Jensch C et al (2022) Development of concepts for a climate-neutral chemical–pharmaceutical industry in 2045. Processes 10:1289. https://doi.org/10.3390/pr10071289
64. Rosengarten JF, Stitz J, Hengelbrock A et al (2022) Digital Twin for HIV-gag VLP production in HEK293 cells. Chemie Ingenieur Technik 94:1280. https://doi.org/10.1002/cite.202255327
65. Helgers H, Hengelbrock A, Schmidt A et al (2022) Process design and optimization towards Digital Twins for HIV-gag VLP production in HEK293 cells, including purification. Processes 10:419. https://doi.org/10.3390/pr10020419
66. Rosengarten JF, Schatz S, Wolf T et al (2022) Components of a HIV-1 vaccine mediate virus-like particle (VLP)-formation and display of envelope proteins exposing broadly neutralizing epitopes. Virology 568:41–48. https://doi.org/10.1016/j.virol.2022.01.008
67. Defense Advanced Research Projects Agency (2017) Removing the viral threat: two months to stop pandemic X from taking hold: DARPA aims to develop an integrated end-to-end platform that uses nucleic acid sequences to halt the spread of viral infections in sixty days or less. https://www.darpa.mil/news-events/2017-02-06a. Accessed 08 Nov 2022

68. Saied AA, Metwally AA, Dhawan M et al (2022) Strengthening vaccines and medicines manufacturing capabilities in Africa: challenges and perspectives. EMBO Mol Med 14: e16287. https://doi.org/10.15252/emmm.202216287
69. Fahr S, Peña-Benavides SA, Thiel L et al (2022) Mobile on demand COVID-19 vaccine production units for developing countries. Ind Eng Chem Res 61:13191–13204. https://doi.org/10.1021/acs.iecr.2c01217
70. Gürsel IV, Hessel V, Wang Q et al (2012) Window of opportunity – potential of increase in profitability using modular compact plants and micro-reactor based flow processing. Green Process Synth 1. https://doi.org/10.1515/gps-2012-0046
71. European Medicines Agency (2021) Assessment report: Comirnaty. https://www.ema.europa.eu/en/documents/assessment-report/comirnaty-epar-public-assessment-report_en.pdf. Accessed 08 Nov 2022
72. Schmidt A, Helgers H, Vetter FL et al (2022) Process automation and control strategy by quality-by-design in total continuous mRNA manufacturing platforms. Processes 10:1783. https://doi.org/10.3390/pr10091783
73. Vetter FL, Zobel-Roos S, Mota JPB et al (2022) Toward autonomous production of mRNA-therapeutics in the light of advanced process control and traditional control strategies for chromatography. Processes 10:1868. https://doi.org/10.3390/pr10091868

Chapter 11
Biopharma 4.0: Digital Technologies Driving Smart Manufacturing

Shilpa Nargund, Klaus Mauch, Sabrina Esswein, Bastian Niebel, Mehran Rafigh, and Pushkar Pendse

Abstract Industry 4.0 is transforming all industrial sectors, and the biopharma industry is no exception. After a cautious start to Industry 4.0 adoption, the biopharma industry has embraced this new paradigm in response to the challenges posed by the COVID-19 pandemic. The Biopharma 4.0 transformation is being driven by three categories of forces: market forces, advancements in life science technologies, and the digital technology revolution. The market forces demand improvements in productivity, quality, speed and flexibility, and the technologies provide the necessary solutions. Manufacturing lies at the center of the any industry's value chain, bridging product lifecycle and supply chain. As a result, our principal focus for this chapter will be on the manufacturing operations and the aspects of product life cycle and supply chain that are closely tied to it. We will begin by exploring why manufacturing is so critical to the biopharma industry and then present a vision of a smart biofactory. We will then delve deeper into how the product lifecycle is being transformed from process development to commercial manufacturing. We will conclude by highlighting the gaps present in our current state of technologies and the innovations required to fill them.

Keywords Digital twins · Smart biomanufacturing · Biopharma 4.0 · Artificial intelligence · Machine learning · Biologics

S. Nargund (✉)
Yokogawa Insilico Biotechnology Pte. Ltd., Singapore, Singapore
e-mail: s.nargund@yokogawa.com

K. Mauch · S. Esswein · B. Niebel · M. Rafigh
Yokogawa Insilico Biotechnology GmbH, Stuttgart, Germany
e-mail: K.Mauch@yokogawa.com; S.Esswein@yokogawa.com; Bastian.Niebel@yokogawa.com; M.Rafigh@yokogawa.com

P. Pendse (✉)
Seemulate Technologies Pvt. Ltd., Bengaluru, India

R. Pörtner (ed.), *Biopharmaceutical Manufacturing*, Cell Engineering 11,
https://doi.org/10.1007/978-3-031-45669-5_11

Abbreviations

AI/ML	Artificial intelligence and machine learning
COVID-19	CoronaVirus Disease of 2019
FDA	U.S. Food and Drug Administration
IIoT	Industrial internet of things
IT	Information technology
mRNA	messenger ribonucleic acid
PAT	Process analytical technology
R&D	Research and development

Introduction

In recent years, the digital technological transformation we have witnessed has had a significant impact on economic and social spheres. Dubbed the fourth industrial revolution or Industry 4.0, this transformation is driving fundamental shifts in how products and services are manufactured and delivered [1]. Industry 4.0 is characterized by two critical phenomena: the exponential growth of innovative technologies and the convergence of two or more technologies to create new applications rapidly.

At first, the biopharmaceutical industry was slow to adopt these advanced technologies and applications, a fact that can be attributed to the presence of stringent regulations and tough legal scrutiny. However, the global COVID-19 pandemic mandated the industry to drastically reduce timelines and the industry turned towards digitalization for this. Today, every domain of the biopharmaceutical value chain, from bench to bedside, has felt the impact of Industry 4.0 technologies [2].

As the biopharmaceutical industry continues to leverage the benefits of Industry 4.0, it is better equipped than ever to meet the healthcare challenges of today and tomorrow. Despite the challenges posed by regulatory constraints and legal scrutiny, the industry has embraced innovative technologies and novel applications at an unprecedented pace. Thanks to the transformative power of Industry 4.0, the biopharmaceutical industry is poised to deliver life-saving solutions that will benefit patients worldwide.

Forces Driving Industry 4.0 Transformation in the Biopharmaceuticals

The Biopharma 4.0 transformation is being propelled by three categories of drivers as illustrated in Fig. 11.1 [3]. The first category comprises market forces that are shaping the industry's economic landscape. Market forces include evolving patient needs, novel treatment options, and embracing personalized medicine, cell and gene

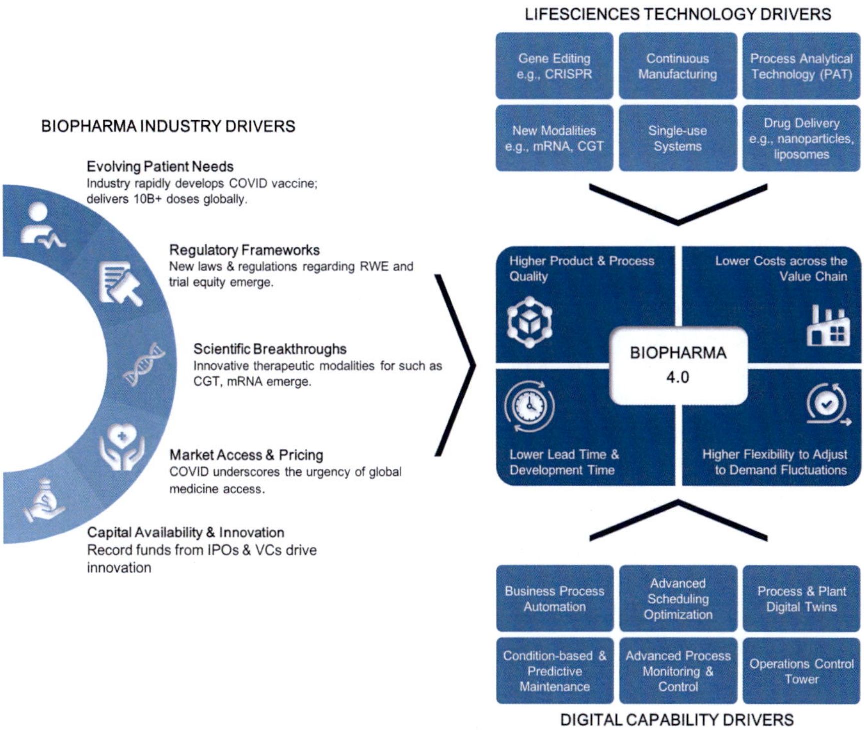

Fig. 11.1 Three categories of drives propelling Biopharma 4.0. Several disparate drivers are coming together to propel the Biopharma 4.0 transformation. This figure summarizes the factors in market forces, advances in life science technologies and advances in digital technologies that are intersecting to create Biopharma 4.0

therapies. Regulatory frameworks, including the FDA's Quality Metrics initiative, are incentivizing companies to adopt digital tools that can meticulously monitor and measure product lifecycles and manufacturing processes. Moreover, market access and pricing pressures are prompting companies to seek out innovative business models and partnerships to maintain their competitiveness. Finally, capital availability and innovation are pushing companies to invest in new technologies and capabilities to keep ahead of the curve.

The advancements in life science technologies comprise the second category of driving forces behind Biopharma 4.0. These innovations, including high throughput screening, genomics, continuous manufacturing, and single-use technology, have streamlined drug development and manufacturing processes, resulting in increased efficiency. But it's the cell and gene therapies (CGT) and messenger RNA (mRNA) vaccines that are truly transforming the field. CGT offers the potential to revolutionize medicine, allowing treatments to be tailored to specific patients and even repair genetic defects. Similarly, mRNA technology has demonstrated remarkable potential in the development of vaccines and therapeutics, as evidenced by the recent breakthrough mRNA COVID-19 vaccines.

The final category of driving forces behind Biopharma 4.0 is the incredible potential of digital technology capabilities. Predictive digital twins are enabling virtual experiments that significantly reduce wet-lab experiments, thereby reducing the laboratory's environmental footprint. The advent of advanced process monitoring and control systems is allowing companies to optimize their manufacturing processes in real-time, leading to greater efficiency and cost savings. Furthermore, predictive maintenance is empowering companies to proactively identify issues with their machines before they arise, which reduces downtime and improves production efficiency. Scheduling optimization is enabling companies to streamline their supply chains and minimize waste. Finally, the implementation of operations control towers provides companies with real-time visibility into their operations, allowing them to make more informed, rational, and faster decisions. The potential of these digital technologies is truly limitless and is set to revolutionize the biopharmaceutical industry.

Biopharma Manufacturing: The Crux of Product Lifecycle and Supply Chain

In the fast-paced world of biopharmaceuticals, companies must continuously improve their operations to stay competitive. To evaluate an organization's competitiveness, it is essential to monitor four key operational performance indicators: quality, productivity, speed, and flexibility. Maintaining high levels of quality is necessary to provide the best drugs to patients consistently. Achieving high levels of productivity is vital for reducing costs and increasing profitability. Speed is crucial to meet demands by patients and to stay ahead of the competition. Flexibility is critical to quickly adapt to volatile market conditions and seize new opportunities. By focusing on these performance indicators, companies can ensure that they are operating at peak efficiency and competitiveness.

Manufacturing plays a vital role in improving the performance indicators. It is the pivotal stage where raw materials are transformed into finished drug products and regulatory compliance is ensured. The manufacturing function is complex, encompassing various stages of the drug product lifecycle and supply chain (Fig. 11.2) [4]. Effective management of these processes is essential to produce high-quality drug products that meet regulatory standards and patient needs while keeping up with the demand. The product life cycle determines how long it takes to develop a drug and bring it market whereas the supply chain determines how long it takes to manufacture a batch of an approved drug. The manufacturing function affects both these timelines greatly. As such, the manufacturing function is critical to the success of biopharma organizations.

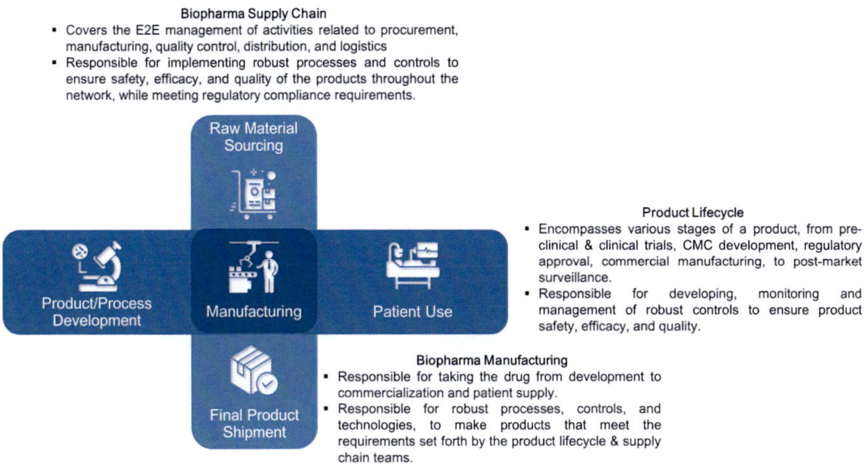

Fig. 11.2 Manufacturing is the crux of the product lifecycles and the supply chain

The digital transformation of the manufacturing function is expected to greatly improve its efficiency and therefore the key operational performance indicators [5, 6]. This chapter focuses on the digital transformation of this crucial function and parts of the product life cycle closely tied to it. We start by outlining a vision of the smart biofactory and the benefits it can provide. From there, we delve into the transformative potential of digital technologies on key workflows. Finally, we analyze the requirements for transitioning from the current state to smart biomanufacturing, taking a comprehensive approach to the path ahead.

Vision of the Smart Biofactory

The biopharmaceutical industry is constantly evolving, with new technologies paving the way for the next generation of manufacturing operations. The "smart" biofactory will use digital tools to seamlessly integrate different areas of a plant to improve efficiency, productivity, and quality while reducing costs. Digitalization will also integrate all stages of the product lifecycle and supple chain. In this section, we envision how digitalization will transform the biofactory.

The transformation of biomanufacturing can be envisioned (Fig. 11.3) by categorizing the operations into seven elements as follows.

1. Digital thread: Data is the prime driver of the digitalization of the biofactory. Therefore, it needs to be captured and integrated across the end-to-end value chain to build the digital thread. A digital thread is a communication framework that allows a connected and integrated view of a product throughout its lifecycle by interweaving all related data and models. This framework connects information from research and development (R&D) laboratories, manufacturing sites,

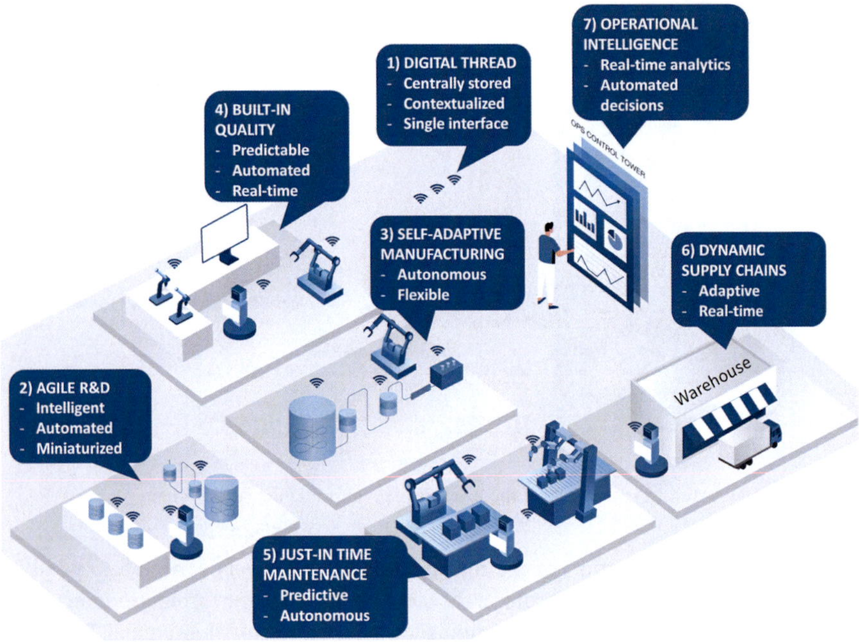

Fig. 11.3 Vision of the smart biofactory. The seven elements that need to be transformed to achieve smart manufacturing

quality management systems, maintenance, warehouse, distributions channels, promoting data flow and continuity. The digital thread drives informed decision-making and optimization of product lifecycles, while also enabling the effective operation of digital twins by providing them with the necessary data to accurately represent their physical counterparts. The digital thread ensures that the data for any vial of a biologic can be traced back to the vial of the master cell bank, working cell bank or research cell bank. In other words, the data generated across the product life cycle and supply chain will be integrated such that it can be viewed from a single interface. It will be easy to provide secure access to the data to people across functions, sites, third party collaborations and regulatory agencies. The data will also be continuously shared with AI-ML models that use it to make predictions on manufacturing processes, maintenance of equipment, production capacity, etc.

2. Agile R&D: The most important facet of the R&D function is agility. The ability to quickly generate data on new biologics and make go/no-go decisions early is critical for speed-to-market and reducing costs. In the smart biofactory, the R&D lab will be intelligent, automated and miniaturized. Artificial intelligence-machine learning (AI-ML) models will optimize processes virtually first. Then a select few experiments will be automatically executed on miniaturized instruments by robotic handlers to validate the predicted solutions. The analytical instruments or sensors will be in-line with process equipment to eliminate

sampling by humans. The data from the analytical instruments will be automatically sent to the AI-ML models to update their learning and iteratively optimize the processes with virtual and physical experiments. The lab footprint and the timelines will be reduced significantly.

3. Self-adaptive manufacturing: The most important facet of the manufacturing function is production of consistent quality biologics. Minimizing deviations or failures of production batches is critical to providing quality therapeutics to patients. In the smart biofactory, the self-adaptive plant will be autonomous and flexible. AI-ML models will enable all unit operations to self-diagnose deviations before they occur and take corrective actions using robots. Transfer of technology across sites will be seamless as AI-ML models across sites will communicate to establish the change in operating parameters. The AI-ML models will continuously learn from data to incorporate any changes to the processes. Moreover, the unit operations will be modular and flexibly operated to easily enable change in production capacity to meet market demand. The interventions by humans in operating the plant will be minimal so much so that there would be no need to turn on the lights in the plant. Therefore, the self-adaptive plant is also called 'Lights-out manufacturing'. Humans will only monitor the manufacturing processes remotely through dashboards that provide operational intelligence. Humans will intervene only in exceptional cases.
4. Built-in quality: The most important facet of the quality function is to ensure that only good quality biologics reach the patient. The manufacturing processes of all therapeutics are prone to variability that lead to variability in the products. In that, biologics are especially prone to variability because they are produced by living cells. It therefore becomes important to control the quality of the biologics at several stages of the manufacturing process. In the smart biofactory, quality will be predictable, automated and real-time. The AI-ML models used in manufacturing for predicting future deviations will also predict the impact on quality attributes if a deviation occurs. Corrective and preventive actions will be taken automatically to avoid the impact on quality. The release testing of biologics from the intelligently controlled processes will be real-time and in-line. Advanced process analytical technologies including sensors will give real-time read-outs on critical process and quality attributes. The data from the release tests will be fed back to the AI-ML models to further improve their predictive accuracy.

   Another important facet of the quality function is to interface with regulatory agencies. The manufacturing batch records, release testing and other relevant information will be automatically shared with regulatory agencies. There would be no need for regulatory officers to visit the sites for inspection.
5. Just-in-time maintenance: To keep a plant running, several systems need to be maintained. These include the manufacturing equipment, analytical instruments, the environmental monitoring systems, utilities and energy management systems. Unexpected breakdowns in any of these systems leads to expensive plant shutdowns and delays in production. It is therefore critical to maintain these systems with just-in-time interventions. In the smart biofactory, maintenance will be predictive and autonomous. Sensors on the equipment or instruments will

continuously report their health metrics and AI-ML models will use these metrics to predict breakdowns. The AI-ML models will direct autonomous robots to carry out the maintenance just-in-time to avoid system breakdowns. Further, the routine maintenance tasks such as filling solvents, replacing filters, cleaning, etc. will be done automatically. Similarly, routine calibration of instruments will also be automated. From all the above advances, unplanned downtime will be eliminated.

6. Dynamically optimized supply chains: The role of supply chain management teams is to make sure that the biologics are received by patients in time, within the shelf-life of the drug while minimizing stockpiling. This requires that information from several nodes of the supply chain i.e. warehouses, market demand, production and distribution is integrated and readily accessible. In the smart biofactory, the supply chain will be adaptive in real-time. The information from all the above supply chain nodes will be available in real-time. This data will be used by AI-ML models to forecast future demand and adapt the capacity of upstream operations in real-time. Inventories of raw materials and drug products become leaner, warehouses have smaller footprints while at the same time the risk of drug shortages are reduced. As supply chain management becomes more precise, the production of drugs will be decentralized among multiple sites to serve patients better.
7. Operational intelligence: In the smart biofactory, every function in the organization will monitor its operations through dashboards that provide analytics in real-time. For the most part, the operations will be automated and even autonomous as AI-ML models will take decisions and actuate them through robots. Human intervention will occur only in exceptional cases wherein the predictive models themselves will alert us about the risks. The role of humans will be largely confined to monitoring the dashboards for alerts and mitigating the risks in exceptional cases.

Thus, the seven elements of a smart biofactory will come together and help it excel at the key performance metrics of a biopharma organization.

A defining feature of each of the seven elements is the use of AI-ML models for making predictions and automating decisions. We want to further highlight the role of these models, also called digital twins. The digital twins put the "smart" in smart biomanufacturing. They serve as the primary vehicle for harnessing the knowledge contained within the data generated during the product lifecycle and supply chain.

First, what is a digital twin. A digital twin is a virtual model or replica of a physical product, process, or system. It is created by integrating data from various sources, including sensors, equipment, and simulations, to provide a comprehensive digital representation of the physical object or system. The digital twin can be used for various purposes, including testing, monitoring, predicting performance, and identifying potential issues or improvements. It can also be used to simulate scenarios and test different strategies before implementing them in the physical world. To make the smart biofactory vision even more concrete, especially in the uses of digital twins, the following table gives examples of how particular workflows will change between the current approach and the smart biomanufacturing approach. We also list the key performance indicators that are impacted by these changes.

Workflow	Functions involved	Traditional approach	Smart biofactory approach	KPIs impacted
Process development	Research and development	Optimize process parameters through targeted or statistically designed lab scale experiments	Utilize digital twins to conduct virtual experiments that intelligently identify optimal process parameters. Only implement a select few experiments in the wet lab	Speed
Manufacturing execution	Manufacturing, quality control, facilities	Operators use manual records and paper-based processes to manage the manufacturing process. They react to deviations only after they have occurred.	A digital twin monitors and controls the manufacturing process in real-time by predicting the future course of the process. Deviations are predicted early and prevented by model-predicted corrective actions	Quality and productivity
Planning	Supply chain, manufacturing	Plan the production schedule based on demand and capacity	Use digital twins to predict demand and optimize production scheduling based on real-time data, available resources and potential disruptions.	Flexibility
Quality control	Quality control, manufacturing	Test samples at several stages as forward processing criteria.	Utilize digital twins for advanced process control systems to predict quality and ensure quality throughout the manufacturing process, from raw material to finished product.	Quality and speed
Tech transfer	Manufacturing, quality control, Research and development	Transfer process knowledge and methods from research and development to manufacturing iteratively through engineering runs	Utilize digital twins to simulate the processes at different sites or scales and minimize the engineering runs during transfer.	Speed and productivity

(continued)

Workflow	Functions involved	Traditional approach	Smart biofactory approach	KPIs impacted
Maintenance	Facilities, manufacturing	Perform regular maintenance based on a pre-defined schedule.	Use predictive maintenance tools and IoT devices to monitor equipment and identify potential issues before they become problems, minimizing downtime and maximizing equipment effectiveness.	Productivity

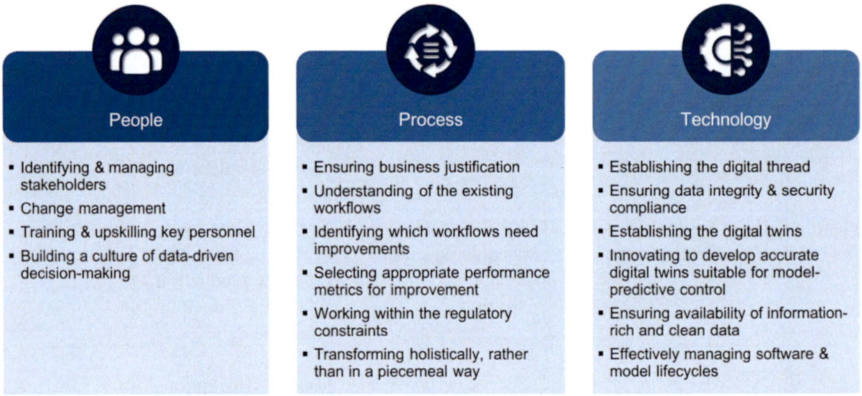

Fig. 11.4 Digital transformation of an organization. The three pillars of an organization that need transformation to achieve digitalization

Getting There: What It Will Take to Achieve Smart Biomanufacturing

To achieve smart biomanufacturing, companies must transform multiple aspects of their organization. To identify gaps systematically, we will use the People, Processes, and Technology framework, which provides a comprehensive view of the key considerations involved. Figure 11.4 illustrates the critical elements in each aspect. Since the scope of this exercise is extensive, we will largely focus on the transformation of digital technologies. By doing so, we can gain a deeper understanding of the critical role that technology plays in the development of smart biomanufacturing.

People

To ensure a successful digital transformation, it's essential that people are at the heart of the process. Identifying the right stakeholders, drafting a comprehensive change management plan, and establishing a culture of data-driven decision-making are all crucial steps to take. To begin with, buy-in from key stakeholders such as senior leadership, information technology (IT), operations, and quality control is vital, as is engaging regulatory agencies from the outset to avoid costly complications. Resistance to change is a common challenge, which is why providing the right training and support is crucial to ensuring a smooth transition. Building a culture of data-driven decision-making requires fostering a collaborative mindset across functions, promoting experimentation and continuous improvement, and ensuring that data is both accessible and actionable. With a focus on people, the digital transformation of biopharmaceutical manufacturing can truly thrive.

Processes

To successfully undergo digital transformation, a multitude of processes across business, R&D, manufacturing, and supply chain operations need to be revamped. In order to achieve this, mapping the flow of data across the product lifecycle is crucial. While R&D and manufacturing functions are already familiar with mapping the flow of materials, mapping the flow of data is equally important to ensure a smooth transition to digitalization. The workflows will undergo significant changes, particularly at the points where data enters models or digital twins and becomes decisions. These decisions or predictions will initially be evaluated manually in pilot studies, but ultimately, they must be automatically applied by coupling the digital twins with their physical counterparts. Achieving this final state of workflows that involve digital twins will require collaboration between industry, technology vendors, and regulatory agencies, especially if the models are driven by artificial intelligence. Therefore, mapping data flow across workflows and fostering collaboration across stakeholders will be critical in driving successful digital transformation.

Technology

The major disruptive force that is driving biopharma 4.0 is technology. First came the advancements in big data analytics from other industries. Then came the advancements in hardware such as high-throughput miniaturized lab equipment, robotics, industrial internet of things (IIoT) sensors, spectrometric probes, etc. And now, advances in AI-ML are enabling us to harness the big data being generated by the advanced hardware. However, we are still at the beginning of the biopharma 4.0

transformation. To achieve the vision of the smart biofactory, many technological advancements are required. We will focus on two aspects that we think will be the make-or-break factors in this transformation – the digital thread and the digital twins [7, 8].

Establishing the Digital Thread

Data is not just an incidental by-product of the manufacturing process; it is a core asset that drives innovation and efficiency. In fact, the industry 4.0 paradigm was born out of the realization that Big Data could be harnessed for intelligent decision-making. Building a digital thread is the first step towards digitalization for any organization. By capturing data across the enterprise and facilitating its flow upstream and downstream, the digital thread enables seamless communication and collaboration between different departments. This is particularly crucial in the pharma world, where traceability of information is a key regulatory requirement. Therefore, investing in a robust digital thread is not just a smart business decision, but a necessary one for staying compliant and competitive in today's market.

The biopharma industry generates massive amounts of data, but this data is often stranded in silos, rendering it useless. This is due to a lack of skills, expertise, and technology to harness the potential of the vast quantities of data being produced. In addition, the process of manual sampling and integration of disparate data sources and formats is both time-consuming and cost-inefficient, leading to errors. According to research, an astonishing 80% of time in current R&D projects is spent on data pre-processing and curation [9].

Compounding this issue is the fact that data is often not stored and managed in a central location, resulting in various data silos within a single company. This limits collaboration and hinders the flexibility and reusability of data across projects. To solve this problem, proper data integration and management require cross-disciplinary team efforts. Unfortunately, the cost savings and probability of success in this area are not always immediately apparent to decision-makers in the company, leading to slow adoption [10]. It's clear that the biopharma industry needs to overcome these challenges and create a robust digital infrastructure to unlock the full potential of its data. To achieve true digital transformation, companies must adopt a data-centric approach, starting with the creation of a digital thread. This thread captures critical data across the entire product lifecycle, from various functions and sites, enabling a seamless flow of information. It's not a one-way street, either. The digital thread also feeds data to digital twins, capturing their insights and decisions. By establishing this foundation, companies can unlock the full potential of their data, empowering them to make informed decisions and drive innovation.

To build a reliable digital thread, it's imperative that every member of an organization places data quality as a top priority. Establishing such a data-driven ecosystem requires the implementation of robust data governance and guidelines, such as the FAIR (findable, accessible, interoperable, reusable) data principles [11]. Responsibility for data ownership, stewardship, and accountability should be ingrained in the fabric of the organization. Regular audits and reviews are necessary to ensure cybersecurity, data integrity, and regulatory compliance. A fundamental prerequisite for effective data usage is to ensure that it is interpretable and contextualized through machine-readable standards like formats, vocabulary, and ontology. The Big Data to Smart Data initiative is a great example of this [12]. A centralized infrastructure coupled with these practices breaks down data silos and fosters seamless internal and external communication.

The implementation of the measures discussed above will not only allow for the establishment of a comprehensive corporate knowledge base, but also facilitate cross-functional collaboration throughout the product lifecycle. By enabling everyone to access and utilize data, both internal and external stakeholders can work together more closely and effectively. Additionally, this centralized approach to data management creates exciting opportunities for innovative business models, where data is a critical strategic asset that is integrated into every aspect of the business process.

Establishing the Digital Twins

Smart manufacturing means that processes become self-adaptive and autonomous. These processes will have the ability to monitor their own performance and adjust their behavior in real-time to optimize their performance. To develop self-adaptive processes, it is critical to build accurate digital twins of these processes. The digital twins should be able to (i) accurately simulate the behavior of the process using a mathematical model, (ii) make decisions on controlling the process based on simulations that are made on real-time data, (iii) take control actions on the physical process such as adjusting process parameters and (iv) continuously learn from the data to improve accuracy of simulations and decision-making. Innovations are required in several spheres to develop such digital twins. In this section we will elaborate on the areas that need innovation and the advancements in them.

Mathematical Modeling Physical systems involved in the production of biologics are complex and have multiple interacting components. Modeling the behavior of such systems is challenging. For example, the cells used to produce the biologic during a cell culture process are sensitive to the presence of over 50 media components, pH, temperature, aeration and agitation. To model the cell culture process, one must consider the dynamically changing biological, chemical and physical phenomena at play. Even then, the model may not be accurate or fast enough to deploy in a manufacturing setting.

Innovative modeling methods that combine mechanistic know-how with AI/ML are necessary to overcome this challenge. The role of AI/ML should be to learn the unknown rules in the data while being bound by the known rules from the mechanistic models. Thus, the AI/ML model is not burdened with learning known mechanisms and can perform well even when trained on small sets of data. For example, the metabolism of cells is understood very well, however the kinetics of metabolic reactions are hard to measure. Therefore, methods that use AI/ML methods to learn only the kinetics of metabolic reactions from data rather than the metabolic pathways themselves will yield superior results [13, 14]. Another important role of the AI/ML model is to transfer learning across data sets because mechanistic models alone are incapable of doing so. Thus, the knowledge from data generated across scales and across projects can be used in perpetuity. The models become more intelligent every time new data is fed and reduce the amount of data required in subsequent projects.

While the AI/ML model makes data a strategic asset, the mechanistic models can ensure that the data being used is of good quality. For instance, biological systems follow mechanistic rules such as metabolic pathways, conservation of mass, etc., and it is necessary to evaluate the feasibility of the data acquired. For example, 1 mole of glucose is converted to exactly 2 moles of pyruvate. However, analytical instruments may introduce errors during measurements of concentrations of glucose and pyruvate which may lead to violation of the above rule. We need to develop methods of cleaning the data so that such errors are minimized. This is another reason why hybrid models of mechanistic know-how and AI/ML are needed. AI/ML models alone cannot distinguish between accurate and inaccurate data whereas, the mechanistic model counterparts can be used to test the data for known rules. Similarly, the mechanistic model is also useful in explaining the outputs of AI/ML model which by nature is a black box model. Furthermore, it is the mechanistic models that enable the digital twin to extrapolate beyond the known solution space. They also enable us to increase process knowledge and do root-cause analysis if deviations were to occur.

Another aspect where development of mathematical modeling methods is required is for making the most of the data acquired. Process analytical technologies (PAT) like Raman spectra produce information rich data that can improve the predictive accuracy of digital twins. However, even though Raman technology has been available for a long time it is not widely used in biopharma because intricate mathematical models need to be developed [15]. Thus, innovation is also required in models for deconvoluting PAT data. Again, use of AI/ML to rapidly build models for inferring Raman data seems like a promising approach [16].

Data Quality The quality of data used to train the digital twins is as critical as the mathematical model itself. If the data is incomplete, unlabeled, or inaccurate, the digital twin will not accurately represent the physical system. Unfortunately, most historical data falls in one of the above categories. Further, the information content in historical data is inadequate for training AI/ML models. It is important that the processes from which the data is gathered are dynamically variable. For this, it may be necessary to design experiments using beta versions of the digital twins themselves. The more variable the data used for training, the more the twin can extrapolate beyond the known solution space. Thus, it is best to develop digital twins at early stages of the process development. The iterative development and utilization of digital twins throughout clone selection, process development, scale-up, and process characterization stages compounds their accuracy by retaining knowledge from each stage.

Another way to get information rich data is to integrate data from process analytical technologies. For example, data from Raman probes can improve the predictive accuracy of digital twins of cell culture processes.

Managing the Software and Model's Lifecycle The key parts of managing the software's lifecycle include qualification and validation. While regulatory agencies are highly supportive of AI/ML technologies [17], the guidelines for use of AI/ML based models in Good Manufacturing Practice settings are still under discussion [18]. A qualified digital twin means that its prediction accuracy is such that any control action derived from the prediction will always lead to product quality attributes and parameters within the acceptance criteria [19]. Several aspects impact whether the digital twin passes qualification. First, the quality of the data affects the accuracy and the precision of the digital twin greatly. As discussed above, the information content of the data needs to be high so that the AI/ML part of the digital twin learns a range of process behaviors.

While the information content in the data affects the accuracy of the digital twin predictions, the precision of predictions is affected by other aspects such as (i) the uncertainties in measured data, (ii) parameter estimations done to get outputs from the mechanistic model and (iii) parameter estimations done while training the AI/ML models. The uncertainties from every step should be propagated forward to derive the uncertainty of the outputs. Thus, the qualification of an AI/ML based digital twin will depend on the data quality and the modeling practices used to estimate the prediction uncertainties.

Validation of AI/ML based digital twins poses a peculiar challenge especially if they are being continuously trained on newly generated data. It may be necessary to qualify the digital twin again if it is re-trained on additional data as it might change the uncertainties coming from stochastic elements of the models. However, this is counterproductive. Therefore, we need to develop methods or agree on guidelines for validation of AI/ML models that are faster and simpler.

Deployment in the Manufacturing Setting The main challenge in deploying a digital twin is that we need to ensure that different systems can communicate and work together. For example, a digital twin of a bioreactor should be able to receive the data from, sensors, probes and analytical instruments that are on-line or off-line with the bioreactor. Typically, all data generated is sent to a central repository and the re-routed to the digital twin from there. Once the twin receives the data and predicts control actions, these decisions need to be relayed back to the bioreactor. Usually, a procedure manager will interface between the digital twin and the process controller to do so. The IT solutions required to accomplish such communication are already present and deployed in other industries. The challenges are mainly in retrofitting these solutions to the biopharma industry. In particular, all software will need to be 21 Code of Federal Regulations Part 11 compliant.

In summary, achieving smart manufacturing requires active collaboration between industry, academia, technology providers and regulators. In that, special focus is required in driving innovations for building digital twins of manufacturing processes.

Outlook

The biopharma industry is at the cusp of a digital transformation. In particular, the digitalization of the manufacturing function promises to greatly impact an organization's performance. Smart manufacturing will improve quality, productivity, speed, and flexibility and ultimately a patient's health. Digitalization will touch all areas of the product lifecycle and the supply chain in the smart biofactory to achieve these improvements. To achieve this vision of smart manufacturing the people, processes and technologies of an organization need to be transformed. The most critical transformations, however, must happen in the technologies. This is because the 'Smart' in smart manufacturing relies heavily on developing the technologies required for making processes self-adaptive. Further, the two pillars of smart manufacturing technologies are the digital thread and the digital twins. The digital thread is the fountainhead of the knowledge that is trapped in data. Once established, it will enable us to tap this knowledge by building digital twins. Digital twins will be the make-or-break technology that will decide if smart manufacturing is achieved. Collaborations between industry, academia, technology providers and regulators will enable us to drive the innovations required and achieve smart manufacturing together.

Funding No funding was received to assist with the preparation of this manuscript.

References

1. Schwab K (2017) The fourth industrial revolution. Penguin, London
2. Levy V (2022) Global life sciences outlook | Digitalization at scale: delivering on the promise of science. Deloitte Global. https://www2.deloitte.com/cn/en/pages/life-sciences-and-healthcare/articles/global-life-sciences-sector-outlook-2022.html. Accessed 22 May 2023
3. Ahlawat H, Graves G, Hou T, Le Deu F, Moss R, Parekh R (2022) Helix: rewiring the DNA for the next wave of impact in biopharma. McKinsey and Company. https://www.ehcpea.org/wp-content/uploads/2022/11/Helix-Rewiring-the-DNA-for-the-next-wave-of-impact-in-biopharma.pdf. Accessed 23 May 2023
4. Rees H (2011) End-to-end pharmaceutical supply chains. In: Supply chain management in the drug industry. Wiley, New York, pp 73–102. https://doi.org/10.1002/9780470920817.ch4
5. Chen W, Park SJ, Cheng TNH, Lau NWH, Khaw LF, Lee-Lane D, Chen X, Sesen M (2022) Pharmaceutical industry. Wiley Online Library. https://onlinelibrary.wiley.com/doi/abs/10.1002/9781119695868.ch12. Accessed 22 May 2023
6. Kumar SH, Talasila D, Gowrav MP, Gangadharappa HV (2020) Adaptations of pharma 4.0 from industry 4.0. Drug Invent Today 14

7. Margaria T, Pesch D, McGibney A (2022) Digital thread in smart manufacturing. In: Margaria T, Steffen B (eds) Leveraging applications of formal methods, verification and validation, Practice, lecture notes in computer science. Springer Nature, Cham, pp 179–183. https://doi.org/10.1007/978-3-031-19762-8_12
8. Onaji I, Tiwari D, Soulatiantork P, Song B, Tiwari A (2022) Digital twin in manufacturing: conceptual framework and case studies. Int J Comput Integr Manuf 35:831–858. https://doi.org/10.1080/0951192X.2022.2027014
9. Zhang S, Zhang C, Yang Q (2003) Data preparation for data mining. Appl Artif Intell 17(5–6): 375–381
10. Mobley I (2021) A guide to digitizing biopharma. Front Line Genomics. https://frontlinegenomics.com/a-guide-to-digitizing-biopharma/. Accessed 22 May 2023
11. Wilkinson MD, Dumontier M, Aalbersberg IJ, Appleton G, Axton M, Baak A, Blomberg N, Boiten J-W, da Silva Santos LB, Bourne PE, Bouwman J, Brookes AJ, Clark T, Crosas M, Dillo I, Dumon O, Edmunds S, Evelo CT, Finkers R, Gonzalez-Beltran A, Gray AJG, Groth P, Goble C, Grethe JS, Heringa J, 't Hoen PAC, Hooft R, Kuhn T, Kok R, Kok J, Lusher SJ, Martone ME, Mons A, Packer AL, Persson B, Rocca-Serra P, Roos M, van Schaik R, Sansone S-A, Schultes E, Sengstag T, Slater T, Strawn G, Swertz MA, Thompson M, van der Lei J, van Mulligen E, Velterop J, Waagmeester A, Wittenburg P, Wolstencroft K, Zhao J, Mons B (2016) The FAIR Guiding Principles for scientific data management and stewardship. Sci Data 3: 160018. https://doi.org/10.1038/sdata.2016.18
12. Digital Technology Roadmap (2022) BioPhorum. https://www.biophorum.com/download/digital-technology-roadmap/. Accessed 22 May 2023
13. Nargund S, Mauch K (2020) Hybrid models for biopharmaceutical process development – making the best of mechanistic know-how and data-driven insights. Biopharma Asia. https://biopharma-asia.com/magazine-articles/hybrid-models-for-biopharmaceutical-process-development-making-the-best-of-mechanistic-know-how-and-data-driven-insights/. Accessed 4 May 2020
14. Sokolov M, von Stosch M, Narayanan H, Feidl F, Butté A (2021) Hybrid modeling – a key enabler towards realizing digital twins in biopharma? Curr Opin Chem Eng 34:100715. https://doi.org/10.1016/j.coche.2021.100715
15. Esmonde-White KA, Cuellar M, Lewis IR (2022) The role of Raman spectroscopy in biopharmaceuticals from development to manufacturing. Anal Bioanal Chem 414:969–991. https://doi.org/10.1007/s00216-021-03727-4
16. Tulsyan A, Garvin C, Undey C (2018) Machine-learning for biopharmaceutical batch process monitoring with limited data. IFAC-PapersOnLine 51:126–131. https://doi.org/10.1016/j.ifacol.2018.09.287
17. Emerging Technology Program (2019) Food and Drug Administration website. https://www.fda.gov/about-fda/center-drug-evaluation-and-research-cder/emerging-technology-program. Accessed 31 Dec 2020
18. Artificial intelligence in drug manufacturing (2023) Food and Drug Administration website. https://www.fda.gov/media/165743/download#:~:text=AI%20methods%20can%20also%20be%20used%20to%20monitor%20product%20quality,a%20product's%20given%20quality%20attribute. Accessed 25 May 2023
19. Manzano T (2022) Qualifying AI algorithms in pharmaceutical manufacturing. BioPharm Int 35(1):30–31, 37

Part II
Manufacturing Aspects of Cell and Gene Therapy

Chapter 12
Essential Aspects of Mesenchymal Stem Cell Manufacturing

Tabea Marie Fleischhammer, Marline Kirsch, Mariia Abyzova, Sandra Dienemann, Iliyana Pepelanova, and Antonina Lavrentieva

Abstract The field of regenerative medicine encompasses various treatment strategies including gene therapy, cell therapy and the use of bioengineered tissue constructs. Although the applications of regenerative medicine involving living cells have demonstrated promising results in clinical trials, challenges in the manufacturing of reproducible cell material are still present. Mesenchymal stem/stromal cells (MSCs) are involved in over a thousand of clinical trials for treating a broad spectrum of conditions and *ex vivo* expansion of MSCs plays a crucial role in the treatment outcome. In this chapter, we discuss key parameters of *in vitro* MSCs culture such as media composition, scale-up strategies, cultivation under physiological oxygen tensions and three-dimensional cultivation. We demonstrate how these parameters influence cell physiology and functioning, and underline the importance of cultivation conditions for the preservation of all possible native functions of these cells during *ex vivo* cultivation.

T. M. Fleischhammer · M. Kirsch · M. Abyzova · S. Dienemann · I. Pepelanova ·
A. Lavrentieva (✉)
Institute of Technical Chemistry, Leibniz University of Hanover, Hanover, Germany
e-mail: fleischhammer@iftc.uni-hannover.de; kirsch@iftc.uni-hannover.de;
abyzova@iftc.uni-hannover.de; dienemann@iftc.uni-hannover.de;
pepelanova@iftc.uni-hannover.de; lavrentieva@iftc.uni-hannover.de

R. Pörtner (ed.), *Biopharmaceutical Manufacturing*, Cell Engineering 11,
https://doi.org/10.1007/978-3-031-45669-5_12

Graphical Abstract

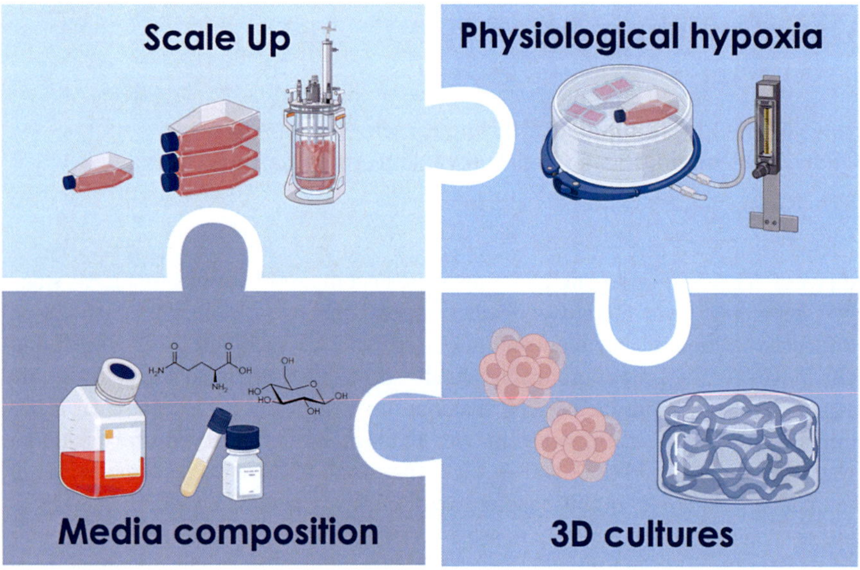

Graphical abstract. (Created with Biorender.com)

Keywords MSC · Manufacturing · Media supplements · 3D cell culture · Hypoxia · Physioxia

Abbreviations

2D	Two-dimensional
3D	Three-dimensional
a-MEM	Alpha minimum essential medium
bFGF	Basic fibroblast growth factor
ChiP	Chromatin immunoprecipitation
cm	Centimetre
DMEM	Dulbecco's modified Eagle's medium
e.g.	exempli gratia
ECM	Extracellular matrix
EGF	Epidermal growth factor
EMA	European Medicines Agency
et al.	*et alii, et aliae, et alia*
etc.	*et cetera*
EV	Extracellular vesicles
FBS	Fetal bovine serum

FCS	Fetal calf serum
g	Gramm
GCCP	Good cell culture practice
GMP	Good manufacturing practise
hBM-MSCs	Human bone marrow derived mesenchymal stem/stromal cells
hPL	Human platelet lysate
HRE	Hypoxia responsive element
HS	Human serum
IGF-1	Insulin-like growth factor 1
IMDM	Iscove's modified Dulbecco's medium
L	Litre
m	Meters
mL	Millilitre
MSC	Mesenchymal stem/stromal cell
PC	Human platelet concentrate
PDGF	Platelet-derived growth factor
pH	Decimal logarithm of the reciprocal of the hydrogen ion activity in a solution
S/D	Solvent/detergent
TGF-β	Transforming growth factor β
VEGF	Vascular endothelial growth factor
vWF	von Willebrand factor
WHO	World Health Organisation

Introduction

Isolation of mesenchymal stem/stromal cells (MSCs) from the bone marrow of guinea pigs was first reported in 1970 by Friedenstein and colleagues [1]. First described as fibroblast-like adherent cells with self-renewal and osteogenic capacities, now MSCs, isolated from different postnatal tissues represent a large area of clinical research. Besides their proliferation and differentiation capacities, MSCs can modulate immune cells, promote neovascularization, support angiogenesis, and protect target cells from apoptosis via direct cell-cell contacts, release of cytokines, as well as secretion of extracellular vesicles (EVs) [2]. Currently, MSCs are involved into over 1450 clinical trials worldwide [3]. However, despite the relatively long post-discovery history, the manufacture of MSCs and MSCs-based products still represents a significant a challenge.

For cellular therapies, depending on the introduction way (arterial, venous or intramuscular) and frequency of injections, millions of cells (in the range 0.5 to 2×10^6 cells per kilogram of patient's body weight) must be available for reaching maximal clinical efficacy. In order to produce tissue engineered constructs of clinically relevant sizes, hundreds of millions of cells must be available. This means that independently of the final application, MSCs must be first expanded *in vitro*. Therefore, the effective and economical expansion of MSCs plays a crucial role for their application in cell-based therapies and tissue engineering.

Although the choice of donor, cell source, isolation protocols, final quality control and implantation strategy also play an essential role in the treatment outcome, in this chapter we will focus on cell manufacturing. The entire pipeline of MSCs manufacturing consists of several steps: cell isolation and creation of a master cell bank, cell expansion, cell harvest, purification, medium removal, formulation and storage. There are several approaches to expand MSCs for clinical applications: (1) conventional static cultivation in cell culture flasks, (2) cultivation on microcarriers – static and dynamic and (3) dynamic cultivation in various bioreactors. Besides the use of different cultivation platforms, further bioprocess-relevant factors must be considered, *e.g.*, the usage of chemically defined media and xeno-free supplements or cultivation under physiologically relevant oxygen concentrations.

In general, one can outline four most important (and most challenging) aspects for the large-scale production of MSC. The **first aspect** is the fact that MSCs are anchorage-dependent cells and thus they cannot be cultivated in conventional shake flasks or stirred-tank reactors like suspension cells (*e.g.*, CHO cells). In this case, several strategies and approaches for the process scale-up are used, which are described below. The **second important parameter** is the cell culture medium composition. Basal cell culture medium provides cells with an energy source, contains physiological salt and pH composition, but lacks adhesion proteins, biological signals for cell survival, growth and differentiation factors. Therefore, basal media are usually supplemented with naturally-derived serum. However, xenogenic proteins from the widely used calf serum can be incorporated into the cultivated cells and cause immune reactions after cell implantation even during autologous applications [4, 5]. Moreover, the cultivated cells can be infected with various known and unidentified viruses, mycoplasma and prions [6]. Furthermore, the use of FCS raises great concerns regarding animal welfare and the cruel methods of its manufacturing. Thus, xeno-free alternatives for media supplementations are intensively studied and will also be discussed in this chapter. The **third essential factor** is cultivation of cells under physiological oxygen concentrations. *In vivo*, MSCs grow and function in much lower oxygen tensions than the ambient 21% O_2 provided by conventional incubators. Depending on the consumption rate and tissue vascularisation, oxygen tensions vary from very low (less that 1% O_2) in the deep zone of articular cartilage [7] to up to 7.8% O_2 in adipose tissue [8–11]. These data indicate that low oxygen tensions represent so-called physiological hypoxia or "physioxia" and that efficient respiratory chain function occurs within a specific narrow range of oxygen cell concentrations for different types of cells [12]. The influence of physiological hypoxia on MSCs is also discussed below. Possible cell expansion in three-dimensional (3D) environment is the **fourth challenge** in MSCs production. In comparison to non-physiological, two-dimensional (2D) cultures, 3D microenvironment allows dynamic intercellular interaction via different types of cell-cell contacts (gap junctions, tight junctions, *etc.*) and bioactive signalling molecules. Cell expansion in a physiological microenvironment improves the quality of the final cell product in terms of stemness, chemotaxis, immunomodulatory properties and angiogenic effects [13–15]. Although expansion of MSCs as cellular aggregates would improve cell properties, cell harvesting via aggregate dissociation requires treatment with more aggressive proteases, which in turn, can affect cellular physiology, *e.g.*, by proteolysis of surface receptors. Advantages and disadvantages of MSCs expansion in 3D systems will also be reviewed in this chapter.

Importance of the Composition of Cell Culture Medium

The choice of a suitable cultivation medium is one of the most important and decisive steps for mesenchymal stem cell manufacturing [16]. Especially in terms of the use of MSCs in regenerative medicine, it is essential to obtain clinically relevant cell numbers in a limited period of time without loss of all original MSC functions (such as retention of all receptors and adhesion molecules, as well as production of extracellular vesicles, chemokines, cytokines and growth factors) [17–22].

In order to achieve successful treatments and reproducible manufacturing of cells under GMP- compliant conditions, the expansion procedures for MSCs must be effective, safe and well-defined, as well as controlled conditions must be developed and established [17, 20, 23]. Nevertheless, to date, there is a lack of general standardised *in vitro* expansion protocols. These cultivation protocols are, however, of crucial importance, as the culture conditions may have an impact on the transcriptome, proteome and cellular organisation of MSCs. Hence, cultivation protocols may consequently influence the engraftment and MSC properties as well as MSC behaviour after transplantation [24]. Accordingly, the choice of a suitable growth medium for the *in vitro* cultivation represents one of the most crucial steps for optimal cell growth and the quality of the final cell product.

A conventional cell culture medium is composed of basal medium consisting of various amino acids, vitamins, inorganic salts, and energy sources (glucose and glutamine), as well as serum as a source of growth factors, hormones, and binding factors (Fig. 12.1) [16, 23, 25–29]. In addition to its purpose to provide nutrients, the medium also contributes to maintain the pH and osmolality maintenance [16]. The requirements on the medium can vary depending on the cell type and therefore result in a large number of media compositions being used [16, 30]. Since each of the components in the basal medium fulfils a specific function, the choice of basal medium alone can be crucial for the cultivation success of MSCs.

For clinical applications of MSCs, the time required for cell expansion is considered as the limiting factor. In addition, various studies have demonstrated negative effects of long-term cultivation on the migration, proliferation, differentiation and genetic stability of MSCs [31, 32]. By using high glucose medium, rapid cell growth can be achieved due to the easy availability of glucose. However, rapid cell expansion still cannot guarantee the required quality of cell-based products. For instance,

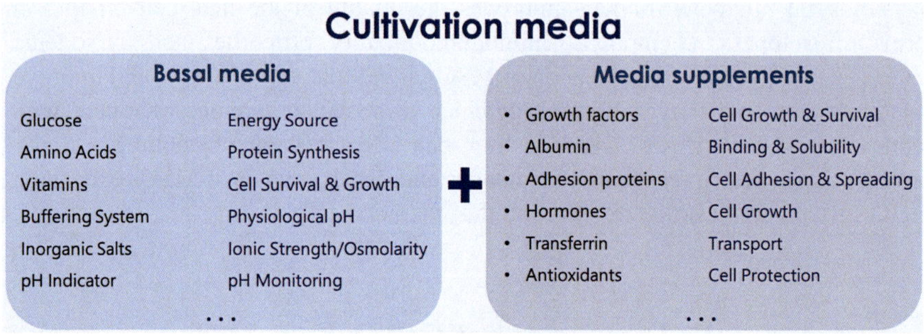

Fig. 12.1 Composition of conventional cultivation media

Zhang et al. and Parsch et al. demonstrated that high glucose concentration (higher than 1 g/L) can result in cellular senescence as well as telomere shortening and genomic instability [33–35]. For human bone marrow derived mesenchymal stem cells (hBM-MSCs), Sotiropoulou et al. showed that low-glucose Minimum Essential Medium Eagle Alpha Modification (α-MEM) (1 g/L glucose) is a more beneficial basal medium than Dulbecco's Modified Eagle's Medium (DMEM) (1–4.5 g/L glucose) [36] or Iscove's Modified Dulbecco's Medium (IMDM) (4.5 g/L) [37].

To provide cells with biological signals and ensure optimal cell growth, media supplements are added to the basal medium. Serum as a medium supplement is considered one of the most important components of the cell culture medium and serves as a source of many factors crucial for cell functions [16, 23]. For instance, serum provides the cells with basic nutrients and molecules involved in growth promotion and specialised cellular functions (growth factors and hormones), as well as binding proteins (*e.g.*, albumin, transferrin) and cell adhesion proteins (*e.g.*, fibronectin). Factors supporting cell spreading as well as protease inhibitors, minerals (*e.g.*, Na^+, K^+, Zn^{2+}, Fe^{2+}), and trace elements are also provided by the serum [38].

The most common and widely used cell culture medium supplement is fetal calf serum (FCS), also known as fetal bovine serum (FBS). To date, it remains widely used as a medium supplement in most cell isolation or expansion protocols as well as for the cultivation in some regenerative medical studies [39–54]. Although there are various disadvantages associated with the use of FCS, there are no widely accepted alternatives for FCS to date [41–46, 48, 49, 55, 56]. A suitable alternative to FCS should ideally ensure a well-defined composition, a preferably reduced level of impurities, low production costs, long shelf life, and good availability [16, 23]. Alternatives to FCS as a medium supplement being studied include human blood derivatives (*e.g.*, human serum (HS) or human platelet lysate (hPL)) and chemically defined media [19].

Conventional Media Supplements

More than 50 years ago, the first isolation and cultivation of MSCs also marked the beginning of the first experiments in which human and xenogeneic sera were used to supplement cells [57, 58]. Since then, FCS obtained from the blood of fetal calf hearts has become an established supplement in cell culture and it is still being used in most isolation and cultivation protocols for clinical studies/cells, although it is no longer considered as the gold standard [39–50]. As a very complex supplement, it provides a variety of growth factors, adhesion proteins, binding and transport proteins, vitamins, amino acids, carbohydrates and hormones essential for cell growth [23, 42, 59, 60]. In addition, FCS is produced in virtually unlimited supply, is very inexpensive in production, and therefore can be easily acquired [43].

Between 500,000 and 800,000 L of FCS are produced per year for cell culture laboratories taken from up to two million calves [61–63]. The serum is taken directly from the foetuses of pregnant slaughter cows by cardiac puncture and without anaesthesia [23, 30, 43, 59, 63, 64]. This ethically controversial serum collection is used due to the higher yield of growth factors and fewer antibodies in the fetal blood, and thus the serum obtained from it, compared to mature cow blood [42]. Despite the low antibody titre, bovine xenogenic compounds such as proteins or polysaccharides can be internalised by the cells during culture. Therefore, immunological reactions can be caused after implantation even when autologous cells are used [4, 65]. Furthermore, cows are exposed to environmental factors and pathogens, which can be directly transferred to the foetus via the placental barrier [59]. Based on this, the use of FCS in the cultivation of human cells can lead to the transmission of contaminants (*e.g.*, fungi, mycoplasmas or endotoxins) or transmissible infectious diseases caused by viruses, bacteria or prions to the patient [42, 43, 55, 56, 59, 60, 66–73].

Besides the ethical and immunological concerns, a large batch variability can significantly affect the quality of the cell product and thus lead to large deviations in the results [23, 59, 74–80]. Despite many years of application and research on FCS, to date, it was not possible to identify all components or to fully understand the molecular mechanism of action behind the positive effect of FCS on cell growth [61].

Based on these multiple issues, the use of FCS for MSC manufacturing is not favoured anymore as a gold standard. Its use is highly controversial and not recommended for clinical applications. In fact, organisations and regulatory guidelines (world health organization (WHO), good manufacturing practice (GMP), European Medicines Agency (EMA)) highly recommend the use of a xeno-free alternative to develop safe and standardised protocols for MSC manufacturing and application for regenerative medicine (*e.g.*, cell-based therapies, tissue engineering) [23, 30, 42, 59, 60].

Derivatives from human blood, such as HS or hPL, represent a possible xeno-free alternative. Due to its human origin (obtained via venepuncture of healthy donors) and the associated lower risk of immune rejection reactions and zoonotic infectious diseases as well as an enhancing effect on cell expansion/proliferation, differentiation and immunophenotype of cells shown in various studies, HS has already been commercialised and used as a possible alternative to FCS [23, 43, 81–84]. However, the production of HS is already declining because of its limited availability (obtained from blood of healthy donors) and its expensive and complex but not standardised production [82, 83, 85, 86]. Therefore, HS is no longer considered as an efficient alternative to FCS for clinical applications [23, 42, 43, 59, 86]. However, until universal chemically defined media are developed and commercially available, it is necessary to evaluate existing alternatives to FCS. Of the available blood derivatives, hPL has so far shown to be a promising alternative to FCS and HS and has also become more commercially available over the last four to 5 years [87–90].

Human Platelet Lysate (hPL)

Pending the development of a universally applicable chemically defined medium, hPL appears to be a promising medium supplement for mesenchymal stem cell manufacturing. It represents an efficient alternative that provides better results in many studies in the field of cell culture compared to the commercial FCS [42–44, 91–95]. Worldwide, blood donation campaigns collect whole blood donations or pure platelets via apheresis donations for transfusion purposes, which are used for the production of human platelet concentrates (PC) [44, 96]. PCs have only a limited shelf life of three to 7 days, so that a large number of PCs are discarded (up to 20% of produced PCs are not transfused) [18, 55, 97]. However, expired PCs are safe and suitable as a raw material for the preparation of hPL while not compromising any subsequent cell cultivation [18, 19, 55, 96]. hPL can be obtained from PCs in accordance with GMP guidelines via various methods (sonication, solvent/detergent (S/D) treatment or via repetitive freeze-thaw cycles) [43, 44, 96, 98].

hPL is a cell-free, protein and growth factor rich medium supplement with many advantages. The storage granules of platelets contain numerous bioactive substances [44, 60, 99–104], which are released into solution after lysis and can thus influence the cells cultured with hPL-supplemented medium [44, 104, 105]. Among albumin, fibrinogen (1–3 mg/mL) [44], the regulatory proteins and factors in hPL are multiple cytokines (*e.g.*, epidermal growth factor (EGF), basic fibroblast growth factor (bFGF), vascular endothelial growth factor (VEGF), transforming growth factor β (TGF-β), insulin-like growth factor 1 (IGF-1), platelet-derived growth factors (PDGFs)) [106], many of which have been shown to have a positive effect on the cultivation of mesenchymal stem cells [43, 106–108]. For instance, Mizuno et al. studied the influence of PDGFs in hPL on the proliferation of MSCs and demonstrated a concentration-dependent increase in proliferation activity [106, 109]. Furthermore, several studies showed enhanced cell adhesion when cells were cultured with hPL, which could be explained by various adhesion factors such as von Willebrand factor (vWF), fibronectin, vitronectin, and thrombospondin [110–113]. In recent years, an increasing number of studies also showed hPL enhanced spreading, viability, proliferation and differentiation of 3D cultured MSCs as well as advantages if MSCs were directly isolated from the tissue in 3D hPL matrices [114–117]. Overall, numerous *in vitro* studies already show clear and promising results on the successful application of hPL. Among them are successful cultivations of fibroblasts, endothelial cells, and cancer cell lines [118–121], the use of hPL for isolation, expansion, and successful differentiation of MSCs, or the *ex vivo* cultivation of MSC-based medical products [42, 44, 94, 122–124].

In conclusion, until the development of efficient and universal chemically defined media, hPL appears to represent an optimal alternative xeno-free medium supplement (compared to conventionally used FCS) for MSC manufacturing and applications in cell-based procedures as well as in tissue engineering [18, 19, 123]. The application of hPL as medium supplement combines modern social principles of ethics, sustainability, recycling, health safety and resource conservation [18, 19, 42, 44, 55, 96, 123, 125]. Although standardised xeno-free cultivation protocols and processes to ensure quality and safety in the handling of hPL are still required [18, 44], the use of hPL as medium supplement represents an important step towards the development of xeno-free humanised cultivation models.

Serum-Free and (Xeno-Free) Chemically Defined Media

The development of serum-free, chemically defined media can improve the reproducibility, upscaling, and cost of mesenchymal stem cell manufacturing [126, 127]. They do not contain proteins, hydrolysates or other components of unknown composition. Most of the hormones or growth factors that were previously isolated and purified from human or animal serum can now be produced by recombinant technology, thus enabling the production of completely xeno-free chemically defined medium with an eliminated risk of transmitting pathogens [30, 127]. Especially for the increasingly important quality control, generally applicable standards for *in vitro* methods are to be developed regarding the good cell culture practice (GCCP). Part of these standards include the introduction of the general use of chemically defined media and the associated replacement of animal-based, undefined supplements with variable compositions [30, 59, 63, 128–130]. Since MSCs can vary greatly depending on their isolation source and the donor, there are several specific requirements for the media additives of an optimal chemically defined medium for the cells [131]. Nevertheless, the respective chemically defined medium can be specifically modified to the characteristics and demands of the MSCs to be cultured. Thus, even MSC selection could be supported during isolation. For some cell types, specifically defined media supplements (serum and/or xeno-free) are already commercially available. However, the development of a universal chemically defined medium for MSC expansion on a clinical scale remains a major challenge [23, 30, 132]. For instance, different studies showed variations with respect to cell doubling times, cell morphologies and sizes, as well as changes in cell vacuoles after cultivation with chemically defined media [23, 127, 133, 134].

To date, another challenge is the often non-xeno-free pre-treatment of cells or cell culture substrates. Many of the currently available chemically defined media require preincubation in serum-containing media or pre-coating of the culture substrates with ECM proteins to support cell adhesion [16, 23, 30, 135–137]. The proteins required for the coating are again often of animal origin, so that no completely xeno-free cultivation is achievable with the chemically defined media [30, 49].

Even if chemically defined media can be specifically tailored to the needs of MSCs, intended clinical objectives [124, 138], and can achieve acceptable cell growth, as well as overcome the risk of pathogen transmission, chemically defined medium cannot yet compete with the performance of media supplements such as FCS or hPL [4]. Furthermore, xeno-free chemically defined media has to be tested and optimised for additional applications such as 3D cultivations and scale up cultivation processes (*e.g.*, attachment and shear stress efficiency) [139–142]. To date, commercially available chemically defined xeno-free media often do not result in optimal cell growth compared to serum or xenogeneic supplemented media, and improvement in the evaluation of functional characteristics of cells cultivated in it is necessary [59, 143]. Therefore, xeno-free chemically defined media is not yet universally applicable for primary adherent cells and MSC cultivation with xeno-free supplements will still be necessary until chemically defined media for MSCs is optimised [144].

Scale-Up Strategies for Production of High Cell Numbers

MSCs belong to an anchorage-dependent cell type, which means that they require a growth surface in order to proliferate *ex vivo*. Traditionally, MSCs are maintained at low confluence in cultural plates or flasks. However, the expansion of cells in these conditions correlates with a higher risk of contamination due to the frequency of open manipulations [145]. Closed robotic systems, which are able to avoid human factors and ensure reproducibility and higher throughputs, are developed to improve two-dimensional cell culture maintenance [146, 147]. Cell expansion in 2D conditions is associated with a series of passages. Due to this process, MSCs can change their phenotype and undergo senescence. Additionally, typical culture flasks require large areas in incubators to obtain therapeutic volumes of cell mass [148]. Thus, several approaches and strategies are developed to improve the in vitro expansion and upscale of MSCs cultivation (Fig. 12.2).

Cultivation in Flasks

To reduce the volume-to-area ratio in planar cell maintenance, multi-layered flasks or cell stackers (or cell factories) can be used. In this case, the “scale-out” method is

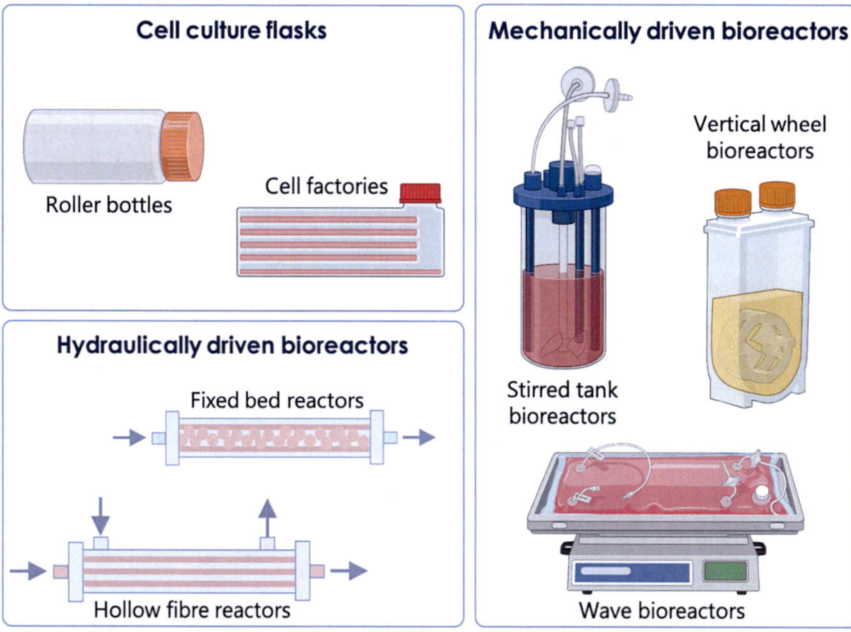

Fig. 12.2 Overview of different methods and reactors for producing high cell numbers of MSCs. (Created with Biorender.com)

implemented when the same surface area is multiplied. However, typical 2D cultivation drawbacks associated with batch-to-batch variability primarily when handled manually are inherent to this method. Cultivation in cell stackers makes it difficult to observe the cells under the microscope, because of the cell factory height. Moreover, achieving homogeneous conditions between layers of a cell stacker is challenging [142].

Another approach to decrease the volume-to-area ratio is performed in roller bottles. A number of bottles with adhesive walls are incorporated in a heating and revolving rack. Adherent cells cover the inner surface of those bottles forming a monolayer [149]. In a rotating bottle, cells are covered with a thin layer of media for a significant period of time that ensures robust gas exchange [150]. However, it was shown that roller bottles demonstrate less efficient MSC expansion capacity compared to traditional culture flasks [151]. The same disadvantages as for culture flasks and cell stackers apply to roller bottles. The risk of contamination due to manual manipulations, batch-to-batch variations, and high labour demand limits the application of this technique in large-scale MSC production. This method is not widely used now and considered mostly as a historical stage.

Expansion in Bioreactors

In contrast, bioreactor systems demonstrate significantly higher scalability capacity that meets the requirements of MSC manufacturing for therapeutical use. Furthermore, bioreactors are considered closed systems, which are only opened for a limited number of manipulations such as loading, vessel connection, and harvesting. In this way, the risk of contamination decreases. With the exception of the most simple models, the majority of bioreactors are instrumented with sensors to monitor nutrient utilisation, metabolite accumulation, optical density, and other desired parameters [152].

Multiple various bioreactor architectures have been developed and can be classified into groups according to the type of power input. Mechanically driven reactors, including stirred tanks, wave-bags, and rotating-wheel reactors enable the cultivation of MSCs adhered to carriers in constantly mixed media and to reach a cell density of 10^5–10^6 cells/mL. This makes them preferable for therapeutic cell manufacturing. Hydraulically driven systems such as fixed-bed and hollow fibre reactors contain fixed scaffolds for cell adhesion while media is pumped to flow along those structures seeded with cells [153].

Stirred tanks are one of the most widely used types of reactors due to their relatively simple structure, scalability and well-characterised properties. In the most primitive case, a stirred tank reactor consists of a reservoir with an impeller inside. Bench-scale reactors with working volumes of up to 250 mL are usually surface-aerate, while larger models may be equipped with two or more impellers and additional aeration devices. Accelerating the impeller rotation speed improves mixing and oxygenation, however, at the same time increases shear forces. Being anchorage-dependent cell types, MSCs can be cultivated in these systems (1) on microcarriers, (2) as cell aggregated and (3) encapsulated in hydrogels. If cultivated on microcarriers, the interaction between mixing particles and media flow prevents the formation of aggregates of several microcarriers and enables cells to migrate from confluent microcarriers to empty ones. At the same time, high mixing speed is not the best option, as MSCs are sensitive to shear stress which can affect immunomodulatory properties, induce differentiation, or damage cells [154–156]. Up to now, according to the relative simplicity of stirred tank bioreactors, numerous publications and protocols have been performed and reported, making this technology preferable for new studies and industrial applications. The different geometry of vertical wheel bioreactors reduces shear stress while achieving effective mixing. Vertical wheel reactors can be scalable up to 100 L according to computational models [157].

In wave bioreactors, cells are loaded inside a plastic bag fixed on a moving and heating platform. The oscillatory movement of the platform facilitates sufficient mass and energy transfer, while limiting foam formation. In this way, antifoam agents can be reduced or even excluded from the media formulation [158–162]. This kind of bioreactor is suitable for shear-sensitive cells due to a more homogeneous shear stress pattern inside the cultivation reservoir. Nevertheless, aggregate formation in this bioreactor type remains challenging.

Independent of architecture, MSC cultivation in the described bioreactors requires usage of carriers for MSCs adhesion. Frequently, small polymer beads or microcarriers are applied for obtaining phenotypically stable cells on a therapeutic scale. Microcarriers can be made of a wide range of materials and coated with various adhesive substances. In general, microcarriers can be divided into two main groups: smooth and macroporous microcarriers. In the first case, cells grow on the surface without penetration into the core. In porous particles, cells are trapped within the particle and grow inside of it. Hereby, microcarriers are providing a larger area than 2D systems within the same volume of growth media [163]. Successful 43-fold MSC expansion on microcarriers (with a yield of 1.28×10^{10} cells) in a 50 L stirred-tank reactor was reported recently [164]. To prevent shear stress during cultivation, some protocols recommend loading cells in hydrogel spheres using special microfluidic-based equipment [165]. Alternatively, cells can be cultivated as self-assembling cell aggregates or spheroids. The possible effects of 3D cultivation in spheroids on MSCs properties will be discussed later in this chapter. Generally, using spheroids mimics natural tissue conditions more closely. At the same time, to facilitate the self-assembly of cellular aggregates, adhesive molecules including recombinant vitronectin, integrins and laminin are required to be present in the growth medium, which makes a large-scale GMP production of MSCs in spheroids relatively expensive [166].

The next important challenge to overcome in MSC expansion on microcarriers or in gel-beads is the process of harvesting. Traditionally, proteolytic enzymes are used for cell detachment, however, prolonged enzymatic treatment can affect cell immunophenotype [167]. To avoid this, it has been suggested to use a mixture of enzymes in combination with reduced exposure time [168, 169]. After detachment, cells can be separated from the microcarriers using a filter system with a suitable pore size which retains microcarriers and allows cells to pass through [164]. In addition, other approaches such as thermosensitive microcarriers can be applied to enable enzyme-free MSCs harvesting [170].

The interaction between MSCs' surface ligands and the microcarriers to which they are attached regulates critical signalling pathways [171]. Natural materials which are components of the extracellular matrix *in vivo*, such as collagen and gelatine provide high biocompatibility. Collagen and gelatine also facilitate cell attachment [172–174]. Moreover, gelatine- and collagen-based microcarriers can be digested enzymatically which improves harvesting [171]. However, since many of these materials are derived from animal sources, they are not recommended for clinical practice [175]. Alternatively, synthetic polymers, usually polystyrene-based, represent a defined and xenogenic-free material for microcarrier production. Their biocompatibility and cell adhesion support can be improved by physical surface treatment or by coating with synthesised polymers that mimic natural ligands and facilitate cell adhesion and proliferation [176]. Mechanically driven rotatory bed bioreactors' structure includes plastic plates as cell carriers. The structure of this type of reactor includes a rotating cylindrical reservoir with numerous adhesive plates inside. This promising approach makes it possible to achieve up to 6000 cm^2 of cultivating area [153].

Hydraulically driven bioreactors provide cell maintenance due to continuous medium flow. For example, hollow fibre reactors consist of numerous semi-permeable hollow fibres placed in a cylindric reservoir. Cells can be seeded inside the fibres while medium constantly flows along their external surface. Alternatively, the opposite seeding principle is available for some models: cells in medium are seeded on the external surface of the hollow fibres while oxygenated medium flows inside them. Hollow fibre reactors are able to obtain cell densities in the order of 10^9 cells/mL [177, 178]. Moreover, a few units can be connected in parallel which enables a scale-up principle [179]. One of the challenges to be overcome when working with hollow fibre bioreactors is the longitudinal concentration gradient forming when medium or dissociation reagent flows along the fibres and thus may result in inconsistent distribution of nutrients or dissociation reagent [180].

Fixed bed bioreactors include an immobilised cell carrier which is made from porous beads, discs or hydrogels perfused with oxygenated medium with nutrients. High cell densities of around 10^8 cells/mL can be reached in this type of reactor [181]. Another advantage of the fixed bed construction is the large cultivation area of around 500 m^2 for industrial cell manufacturing [182]. On a large scale, axial and radial concentration gradients can occur due to the reactor structure. Similar to microcarrier-based cultivations, cell harvesting remains challenging because of the complicated distribution of the dissociation reagent and also because of high cell packaging density in the fixed bed [176].

In all dynamic systems, shear stress is considered a limitation for the production of large amounts of phenotypically stable cell mass with a high proliferation rate. There are controversial findings regarding the effect of shear stress on the proliferation rate [183–189]; however, the majority of authors agree on the fact that shear stress induces osteogenic and chondrogenic differentiation [184, 190–193]. Shear forces affect expanded MSCs not only during their growth in bioreactors but also in the consequent downstream process. Usually, the downstream process includes cell detachment, concentration, washing, formulation and cryopreservation. The principles for downstream process development are aiming to preserve cell viability and function. Downstream strategies need to meet GMP requirements, be scalable, automated, closed and be performed in a short time [194]. Also, omitting some process steps, according to cultivation technology can be an advantage. For example, growing cells in self-assembling aggregates does not only mimic *in vivo* conditions, but at the same time does not require cell separation from microcarriers.

Influence of Physiological Hypoxia on MSC Production

Oxygen is a key element for the survival of higher organisms. It has long been known that too high or too low oxygen concentrations can both act as biological stressors. Accordingly, cultivation under physiologically relevant, reduced oxygen conditions has been shown to be beneficial for some applications in the field of MSC manufacturing. Understanding the divergent effects of hypoxia is necessary for producing clinically relevant cell numbers, MSC-derived products or transplantation constructs.

Biochemical Regulation of Oxygen Homeostasis

For the maintenance of cellular and tissue homeostasis, the adaptation of cellular processes to low oxygen availability is essential for all cells of vertebrates and larger invertebrates [195]. The so-called hypoxia inducible factors (HIFs) have been identified as the main biochemical regulator of oxygen homeostasis [195]. HIF-1 was found to be a heterodimeric protein consisting of HIF-1α and HIF-1β subunits that interact directly with DNA in the major groove [196, 197]. HIF-1α and HIF-1β mRNA is expressed in most, if not all, human and murine tissues and a conserved homolog has even been found in *Caenorhabditis elegans* [198–200]. While HIF-1 appears to have a very basic role in hypoxia-induced signalling to the transcriptional machinery in the nucleus of all cells, its other two isoforms show more specialised or limited roles in oxygen homeostasis [195].

Under non-hypoxic conditions, the constitutively translated [201, 202] HIF-1α subunit undergoes proteasomal degradation, but it is stabilised during hypoxic conditions (cf. Fig. 12.3) [203–207]. HIF-1α is subsequently translocated to the nucleus where it dimerises with HIF-1β [205, 208, 209]. HIF-1 and its co-activators bind to consensus sequences in DNA found in hypoxia responsive elements (HRE) which have a conserved sequence of five nucleotides [195, 209, 210]. ChiP-based analyses by Schödel et al. revealed over 400 of potential HIF-1 binding sites with high stringency at over 500 gene loci [211]. Among the genes controlled by HIF-1 are mainly genes for cell growth and apoptosis, genes of energy and iron metabolism, and genes involved in pH control and angiogenesis or erythropoiesis [212]. HIF-1 is also capable of activating other transcription factors, such as SOX9, which has a core function in chondrogenesis of MSCs [213–215]. Stabilisation of HIF-1α can also occur under normoxic conditions: Growth factors [216–219] and cytokines [219–221] play a major role in this process.

Influence of Oxygen on MSC Physiology

In vivo, MSCs can be found in many different tissue types such as adipose tissue, bone marrow or umbilical cord blood, amongst others. Oxygen concentrations lower than the atmospheric ones are present in these tissues: for example, the oxygen concentration of adipose tissue varies between 5.3% and 7.3% [8–11] and for bone marrow it is estimated to range between 1% and 7% [222–225]. Thus, the physiological oxygen concentrations (often also referred to as "physiological hypoxia" or

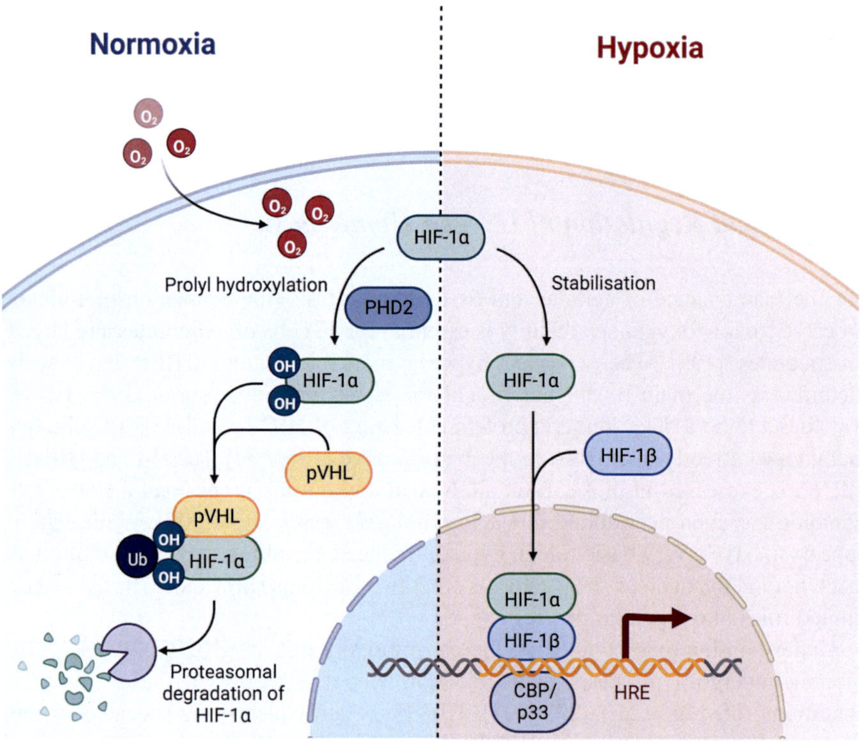

Fig. 12.3 Major regulatory mechanisms of HIF-1 during normoxia and hypoxia. (Adapted from "Organoid Production Methods Overview", by BioRender.com (2022). Retrieved from https://app.biorender.com/biorender-templates)

"physioxia") are generally much lower than the ambient 21% oxygen that are conventionally applied in cell culture. This implies that the cultivation of MSC under physiologically relevant oxygen tensions can recapitulate the influence of native local signals on differentiation or proliferation of these cells more closely and thus might have a positive effect on their cultivation. Moreover, cultivation under hypoxia can also be important in subsequent cell survival after transplantation of MSCs in ischemic, injured tissues [226–228].

The effects of hypoxia on MSC biology have been studied and extensively discussed in the literature [229–231]. Hypoxia has been reported to show numerous positive effects on MSCs. When cultured under hypoxic conditions, MSCs have been reported to show increased cell proliferation [232–239] and preservation of stemness [240, 241], which are key features of a successful expansion of MSCs. Additionally, a hypoxic pre-conditioning stimulus potentially increases the regenerative potential of MSCs via their secretory functions: in various studies, the secretion of pro-angiogenic and pro-survival factors was increased after hypoxic preconditioning of MSCs [226, 242]. Hypoxic pre-conditioning also lead to an enhanced cell survival after implantation into ischaemic heart tissue [226], potentially evading hypoxia-induced apoptosis via the preceding priming of the cells. Depending on the targeted application of MSCs, the positive influence of hypoxia on chondrogenic differentiation in different 3D cell culture applications could also be exploited [243–245]. Conversely, there is conflicting data for osteogenic and adipogenic differentiation: While an enhanced osteogenic and chondrogenic differentiation was reported in some studies [239, 243–245], others report a diminished adipogenicity [246–248] or impaired osteogenic differentiation [246, 249–251]. A recent review on these discrepancies by Samal et al. provides a comprehensive overview of the used culture conditions and the observed effects on MSC fate in over 30 studies [229].

Despite the aforementioned positive effects of hypoxia on MSCs, there is conflicting evidence, however, if cultivation of MSCs in hypoxic conditions has universally positive effects. These diverging findings on the effect of hypoxia may be due to the extreme variability of the experimental set-ups under which the effect of hypoxia has been examined: Firstly, the oxygen tensions that authors consider as "hypoxic" diverge between 1% and 10% and the timing and duration of the exposure to reduced oxygen concentrations differs widely between studies, too. Exposure to hypoxia ranges from a few hours up to weeks. Secondly, the examined MSCs may be originating not only from human tissue, but are also often derived from a variety of other species (*e.g.*, mouse, rat, pig, rabbit *etc.*). Additionally, the multitude of possible tissues of origin for MSCs creates even more variability. Cultivation methods in these studies range between conventional 2D cultivation and a diverse spectrum of different 3D cultivation systems. Moreover, in some studies hypoxia mimicking reagents are used instead of decreasing the oxygen concentration itself. Lastly, different media and supplement combinations are used, too. The abovementioned aspects of the influence of hypoxia on proliferation behaviour, differentiation and survival rate of MSCs show how essential it is to tailor the cultivation conditions to the respective combination of cultivation method and target tissue in order to obtain the best results during MSC cultivation.

Taking MSC Production to the Third Dimension

Another important physiological aspect in the cultivation of MSCs is the composition of their microenvironment. The cultivation of cells in 2D has been routinely performed in thousands of laboratories worldwide for over 100 years [252]. Although 2D cultivation is less labour-intensive and less costly, cultivation in a monolayer barely, if at all, reproduces the physiological conditions that cells would encounter in their tissue of origin [253–256]. In 2D culture, cell-cell interactions are fundamentally different than in the physiological environment [253]. Here, only the homogeneous contact surface of the cultivation vessel is provided (Fig. 12.4). Generally, cell-cell and cell-matrix interactions are, among other things, crucial for the maintenance and progression of cell differentiation [253, 255, 257–259], the resulting

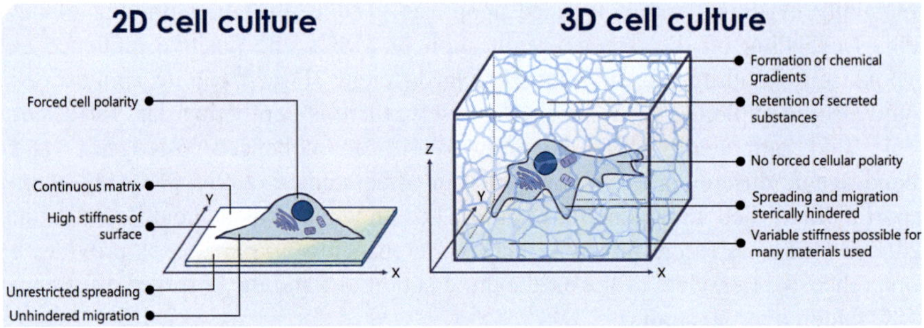

Fig. 12.4 Comparison of 2D and 3D cell cultivation conditions

expression profiles [260–262] as well as migration [263–266] and thus also for the cellular response to stimuli.

Three-dimensional (3D) cultivation of cells provides a solution to many of the aforementioned problems for recapitulating a more *in vivo*-like microenvironment. The different techniques in 3D cell cultivation can be classified as either an *ex vivo* cultivation of grafts, the cultivation as 3D cell aggregates (spheroids) or the cultivation in/on different matrices, *i.e.*, using decellularised extracellular matrix, concentrated medium, gel-based substances or scaffolds [253, 254].

MSC Cultivation in Spheroids

Spheroids have been extensively studied for the expansion of MSCs and MSC-derived applications due to their ease of fabrication, scalability and their potential positive effects on MSC biology. The production of spheroids is based on the natural tendency of many adherent cells to form aggregates with one another [255]. This leads to the development of spherical, cellular aggregates in the absence of a surface to which they can adhere to. An apparent advantage of spheroids when compared to other 3D cultivation methods is that no adhesion matrix is required for the creation of spheroids. Also, spheroids are potentially well-suited for the application in high-throughput manufacturing, depending on the method of producing spheroids [267]. As already mentioned above, the cultivation of MSCs in cell aggregates also allows the involvement of different types of bioreactors in cell production.

There are different methods for obtaining cell aggregates: Historically, spheroids have been obtained via the liquid overlay or the spinner flask methods which both produce a comparatively heterogeneous population of cellular aggregates, but are able to yield large numbers of spheroids [14, 268–270]. For fabricating spheroids via the spinner flask technique, cell-surface contacts are minimised via agitation of the flask while aiming for maximising cell-cell contacts that are facilitated in the high-density cell suspension [14]. Although a high stirring rate is crucial for the formation of spheroids, it can induce damage to the cells [271]. The liquid overlay method prevents cellular adhesion to the surface with a layer of agarose gel [14]. Cell culture flasks coated with cell-repellent agents allow the production of larger quantities of spheroids, but with varying spheroid sizes [14, 272]. Other currently used approaches are the generation of cell aggregates via the hanging drop technique or via using low attachment microplates [14, 272–274]. Both of these methods generate uniformly sized cell aggregates but production of large quantities is more difficult [272–274]. A more modern approach is the generation of spheroids with so called microstructured plates: These are able to produce large amounts of spheroids but have been shown to have an inequal distribution of oxygen concentration throughout the plate [275]. Lastly, there are other methods that require more specialised equipment: For example, rotating wall vessels [276] or magnetic levitation [277] as well as microfluidic systems [278] can also be applied for the purpose of creating spheroids (Fig. 12.5).

Although the spheroid manufacturing method seems to influence the exact experimental outcome to some extent, it has been shown that cultivating MSCs in spheroids can have numerous positive effects. Firstly, MSCs can form cell-cell interactions and show cell polarity which is a key advantage of spheroids over traditional 2D cell culture [272, 276]. Moreover, spheroid culture leads to a maintained stemness [13, 279] as well as differentiation potential [279] of MSCs and can also positively influence their paracrine capabilities: anti-inflammatory

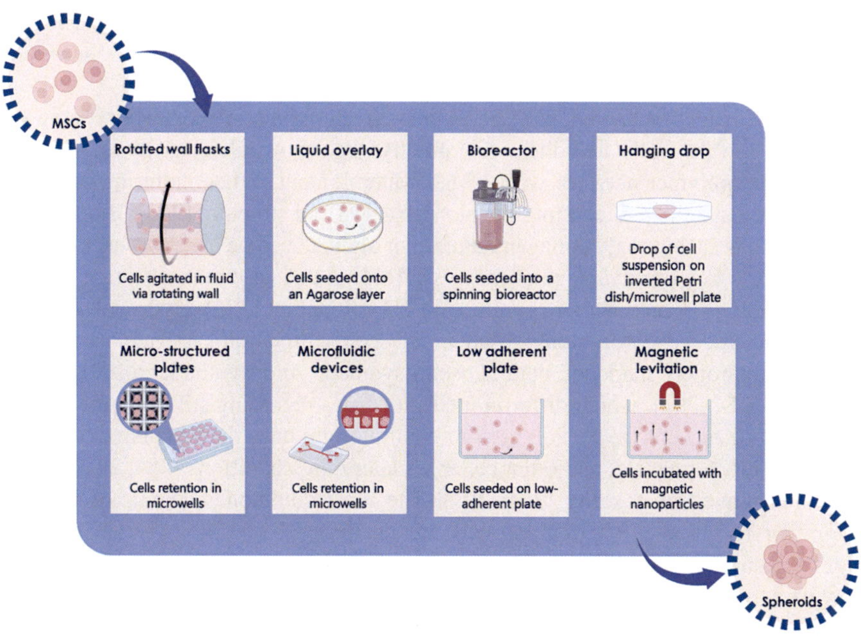

Fig. 12.5 Different methods for generating spheroids from MSCs. (Adapted from "Organoid Production Methods Overview", by BioRender.com (2022). Retrieved from https://app.biorender.com/biorender-templates)

[280–282], angiogenic [13, 283, 284], and immunomodulatory [281, 285] potentials were positively influenced when MSCs were cultured in microspheres. Moreover, MSCs that have been cultured in spheroids over a short period of time and subsequently expanded as monolayers show increased regenerative potential, greater expansion capacity with decreased senescence, and improved *in vitro* migration as well as *in vivo* wound healing [13].

As it is the case for most, if not all, cell cultivation platforms, it has also been shown that the composition of the cell culture medium has a non-negligible effect on the formed spheroids. Even though MSC expansion under serum-free conditions is possible [286], Ylostalo et al. showed that a chemically defined, xeno-free medium supplemented with blood-derived or recombinant human serum albumin resulted in tighter spheroids with a high expression of anti-inflammatory and anti-cancer molecules in comparison to spheroids that were cultivated in several other commercially available stem-cell media, *i.e.*, α-MEM [287].

Matrix-Based Cultivation

Since the first 3D cell cultures in 1911 by Carrel and Burrows [288], the range of matrix-based solutions for 3D cell cultivation has grown massively. The key advantages of culturing in matrix-based methods are that cell-cell and cell-ECM contacts can be formed, resembling the *in vivo* microenvironment more closely [253]. The 3D matrix also leads to higher local concentrations of effector proteins, *i.e.* secreted growth factors, due to diffusion limitations [289].

A basic distinction can be made between cultivation on scaffolds, which serve as a surface for the cells to adhere to, and hydrogels, which consist of a three-dimensional polymer network with a high water content. When culturing on scaffolds, cells can adhere to and migrate along the scaffold structure [253]. Biodegradable materials such as silk, collagen, laminin or alginate can be used for this purpose [290–292]. A key disadvantage of many scaffolds is that adhered cells often have a flat morphology, which is more common in 2D cultivation [253]. Hydrogels, on the other hand, are very commonly used in 3D cell cultivation. A wide variety of different hydrogels made of natural, semi-synthetic and synthetic materials are now used [254, 293], which differ in their properties regarding stiffness, mesh size and influence on cells [294]. Polymers of natural origin have the advantage of being abundant, biodegradable, and often gellable under mild conditions [254, 294]. They also often have low toxicity [254, 294]. The most common materials of natural origin for the production of hydrogels are gelatine, glycosaminoglycans, fibrin, alginate and collagen [295]. Synthetic materials also can be used as scaffolds, but they usually lack biological clues. Due to possible batch-to-batch variations and in the absence of the desired mechanical properties, starting materials of natural origin are often additionally chemically modified or combined with synthetic components to obtain improved, controllable properties [295, 296]. Gaining control over the mechanical stiffness of a matrix is a major advantage as MSCs - given their plasticity and thus remarkable ability to adapt to different cellular contexts in different ways – get mechanical cues from their microenvironment and commit to the lineage specified by matrix elasticity, as highlighted by the findings of Engler et al. [297]. The authors were able to show that Collagen I coated polyacrylamide gels of different stiffnesses stimulate neurogenic differentiation on soft gels even in the absence of differentiation-inducing soluble factors. Medium matrix elasticity proved myogenic and a rigid matrix led to an osteogenic differentiation of MSCs [297]. Similar influences of matrix elasticity of other cultivation substrates on MSC differentiation of other cultivation substrates [298–304], including topography-induced forced cell geometry [305–309] and the interplay of seeding density and matrix stiffness [310] have also been investigated.

Limitations of 3D Cell Culture

As with any other *in vitro* cultivation technique, the 3D cultivation of MSCs also has limitations that have to be considered when choosing a cultivation platform. First of all, the use of MSCs cultured in spheroids or in matrices in regenerative medicine is limited to site-directed injection and cannot be directly applied for systemic diseases that would need intravenous or intraarterial injections without dissociation. For obtaining cell suspensions from 3D cultures, however, the cell-cell and cell-matrix connections must be proteolytically removed first: This can differ from detaching cells that have been grown in monolayers, *e.g.*, due to diffusion limitations of the enzyme into the 3D construct. As discussed before, this enzymatic step may decrease cell viability and overall cell number, and can also lead to proteolysis of surface receptors, thus changing cell physiology [311, 312]. Secondly, there will be mechanical and chemical gradients (including waste product and nutrient gradients) that exist from the centre of the 3D construct to its outer layers. These concentration gradients are likely to show spatial-temporal changes which can potentially alter cell behaviour and will expose the cells to some degree of heterogeneity and are likely to show spatial-temporal changes. Third, the cultivation in 3D will also lead to the formation of oxygen gradients within the construct that can either have beneficial or unwanted effects, as already discussed in the last chapter. Static cultivation of 3D constructs in tissue engineering often does not use active forms of supplemental oxygenation beyond diffusion and *in vitro* vascularisation of artificial tissues remains a major challenge [313]. This limits, for example, the maximum thickness for a viable tissue construct [313]. When developing a 3D cell cultivation system, it is therefore necessary to find a balance between the specific demands of the cells on their environment and the selected cultivation conditions [313].

Concluding Remarks and Future Perspective

In the present chapter we discussed key parameters for MSCs *ex vivo* cultivation. The choice of right media supplements, scale-up strategies, cultivation in adjustable 3D systems and implementation of hypoxia during cultivation of 2D and 3D cell cultures can tremendously improve the quality of final MSCs-based products for regenerative medicine. Until efficient chemically defined serum-free media are successfully established and approved for MSC production and differentiation, hPL can serve as a suitable supplement for xeno-free 2D and 3D cell cultivation. Commercially available chemically defined media will be the best choice in the future, but to date, cultivation in such media needs additional coating of the cell culture surface with proteins – collagen or fibronectin, in order to support cell adhesion.

Bioreactors have high potential for MSCs expansion for future clinical applications since they offer reproducible cultivation conditions and provide a controllable environment. Moreover, process scale up is not possible without the implementation of bioreactor systems. Challenges in the culturing of anchorage-dependent cells in stirred tanks and wave-mixed bioreactors remain to be overcome. The expansion of MSCs under hypoxic conditions can be beneficial to the cells, but the precise oxygen tensions beneficial to MSCs isolated from different sources need to be determined.

A growing body of research supports the theory that cell expansion in 3D systems increases their therapeutic potential. The choice of cultivation conditions should also take into account secondary effects such as an additional limitation of oxygen diffusion due to an accumulation of produced ECM in the tissue construct or an increase in cell numbers over the course of the cultivation. To exploit the positive effects of 3D cultivation options, it is essential to actively monitor of oxygen distribution within the designed 3D constructs. In addition, cell harvesting from 3D culture systems remains a challenge when cells are to be used for cell therapy.

Last but not least, quality control of MSCs for therapeutic use must be considered. In the most common cases, cell morphology, growth rate, karyotype, and sterility are evaluated at the final product stage and during the whole production process [314, 315]. Besides the above listed parameters, valid potency assays must be included in the quality control of MSCs [21]. With all of the above parameters in mind, standard protocols and guidelines for the manufacture of MSCs need to be developed that will enable further progress towards the delivery of truly working advanced therapies.

Funding and Acknowledgement Statement This work was supported by the German Research Foundation (DFG project 398007461 "BiomolecularSensor Platform for Elucidating Hypoxic Signatures in 2D and 3D in vitro culture Systems" and DFG project 464596535 "Ex-vivo modelling of the lymph node microenvironment and its impact on drug response in lymphoma") and by the SMART BIOTECS initiative, which is financially supported by the Lower Saxony Ministry of Science and Culture.

Conflicts of Interest Statement All authors declare that they have no conflicts of interest.

References

1. Friedenstein AJ, Chailakhjan RK, Lalykina KS (1970) The development of fibroblast colonies in monolayer cultures of Guinea-pig bone marrow and spleen cells. Cell Prolif 3:393–403. https://doi.org/10.1111/j.1365-2184.1970.tb00347.x
2. Fan X-LL, Zhang Y, Li X, Fu Q-LL (2020) Mechanisms underlying the protective effects of mesenchymal stem cell-based therapy. Cell Mol Life Sci 77:2771–2794. https://doi.org/10.1007/s00018-020-03454-6
3. U.S. National Library of Medicine (2022) ClinicalTrials.gov – mesenchymal stem cells – list results. https://clinicaltrials.gov/ct2/results?cond=&term=mesenchymal+stem+cells&cntry=&state=&city=&dist=. Accessed 15 Nov 2022
4. Spees JL, Gregory CA, Singh H et al (2004) Internalized antigens must be removed to prepare hypoimmunogenic mesenchymal stem cells for cell and gene therapy. Mol Ther 9:747–756. https://doi.org/10.1016/j.ymthe.2004.02.012
5. Sundin M, Ringdén O, Sundberg B et al (2007) No alloantibodies against mesenchymal stromal cells, but presence of anti-fetal calf serum antibodies, after transplantation in allogeneic hematopoietic stem cell recipients. Haematologica 92:1208–1215. https://doi.org/10.3324/haematol.11446
6. Taylor MW (2015) Viruses and man: a history of interactions. Springer, Cham
7. Silver IA (1975) Measurement of pH and ionic composition of pericellular sites. Philos Trans R Soc Lond B Biol Sci 271:261–272. https://doi.org/10.1098/RSTB.1975.0050
8. Pasarica M, Sereda OR, Redman LM et al (2009) Reduced adipose tissue oxygenation in human obesity evidence for rarefaction, macrophage chemotaxis, and inflammation without an angiogenic response. Diabetes 58:718–725. https://doi.org/10.2337/db08-1098
9. Goossens GH, Bizzarri A, Venteclef N et al (2011) Increased adipose tissue oxygen tension in obese compared with lean men is accompanied by insulin resistance, impaired adipose tissue capillarization, and inflammation. Circulation 124:67–76. https://doi.org/10.1161/CIRCULATIONAHA.111.027813
10. Lawler HM, Underkofler CM, Kern PA et al (2016) Adipose tissue hypoxia, inflammation, and fibrosis in obese insulin-sensitive and obese insulin-resistant subjects. J Clin Endocrinol Metab 101:1422. https://doi.org/10.1210/JC.2015-4125
11. Goossens GH, Vogel MAA, Vink RG et al (2018) Adipose tissue oxygenation is associated with insulin sensitivity independently of adiposity in obese men and women. Diabetes Obes Metab 20:2286–2290. https://doi.org/10.1111/DOM.13329
12. Semenza GL (2007) Life with oxygen. Science 318:62–64
13. Cheng N, Chen S, Li J-R, Young T (2013) Short-term spheroid formation enhances the regenerative capacity of adipose-derived stem cells by promoting stemness, angiogenesis, and chemotaxis. Stem Cells Transl Med 2:584–594. https://doi.org/10.5966/sctm.2013-0007
14. Cesarz Z, Tamama K (2016) Spheroid culture of mesenchymal stem cells. Stem Cell Int 2016:9176357
15. Noronha Nc NDC, Mizukami A, Caliári-Oliveira C et al (2019) Priming approaches to improve the efficacy of mesenchymal stromal cell-based therapies. Stem Cell Res Ther 10(1):132
16. Arora M (2013) Cell culture media: a review. Mater Methods 3. https://doi.org/10.13070/MM.EN.3.175
17. Bieback K, Hecker A, Schlechter T et al (2012) Replicative aging and differentiation potential of human adipose tissue-derived mesenchymal stromal cells expanded in pooled human or fetal bovine serum. Cytotherapy 14:570–583. https://doi.org/10.3109/14653249.2011.652809
18. Schallmoser K, Henschler R, Gabriel C et al (2020) Production and quality requirements of human platelet lysate: a position statement from the working party on cellular therapies of the International Society of Blood Transfusion. Trends Biotechnol 38:13–23. https://doi.org/10.1016/J.TIBTECH.2019.06.002

19. Henschler R, Gabriel C, Schallmoser K et al (2019) Human platelet lysate current standards and future developments. Transfusion 59:1407–1413. https://doi.org/10.1111/trf.15174
20. Lavrentieva A, Hatlapatka T, Winkler R et al (2011) Strategies in umbilical cord-derived mesenchymal stem cells expansion: influence of oxygen, culture medium and cell separation. BMC Proc 5:1–2. https://doi.org/10.1186/1753-6561-5-s8-p88
21. Lavrentieva A, Hoffmann A, Lee-Thedieck C (2020) Limited potential or unfavorable manipulations? Strategies toward efficient mesenchymal stem/stromal cell applications. Front Cell Dev Biol 8:316. https://doi.org/10.3389/fcell.2020.00316
22. Mizukami A, Swiech K (2018, 2018) Mesenchymal stromal cells: from discovery to manufacturing and commercialization. Stem Cells Int. https://doi.org/10.1155/2018/4083921
23. Cimino M, Gonçalves RM, Barrias CC, Martins MCLL (2017) Xeno-free strategies for safe human mesenchymal stem/stromal cell expansion: supplements and coatings. Stem Cells Int 2017:6597815. https://doi.org/10.1155/2017/6597815
24. Tonti GA, Mannello F (2008) From bone marrow to therapeutic applications: different behaviour and genetic/epigenetic stability during mesenchymal stem cell expansion in autologous and foetal bovine sera? Int J Dev Biol 52:1023–1032. https://doi.org/10.1387/ijdb.082725gt
25. Berthois Y, Katzenellenbogen JA, Katzenellenbogen BS (1986) Phenol red in tissue culture media is a weak estrogen: implications concerning the study of estrogen-responsive cells in culture. Proc Natl Acad Sci U S A 83:2496–2500. https://doi.org/10.1073/pnas.83.8.2496
26. Howorth PJN (1975) The physiological assessment of acid-base balance. Br J Dis Chest 69:75–102. https://doi.org/10.1016/0007-0971(75)90061-3
27. Zigler JS, Lepe-Zuniga JL, Vistica B, Gery I (1985) Analysis of the cytotoxic effects of light-exposed HEPES-containing culture medium. Vitr Cell Dev Biol Anim 21:282–287. https://doi.org/10.1007/BF02620943
28. Shipman C (1969) Evaluation of 4-(2-hydroxyethyl)-1-piperazineëthanesulfonic acid (HEPES) as a tissue culture buffer. Proc Soc Exp Biol Med 130:305–310. https://doi.org/10.3181/00379727-130-33543
29. Reznikov BF (1972) Vyrashchivanie brutsell na tverdykh pitatel'nykh sredakh s indikatorom fenolovym krasnym. Veterinariia 7:109–110
30. van der Valk J, Brunner D, De Smet K et al (2010) Optimization of chemically defined cell culture media – replacing fetal bovine serum in mammalian in vitro methods. Toxicol Vitr 24:1053–1063. https://doi.org/10.1016/j.tiv.2010.03.016
31. Wagner W, Bork S, Horn P et al (2009) Aging and replicative senescence have related effects on human stem and progenitor cells. PLoS One 4:e5846. https://doi.org/10.1371/journal.pone.0005846
32. Hladik D, Höfig I, Oestreicher U et al (2019) Long-term culture of mesenchymal stem cells impairs ATM-dependent recognition of DNA breaks and increases genetic instability. Stem Cell Res Ther 10:1–12. https://doi.org/10.1186/s13287-019-1334-6
33. Nuschke A, Rodrigues M, Wells AAW et al (2016) Mesenchymal stem cells/multipotent stromal cells (MSCs) are glycolytic and thus glucose is a limiting factor of in vitro models of MSC starvation. Stem Cell Res Ther 7:1–9
34. Zhang D, Lu H, Chen Z et al (2017) High glucose induces the aging of mesenchymal stem cells via Akt/mTOR signaling. Mol Med Rep 16:1685–1690. https://doi.org/10.3892/mmr.2017.6832
35. Parsch D, Fellenberg J, Brümmendorf TH et al (2004) Telomere length and telomerase activity during expansion and differentiation of human mesenchymal stem cells and chondrocytes. J Mol Med (Berl) 82:49–55. https://doi.org/10.1007/s00109-003-0506-z
36. Eagle H (1955) Nutrition needs of mammalian cells in tissue culture. Science 122:501–504. https://doi.org/10.1126/science.122.3168.501
37. Sotiropoulou PA, Perez SA, Salagianni M et al (2006) Characterization of the optimal culture conditions for clinical scale production of human mesenchymal stem cells. Stem Cells 24:462–471. https://doi.org/10.1634/stemcells.2004-0331

38. Weller TH, Wheeldon SK (1982) The cultivation in vitro of cells derived from adult Schistosoma mansoni. I. Methodology; criteria for evaluation of cultures; and development of media. Am J Trop Med Hyg 31:335–348. https://doi.org/10.4269/AJTMH.1982.31.335
39. Schrödel A (2007) Die Rolle des fetalen Kälberserums in Zellkulturmedien. Biol Unserer Zeit 37:289. https://doi.org/10.1002/biuz.200790079
40. Wagner M, Yoshihara M, Douagi I et al (2020) Single-cell analysis of human ovarian cortex identifies distinct cell populations but no oogonial stem cells. Nat Commun 11:1147. https://doi.org/10.1038/s41467-020-14936-3
41. Lindroos B, Boucher S, Chase L et al (2009) Serum-free, xeno-free culture media maintain the proliferation rate and multipotentiality of adipose stem cells in vitro. Cytotherapy 11:958–972. https://doi.org/10.3109/14653240903233081
42. Bieback K (2013) Platelet lysate as replacement for fetal bovine serum in mesenchymal stromal cell cultures. Transfus Med Hemother 40:326–335. https://doi.org/10.1159/000354061
43. Hemeda H, Giebel B, Wagner W (2014) Evaluation of human platelet lysate versus fetal bovine serum for culture of mesenchymal stromal cells. Cytotherapy 16:170–180. https://doi.org/10.1016/J.JCYT.2013.11.004
44. Burnouf T, Strunk D, Koh MBCC, Schallmoser K (2016) Human platelet lysate: replacing fetal bovine serum as a gold standard for human cell propagation? Biomaterials 76:371–387. https://doi.org/10.1016/j.biomaterials.2015.10.065
45. Monsanto MM, White KS, Kim T et al (2017) Concurrent isolation of 3 distinct cardiac stem cell populations from a single human heart biopsy. Circ Res 121:113–124. https://doi.org/10.1161/CIRCRESAHA.116.310494
46. Motedayyen H, Esmaeil N, Tajik N et al (2017) Method and key points for isolation of human amniotic epithelial cells with high yield, viability and purity. BMC Res Notes 10:1–8. https://doi.org/10.1186/s13104-017-2880-6
47. Araújo AB, Furlan JM, Salton GD et al (2018) Isolation of human mesenchymal stem cells from amnion, chorion, placental decidua and umbilical cord: comparison of four enzymatic protocols. Biotechnol Lett 40:989–998. https://doi.org/10.1007/s10529-018-2546-z
48. Lee MS, Wang J, Yuan H et al (2019) Endothelin-1 differentially directs lineage specification of adipose- and bone marrow–derived mesenchymal stem cells. FASEB J 33:996–1007. https://doi.org/10.1096/FJ.201800614R
49. Cherian DS, Bhuvan T, Meagher L, Heng TSP (2020) Biological considerations in scaling up therapeutic cell manufacturing. Front Pharmacol 11:654
50. Ghamari SH, Abbasi-Kangevari M, Tayebi T et al (2020) The bottlenecks in translating placenta-derived amniotic epithelial and mesenchymal stromal cells into the clinic: current discrepancies in marker reports. Front Bioeng Biotechnol 8:180
51. Brooke G, Rossetti T, Pelekanos R et al (2009) Manufacturing of human placenta-derived mesenchymal stem cells for clinical trials. Br J Haematol 144:571–579. https://doi.org/10.1111/j.1365-2141.2008.07492.x
52. Hesari R, Keshvarinia M, Kabiri M et al (2020) Comparative impact of platelet rich plasma and transforming growth factor-β on chondrogenic differentiation of human adipose derived stem cells. Bioimpacts 10:37–43. https://doi.org/10.15171/bi.2020.05
53. Kotobuki N, Hirose M, Takakura Y, Ohgushi H (2004) Cultured autologous human cells for hard tissue regeneration: preparation and characterization of mesenchymal stem cells from bone marrow. Artif Organs 28:33–39. https://doi.org/10.1111/j.1525-1594.2004.07320.x
54. Ling L, Ren X, Cao X et al (2020) Enhancing the efficacy of stem cell therapy with glycosaminoglycans. Stem Cell Rep 14:105–121. https://doi.org/10.1016/j.stemcr.2019.12.003
55. Jonsdottir-Buch SM, Lieder R, Sigurjonsson OE (2013) Platelet lysates produced from expired platelet concentrates support growth and osteogenic differentiation of mesenchymal stem cells. PLoS One 8:e68984. https://doi.org/10.1371/journal.pone.0068984

56. Mastrolia I, Foppiani EM, Murgia A et al (2019) Challenges in clinical development of mesenchymal stromal/stem cells: concise review. Stem Cells Transl Med 8:1135–1148. https://doi.org/10.1002/SCTM.19-0044
57. Friedenstein AJ, Piatetzky-Shapiro II, Petrakova KV (1966) Osteogenesis in transplants of bone marrow cells. J Embryol Exp Morphol 16:381–390
58. Gstraunthaler G, Lindl T, van der Valk J (2013) A plea to reduce or replace fetal bovine serum in cell culture media. Cytotechnology 65:791–793. https://doi.org/10.1007/s10616-013-9633-8
59. Lee M-SS, Youn C, Kim JH et al (2017) Enhanced cell growth of adipocyte-derived mesenchymal stem cells using chemically-defined serum-free media. Int J Mol Sci 18:1779. https://doi.org/10.3390/ijms18081779
60. Astori G, Amati E, Bambi F et al (2016) Platelet lysate as a substitute for animal serum for the ex-vivo expansion of mesenchymal stem/stromal cells: present and future. Stem Cell Res Ther 7:93. https://doi.org/10.1186/s13287-016-0352-x
61. Rauch C, Feifel E, Amann E-MM et al (2011) Alternatives to the use of fetal bovine serum: human platelet lysates as a serum substitute in cell culture media. ALTEX 28:305–316. https://doi.org/10.14573/altex.2011.4.305
62. Pittenger MF, Mackay AM, Beck SC et al (1999) Multilineage potential of adult human mesenchymal stem cells. Science 284:143–147. https://doi.org/10.1126/science.284.5411.143
63. Van Der Valk J, Mellor D, Brands R et al (2004) The humane collection of fetal bovine serum and possibilities for serum-free cell and tissue culture. Toxicol In Vitro 18:1–12
64. Jochems CEA, Van der Valk JBF, Stafleu FR, Baumans V (2002) The use of fetal bovine serum: ethical or scientific problem? Altern Lab Anim 30:219–227. https://doi.org/10.1177/026119290203000208
65. Lepperdinger G, Brunauer R, Jamnig A et al (2008) Controversial issue: is it safe to employ mesenchymal stem cells in cell-based therapies? Exp Gerontol 43:1018–1023. https://doi.org/10.1016/J.EXGER.2008.07.004
66. Martin MJ, Muotri A, Gage F, Varki A (2005) Human embryonic stem cells express an immunogenic nonhuman sialic acid. Nat Med 11:228–232. https://doi.org/10.1038/nm1181
67. Heiskanen A, Satomaa T, Tiitinen S et al (2007) N-glycolylneuraminic acid xenoantigen contamination of human embryonic and mesenchymal stem cells is substantially reversible. Stem Cells 25:197–202. https://doi.org/10.1634/stemcells.2006-0444
68. Nikfarjam L, Farzaneh P (2012) Prevention and detection of mycoplasma contamination in cell culture. Cell J 13:203–212
69. Pinheiro de Oliveira TF, Fonseca AÔA, Camargos MF et al (2013) Detection of contaminants in cell cultures, sera and trypsin. Biologicals 41:407–414
70. Rottem S, Barile MF (1993) Beware of mycoplasmas. Trends Biotechnol 11(4):143–151
71. De Sousa PA, Galea G, Turner M (2006) The road to providing human embryo stem cells for therapeutic use: the UK experience. Reproduction 132:681–689. https://doi.org/10.1530/rep.1.01080
72. Windsor HM, Windsor GD, Noordergraaf JH (2010) The growth and long term survival of Acholeplasma laidlawii in media products used in biopharmaceutical manufacturing. Biologicals 38:204–210. https://doi.org/10.1016/j.biologicals.2009.11.009
73. Wessmann SJ, Levings RLL, Wessman SJ, Levings RLL (1999) Benefits and risks due to animal serum used in cell culture production. Dev Biol Stand 99:3–8
74. Shih DT-BB, Burnouf T (2015) Preparation, quality criteria, and properties of human blood platelet lysate supplements for ex vivo stem cell expansion. New Biotechnol 32:199–211. https://doi.org/10.1016/j.nbt.2014.06.001
75. Bjare U (1992) Serum-free cell culture. Pharmacol Ther 53:355–374. https://doi.org/10.1016/0163-7258(92)90056-6
76. Kyllönen L, Haimi S, Mannerström B et al (2013) Effects of different serum conditions on osteogenic differentiation of human adipose stem cells in vitro. Stem Cell Res Ther 4:1–15. https://doi.org/10.1186/scrt165

77. Zheng X, Baker H, Hancock WS et al (2006) Proteomic analysis for the assessment of different lots of fetal bovine serum as a raw material for cell culture. Part IV. Application of proteomics to the manufacture of biological drugs. Biotechnol Prog 22:1294–1300. https://doi.org/10.1021/bp060121o
78. Knepper PA, Mayanil CS, Goossens W et al (1998) The presence of transcription factors in fetal bovine sera. Vitr Cell Dev Biol Anim 34:170–173. https://doi.org/10.1007/s11626-998-0101-2
79. Wappler J, Rath B, Läufer T et al (2013) Eliminating the need of serum testing using low serum culture conditions for human bone marrow-derived mesenchymal stromal cell expansion. Biomed Eng Online 12:1–10. https://doi.org/10.1186/1475-925X-12-15
80. Stein A (2007) Decreasing variability in your cell culture. BioTechniques 43:228–229. https://doi.org/10.2144/000112561
81. Mannello F, Tonti GA (2007) Concise review: no breakthroughs for human mesenchymal and embryonic stem cell culture: conditioned medium, feeder layer, or feeder-free; medium with fetal calf serum, human serum, or enriched plasma; serum-free, serum replacement nonconditioned medium. Stem Cells 25:1603–1609. https://doi.org/10.1634/stemcells.2007-0127
82. Tateishi K, Ando W, Higuchi C et al (2008) Comparison of human serum with fetal bovine serum for expansion and differentiation of human synovial MSC: potential feasibility for clinical applications. Cell Transplant 17:549–557. https://doi.org/10.3727/096368908785096024
83. Aldahmash A, Haack-Sørensen M, Al-Nbaheen M et al (2011) Human serum is as efficient as fetal bovine serum in supporting proliferation and differentiation of human multipotent stromal (mesenchymal) stem cells in vitro and in vivo. Stem Cell Rev Rep 7:860–868. https://doi.org/10.1007/s12015-011-9274-2
84. Müller I, Kordowich S, Holzwarth C et al (2006) Animal serum-free culture conditions for isolation and expansion of multipotent mesenchymal stromal cells from human BM. Cytotherapy 8:437–444. https://doi.org/10.1080/14653240600920782
85. Kocaoemer A, Kern S, Klüter H et al (2007) Human AB serum and thrombin-activated platelet-rich plasma are suitable alternatives to fetal calf serum for the expansion of mesenchymal stem cells from adipose tissue. Stem Cells 25:1270–1278. https://doi.org/10.1634/stemcells.2006-0627
86. Bieback K, Hecker A, Kocaömer A et al (2009) Human alternatives to fetal bovine serum for the expansion of mesenchymal stromal cells from bone marrow. Stem Cells 27:2331–2341. https://doi.org/10.1002/stem.139
87. Lensch M, Muise A, White L et al (2018) Comparison of synthetic media designed for expansion of adipose-derived mesenchymal stromal cells. Biomedicine 6:54. https://doi.org/10.3390/biomedicines6020054
88. Karadjian M, Senger A-SS, Essers C et al (2020) Human platelet lysate can replace fetal calf serum as a protein source to promote expansion and osteogenic differentiation of human bone-marrow-derived mesenchymal stromal cells. Cells 9:918. https://doi.org/10.3390/cells9040918
89. Kakudo N, Morimoto N, Ma Y, Kusumoto K (2019) Differences between the proliferative effects of human platelet lysate and fetal bovine serum on human adipose-derived stem cells. Cells 8:1218. https://doi.org/10.3390/cells8101218
90. Bandeiras C, Koc JRR, Ma Y et al (2018) Cost effectiveness analysis of allogeneic, just-in-time expansion of mesenchymal stem cells with PLUS™ human platelet lysate for a clinical trial. Cytotherapy 20:S60. https://doi.org/10.1016/j.jcyt.2018.02.166
91. Palombella S, Guiotto M, Higgins GC et al (2020) Human platelet lysate as a potential clinical-translatable supplement to support the neurotrophic properties of human adipose-derived stem cells. Stem Cell Res Ther 11:432. https://doi.org/10.1186/s13287-020-01949-4

92. Escobar CH, Chaparro O (2016) Xeno-free extraction, culture, and cryopreservation of human adipose-derived mesenchymal stem cells. Stem Cells Transl Med 5:358–365. https://doi.org/10.5966/SCTM.2015-0094
93. Guiotto M, Raffoul W, Hart AM et al (2020) Human platelet lysate to substitute fetal bovine serum in hMSC expansion for translational applications: a systematic review. J Transl Med 18: 1–14
94. Doucet C, Ernou I, Zhang Y et al (2005) Platelet lysates promote mesenchymal stem cell expansion: a safety substitute for animal serum in cell-based therapy applications. J Cell Physiol 205:228–236. https://doi.org/10.1002/jcp.20391
95. Mojica-Henshaw MP, Jacobson P, Morris J et al (2013) Serum-converted platelet lysate can substitute for fetal bovine serum in human mesenchymal stromal cell cultures. Cytotherapy 15: 1458–1468. https://doi.org/10.1016/J.JCYT.2013.06.014
96. Strunk D, Lozano M, Marks DC et al (2018) International forum on GMP-grade human platelet lysate for cell propagation. Vox Sang 113:e1–e25. https://doi.org/10.1111/vox.12594
97. Whitaker B, Rajbhandary S, Kleinman S et al (2016) Trends in United States blood collection and transfusion: results from the 2013 AABB Blood Collection, Utilization, and Patient Blood Management Survey. Transfusion 56:2173–2183. https://doi.org/10.1111/TRF.13676
98. Schallmoser K, Strunk D (2013) Generation of a pool of human platelet lysate and efficient use in cell culture. Methods Mol Biol 946:349–362. https://doi.org/10.1007/978-1-62703-128-8_22
99. Lesurtel M, Graf R, Aleil B et al (2006) Platelet-derived serotonin mediates liver regeneration. Science 312:104–107. https://doi.org/10.1126/science.1123842
100. Lesurtel M, Clavien PA (2014) Platelet-derived serotonin: translational implications for liver regeneration. Hepatology 60:30–33. https://doi.org/10.1002/hep.27067
101. Myronovych A, Murata S, Chiba M et al (2008) Role of platelets on liver regeneration after 90% hepatectomy in mice. J Hepatol 49:363–372. https://doi.org/10.1016/J.JHEP.2008.04.019
102. Blair P, Flaumenhaft R (2009) Platelet alpha-granules: basic biology and clinical correlates. Blood Rev 23:177–189. https://doi.org/10.1016/J.BLRE.2009.04.001
103. Semple JW, Italiano JE, Freedman J (2011) Platelets and the immune continuum. Nat Rev Immunol 11:264–274. https://doi.org/10.1038/nri2956
104. Italiano JE, Richardson JL, Patel-Hett S et al (2008) Angiogenesis is regulated by a novel mechanism: pro- and antiangiogenic proteins are organized into separate platelet alpha granules and differentially released. Blood 111:1227–1233. https://doi.org/10.1182/blood-2007-09-113837
105. Kamykowski J, Carlton P, Sehgal S, Storrie B (2011) Quantitative immunofluorescence mapping reveals little functional coclustering of proteins within platelet α-granules. Blood 118:1370–1373. https://doi.org/10.1182/blood-2011-01-330910
106. Scully D, Naseem KM, Matsakas A (2018) Platelet biology in regenerative medicine of skeletal muscle. Acta Physiol 223. https://doi.org/10.1111/apha.13071
107. Yang J, Zhang YS, Yue K, Khademhosseini A (2017) Cell-laden hydrogels for osteochondral and cartilage tissue engineering. Acta Biomater 57:1–25. https://doi.org/10.1016/j.actbio.2017.01.036
108. Reible B, Schmidmaier G, Moghaddam A, Westhauser F (2018) Insulin-like growth factor-1 as a possible alternative to bone morphogenetic protein-7 to induce osteogenic differentiation of human mesenchymal stem cells in vitro. Int J Mol Sci 19:1674. https://doi.org/10.3390/ijms19061674
109. Mizuno M, Katano H, Otabe K et al (2015) Platelet-derived growth factor (PDGF)-AA/AB in human serum are potential indicators of the proliferative capacity of human synovial mesenchymal stem cells. Stem Cell Res Ther 6:1–11. https://doi.org/10.1186/s13287-015-0239-2
110. Kore-Grodzicki B, Tauber-Finkelstein M, Chain D, Shaltiel S (1988) Vitronectin is phosphorylated by a cAMP-dependent protein kinase released by activation of human platelets with

thrombin. Biochem Biophys Res Commun 157:1131–1138. https://doi.org/10.1016/S0006-291X(88)80991-4
111. Sander HJ, Slot JW, Bouma BN et al (1983) Immunocytochemical localization of fibrinogen, platelet factor 4, and beta thromboglobulin in thin frozen sections of human blood platelets. J Clin Invest 72:1277–1287. https://doi.org/10.1172/JCI111084
112. Heijnen H, van der Sluijs P (2015) Platelet secretory behaviour: as diverse as the granules . . . or not? J Thromb Haemost 13:2141–2151. https://doi.org/10.1111/jth.13147
113. Wencel-Drake JD, Painter RG, Zimmerman TS, Ginsberg MH (1985) Ultrastructural localization of human platelet thrombospondin, fibrinogen, fibronectin, and von Willebrand factor in frozen thin section. Blood 65:929–938. https://doi.org/10.1182/blood.v65.4.929.929
114. Egger D, Oliveira AC, Mallinger B et al (2019) From 3D to 3D: isolation of mesenchymal stem/stromal cells into a three-dimensional human platelet lysate matrix. Stem Cell Res Ther 10:248. https://doi.org/10.1186/s13287-019-1346-2
115. Kirsch M, Herder A-CC, Boudot C et al (2020) Xeno-free in vitro cultivation and osteogenic differentiation of hAD-MSCs on resorbable 3D printed RESOMER®. Materials (Basel) 13: 3399. https://doi.org/10.3390/ma13153399
116. Kirsch M, Rach J, Handke W et al (2020) Comparative analysis of mesenchymal stem cell cultivation in fetal calf serum, human serum, and platelet lysate in 2D and 3D systems. Front Bioeng Biotechnol 8:598389. https://doi.org/10.3389/fbioe.2020.598389
117. Kirsch M, Birnstein L, Pepelanova I et al (2019) Gelatin-methacryloyl (GelMA) formulated with human platelet lysate supports mesenchymal stem cell proliferation and differentiation and enhances the hydrogel's mechanical properties. Bioengineering 6:76. https://doi.org/10.3390/bioengineering6030076
118. Hara Y, Steiner M, Baldini MG (1980) Platelets as a source of growth-promoting factor(s) for tumor cells. Cancer Res 40:1212–1216
119. Umeno Y, Okuda A, Kimura G (1989) Proliferative behaviour of fibroblasts in plasma-rich culture medium. J Cell Sci 94(Pt 3):567–575. https://doi.org/10.1242/JCS.94.3.567
120. King GL, Buchwald S (1984) Characterization and partial purification of an endothelial cell growth factor from human platelets. J Clin Invest 73:392–396. https://doi.org/10.1172/JCI111224
121. Aldén A, Gonzalez L, Persson A et al (2007) Porcine platelet lysate as a supplement for animal cell culture. Cytotechnology 55:3–8. https://doi.org/10.1007/s10616-007-9097-9
122. Crespo-Diaz R, Behfar A, Butler GW et al (2011) Platelet lysate consisting of a natural repair proteome supports human mesenchymal stem cell proliferation and chromosomal stability. Cell Transplant 20:797–811. https://doi.org/10.3727/096368910X543376
123. Bieback K, Fernandez-Muñoz B, Pati S, Schäfer R (2019) Gaps in the knowledge of human platelet lysate as a cell culture supplement for cell therapy: a joint publication from the AABB and the International Society for Cell & Gene Therapy. Transfusion 59:3448–3460. https://doi.org/10.1111/trf.15483
124. Stanco D, Caprara C, Ciardelli G et al (2019) Tenogenic differentiation protocol in xenogenic-free media enhances tendon-related marker expression in ASCs. PLoS One 14:e0212192. https://doi.org/10.1371/journal.pone.0212192
125. Altaie A, Owston H, Jones E (2016) Use of platelet lysate for bone regeneration – are we ready for clinical translation? World J Stem Cells 8:47–55. https://doi.org/10.4252/wjsc.v8.i2.47
126. Hartung T (2007) Food for thought... on cell culture. ALTEX 24:143–152. https://doi.org/10.14573/altex.2007.3.143
127. Chase LG, Lakshmipathy U, Solchaga LA et al (2010) A novel serum-free medium for the expansion of human mesenchymal stem cells. Stem Cell Res Ther 1:8. https://doi.org/10.1186/scrt8
128. Coecke S, Balls M, Bowe G et al (2005) Guidance on good cell culture practice: a report of the second ECVAM task force on good cell culture practice. Altern Lab Anim 33:261–287. https://doi.org/10.1177/026119290503300313/ASSET/026119290503300313.FP.PNG_V03

129. Hartung T, Balls M, Bardouille C et al (2002) Good cell culture practice: ECVAM good cell culture practice task force report 1. Altern Lab Anim 30:407–414. https://doi.org/10.1177/026119290203000404
130. Gupta K, Rispin A, Stitzel K et al (2005) Ensuring quality of in vitro alternative test methods: issues and answers. Regul Toxicol Pharmacol 43:219–224. https://doi.org/10.1016/J.YRTPH.2005.03.010
131. Tekkatte C, Gunasingh GP, Cherian KM, Sankaranarayanan K (2011) "Humanized" stem cell culture techniques: the animal serum controversy. Stem Cells Int 2011. https://doi.org/10.4061/2011/504723
132. Brunner D, Frank J, Appl H et al (2010) The serum-free media interactive online database. ALTEX 27:53–62. https://doi.org/10.14573/ALTEX.2010.1.53
133. Dolley-Sonneville PJ, Romeo LE, Melkoumian ZK (2013) Synthetic surface for expansion of human mesenchymal stem cells in xeno-free, chemically defined culture conditions. PLoS One 8:e70263. https://doi.org/10.1371/journal.pone.0070263
134. Mark P, Kleinsorge M, Gaebel R et al (2013) Human mesenchymal stem cells display reduced expression of CD105 after culture in serum-free medium. Stem Cells Int 2013:698076. https://doi.org/10.1155/2013/698076
135. Jung S, Sen A, Rosenberg L, Behie LA (2010) Identification of growth and attachment factors for the serum-free isolation and expansion of human mesenchymal stromal cells. Cytotherapy 12:637–657. https://doi.org/10.3109/14653249.2010.495113
136. Barrias CC, Martins MCL, Almeida-Porada G et al (2009) The correlation between the adsorption of adhesive proteins and cell behaviour on hydroxyl-methyl mixed self-assembled monolayers. Biomaterials 30:307–316. https://doi.org/10.1016/j.biomaterials.2008.09.048
137. Dånmark S, Finne-Wistrand A, Albertsson A-CC et al (2012) Integrin-mediated adhesion of human mesenchymal stem cells to extracellular matrix proteins adsorbed to polymer surfaces. Biomed Mater 7:35011. https://doi.org/10.1088/1748-6041/7/3/035011
138. Esch MB, Ueno H, Applegate DR, Shuler ML (2016) Modular, pumpless body-on-a-chip platform for the co-culture of GI tract epithelium and 3D primary liver tissue. Lab Chip 16: 2719–2729. https://doi.org/10.1039/c6lc00461j
139. Tan KY, Reuveny S, Oh SKW (2016) Recent advances in serum-free microcarrier expansion of mesenchymal stromal cells: parameters to be optimized. Biochem Biophys Res Commun 473:769–773. https://doi.org/10.1016/j.bbrc.2015.09.078
140. Leber J, Barekzai J, Blumenstock M et al (2017) Microcarrier choice and bead-to-bead transfer for human mesenchymal stem cells in serum-containing and chemically defined media. Process Biochem 59:255–265. https://doi.org/10.1016/J.PROCBIO.2017.03.017
141. Heathman TR, Nienow AW, McCall MJ et al (2015) The translation of cell-based therapies: clinical landscape and manufacturing challenges. Regen Med 10:49–64
142. Jossen V, van den Bos C, Eibl R, Eibl D (2018) Manufacturing human mesenchymal stem cells at clinical scale: process and regulatory challenges. Appl Microbiol Biotechnol 102: 3981–3994. https://doi.org/10.1007/s00253-018-8912-x
143. Al-Saqi SH, Saliem M, Asikainen S et al (2014) Defined serum-free media for in vitro expansion of adipose-derived mesenchymal stem cells. Cytotherapy 16:915–926. https://doi.org/10.1016/J.JCYT.2014.02.006
144. Gottipamula S, Muttigi MS, Chaansa S et al (2016) Large-scale expansion of pre-isolated bone marrow mesenchymal stromal cells in serum-free conditions. J Tissue Eng Regen Med 10: 108–119. https://doi.org/10.1002/term.1713
145. Martin I, Wendt D, Heberer M (2004) The role of bioreactors in tissue engineering. Trends Biotechnol 22:80–86
146. Thomas RJ, Chandra A, Liu Y et al (2007) Manufacture of a human mesenchymal stem cell population using an automated cell culture platform. Cytotechnology 55:31–39. https://doi.org/10.1007/s10616-007-9091-2

147. Thomas RJ, Anderson D, Chandra A et al (2009) Automated, scalable culture of human embryonic stem cells in feeder-free conditions. Biotechnol Bioeng 102:1636–1644. https://doi.org/10.1002/bit.22187
148. Campbell A, Brieva T, Raviv L et al (2015) Concise review: process development considerations for cell therapy. Stem Cells Transl Med 4:1155–1163. https://doi.org/10.5966/sctm.2014-0294
149. Jung S, Panchalingam KM, Wuerth RD et al (2012) Large-scale production of human mesenchymal stem cells for clinical applications. Biotechnol Appl Biochem 59:106–120. https://doi.org/10.1002/bab.1006
150. Bodiou V, Moutsatsou P, Post MJ (2020) Microcarriers for upscaling cultured meat production. Front Nutr 7:10. https://doi.org/10.3389/FNUT.2020.00010/BIBTEX
151. Tozetti PA, Caruso SR, Mizukami A et al (2017) Expansion strategies for human mesenchymal stromal cells culture under xeno-free conditions. Biotechnol Prog 33:1358–1367. https://doi.org/10.1002/btpr.2494
152. Jossen V, Pörtner R, Kaiser SC et al (2014) Mass production of mesenchymal stem cells—impact of bioreactor design and flow conditions on proliferation and differentiation. In: Eberli D (ed) Cells and biomaterials in regenerative medicine. IntechOpen, Rijeka
153. Reichardt A, Polchow B, Shakibaei M et al (2013) Large scale expansion of human umbilical cord cells in a rotating bed system bioreactor for cardiovascular tissue engineering applications. Open Biomed Eng J 7:50–61. https://doi.org/10.2174/1874120701307010050
154. Das R, Roosloot R, van Pel M et al (2019) Preparing for cell culture scale-out: establishing parity of bioreactor- and flask-expanded mesenchymal stromal cell cultures. J Transl Med 17: 241. https://doi.org/10.1186/s12967-019-1989-x
155. Stathopoulos NA, Hellums JD (1985) Shear stress effects on human embryonic kidney cells in vitro. Biotechnol Bioeng 27:1021–1026. https://doi.org/10.1002/bit.260270713
156. dos Santos F, Campbell A, Fernandes-Platzgummer A et al (2014) A xenogeneic-free bioreactor system for the clinical-scale expansion of human mesenchymal stem/stromal cells. Biotechnol Bioeng 111:1116–1127
157. Borys BS, Dang T, So T et al (2021) Overcoming bioprocess bottlenecks in the large-scale expansion of high-quality hiPSC aggregates in vertical-wheel stirred suspension bioreactors. Stem Cell Res Ther 12:1–19. https://doi.org/10.1186/s13287-020-02109-4
158. Hundt B, Best C, Schlawin N et al (2007) Establishment of a mink enteritis vaccine production process in stirred-tank reactor and wave® bioreactor microcarrier culture in 1-10 L scale. Vaccine 25:3987–3995. https://doi.org/10.1016/j.vaccine.2007.02.061
159. Genzel Y, Dietzsch C, Rapp E et al (2010) MDCK and Vero cells for influenza virus vaccine production: a one-to-one comparison up to lab-scale bioreactor cultivation. Appl Microbiol Biotechnol 88:461–475. https://doi.org/10.1007/s00253-010-2742-9
160. Slivac I, Gaurina Srček V, Radošević K et al (2006) Aujeszky's disease virus production in disposable bioreactor. J Biosci 31:363–368. https://doi.org/10.1007/BF02704109
161. Hami LS, Green C, Leshinsky N et al (2004) GMP production and testing of Xcellerated T Cells™ for the treatment of patients with CLL. Cytotherapy 6:554–562. https://doi.org/10.1080/14653240410005348
162. Genzel Y, Behrendt I, König S et al (2004) Metabolism of MDCK cells during cell growth and influenza virus production in large-scale microcarrier culture. Vaccine 22:2202–2208. https://doi.org/10.1016/j.vaccine.2003.11.041
163. Rafiq QA, Brosnan KM, Coopman K et al (2013) Culture of human mesenchymal stem cells on microcarriers in a 5 l stirred-tank bioreactor. Biotechnol Lett 35:1233–1245. https://doi.org/10.1007/s10529-013-1211-9
164. Lawson T, Kehoe DE, Schnitzler AC et al (2017) Process development for expansion of human mesenchymal stromal cells in a 50 L single-use stirred tank bioreactor. Biochem Eng J 120:49–62. https://doi.org/10.1016/j.bej.2016.11.020
165. Freimark D, Pino-Grace PP, Pohl S et al (2010) Use of encapsulated stem cells to overcome the bottleneck of cell availability for cell therapy approaches. Transfus Med Hemother 37:66–73

166. Badenes SM, Fernandes TG, Rodrigues CAV et al (2016) Microcarrier-based platforms for in vitro expansion and differentiation of human pluripotent stem cells in bioreactor culture systems. J Biotechnol 234:71–82
167. Potapova IA, Brink PR, Cohen IS, Doronin SV (2008) Culturing of human mesenchymal stem cells as three-dimensional aggregates induces functional expression of CXCR4 that regulates adhesion to endothelial cells. J Biol Chem 283:13100–13107. https://doi.org/10.1074/jbc.M800184200
168. Sart S, Schneider YJ, Agathos SN (2009) Ear mesenchymal stem cells: an efficient adult multipotent cell population fit for rapid and scalable expansion. J Biotechnol 139:291–299. https://doi.org/10.1016/j.jbiotec.2008.12.011
169. Rubin JP, Bennett JM, Doctor JS et al (2007) Collagenous microbeads as a scaffold for tissue engineering with adipose-derived stem cells. Plast Reconstr Surg 120:414–424. https://doi.org/10.1097/01.prs.0000267699.99369.a8
170. Yang HS, Jeon O, Bhang SH et al (2010) Suspension culture of mammalian cells using thermosensitive microcarrier that allows cell detachment without proteolytic enzyme treatment. Cell Transplant 19:1123–1132. https://doi.org/10.3727/096368910X516664
171. Berrier AL, Yamada KM (2007) Cell-matrix adhesion. J Cell Physiol 213:565–573
172. Kleinman HK, McGoodwin EB, Rennard SI, Martin GR (1979) Preparation of collagen substrates for cell attachment: effect of collagen concentration and phosphate buffer. Anal Biochem 94:308–312. https://doi.org/10.1016/0003-2697(79)90365-8
173. Harunaga JS, Yamada KM (2011) Cell-matrix adhesions in 3D. Matrix Biol 30:363–368. https://doi.org/10.1016/j.matbio.2011.06.001
174. Davidenko N, Schuster CF, Bax DV et al (2016) Evaluation of cell binding to collagen and gelatin: a study of the effect of 2D and 3D architecture and surface chemistry. J Mater Sci Mater Med 27:148. https://doi.org/10.1007/s10856-016-5763-9
175. Halme DG, Kessler DA (2006) FDA regulation of stem-cell-based therapies. N Engl J Med 355:1730–1735. https://doi.org/10.1056/nejmhpr063086
176. Rafiq QA, Coopman K, Nienow AW, Hewitt CJ (2016) Systematic microcarrier screening and agitated culture conditions improves human mesenchymal stem cell yield in bioreactors. Biotechnol J 11:473–486. https://doi.org/10.1002/biot.201400862
177. Chen AKL, Reuveny S, Oh SKW (2013) Application of human mesenchymal and pluripotent stem cell microcarrier cultures in cellular therapy: achievements and future direction. Biotechnol Adv 31:1032–1046
178. Elseberg CL, Leber J, Salzig D et al (2012) Microcarrier-based expansion process for hMSCs with high vitality and undifferentiated characteristics. Int J Artif Organs 35:93–107. https://doi.org/10.5301/ijao.5000077
179. Thanos C, Emerich D (2008) On the use of hydrogels in cell encapsulation and tissue engineering systems. Recent Pat Drug Deliv Formul 2:19–24. https://doi.org/10.2174/187221108783331384
180. Martin Y, Vermette P (2005) Bioreactors for tissue mass culture: design, characterization, and recent advances. Biomaterials 26:7481–7503. https://doi.org/10.1016/j.biomaterials.2005.05.057
181. Warnock JN, Bratch K, Al-Rubeai M (2005) Packed bed bioreactors. In: Bioreactors for tissue engineering: principles, design and operation. Springer, Dordrecht, pp 87–113
182. Cunha B, Aguiar T, Carvalho SB et al (2017) Bioprocess integration for human mesenchymal stem cells: from up to downstream processing scale-up to cell proteome characterization. J Biotechnol 248:87–98. https://doi.org/10.1016/j.jbiotec.2017.01.014
183. Higuera GA, van Boxtel A, van Blitterswijk CA, Moroni L (2012) The physics of tissue formation with mesenchymal stem cells. Trends Biotechnol 30:583–590
184. Yeatts AB, Choquette DT, Fisher JP (2013) Bioreactors to influence stem cell fate: augmentation of mesenchymal stem cell signaling pathways via dynamic culture systems. Biochim Biophys Acta 1830:2470–2480. https://doi.org/10.1016/j.bbagen.2012.06.007

185. Adamo L, García-Cardeña G (2011) Directed stem cell differentiation by fluid mechanical forces. Antioxid Redox Signal 15:1463–1473. https://doi.org/10.1089/ars.2011.3907
186. Prockop DJ, Brenner M, Fibbe WE et al (2010) Defining the risks of mesenchymal stromal cell therapy. Cytotherapy 12:576–578. https://doi.org/10.3109/14653249.2010.507330
187. Prockop DJ, Oh JY (2012) Medical therapies with adult stem/progenitor cells (MSCs): a backward journey from dramatic results in vivo to the cellular and molecular explanations. J Cell Biochem 113:1460–1469. https://doi.org/10.1002/jcb.24046
188. Luo W, Xiong W, Zhou J et al (2011) Laminar shear stress delivers cell cycle arrest and anti-apoptosis to mesenchymal stem cells. Acta Biochim Biophys Sin Shanghai 43:210–216. https://doi.org/10.1093/abbs/gmr004
189. Maul TM, Chew DW, Nieponice A, Vorp DA (2011) Mechanical stimuli differentially control stem cell behavior: morphology, proliferation, and differentiation. Biomech Model Mechanobiol 10:939–953. https://doi.org/10.1007/s10237-010-0285-8
190. Barron MJ, Tsai CJ, Donahue SW (2010) Mechanical stimulation mediates gene expression in MC3T3 osteoblastic cells differently in 2D and 3D environments. J Biomech Eng 132. https://doi.org/10.1115/1.4001162
191. Grellier M, Bareille R, Bourget C, Amédée J (2009) Responsiveness of human bone marrow stromal cells to shear stress. J Tissue Eng Regen Med 3:302–309. https://doi.org/10.1002/term.166
192. Alves da Silva ML, Martins A, Costa-Pinto AR et al (2011) Chondrogenic differentiation of human bone marrow mesenchymal stem cells in chitosan-based scaffolds using a flow-perfusion bioreactor. J Tissue Eng Regen Med 5:722–732. https://doi.org/10.1002/term.372
193. Wescoe KE, Schugar RC, Chu CR, Deasy BM (2008) The role of the biochemical and biophysical environment in chondrogenic stem cell differentiation assays and cartilage tissue engineering. Cell Biochem Biophys 52:85–102
194. Hassan S, Simaria AS, Varadaraju H et al (2015) Allogeneic cell therapy bioprocess economics and optimization: downstream processing decisions. Regen Med 10:591–609. https://doi.org/10.2217/rme.15.29
195. Semenza GL (2000) HIF-1: mediator of physiological and pathophysiological responses to hypoxia. J Appl Physiol 88:1474–1480
196. Wang GL, Semenza GL (1995) Purification and characterization of hypoxia-inducible factor. J Biol Chem 270:1230–1237. https://doi.org/10.1074/jbc.270.3.1230
197. Wang GL, Semenza GL (1993) Characterization of hypoxia-inducible factor 1 and regulation of DNA binding activity by hypoxia. J Biol Chem 268:21513–21518. https://doi.org/10.1016/s0021-9258(20)80571-7
198. Wiener CM, Booth G, Semenza GL (1996) In vivo expression of mRNAs encoding hypoxia-inducible factor 1. Biochem Biophys Res Commun 225:485–488. https://doi.org/10.1006/bbrc.1996.1199
199. Jain S, Maltepe E, Lu MM et al (1998) Expression of ARNT, ARNT2, HIF1α, HIF2α and Ah receptor mRNAs in the developing mouse. Mech Dev 73:117–123. https://doi.org/10.1016/S0925-4773(98)00038-0
200. Shen C, Nettleton D, Jiang M et al (2005) Roles of the HIF-1 hypoxia-inducible factor during hypoxia response in Caenorhabditis elegans. J Biol Chem 280:20580–20588. https://doi.org/10.1074/JBC.M501894200/ATTACHMENT/479A7C9B-DF5B-459A-A597-71BF5D0986C1/MMC1.PDF
201. Jewell UR, Kvietikova I, Scheid A et al (2001) Induction of HIF–1α in response to hypoxia is instantaneous. FASEB J 15:1312–1314. https://doi.org/10.1096/fj.00-0732fje
202. Görlach A, Camenisch G, Kvietikova I et al (2000) Efficient translation of mouse hypoxia-inducible factor-1α under normoxic and hypoxic conditions. Biochim Biophys Acta 1493:125–134. https://doi.org/10.1016/S0167-4781(00)00172-X
203. Huang LE, Arany Z, Livingston DM, Franklin Bunn H (1996) Activation of hypoxia-inducible transcription factor depends primarily upon redox-sensitive stabilization of its α subunit. J Biol Chem 271:32253–32259. https://doi.org/10.1074/jbc.271.50.32253

204. Huang LE, Gu J, Schau M, Bunn HF (1998) Regulation of hypoxia-inducible factor 1α is mediated by an O_2-dependent degradation domain via the ubiquitin-proteasome pathway. Proc Natl Acad Sci 95:7987–7992. https://doi.org/10.1073/pnas.95.14.7987
205. Salceda S, Caro J (1997) Hypoxia-inducible factor 1α (HIF-1α) protein is rapidly degraded by the ubiquitin-proteasome system under normoxic conditions. Its stabilization by hypoxia depends on redox-induced changes. J Biol Chem 272:22642–22647. https://doi.org/10.1074/jbc.272.36.22642
206. Kallio PJ, Wilson WJ, O'Brien S et al (1999) Regulation of the hypoxia-inducible transcription factor 1α by the ubiquitin-proteasome pathway. J Biol Chem 274:6519–6525. https://doi.org/10.1074/jbc.274.10.6519
207. Loboda A, Jozkowicz A, Dulak J (2010) HIF-1 and HIF-2 transcription factors-similar but not identical. Mol Cells 29:435–442
208. Chilov D, Camenisch G, Kvietikova I et al (1999) Induction and nuclear translocation of hypoxia-inducible factor-1 (HIF-1): Heterodimerization with ARNT is not necessary for nuclear accumulation of HIF-1α. J Cell Sci 112:1203–1212. https://doi.org/10.1242/jcs.112.8.1203
209. Semenza GL, Agani F, Booth G et al (1997) Structural and functional analysis of hypoxia-inducible factor 1. Kidney Int 51(2):553–555
210. Semenza GL, Wang GL (1992) A nuclear factor induced by hypoxia via de novo protein synthesis binds to the human erythropoietin gene enhancer at a site required for transcriptional activation. Mol Cell Biol 12(12):5447–5454
211. Schödel J, Oikonomopoulos S, Ragoussis J et al (2011) High-resolution genome-wide mapping of HIF-binding sites by ChIP-seq. Blood 117. https://doi.org/10.1182/blood-2010-10-314427
212. Semenza GL (2002) Signal transduction to hypoxia-inducible factor 1. Biochem Pharmacol 64:993–998. https://doi.org/10.1016/S0006-2952(02)01168-1
213. Akiyama H, Chaboissier MC, Martin JF et al (2002) The transcription factor Sox9 has essential roles in successive steps of the chondrocyte differentiation pathway and is required for expression of Sox5 and Sox6. Genes Dev 16:2813–2828. https://doi.org/10.1101/gad.1017802
214. Amarilio R, Viukov SV, Sharir A et al (2007) HIF1α regulation of Sox9 in necessary to maintain differentiation of hypoxic prechondrogenic cells during early skeletogenesis. Development 134:3917–3928. https://doi.org/10.1242/dev.008441
215. Robins JC, Akeno N, Mukherjee A et al (2005) Hypoxia induces chondrocyte-specific gene expression in mesenchymal cells in association with transcriptional activation of Sox9. Bone 37:313–322. https://doi.org/10.1016/j.bone.2005.04.040
216. Zelzer E, Levy Y, Kahana C et al (1998) Insulin induces transcription of target genes through the hypoxia-inducible factor HIF-1α/ARNT. EMBO J 17:5085–5094. https://doi.org/10.1093/emboj/17.17.5085
217. Feldser D, Agani F, Iyer NV et al (1999) Reciprocal positive regulation of hypoxia-inducible factor 1α and insulin-like growth factor 2. Cancer Res 59:3915–3918
218. Tacchini L, Dansi P, Matteucci E, Desiderio MA (2001) Hepatocyte growth factor signalling stimulates hypoxia inducible factor-1 (HIF-1) activity in HepG2 hepatoma cells. Carcinogenesis 22:1363–1371. https://doi.org/10.1093/carcin/22.9.1363
219. Stiehl DP, Jelkmann W, Wenger RH, Hellwig-Bürgel T (2002) Normoxic induction of the hypoxia-inducible factor 1α by insulin and interleukin-1β involves the phosphatidylinositol 3-kinase pathway. FEBS Lett 512:157–162. https://doi.org/10.1016/S0014-5793(02)02247-0
220. Hellwig-Bürgel T, Rutkowski K, Metzen E et al (1999) Interleukin-1β and tumor necrosis factor-α stimulate DNA binding of hypoxia-inducible factor-1. Blood 94:1561–1567. https://doi.org/10.1182/blood.v94.5.1561
221. Thornton RD, Lane P, Borghaei RC et al (2000) Interleukin 1 induces hypoxia-inducible factor 1 in human gingival and synovial fibroblasts. Biochem J 350:307–312. https://doi.org/10.1042/0264-6021:3500307

222. Ishikawa Y, Ito T (1988) Kinetics of hemopoietic stem cells in a hypoxic culture. Eur J Haematol 40:126–129. https://doi.org/10.1111/J.1600-0609.1988.TB00808.X
223. Chow DC, Wenning LA, Miller WM, Papoutsakis ET (2001) Modeling pO_2 distributions in the bone marrow hematopoietic compartment. Biophys J 81:675–684. https://doi.org/10.1016/S0006-3495(01)75732-3
224. Antoniou ES, Sund S, Homsi EN et al (2004) A theoretical simulation of hematopoietic stem cells during oxygen fluctuations: prediction of bone marrow responses during hemorrhagic shock. Shock 22:415–422. https://doi.org/10.1097/01.SHK.0000142185.88094.88
225. Harrison JS, Rameshwar P, Chang V, Bandari P (2002) Oxygen saturation in the bone marrow of healthy volunteers. Blood 99:394
226. Hu X, Yu SP, Fraser JL et al (2008) Transplantation of hypoxia-preconditioned mesenchymal stem cells improves infarcted heart function via enhanced survival of implanted cells and angiogenesis. J Thorac Cardiovasc Surg 135:799–808. https://doi.org/10.1016/j.jtcvs.2007.07.071
227. Leroux L, Descamps B, Tojais NF et al (2010) Hypoxia preconditioned mesenchymal stem cells improve vascular and skeletal muscle fiber regeneration after ischemia through a Wnt4-dependent pathway. Mol Ther 18:1545–1552. https://doi.org/10.1038/MT.2010.108
228. Jaussaud J, Biais M, Calderon J et al (2013) Hypoxia-preconditioned mesenchymal stromal cells improve cardiac function in a swine model of chronic myocardial ischaemia. Eur J Cardiothorac Surg 43:1050–1057. https://doi.org/10.1093/ejcts/ezs549
229. Samal JRK, Rangasami VK, Samanta S et al (2021) Discrepancies on the role of oxygen gradient and culture condition on mesenchymal stem cell fate. Adv Healthc Mater 10:2002058. https://doi.org/10.1002/adhm.202002058
230. Ejtehadifar M, Shamsasenjan K, Movassaghpour A et al (2015) The effect of hypoxia on mesenchymal stem cell biology. Adv Pharm Bull 5:141–149. https://doi.org/10.15171/apb.2015.021
231. Ma T, Grayson WL, Fröhlich M, Vunjak-Novakovic G (2009) Hypoxia and stem cell-based engineering of mesenchymal tissues. Biotechnol Prog 25:32–42. https://doi.org/10.1002/btpr.128
232. Boyette LB, Creasey OA, Guzik L et al (2014) Human bone marrow-derived mesenchymal stem cells display enhanced Clonogenicity but impaired differentiation with hypoxic preconditioning. Stem Cells Transl Med 3:241–254. https://doi.org/10.5966/sctm.2013-0079
233. Grayson WL, Zhao F, Izadpanah R et al (2006) Effects of hypoxia on human mesenchymal stem cell expansion and plasticity in 3D constructs. J Cell Physiol 207:331–339. https://doi.org/10.1002/jcp.20571
234. Krinner A, Zscharnack M, Bader A et al (2009) Impact of oxygen environment on mesenchymal stem cell expansion and chondrogenic differentiation. Cell Prolif 42:471–484. https://doi.org/10.1111/j.1365-2184.2009.00621.x
235. Hu Y, Lou B, Wu X et al (2018) Comparative study on in vitro culture of mouse bone marrow mesenchymal stem cells. Stem Cells Int 2018. https://doi.org/10.1155/2018/6704583
236. Antebi B, Rodriguez LA, Walker KP et al (2018) Short-term physiological hypoxia potentiates the therapeutic function of mesenchymal stem cells. Stem Cell Res Ther 9. https://doi.org/10.1186/S13287-018-1007-X
237. Lavrentieva A, Majore I, Kasper C, Hass R (2010) Effects of hypoxic culture conditions on umbilical cord-derived human mesenchymal stem cells. Cell Commun Signal 8:1–9. https://doi.org/10.1186/1478-811X-8-18
238. Grayson WL, Zhao F, Bunnell B, Ma T (2007) Hypoxia enhances proliferation and tissue formation of human mesenchymal stem cells. Biochem Biophys Res Commun 358:948–953. https://doi.org/10.1016/j.bbrc.2007.05.054
239. Hung SP, Ho JH, Shih YRV et al (2012) Hypoxia promotes proliferation and osteogenic differentiation potentials of human mesenchymal stem cells. J Orthop Res 30:260–266. https://doi.org/10.1002/jor.21517

240. Saller MM, Prall WC, Docheva D et al (2012) Increased stemness and migration of human mesenchymal stem cells in hypoxia is associated with altered integrin expression. Biochem Biophys Res Commun 423:379–385. https://doi.org/10.1016/j.bbrc.2012.05.134
241. Yamamoto Y, Fujita M, Tanaka Y et al (2013) Low oxygen tension enhances proliferation and maintains stemness of adipose tissue-derived stromal cells. Biores Open Access 2:199–205. https://doi.org/10.1089/biores.2013.0004
242. Bader AM, Klose K, Bieback K et al (2015) Hypoxic preconditioning increases survival and pro-angiogenic capacity of human cord blood mesenchymal stromal cells in vitro. PLoS One 10:e0138477. https://doi.org/10.1371/JOURNAL.PONE.0138477
243. Zscharnack M, Poesel C, Galle J, Bader A (2009) Low oxygen expansion improves subsequent chondrogenesis of ovine bone-marrow-derived mesenchymal stem cells in collagen type I hydrogel. Cells Tissues Organs 190:81–93. https://doi.org/10.1159/000178024
244. Adesida AB, Mulet-Sierra A, Jomha NM (2012) Hypoxia mediated isolation and expansion enhances the chondrogenic capacity of bone marrow mesenchymal stromal cells. Stem Cell Res Ther 3:9. https://doi.org/10.1186/scrt100
245. Xu Y, Malladi P, Chiou M et al (2007) In vitro expansion of adipose-derived adult stromal cells in hypoxia enhances early chondrogenesis. Tissue Eng 13:2981–2993. https://doi.org/10.1089/ten.2007.0050
246. Fehrer C, Brunauer R, Laschober G et al (2007) Reduced oxygen tension attenuates differentiation capacity of human mesenchymal stem cells and prolongs their lifespan. Aging Cell 6:745–757. https://doi.org/10.1111/J.1474-9726.2007.00336.X
247. Holzwarth C, Vaegler M, Gieseke F et al (2010) Low physiologic oxygen tensions reduce proliferation and differentiation of human multipotent mesenchymal stromal cells. BMC Cell Biol 11:11. https://doi.org/10.1186/1471-2121-11-11
248. Ding H, Chen S, Yin J-H et al (2014) Continuous hypoxia regulates the osteogenic potential of mesenchymal stem cells in a time-dependent manner. Mol Med Rep 10:2184–2190. https://doi.org/10.3892/mmr.2014.2451
249. Hsu S-H, Chen C-T, Wei Y-H (2013) Inhibitory effects of hypoxia on metabolic switch and osteogenic differentiation of human mesenchymal stem cells. Stem Cells 31:2779–2788. https://doi.org/10.1002/stem.1441
250. Xu N, Liu H, Qu F et al (2013) Hypoxia inhibits the differentiation of mesenchymal stem cells into osteoblasts by activation of notch signaling. Exp Mol Pathol 94:33–39. https://doi.org/10.1016/j.yexmp.2012.08.003
251. Potier E, Ferreira E, Andriamanalijaona R et al (2007) Hypoxia affects mesenchymal stromal cell osteogenic differentiation and angiogenic factor expression. Bone 40:1078–1087. https://doi.org/10.1016/j.bone.2006.11.024
252. Freshney IR (2005) Culture of animal cells: a manual of basic technique. Wiley-Liss, New York
253. Kapałczyńska M, Kolenda T, Przybyła W et al (2018) 2D and 3D cell cultures – a comparison of different types of cancer cell cultures. Arch Med Sci 14:910–919. https://doi.org/10.5114/aoms.2016.63743
254. Dutta RC, Dutta AK (2018) 3D cell culture. Jenny Stanford Publishing, Singapore
255. Pampaloni F, Reynaud EG, Stelzer EHK (2007) The third dimension bridges the gap between cell culture and live tissue. Nat Rev Mol Cell Biol 8:839–845. https://doi.org/10.1038/nrm2236
256. Baker BM, Chen CS (2012) Deconstructing the third dimension-how 3D culture microenvironments alter cellular cues. J Cell Sci 125:3015–3024
257. Benya PD, Shaffer JD (1982) Dedifferentiated chondrocytes reexpress the differentiated collagen phenotype when cultured in agarose gels. Cell 30:215–224. https://doi.org/10.1016/0092-8674(82)90027-7
258. Gómez-Lechón MJ, Jover R, Donato T et al (2000) Expression of liver specific-genes in hepatocytes cultured in collagen gel matrix. Prog Mol Subcell Biol 25:89–104

259. Von Der Mark K, Gauss V, Von Der Mark H, Müller P (1977) Relationship between cell shape and type of collagen synthesised as chondrocytes lose their cartilage phenotype in culture. Nature 267:531–532
260. Delcommenne M, Streuli CH (1995) Control of integrin expression by extracellular matrix. J Biol Chem 270:26794–26801. https://doi.org/10.1074/jbc.270.45.26794
261. Ghosh S, Spagnoli GC, Martin I et al (2005) Three-dimensional culture of melanoma cells profoundly affects gene expression profile: a high density oligonucleotide array study. J Cell Physiol 204:522–531. https://doi.org/10.1002/jcp.20320
262. Bissell MJ, Hall HG, Parry G (1982) How does the extracellular matrix direct gene expression? J Theor Biol 99:31–68. https://doi.org/10.1016/0022-5193(82)90388-5
263. Zaman MH, Trapani LM, Sieminski AL et al (2006) Migration of tumor cells in 3D matrices is governed by matrix stiffness along with cell-matrix adhesion and proteolysis. Proc Natl Acad Sci 103:10889–10894. https://doi.org/10.1073/pnas.0604460103
264. Meshel AS, Wei Q, Adelstein RS, Sheetz MP (2005) Basic mechanism of three-dimensional collagen fibre transport by fibroblasts. Nat Cell Biol 7:157–164. https://doi.org/10.1038/ncb1216
265. Cukierman E, Pankov R, Stevens DR, Yamada KM (2001) Taking cell-matrix adhesions to the third dimension. Science 294:1708–1712. https://doi.org/10.1126/science.1064829
266. Walpita D, Hay E (2002) Studying actin-dependent processes in tissue culture. Nat Rev Mol Cell Biol 3:137–141. https://doi.org/10.1038/nrm727
267. Kunz-Schughart LA, Freyer JP, Hofstaedter F, Ebner R (2004) The use of 3-D cultures for high-throughput screening: the multicellular spheroid model. J Biomol Screen 9:273–285
268. Mueller-Klieser W (1987) Multicellular spheroids – a review on cellular aggregates in cancer research. J Cancer Res Clin Oncol 113:101–122. https://doi.org/10.1007/BF00391431
269. Costa EC, de Melo-Diogo D, Moreira AF et al (2018) Spheroids formation on non-adhesive surfaces by liquid overlay technique: considerations and practical approaches. Biotechnol J 13:1700417. https://doi.org/10.1002/biot.201700417
270. Lin RZ, Chang HY (2008) Recent advances in three-dimensional multicellular spheroid culture for biomedical research. Biotechnol J 3:1285–1285. https://doi.org/10.1002/BIOT.1285
271. Achilli TM, Meyer J, Morgan JR (2012) Advances in the formation, use and understanding of multi-cellular spheroids. Expert Opin Biol Ther 12:1347–1360
272. Egger D, Tripisciano C, Weber V et al (2018) Dynamic cultivation of mesenchymal stem cell aggregates. Bioengineering 5:48. https://doi.org/10.3390/bioengineering5020048
273. Foty R (2011) A simple hanging drop cell culture protocol for generation of 3D spheroids. J Vis Exp. https://doi.org/10.3791/2720
274. Howes AL, Richardson RD, Finlay D, Vuori K (2014) 3-dimensional culture systems for anti-cancer compound profiling and high-throughput screening reveal increases in EGFR inhibitor-mediated cytotoxicity compared to monolayer culture systems. PLoS One 9:e108283. https://doi.org/10.1371/JOURNAL.PONE.0108283
275. Schmitz C, Potekhina E, Belousov VV et al (2021) Hypoxia onset in mesenchymal stem cell spheroids: monitoring with hypoxia reporter cells. Front Bioeng Biotechnol 9:23. https://doi.org/10.3389/fbioe.2021.611837
276. Frith JE, Thomson B, Genever PG (2010) Dynamic three-dimensional culture methods enhance mesenchymal stem cell properties and increase therapeutic potential. Tissue Eng Part C Methods 16:735–749. https://doi.org/10.1089/ten.TEC.2009.0432
277. Lewis NS, Lewis EEL, Mullin M et al (2017) Magnetically levitated mesenchymal stem cell spheroids cultured with a collagen gel maintain phenotype and quiescence. J Tissue Eng:8. https://doi.org/10.1177/2041731417704428
278. Aijian AP, Garrell RL (2015) Digital microfluidics for automated hanging drop cell spheroid culture. SLAS Technol 20:283–295. https://doi.org/10.1177/2211068214562002

279. Lee SI, Ko Y, Park JB (2017) Evaluation of the maintenance of stemness, viability, and differentiation potential of gingiva-derived stem-cell spheroids. Exp Ther Med 13:1757–1764. https://doi.org/10.3892/etm.2017.4194
280. Bartosh TJ, Ylöstalo JH, Mohammadipoor A et al (2010) Aggregation of human mesenchymal stromal cells (MSCs) into 3D spheroids enhances their antiinflammatory properties. Proc Natl Acad Sci 107:13724–13729. https://doi.org/10.1073/pnas.1008117107
281. Bartosh TJ, Ylöstalo JH, Bazhanov N et al (2013) Dynamic compaction of human mesenchymal stem/precursor cells into spheres self-activates caspase-dependent IL1 signaling to enhance secretion of modulators of inflammation and immunity (PGE2, TSG6, and STC1). Stem Cells 31:2443–2456. https://doi.org/10.1002/STEM.1499
282. Silva-Carvalho AÉ, da Silva IGM, Corrêa JR, Saldanha-Araujo F (2022) Regulatory T-cell enhancement, expression of adhesion molecules, and production of anti-inflammatory factors are differentially modulated by spheroid-cultured mesenchymal stem cells. Int J Mol Sci 23. https://doi.org/10.3390/IJMS232214349
283. Lee JH, Han YS, Lee SH (2016) Long-duration three-dimensional spheroid culture promotes angiogenic activities of adipose-derived mesenchymal stem cells. Biomol Ther (Seoul) 24(3): 260–267
284. Miceli V, Pampalone M, Vella S et al (2019) Comparison of immunosuppressive and Angiogenic properties of human amnion-derived mesenchymal stem cells between 2D and 3D culture systems. Stem Cells Int 2019. https://doi.org/10.1155/2019/7486279
285. Follin B, Juhl M, Cohen S et al (2016) Increased paracrine immunomodulatory potential of mesenchymal stromal cells in three-dimensional culture. Tissue Eng Part B Rev 22:322–329
286. Alimperti S, Lei P, Wen Y et al (2014) Serum-free spheroid suspension culture maintains mesenchymal stem cell proliferation and differentiation potential. Biotechnol Prog 30:974–983. https://doi.org/10.1002/btpr.1904
287. Ylostalo JH, Bartosh TJ, Tiblow A, Prockop DJ (2014) Unique characteristics of human mesenchymal stromal/progenitor cells pre-activated in 3-dimensional cultures under different conditions. Cytotherapy 16:1486–1500. https://doi.org/10.1016/j.jcyt.2014.07.010
288. Carrel A, Burrows MT (1911) An addition to the techniqie of the cultivation of tissues in vitro. J Exp Med 14:244–247. https://doi.org/10.1084/jem.14.3.244
289. Griffith LG, Swartz MA (2006) Capturing complex 3D tissue physiology in vitro. Nat Rev Mol Cell Biol 7:211–224. https://doi.org/10.1038/nrm1858
290. Sourla A, Doillon C, Koutsilieris M (1996) Three-dimensional type I collagen gel system containing MG-63 osteoblasts-like cells as a model for studying local bone reaction caused by metastatic cancer cells. Anticancer Res 16:2773–2780
291. Glicklis R, Shapiro L, Agbaria R et al (2000) Hepatocyte behavior within three-dimensional porous alginate scaffolds. Biotechnol Bioeng 67:344–353. https://doi.org/10.1002/(SICI)1097-0290(20000205)67:3<344::AID-BIT11>3.0.CO;2-2
292. Jastrzebska K, Kucharczyk K, Florczak A et al (2015) Silk as an innovative biomaterial for cancer therapy. Rep Pract Oncol Radiother 20:87–98. https://doi.org/10.1016/j.rpor.2014.11.010
293. Ruedinger F, Lavrentieva A, Blume C et al (2015) Hydrogels for 3D mammalian cell culture: a starting guide for laboratory practice. Appl Microbiol Biotechnol 99:623–636
294. Malda J, Visser J, Melchels FP et al (2013) 25th anniversary article: engineering hydrogels for biofabrication. Adv Mater 25:5011–5028. https://doi.org/10.1002/adma.201302042
295. Catoira MC, Fusaro L, Di Francesco D et al (2019) Overview of natural hydrogels for regenerative medicine applications. J Mater Sci Mater Med 30:1–10. https://doi.org/10.1007/s10856-019-6318-7
296. Pepelanova I, Kruppa K, Scheper T, Lavrentieva A (2018) Gelatin-methacryloyl (GelMA) hydrogels with defined degree of functionalization as a versatile toolkit for 3D cell culture and extrusion bioprinting. Bioengineering 5:55. https://doi.org/10.3390/bioengineering5030055
297. Engler AJ, Sen S, Sweeney HL et al (2006) Matrix elasticity directs stem cell lineage specification. Cell 126:677–689. https://doi.org/10.1016/J.CELL.2006.06.044

298. Janson IA, Putnam AJ (2015) Extracellular matrix elasticity and topography: material-based cues that affect cell function via conserved mechanisms. J Biomed Mater Res A 103(3): 1246–1258
299. Reilly GC, Engler AJ (2010) Intrinsic extracellular matrix properties regulate stem cell differentiation. J Biomech 43:55–62
300. Yao S, Liu X, Yu S et al (2016) Co-effects of matrix low elasticity and aligned topography on stem cell neurogenic differentiation and rapid neurite outgrowth. Nanoscale 8:10252–10265. https://doi.org/10.1039/C6NR01169A
301. Pek YS, Wan ACA, Ying JY (2010) The effect of matrix stiffness on mesenchymal stem cell differentiation in a 3D thixotropic gel. Biomaterials 31:385–391. https://doi.org/10.1016/J.BIOMATERIALS.2009.09.057
302. Rowlands AS, George PA, Cooper-White JJ (2008) Directing osteogenic and myogenic differentiation of MSCs: interplay of stiffness and adhesive ligand presentation. Am J Physiol Cell Physiol 295:1037–1044. https://doi.org/10.1152/AJPCELL.67.2008/ASSET/IMAGES/LARGE/ZH00100857480008.JPEG
303. Park JS, Chu JS, Tsou AD et al (2011) The effect of matrix stiffness on the differentiation of mesenchymal stem cells in response to TGF-β. Biomaterials 32:3921–3930. https://doi.org/10.1016/J.BIOMATERIALS.2011.02.019
304. Wingate K, Bonani W, Tan Y et al (2012) Compressive elasticity of three-dimensional nanofiber matrix directs mesenchymal stem cell differentiation to vascular cells with endothelial or smooth muscle cell markers. Acta Biomater 8:1440–1449. https://doi.org/10.1016/J.ACTBIO.2011.12.032
305. Kilian KA, Bugarija B, Lahn BT, Mrksich M (2010) Geometric cues for directing the differentiation of mesenchymal stem cells. Proc Natl Acad Sci 107:4872–4877. https://doi.org/10.1073/pnas.0903269107
306. McBeath R, Pirone DM, Nelson CM et al (2004) Cell shape, cytoskeletal tension, and RhoA regulate stem cell lineage commitment. Dev Cell 6:483–495. https://doi.org/10.1016/S1534-5807(04)00075-9
307. Kim SJ, Lee JK, Kim JW et al (2008) Surface modification of polydimethylsiloxane (PDMS) induced proliferation and neural-like cells differentiation of umbilical cord blood-derived mesenchymal stem cells. J Mater Sci Mater Med 19:2953–2962. https://doi.org/10.1007/s10856-008-3413-6
308. Song W, Lu H, Kawazoe N, Chen G (2011) Adipogenic differentiation of individual mesenchymal stem cell on different geometric micropatterns. Langmuir 27:6155–6162. https://doi.org/10.1021/la200487w
309. Seo CH, Furukawa K, Suzuki Y et al (2011) A topographically optimized substrate with well-ordered lattice micropatterns for enhancing the osteogenic differentiation of murine mesenchymal stem cells. Macromol Biosci 11:938–945. https://doi.org/10.1002/MABI.201000477
310. Xue R, Li JYS, Yeh Y et al (2013) Effects of matrix elasticity and cell density on human mesenchymal stem cells differentiation. J Orthop Res 31:1360–1365. https://doi.org/10.1002/jor.22374
311. Tsuji K, Ojima M, Otabe K et al (2017) Effects of different cell-detaching methods on the viability and cell surface antigen expression of synovial mesenchymal stem cells. Cell Transplant 26:1089–1102. https://doi.org/10.3727/096368917X694831/ASSET/IMAGES/LARGE/10.3727_096368917X694831-FIG2.JPEG
312. Weber C, Pohl S, Pörtner R et al (2007) Expansion and harvesting of hMSC-TERT. Open Biomed Eng J 1:38. https://doi.org/10.2174/1874120700701010038
313. Malda J, Klein TJ, Upton Z (2007) The roles of hypoxia in the in vitro engineering of tissues. Tissue Eng 13:2153–2162. https://doi.org/10.1089/ten.2006.0417

314. Torre ML, Lucarelli E, Guidi S et al (2015) Ex vivo expanded mesenchymal stromal cell minimal quality requirements for clinical application. Stem Cells Dev 24:677–685. https://doi.org/10.1089/scd.2014.0299
315. Dominici M, Le Blanc K, Mueller I et al (2006) Minimal criteria for defining multipotent mesenchymal stromal cells. The International Society for Cellular Therapy position statement. Cytotherapy 8:315–317. https://doi.org/10.1080/14653240600855905

Chapter 13
Towards a Continuous Production of Human Mesenchymal Stromal Cells in a Chemically Defined Medium: Opportunities and Challenges for a Robust and Scalable Expansion Process

Lars Tasto and Denise Salzig

Abstract Human mesenchymal stromal cells (hMSCs) are one of the most prominent cell types used for cell therapy application. Thereby the current clinical studies focus more and more on juvenile hMSCs (e.g. isolated from the umbilical cord) as these show unique therapeutic properties. For their therapeutic application, a complex manufacturing process is required to produce the high hMSC yield in the therapeutic quality as it is need for the indication. This requires a strong process knowledge given by process development after quality-by-design (QbD) principle. Chemically defined media (CDM) are a critical component to gain process stability and fulfill the requirement of Good Manufacturing Practice (GMP) and Process Analytical Technology (PAT). Another factor are advanced process modes which implements in-line process control to stabilize the process outcome. Within this chapter, we discuss the use of umbilical cord hMSCs and their manufacturing thereby focusing on the cell expansion, the use of chemically defined media, design-of-experiment (DoE) approaches for process development and intensified bioreactor processes.

Keywords ATMP · Perfusion · Fed-batch · Microenvironment · Critical process parameters · Stirred-tank bioreactor · Microcarrier · Design-of-experiments (DoE)

Abbreviations

AT	Adipose tissue
ATMPs	Advanced therapy medicinal products
BM	Bone marrow

L. Tasto · D. Salzig (✉)
Institute of Bioprocess Engineering and Pharmaceutical Technology (IBPT), University of Applied Sciences Mittelhessen (THM), Gießen, Germany
e-mail: lars.tasto@lse.thm.de; denise.salzig@lse.thm.de

R. Pörtner (ed.), *Biopharmaceutical Manufacturing*, Cell Engineering 11,
https://doi.org/10.1007/978-3-031-45669-5_13

CD	Cluster of differentiation
CDM	Chemically defined medium
CPP	Critical process parameter
CQA	Critical quality attribute
DO	Dissolved oxygen
DoE	Design-of-experiments
ECM	Extracellular matrix
EMA	European medicines agency
EVs	Extracellular vesicles
FCS	Fetal calf serum
FDA	Food and drug administration
GMP	Good manufacturing practice
hMSCs	Human mesenchymal stromal cells
IL	Interleukin
INF	Interferon
ISCT	International society of cell therapy
PAT	Process analytical technology
QbD	Quality by design
QTPP	Quality target product profile
R&D	Research and development
SFM	Serum-free medium
STR	Stirred-tank bioreactor
TNF	Tumor necrosis factor
UC	Umbilical cord

Introduction

Human mesenchymal stromal cells (hMSCs) were first isolated from bone marrow [1]. Today we know that hMSCs are found in many human tissues beside bone marrow (BM), e.g., adipose tissue (AD), umbilical cord blood (UC) or cord matrix [2, 3], liver [4], dental pulp [5], human placenta [6], and menstrual blood [7]. Although conflicting results regarding the differentiation potency of hMSCs are reported [8], they are generally considered to be multipotent and capable of differentiating *in vitro* into tissues of mesenchymal lineage. Depending on age, gender, the health status of the donor, and on the tissue origin, hMSCs show different characteristics. hMSCs, which are derived from an impaired patient, may be challenging to adapt to existing *in vitro* expansion protocols. Moreover, hMSCs are different in growth rates, surface marker expression, differentiation potential, and immunomodulatory potency depending on their tissue source. AD-hMSCs proliferate faster than BM-hMSCs [9], and juvenile UC-hMSCs proliferate faster than adult hMSCs [10]. Integrin expression, which is crucial for extracellular matrix connections and adhesion, varies between hMSCs derived from different tissues [11]. Even hMSCs from the same tissue source, but different donors, showed a different growth performance [12, 13]. Further, the *in vitro* culture conditions, e.g., the expansion in different culture media, also affect the expression of immunosuppressive and adhesion-related molecules [14]. hMSCs are a heterogeneous cell pool sensitive to microenvironmental changes. To generate a reliable hMSC drug product, a well-defined, well understood, and robust bioreactor production process is a prerequisite.

hMSCs are one of the most promising cell types for cell therapy applications. Clinical treatment often focuses on immunomodulatory hMSCs, which are either autologous cells isolated from the patient and used for personalized therapy, or allogeneic cells isolated from another individual. The latter option provides the possibility of treating diseases affecting a large number of patients. In 2023, clinicaltrial.gov identified around 298 of hMSC-based clinical trials that were either completed, terminated, ongoing, or recruiting worldwide. These cases were mainly concentrated in the United States, Europe, and China. A minimum dose of approximately $20 \cdot 10^6$ cells was required for hMSC clinical trials, with a therapeutic dose typically containing $2–3 \cdot 10^6$ MSCs $\cdot$ kg^{-1} [15]. Often repeated doses are needed, depending on the therapeutic indication. The demand for cells required for clinical trials and conventional treatments will be high, making *in vitro* expansion mandatory, as naturally occurring hMSCs represent only a small percentage of the total cell population in the body.

Although some companies, such as Osiris Therapeutics (*Prochymal* for GvHD using allogenic BM-MSCs) reported some failures in the phase 3 trials of their products, there are approved hMSC-based products on the global market and under development (Table 13.1), underlining the importance of a large-scale production of hMSCs [16]. hMSCs are classified as an Advanced Therapy Medicinal Product (ATMP) and their production is subject to a number of regulations and guidelines. It is essential to exercise extreme caution during the production process in order to produce the highest quality cells. Each step of the process must be carefully controlled, regulated, and checked according to Good Manufacturing Practice (GMP) guidelines [17].

From Waste to Wonder: Harnessing the Power of UC-hMSCs for Therapeutic Applications

Among hMSCs, especially the cells isolated from the umbilical cord (UC) may have advantageous therapeutic benefits over the others [18]. Thomas Wharton, an English physician and anatomist, first discovered and described the umbilical cord stroma (=Wharton's jelly) [19]. The cell population within Wharton's jelly (WJ) comprises fibroblasts, myofibroblasts, smooth muscle cells, and hMSCs, which possess a unique combination of prenatal and postnatal stromal cell properties [19]. UC-hMSCs have gained attention due to their abundance in comparison to BM-hMSCs, as the procedure for collecting BM-hMSCs is invasive and painful.

Table 13.1 Commercialized hMSC products and products currently under development. Unless stated otherwise, allogeneic cells were used

Drug name	Treated disease	hMSC source	Country	Company
Amesanar®	Chronic wounds	ABCB5+ hMSCs	Europe	Rheacell GmbH & Co. KG
Alofisel®	Complex perianal fistulas in Crohn's disease	AT-hMSC	Europe, Japan	TiGenix/Takeda
Cartistem®	Knee articular cartilage defect	UC-hMSC	South Korea	Medipost Co. Ltd.
Cellgram-AMI®	Acute myocardial infarction	Autologous BM-hMSC	South Korea	Pharmicell Co. Ltd.
Cupistem®	Crohn's fistula	Autologous AT-hMSC	South Korea	Anterogen Co. Ltd.
HeartiCell gram-AMlis	Acute myocardial infarction	UC-hMSC	South Korea	FCB-Pharmicell
Neuronata-R®	Amyotrophic lateral sclerosis (ALS) or Lou Gehrig's disease	Autologous BM-hMSC	South Korea	Corestem Inc.
Neurostem®	Alzheimer disease	UC-hMSC	South Korea	Medipost Co. Ltd.
Pneumostem®	Broncho-pulmonary dysplasia	UC-hMSC	South Korea	Medipost Co. Ltd.
Prochymal®, Remestemcel-L[a]	*GvHD*	*BM-hMSC*	*Canada, New Zealand*	*Osiris Therapeutics Inc., Mesoblast Ltd.*
SMUP-IA-01[b]	*Diabetic neuropathy*	*UC-hMSC*	*South Korea*	*Medipost Co. Ltd.*
SMUP-IV-01[b]	*Osteoarthritis*	*UC-hMSC*	*South Korea*	*Medipost Co. Ltd.*
Stemirac®	Spinal cord injury	Autologous BM-hMSC	Japan	Nipro Corp
Stempeucel®	Critical limb ischemia	BM-hMSC	India	Stempeutics Research PVT
TemCell® HS	GvHD	BM-hMSC	Japan	JCR Pharmaceutical
Queencell®	Regeneration of subcutaneous tissue	Autologous AT-hMSC	South Korea	Anterogen Co. Ltd.

[a] hMSC product has been withdrawn from the market
[b] hMSC product currently under development for treatment and prevention

UC-hMSCs can be easily obtained without ethical concerns, as extra-embryonic tissues are typically discarded after birth while still exhibiting the desired characteristics of young cells. Moreover, the amount of hMSCs that can be isolated from umbilical cord tissue is generally higher compared to that obtained from other sources. UC-hMSCs further showed a promising proliferation performance [20], indicating their relatively primitive nature in comparison to adult hMSCs [21], as an inverse correlation between age and the growth potential of adult hMSCs [22, 23] was demonstrated. UC-MSCs have been shown to elicit a lower immune response compared to hMSCs derived from other sources. This makes them a potentially safer and more effective option for cell therapies.

The concept of the UC as a source of stem cells emerged in 1974, when progenitors and hematopoietic stem cells were first extracted from umbilical cord blood [24]. Prior to this discovery, the remaining UC tissue was considered as medical waste. However, the discovery of UC as a source of hMSC was later, with McElreavey et al. (1991) describing fibroblast-like cells from WJ, [25] and Wang et al. (2004) identifying that WJ-hMSCs express matrix receptors (CD44, CD105) and integrin markers (CD29, CD51), but not hematopoietic lineage markers (CD34, CD45) [26, 27]. The origin of UC-hMSCs is still debated with two hypotheses, suggesting fetal hMSC migration and colonization during embryonic development, or primitive hMSC presence since cord formation [28].

For cell isolation, the sterile UC is collected and processed within 12–24 h after birth [29]. The explant/outgrowth method [25] capitalizes on the inherent ability of hMSCs to adhere to plastic surfaces and migrate from explants. After the removal of blood vessels and epithelium, the WJ is dissected into small pieces and placed into 6-well plates with a culture medium. Over time, UC-hMSCs spontaneously migrate and colonize the plastic surfaces of the well plates [30]. Alternatively, WJ pieces are incubated with enzyme cocktails (e.g., collagenase, hyaluronidase, and trypsin) at 37 °C for 30 min up to 16 h [26]. The resulting suspension is filtered, washed, and centrifuged, and the remaining cells are cultured. Recently, a new technique has been developed, which combines enzymatic digestion with mechanical dissociation. After enzymatic digestion, the mixture is mechanically homogenized, leading to an increase in cell extraction yield by a factor of 160 [31]. To ensure the suitability of hMSCs for clinical trials, the cell culture media must be serum-free, and the isolation procedure should involve minimal tissue handling and be less time-consuming with a typical processing time of 3 to 4 h per cord.

Therapeutic applications of UC-hMSCs have focused mainly on two distinct areas: tissue engineering and cell therapy (Table 13.2). For tissue engineering, UC-hMSCs have been particularly effective in bone regeneration, such as in the development of cartilage [32–35], as well as in other areas such as skin and cornea regeneration. In cell therapy approaches, transplanted hMSCs can act over two mechanisms: (i) regeneration of the pathological tissue via an activation of an intrinsic differentiation of tissue-specific stem cells and cytokine and growth factor secretion properties of hMSCs, or (ii) through the induction of an inflammatory response mediated by the immunomodulatory properties of hMSCs. In 2006, Weiss et al. transplanted UC-hMSCs into the brains of immunosuppressed rats with hemi parkinsonism [3]. The treatment showed an improvement in apomorphine-induced rotations, a standard test for Parkinson's disease models. In that same year, Conconi et al. investigated the injection of UC-hMSCs into damaged skeletal muscles, which supported the muscle regenerative process by spontaneously differentiating into the myogenic lineage [36]. In 2007, Friedman et al. demonstrated the potential of using UC-hMSCs as adjuvants for transplantation [37]. In 2011, Peng et al. identified UC-hMSCs as a promising cell source for nerve tissue engineering, as they were able to differentiate into Schwan-cell lineage [38]. In accordance with the International

Table 13.2 Active, terminated or completed clinical studies regarding the use of allogeneic UC-hMSCs in bone, cartilage, and cardiovascular diseases

Disease	Enrollment	Study design/effects	Clinical phase	Cell mass per patient	NCT	References
Bone and cartilage						
Articular cartilage diseases/OA	20	UC-MSC implantation in the treatment of articular cartilage defect of knee to find alternative concept of autologous chondrocyte implantation	Phase I	$20 \cdot 10^6$	02291926	/
OA of knee	30	Investigating the safety and efficacy of UC-MSCs in patients with knee OA for reduction in cartilage degradation, the attenuation of bone sclerosis, and an effective anti-inflammatory response	Phase I-II	$20 \cdot 10^6$	02580695	/
Cardiovascular						
Chronic ischemic cardiomyopathy	50	UC-MSC with injectable collagen scaffold transplantation for chronic ischemic cardiomyopathy. Testing coronary artery bypass surgery (CABG) with allogeneic MSCs w or w/o the injection into the infarct region and w or w/o collagen scaffold	Phase I-II	$100 \cdot 10^6$	02635464	[47]
ST-elevation myocardial infarction	160	Investigating the safety and efficacy of intracoronary human UC-MSC transfer in patients with ST-segment elevation acute myocardial infarction. Testing efficacy by intracoronary infusion of UC-MSCs or placebo medium into the infarct artery 4–7 days after successful reperfusion therapy	Phase II	$348 \cdot 10^6$	01291329	[20, 48–53]
Acute myocardial infarction	4	UC-MSCs from a donor will be cultured in a clinical grade laboratory with xeno-free medium. Maximum passage was 6 and doubling population is less than 30. To assure the quality of the expanded-UC MSCs at ProSTEM the following tests are done: cell adherence, cell surface marker, *in vitro* differentiation, cell viability, sterility, mycoplasma, endotoxin, and karyotyping	Phase I-II	$50–170 \cdot 10^6$	04340609	/

Heart failure with reduced ejection fraction(PRIME-HFrEF study)	40	Exploratory clinical study to observe the safety and efficacy of the intravenous infusion of UC-MSCs in the heart failure patients with reduced ejection fraction. The study was a 12 month single center, randomized, double-blind, placebo-controlled trial that included 12 weeks of treatment, and 9 months of follow-up	Phase I-II	$225 \cdot 10^6$	04992832	/
Non-ischemic heart failure (RegenHeart)	30	Optimization of the complex treatment of non-ischemic dilated cardiomyopathy due to the addition to the standard drug therapy of intracoronary administration of UC-MSCs	Phase II	$10 \cdot 10^6$	04325594	/
Cardiopathy (RIMECARD)	30	Determine the safety and clinical effectiveness of UC-MSCs transplanted by intravenous infusion in patients with heart failure in dilated stage, of different etiology	Phase I-II	$225 \cdot 10^6$	01739777	[54, 55]
Myocardial infarction (HUC-HEART)	46	All subjects will be taken into the bypass coronary surgery prior to the cell administration. This study comprises three independent groups, where the first group (n = 20) will take no cells, second group will take autologous BM-MNCs (n = 20), and third group (n = 39) will be receiving allogeneic hUC-MSCs. In all transplantations cells will be administered to the approximately 10 peri-infarct areas at one time.	Phase I-II	/	02323477	[56, 57]
Acute myocardial infarction (AMI)	10	This study is a first-in-human assessment of safety of using UC-MSC in patients with AMI via a combination of IC and IV stromal cell administration. The novelty of the current study is the dual route of administration. Since dual administration had never been conducted in humans, there may be unknown risks to humans not predicted from the preclinical studies. However, the risk to patients in this trial will	Phase I	/	04056819	/

(continued)

Table 13.2 (continued)

Disease	Enrollment	Study design/effects	Clinical phase	Cell mass per patient	NCT	References
		be minimized by rigorous adherence to the eligibility criteria, use of appropriate dose and concentration of stromal cells, standardized techniques of infusion, and intensive patient monitoring during and after infusion				

http://www.clinicaltrials.gov, May 2023 – The number of recruitment studies is not on the list but is growing rapidly in other disease areas

Society of Cell Therapy (ISCT) guidelines, UC-hMSCs have the ability to differentiate *in vitro* into various cell types, including chondrocytes, adipocytes, osteoblasts, neuroglia cells, hepatocyte-like cells, dermal fibroblasts, smooth muscle cells, cardiomyocytes, odontogenic-like cells, skeletal muscle cells, insulin-producing cells, endothelial cells, and dopaminergic neurons [19]. However, whether hMSCs also show these differentiation properties *in vivo* is an area of some debate. The plasticity of UC-hMSCs is known to be influenced by the conditions of the pregnancy, as evidenced by studies showing that UC-hMSCs derived from preeclamptic patients, a condition characterized by high blood pressure and excessive proteins in the urine, exhibit an increased propensity towards neuroglial differentiation [19]. UC-hMSCs obtained from patients with gestational diabetes have been observed to exhibit a decreased cell growth, earlier cell senescence, a lower adipogenic and osteogenic differentiation potential, and a reduced mitochondrial activity [19]. Furthermore, UC-hMSCs obtained from preterm pregnancies have been shown to possess reduced osteogenic potential in comparison to cells derived from term pregnancies.

The secretory properties and thus the wide range of secreted bioactive compounds are a unique therapeutic feature of hMSCs. Quantitative assessments of the proteome and transcriptome of hMSCs from various sources have highlighted shared characteristics. UC-hMSCs secrete numerous factors involved in processes [39] such as angiogenesis (e.g., CXCL-1, CXCL, CXCL-5, CXCL-6, and CXCL-8), vascularization (via the expression of genes for CXCL-2, MDK, EGF, and FGF-9), migration of neurons (HBEGF, MDK, EGF) and outgrowth of neurites (NTF-3, MDK). Additionally, UC-hMSCs are capable of secreting cytokines and hematopoietic factors, including G-CSF, GM-CSF, LIF, IL-1, IL-6, IL-8, and IL-11, rendering them a suitable option for hematopoietic stem cell expansion [19]. Furthermore UC-hMSCs can secrete specific factors, such as human insulin, in response to physiological glucose levels following a differentiation into mature islet-like cell clusters [40, 41]. UC-hMSCs are capable of secreting cytokines similar to those secreted by BM-hMSCs. Additionally, UC-hMSCs can secrete the granulocyte-macrophage colony-stimulating factor (GM-CSF) and the granulocyte colony-stimulating factor (G-CSF), which promote the proliferation of white blood cells [42]. This is of therapeutic significance in chemotherapy, in bone marrow transplantation, or for the treatment of human immunodeficiency virus (HIV). UC-hMSCs are primarily used for allogeneic transplantation, and their low immunogenicity is a crucial requirement. They express low levels of human leukocyte antigen class I (HLA-I), which protects them from natural killer cell-mediated lysis [43]. Moreover, UC-hMSCs do not express HLA-II and co-stimulatory antigens, such as CD80 and CD86, which are involved in the activation of T and B cell responses [43–45]. Hence, UC-hMSCs suppress the proliferation of stimulated lymphocytes and have a reduced immunogenicity. T-lymphocytes, by secreting INF-γ and TNF-α, are the primary immune response elements in an allogeneic situation [46].

A Quality Target Product Profile (QTPP) That Identifies the Critical Quality Attributes (CQAs) of the hMSC Product

The heterogeneity of hMSCs and their complex interactions with the microenvironment pose significant challenges in the transition from laboratory-scale experiments to industrial biomanufacturing processes. To overcome this challenge and ensure a consistent and standardized manufacturing process, the hMSC expansion should be developed based on the quality-by-design (QbD) principle [58]. QbD is a scientific and risk-based approach to product development that is currently flourishing in the pharmaceutical industry [59, 60]. This principle integrates scientific knowledge from both biological and engineering perspectives and provides a rational framework for quality assurance through risk analysis. The QbD process (Fig. 13.1) begins by a clear definition of the quality target product profile (QTPP) that outlines the desired product quality characteristics [61]. The QTPP comprises critical quality attributes (CQAs) that specify the physical, chemical, and biological properties of the product. Consequently, the CQAs, e.g., the identity, potency, sterility, purity, and safety of each hMSC product are tightly controlled, but vary across individual hMSC products and their intended therapeutic application, which cannot be transferred from one product to another [62, 63].

Identity: Challenging due to missing unique hMSC-markers

hMSC identity is determined by morphological and phenotypic analysis, however hMSCs from different tissues vary in phenotypic surface marker and gene expression levels [2, 64, 65]. These tissue-source-dependent differences are largely due to variations in the environmental conditions imposed by the specific cellular niche (e.g., oxygen levels, tissue stiffness). As mentioned before, donor age and health status further impact hMSC characteristics [19, 66].

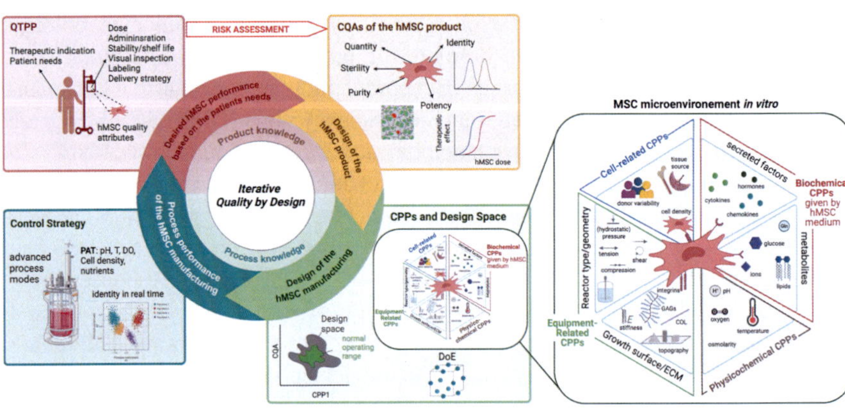

Fig. 13.1 Quality-by-Design Principle for hMSC manufacturing, created with BioRender

As hMSCs lack a distinct surface marker for reliable identification, researchers rely on a combination of surface markers and phenotypical characteristics. Various societies and institutes have outlined hMSC characteristics [67]. The ISCT provides a definition of MSCs in their "minimal criteria for defining multipotent mesenchymal stromal cells" [27, 68]. This definition specifies the identity of cells based on their surface markers (cluster-of-differentiation), which should exhibit positive expression (>95%) of CD105, CD73, CD90, additionally CD166 [69–72] and negative expression (<2%) of CD45, CD34, CD14 or CD11b, CD79α or CD19, and HLA class II. Other criteria in the ISCT definition include the adherence to plastic surfaces when cultured *in vitro* under standard conditions, and the ability to differentiate *in vitro* into osteoblasts, adipocytes, and chondroblasts [2]. These characteristics must be determined and tailored for each hMSC line depending on its source [73]. Gene expression levels impact protein transcription and surface marker expression, which directly affects the characterization of hMSCs. Despite ongoing discussions, the precise phenotype of naturally occurring and/or *in vitro* cultured hMSCs remains unclear. Using antibodies against Syndecan-2 (CD362) [74] and stromal precursor antigen-1 (STRO-1) [75] can enrich specific hMSC subpopulations to enhance the therapeutic potential of these cells, by selecting those with desired properties, such as an increased immunomodulatory capacity or an enhanced regenerative potential.

A CQA of hMSCs can be their frequency of colony-forming units (CFU-F). The CFU-F assay involves seeding 100–500 cells per 10 cm^2 dish, and allowing colonies to form over a period of 14 days, followed by staining with crystal violet to visualize and count the colonies [76]. Typical CFU-F values range from 15% to 25% for AT-hMSC [77], 22% to 38% for BM-hMSC [78], and 22% to 30% for UC-hMSCs [79].

Potency: from viability and differentiation capacities to secretome analysis – quantifiable metrics for predicting therapeutic efficacy for clinical applications

The potency of hMSCs usually depends on the specific therapeutic indication for which the product is intended. While several potency assays are available, only specific assays are suitable for individual hMSC products. For instance, a differentiation potency assay is only appropriate for hMSCs that develop their therapeutic mechanisms based on tissue formation. Due to their mechano-sensitivity, the differentiation process of MSCs can be significantly impacted by various environmental conditions, including shear stress and oxygen concentration [80–82]. Therefore, potency tests for hMSCs should be specific, accurate, robust, and suitable, with clear acceptance and rejection criteria, incorporate relevant standards and controls, and produce quantitative data [83] to ensure that the results obtained in various laboratories can be compared and analyzed accurately. Potency is a measurable property that serve as an important indicator of manufacturing consistency. During the hMSC manufacturing process, it is important to monitor the cell viability, metabolic activity, and growth rates at each step, such as after isolation, expansion, harvest and clarification, and final formulation. Viability is a potential indicator of potency because only living cells can function as a therapeutic unit. To perform this monitoring, viability testing and cell counting are conducted using techniques such as flow cytometry, which includes dye exclusion. Various dyes are used to differentiate between viable and non-viable cells, even in encapsulated formulations [84].

If the therapeutic aim is cell engraftment or tissue formation, an assay that measures differentiation potential may be appropriate to describe hMSC potency. To comply with FDA regulations, quantitative biological assays are required to measure potency. As mentioned before, one standard approach is to differentiate MSCs *in vitro* by culturing them in differentiation medium for 21 days and testing them thereafter. However, non-quantitative staining for differentiation markers may not be sufficient, and alternative methods such as a post-differentiation RNA/protein analysis, or online monitoring of differentiation by Raman spectroscopy [85] may be more suitable.

hMSCs in their undifferentiated state secrete various factors that possess immunomodulatory, anti-inflammatory, anti-apoptotic, antioxidant, antimicrobial properties, and promote angiogenesis. If the therapeutic efficacy of hMSCs is mediated by their secreted factors, the differentiation potential of the cells may not be the primary determinant of their potency [63]. The ISCT recommends a matrix assay approach that includes the use of a quantitative RNA analysis to evaluate selected gene products, flow cytometry to detect functionally relevant surface markers, and protein-based assays to map the secretome and determine the immunomodulatory potency of hMSCs [68]. Using advanced proteomic and physiochemical techniques, such as multiplex enzyme-linked immunosorbent assays (ELISAs) or mass spectroscopy [86], to identify and quantify the secretome of hMSC and improve the understanding of the therapeutic potential of these cells. To further characterize the secretome of hMSCs and evaluate their immunomodulatory properties, *in vitro* functional assays were developed. These assays can directly or indirectly measure the activity of hMSCs and are more suitable for assessing their potency. Several potency assays are available, but it is essential to select specific assays that are applicable to individual hMSC products.

The measurement of the indoleamine 2,3-dioxygenase (IDO) activity may provide insight into the immunomodulatory activity of hMSCs [87–91]. IDO is the initial and rate-limiting enzyme in the breakdown of tryptophan, an amino acid that promotes T-cell activity, to L-kynurenine. Through this mechanism, IDO has been identified as a key contributor to hMSC-mediated suppression of T-cell proliferation.

Furthermore, the scratch or migration assay is a widely used technique that measures the migratory capacity of cells. *In vivo*, cell migration plays a critical role in wound healing, as cells are recruited to the site of injury, where they differentiate into specialized cell types, and secrete cytokines and chemokines to support the regeneration of the damaged tissue [92]. *In vitro*, this process can be simulated by creating a "scratch" or gap in a confluent layer of cells that is then treated with hMSCs or the secreted factors they produce. The subsequent ingrowth or closure of the scratch is monitored over time as an indicator of the migratory and regenerative potential of the treated cells [93, 94].

For instance, the mixed lymphocyte reaction (MLR) is also an appropriate assay for some hMSC products [95–98]. The MLR utilizes the ability of hMSCs to suppress T-cell proliferation. This suppression occurs *in vivo* through two pathways. The first pathway involves promoting T-lymphocyte apoptosis via the Fas/Fas ligand pathway. Specifically, hMSCs express the Fas ligand, while T-lymphocytes express the Fas receptor. The binding of the two induces apoptosis in the T-lymphocytes, effectively eliminating them [99]. The second pathway involves the release of factors such as B7-H4, indoleamine-2,3-dioxygenase (IDO), and haem oxygenase-1 (HO-1) [100–102] by hMSCs. These factors induce cell cycle arrest in T cells [102], and their release can be further enhanced by stimulating MSCs with TNF-α and IFN-γ [103]. *In vivo*, a regulatory mechanism exists to prevent an excessive inhibition of T-lymphocytes, which can lead to a susceptibility to infectious agents that damage the immune system [104].

These assays can help determine the effects of hMSCs on immune cells, provide insight into their mechanism of action and will aid in the development of standardized and validated assays for hMSCs. But obtaining robust and accurate results from these assays is challenging due to their variability [105]. However, before animal models or cell-based assays can be used to determine the limits of relevant factors, it is critical to establish a clear connection between the secretome profile and *in vivo* efficacy, because these modes of action are clinically relevant. The transferability of all tests to *in vivo* conditions must be evaluated. To move from complex and highly variable *in vivo* assays to a multi-assay approach, researchers could use a combination of techniques, including quantification of viability, target-specific cytotoxicity or cytokine release, surrogate biomarkers (morphological phenotype or released factors that correlate with function), bioactivity (e.g., presentation of surface markers), cell-based assays, and genomic, transcriptomic, and proteomic profiles [106].

Sterility, purity, and safety: A must-have for each hMSC product.

The sterility of hMSC products is a critical aspect that requires adherence to stringent safety and purity guidelines to minimize the risk of unintended substances, contaminants, and microbial infections. Impurities are unwanted components that arise during the manufacturing process, such as unwanted cell types, particles from manufacturing equipment and materials, and components of culture medium. On the other hand, contaminants come from external sources, such as bacteria, fungi, viruses, endotoxins, and mycoplasma [63]. Their microbial examination is regulated by the European Pharmacopoeia (EuPh) chapters 2.6.2, 2.6.14, and 2.6.27. Purity testing confirms the absence of contaminants such as residual process reagents or cell debris, while safety testing ensures the absence of microbial contaminants and the sterility of the product. However, classical microbiological methods may not be suitable for products with a limited shelf-life. Therefore, alternative tests are covered in the EuPh chapter on microbiological examination [67]. ELISA-based tests are commonly used to ensure the absence of unwanted biological agents, such as endotoxin, residual proteins, or other contaminants [107]. Multi-target flow cytometry or gene arrays can be employed to better characterize the purity of the product [107]. Genomic screening methods are available to detect the presence of viruses, bacteria, or mycoplasma in hMSC products. Since the product consists of living cells, a final sterilization step is not possible, so the entire hMSC production process must be carried out under aseptic conditions. As hMSCs are heterogeneous, it can be challenging to identify and eliminate unwanted cell types. Ideally, hMSC preparations should be pure, but impurities such as fibroblasts are often present. To achieve sufficient purity, cell-specific sorting, based on markers such as CD166 (which is expressed at higher levels on MSCs) and CD9 (which is expressed at higher levels on fibroblasts), may be used [72]. Alternatively, if most cells in the final product (>98%) meet the minimal criteria based on MSC surface markers, it may be considered sufficient. A risk assessment should be performed to prioritize the most influential attributes on the CQA list, based on their impact and certainty. The goal is to ensure that the final product meets the desired specifications and fulfills its intended therapeutic purpose [63].

The potential tumorigenicity/malignancy of therapeutic hMSC products is a safety issue [108]. According to Prockop et al. (2010), the probability of freshly isolated primary hMSCs becoming tumorigenic is extremely low, with a frequency of less than 10^{-9} per cell division cycle [109]. Nonetheless, primary hMSC populations are heterogeneous and certain parameters may promote tumorigenicity, including cell culture conditions, the donor, or cells of different ages and divisions, e.g., during long-term cell expansion. While abnormalities typically only result in senescence, it is challenging to eliminate the risk of cell transformation due to DNA damage, which is a critical process in tumor formation [108]. A study of 92 clinical-grade BM-MSCs in 2014 showed clonal mutations in only 3 of 86 cases, and none of these mutations caused a malignant transformation or phenotypic changes [110]. To exclude the possibility of abnormal cells, G-band karyotyping, and fluorescence *in situ* hybridization (FISH) can be employed, as recommended by the EMA Cell Products Working Party (CPWP) and the Committee for Advanced Therapies (CAT) [108]. If no chromosomal abnormalities are observed, it is unnecessary to conduct cytogenetic testing for each batch as a release criterion. However, cryopreservation of batch samples during manufacturing is useful for future testing if needed [17].

Enhancing hMSC Manufacturing Through Defining Critical Process Parameters (CPPs)

Every process parameter that influences the CQAs of an hMSC product is a critical process parameter (CPP). Each process step involved in the manufacturing of hMSCs has several CPPs that should be controlled. During the *in vitro* expansion step, CPPs are cell density and cell age, raw material attributes, such as growth factors and culture medium, and operational features of the culture vessel or bioreactor system, such as pH, temperature, dissolved oxygen, and agitation. The impact of each CPP on the CQAs must be quantified in a design space (Fig. 13.1). By adopting a suitable control strategy, CPPs can be maintained within their optimal operating range, ensuring the production of high-quality hMSCs that meet all the required CQAs. Due to the complexity and heterogeneity of hMSCs, the CQAs of each hMSC product are unique, and the corresponding CPPs must be identified on a case-by-case basis. However, depending on the bioreactor design, some CPPs may be challenging to access. Therefore, it is crucial to understand how the CQAs of an hMSC product are affected by different CPPs and to optimize the manufacturing process accordingly. Ultimately, adherence to QbD principles and the control of CPPs are critical to ensure the quality, safety, and efficacy of hMSC products [62, 63]. Typical CPP during the hMSC expansion have been reviewed in Barekzai et al. (2021) [62] and are only summarized below:

- **Cell-Related Parameters**
  Morphology, metabolic behaviour, proliferation rate, viability, and seeding cell density [111–116]
  Cell age, tissue source, and donor variability [12, 117, 118]
- **Biochemical Parameters** – *discussed more in Chap. 3.*

  Metabolites:
  glucose [119–124], ions, lipids, amino acids and peptides, vitamins, trace elements [125] and growth factors
  Secreted factors:
  hormones, chemokines, and cytokines [126–128]

- **Physicochemical Parameters**
  dissolved oxygen conditions (e.g., hypoxia) [129–141]
  temperature, pH and osmolarity [124, 142]
- **Equipment-Related Parameters** – *discussed more in Chap. 4.*

  Growth surface and extracellular matrix (ECM) components:
  stiffness [143], topography [144], integrins [11], coatings (10–20 μm thick) like glycosaminoglycans (GAG), collagen (COL) and fibronectin (FN).
  Reactor type/geometry:
  reactor design (e.g. STR $d_S/d_T \geq 0.4$) [145], aeration [146–150], hydrostatic pressure, compression, tension, shear stress, and agitation [97, 145, 147, 151–155].

- **Parameters in Downstream Processing (DSP)**
  Harvest [84], separation, clarification, concentration, formulation, and storage.

More CPPs are associated with storage, as fresh cells have higher metabolic fitness and replication capacity compared to cryobanked off-the-shelf cells [156]. However, the fresh approach presents logistical challenges, while cryopreservation can lead to short-term effects on hMSC viability, functionality, and *in vivo* persistence [157]. Nevertheless, the theory of efferocytosis suggests that apoptotic hMSCs may be engulfed by phagocytic macrophages, which could explain hMSC-mediated immune suppression, making the concept of hMSC fitness less relevant [158]. Overall, it is important to provide optimal culture conditions, avoiding cell stress, monitoring stem cell quality and genetic modification, and to ensure proper storage. When utilizing an off-the-shelf approach for hMSC therapy, it is essential to use cryoprotectant formulations that are devoid of xenogeneic contaminants and consist of chemically defined components that exclude dimethyl sulfoxide (DMSO). Furthermore, these cryoprotectants should possess the ability to be delivered directly to the patient's bedside without the need for any further manipulation [156].

Finding the Optimal hMSC Expansion Medium: Characterization and Optimization Using Design-of-Experiment (DoE) Approaches

The selection of the optimal growth medium, along with other CPPs, is one of the main tasks of cell culture technology in the field of *in vitro* cultivation [159]. After all, the medium serves as a source of energy and provides cells with all the components they need to survive, proliferate, and maintain cell functions [160]. Current cell culture media formulations for the isolation and expansion of hMSCs include fetal calf serum (FCS), human AB serum (ABS), or human platelet lysate (PL) as supplements. However, these established supplements are inherently ill-defined formulations containing a variety of bioactive molecules in varying compositions from batch to batch, and carry the risk of introducing pathogens such as viruses, prions, or other zoonotic agents that escape routine screening procedures [161].

In most cases, a basal medium with a buffer system is used that already contains essential amino acids, salts, vitamins, and substrates such as glucose, to which up to 20% FCS is added to provide various growth and adhesion factors, fatty acids and lipids [162, 163]. In addition to the risk of introducing pathogens into the hMSC process, the presence of immunogenic bovine proteins can trigger unwanted immune responses [164, 165]. FCS has a highly variable composition with seasonal and geographical fluctuations, making a reproducibility and standardization of the hMSC expansion process almost impossible, and a subsequent purification difficult [165]. Furthermore, FCS is not suitable for manufacturing hMSCs designated as ATMPs under GMP guidelines, as the absence of animal materials must be guaranteed [166]. Therefore, there is a need to find a substitute or alternative to a serum-based hMSC expansion. Divided into the four groups: (i) serum-free media, (ii) protein-free media, (iii) xeno-free media, and (iv) chemically defined media (CDM), there are differences in the permitted supplementary products. In serum-free media, crude protein fractions such as bovine serum albumin can be used, whereas in protein-free media only peptide fractions, i.e., the hydrolysates of the proteins, can be used. In the case of xeno-free media, animal components are completely excluded

Table 13.3 Currently available serum-free media for the cultivation of hMSCs

Condition	Brand	Manufacturer
Serum-free (SF)	StemPro MSC-SFM	Gibco/Thermo Fisher Scientific
	CellCor™ SFD MSC	Xcell Therapeutics/PeloBiotech
	StemXVivo SFM	R&D Bio-Techne
	TheraPEAK MSC GM	LONZA
	Kubota's SC GM	Phoenix songs biologicals
Xeno-free (XF)	MSC Medium Kit XF	PeloBiotech
	StemPro MSC XF	Gibco/Thermo Fisher Scientific
Xeno-/serum-free (XF/SF)	MSC GM XF	PromoCell
	Prime-XV	Irvine scientific (FujiFilm)
	MSC NutriStem® XF	Sartorius
	StemMACS™ XF	Miltenyi Biotec
	Mesencult – ACF Plus	StemCell
	PowerStem MSC1	PanBiotech
	Stemline XF MSC	Merck/Sigma Aldrich
Chemically defined (CDM)	CellCor™ CDM	Xcell Therapeutics/PeloBiotech
	StemFit CDM MSC	Ajinomoto
	SuperCult® CDM	Creative Bioarray
	SteMaxOne	Cell Culture Technologies

and only additives such as human serum albumin are allowed. The clearest criteria apply to chemically defined media, where only high-purity components, such as recombinant proteins, are allowed in precisely defined amounts [160].

However, the composition of many commercially available hMSC media (Table 13.3) is not disclosed [166], meaning that it is not possible to ensure whether hormones or larger molecules are present. This often requires the custom development of hMSC expansion media, usually based on a commercially available basal medium to which various growth and attachment factors are added. An optimal medium results in the best hMSC performance as defined by the chosen CQAs [162].

Chemically Defined Additives for hMSC Expansion

Different growth factors and additives are known to be essential for hMSC expansion (Table 13.4). In some cases, the influence of individual growth factors has been specifically discussed, but in other cases, complete formulations with many different growth factors have been tested, making it impossible to draw precise conclusions about the individual influences [165, 167, 168].

The basic fibroblast growth factor (bFGF), which has proliferation-promoting properties, is often mentioned [169, 170]. Positive effects have also been observed for the platelet-derived growth factor (PDGF) and transforming growth factor-β (TGF-β1) [170, 171]. In particular, a positive synergistic effect of these three factors

Table 13.4 Commonly used substances for enhancing *in vitro* proliferation of hMSCs

Class	Substances	Function	Concentration	References
Amino acids and peptides	Essential branched chain amino acids (BCAA): valine (Val), leucine (Leu) and isoleucine (Ile)	Increases S/G2/M cell cycle phases of MSCs; modulates the synthesis of TGF-β, PGE_2 and the transcription factors NFκB and PPAR-γ; expand the modulating capacity and increase metabolic activity	2 mM	[189]
Growth factors and cytokines	Epidermal growth factor (EGF)	Activates several signaling pathways, including the MAPK/ERK and PI3K/AKT pathways, which are involved in regulating cell proliferation and survival. EGF signaling activates the transcription factor SOX2, which is involved in the regulation of stem cell self-renewal and pluripotency. EGF is often used in combination with other growth factors, such as FGF-2 and platelet-derived growth factor (PDGF), to optimize the expansion of hMSCs in culture	10–20 ng·mL^{-1}	[177, 181]
	Basic fibroblast growth factor-2 (bFGF/FGF-2)	Stimulates the cell cycle progression of hMSCs by activating cyclin-dependent kinases (CDKs) and upregulating the expression of cyclins, which are key regulators of the cell cycle. FGF-2 also maintains the stemness of hMSCs by preventing their differentiation. FGF-2 signaling activates multiple transcription factors, including OCT4 and NANOG, which are involved in the maintenance of pluripotency and the suppression of differentiation	17–20 ng·mL^{-1}	[166, 169, 170, 172, 173, 190–193]
	Hepatocyte growth factor (HGF)	The effect of HGF occurred via the active FAK-AKT-mTOR pathway and increased the synthesis of the cyclin D1 protein, accelerating the cell cycle from G1 phase to S phase progression, and promoting the proliferation of MSCs. HGF could enhance the expression of integrin, which is helpful to signaling transduction. After adding the inhibitors of FAK and AKT, the phosphorylation of the corresponding protein decreased. Finally, after adding HGF, MSCs could still express surface markers including CD44, CD73, CD90, and	10–20 or 100 ng·mL^{-1}	[186–188, 192, 194–196]

		CD105, and led to osteogenetic and adipogenetic differentiation		
	Insulin-like growth factor-1 (IGF-1)	The signaling pathway is not fully understood, but in osteosarcoma cells, it has been linked to the ERK signaling pathway. This MAPK pathway is also characterized by the binding of growth factors to membrane receptors, which then initiate a three-step kinase cascade. The final kinases activate ERK 1 and 2, which in turn phosphorylate proteins in the nucleus involved in proliferation regulation	20 ng·mL^{-1}	[165, 170, 179, 180, 183, 184, 197, 198]
	Interleukin-3 (IL-3)	Enhances *in-vitro* migration, motility, and wound healing abilities of hMSCs	20 ng·mL^{-1}	[170, 199]
	Interleukin-6 (IL-6)	Essential for the proliferation of stromal cells and their immunosuppression capacity	20 ng·mL^{-1}	[170, 200]
	Leukemia-inhibitory factor (LIF)	Prevents differentiation of the embryonic stem cells	20 ng·mL^{-1}	[170, 201]
	Platelet-derived growth factor PDGF-(AB/BB)	Dimeric protein composed of two polypeptide chains (A and B) that bind to specific receptors on the surface of hMSCs and activates several signaling pathways, including the MAPK/ERK and PI3K/AKT pathways	20 ng·mL^{-1}	[165, 170, 172, 174, 193, 202]
	Putrescine	Involved in the regulation of the cell cycle, specifically in the G1 phase, where it promotes the transition from G1 to S phase, which is the phase of DNA replication. Putrescine promotes stem cell proliferation by activating the AKT and ERK signaling pathways.	1.61 mg·mL^{-1}; 55.9 μM	[170, 203]
	Stem cell factor (SCF)	Binds to and activates the c-Kit receptor, a transmembrane receptor tyrosine kinase that is expressed on the surface of stem cells. The activation of c-Kit by SCF triggers a downstream signaling cascade that leads to the activation of various pathways involved in cell growth and division. SCF has been shown to inhibit stem cell differentiation by regulating the activity of various transcription factors, such as Oct4 and Nanog, which are involved in the maintenance of pluripotency	10–20 ng·mL^{-1}	[195]

(continued)

Table 13.4 (continued)

Class	Substances	Function	Concentration	References
	Transforming growth factor-β1 (TGF-β1)	TGF-β1 can enhance the proliferation of MSCs by activating the MAPK and PI3K/Akt signaling pathways. In some cases, TGF-β1 has been shown to inhibit stem cell proliferation by inducing cell cycle arrest and promoting apoptosis. This effect has been observed in various stem cell types, including embryonic stem cells, neural stem cells, and hematopoietic stem cells	20 ng·mL^{-1}	[170–172, 183, 202, 204]
	Vascular endothelial growth factor (VEGF)	Activates various downstream signaling pathways, including the PI3K/Akt and MAPK pathways. VEGF has been shown to regulate the activity of various transcription factors, such as Oct4 and Nanog	20 ng·mL^{-1}	[170, 182, 183, 191]
Hormones	Corticosterone	Activates the MAPK/ERK and PI3K/Akt signaling pathways	1 µg·mL^{-1}	[203]
	Hydrocortisone	Activate the PI3K/Akt signaling pathway. It has been shown to induce cell cycle arrest and apoptosis in MSCs, particularly at high concentrations. Moreover, prolonged exposure to hydrocortisone has been shown to promote MSC differentiation towards adipogenic or osteogenic lineages, rather than self-renewal and proliferation	27.65 µM	[163, 170, 190, 205]
	Human recombinant insulin (HRI)	Activate the PI3K/Akt signaling pathway. HRI promotes MSC proliferation by stimulating the expression of CDKs, which are key regulators of cell cycle progression. HRI also upregulates the expression of various growth factors and cytokines that promote MSC proliferation, such as IGF-1 and TGF-ß	410 µg·mL^{-1}; 4.01 µM	[170, 203]
	Progesterone	Mainly by activating the PI3K/Akt signaling pathway and also upregulating the expression of several genes involved in cell proliferation and differentiation, including cyclin D1 and also has been reported to have anti-inflammatory effects	640 ng·mL^{-1}; 0.018 µM	[170, 203]

	Linoleic acid and/or Linolenic acid	Modulate the activity of key signaling pathways involved in cell proliferation, such as the PI3K/Akt and ERK/MAPK pathways	50 $\mu g \cdot mL^{-1}$	[203]
Trace elements	Human transferrin	Activate the PI3K/Akt signaling pathway and enhance the phosphorylation of Akt	5 $mg \cdot mL^{-1}$; 0.318 mM	[203]
	Sodium selenite	Increasing the expression of various antioxidant enzymes, such as glutathione peroxidase, which can help protect MSCs from oxidative stress. Sodium selenite also activate the PI3K/Akt and MAPK/ERK signaling pathways	900 $ng \cdot mL^{-1}$; 27 nM	[170, 203]

on the growth of BM-hMSCs has been described [172]. The signaling pathway that is induced is important in assessing the influence of these growth factors. For example, PDGF and bFGF induce proliferation through the c-Jun N-terminal kinase (JNK) signaling pathway [173, 174]. This pathway is a member of the mitogen-activated protein kinase (MAPK) signaling pathway and is based on a three-step kinase cascade. JNK is the kinase that then interacts with or phosphorylates the final effector, c-Jun, a component of the transcription factor AP-1, and thus intervenes in proliferation since AP-1-regulated genes are involved in cell division [175]. TGF-β1 contributes to proliferation by signaling through the intracellular protein SMAD-3 [173]. Receptor binding of TGF-β1 leads to phosphorylation of SMAD-3, which accumulates after complex formation and translocates to the nucleus, thereby activating or repressing downstream genes. A link has also been established between this signaling pathway and the Wnt signaling pathway [176].

In publications looking at the overall effect of different growth factors [165], the epidermal growth factor (EGF) is one of the factors mentioned [170]. Sabri et al. (2011) showed that EGF drives hMSC proliferation via four signaling pathways [177]. These include the three MAPK signaling pathways and the PI3K-AKT signaling pathway. In the PI3K-AKT pathway, EGF binding to the tyrosine kinase receptor activates PI3 kinase (phosphatidylinositol 3-kinase), which phosphorylates the cell membrane phospholipid component PIP2 to PIP3. PIP3 then binds to the kinases PDK1 and AKT, which, in the final step, control proliferation by phosphorylating the proteins involved [178].

Other growth factors associated with a positive proliferative effect are the insulin-like growth factor-1 (IGF-1) and the vascular epidermal growth factor (VEGF). IGF-1 is one of the most abundant growth factors in bone and, according to Doorn et al. (2013), is also classified as pro-proliferative [179]. The signaling pathway by which IGF-1 promotes proliferation is not fully understood, but in osteosarcoma cells, it has been linked to the ERK signaling pathway [180]. This MAPK pathway is also characterized by the binding of growth factors to membrane receptors, which then initiate a three-step kinase cascade. The final kinases activate extracellular signal-regulated kinases (ERK) 1 and 2, which in turn phosphorylate proteins in the nucleus involved in proliferation regulation [181]. For VEGF, Kong et al. (2010) also discovered a link to the ERK signaling pathway [182]. However, Hefka Blahnova et al. (2020) did not find a positive effect of VEGF on hMSC proliferation [183]. Similarly, Collins et al. (2010) conclude that IGF-1 does not promote the full proliferation of an hMSC population [184]. This controversy leads to the need to develop a medium that is individually adapted to the specific hMSC type. For example, the different isolation sources and tissue structures of hMSCs may be a reason for differences in the proliferation effect. In addition, such a medium must be tested for its specific use. For example, it has already been shown that a customized serum-free formulation supports the growth of hMSCs in static culture, but had to be adapted for cultivation on microcarriers [185].

Forte et al. (2006) showed that a short-term exposure of hMSCs to the hepatocyte growth factor (HGF) can induce an activation of the cognate Met receptor and the downstream effectors ERK1/2, p38MAPK, and PI3K/Akt, while a long-term exposure to HGF resulted in cytoskeletal rearrangement, cell migration, and a marked inhibition of proliferation by the arrest in the G1-S checkpoint [186]. The concentration-dependent effect of HGF was further described as Osteogenic Differentiation Medium (DM) with a low concentration of HGF (20 ng/ml) preferentially promoting a MSC osteogenic differentiation through an increased c-Met expression and phosphorylation, Akt pathway activation, and an increased expression of p27, Runx2 and Osterix. In contrast, a high concentration of HGF (100 ng/ml) strongly induced a proliferation by inducing a strong activation of the ERK1/2 signaling pathway [187]. In a serum-free medium, a low HGF concentration (10 ng/mL) increased the hMSCs proliferation and viability [188].

Optimization of Chemically Defined Media by Statistical Design-of-Experiments to Enhance Cell Proliferation

The shift towards utilizing CDM and avoiding serum-containing complex media highlights the necessity for an independent optimization of culture media to ensure a precise and tailored supply for different hMSC types. Medium development for hMSCs is a multifaceted process that requires a fine-tuning of various interacting components to support optimal cell growth and potency. To determine the optimal composition and concentration of these components for specific hMSCs and their microenvironment, appropriate experimental setups must be employed.

The traditional method of medium development runs one-factor-at-a-time (OFAT) experiments [206]. This includes a series of titrations in which single medium components are adjusted to their final concentrations while other components are held constant. One problem with this approach is the excessive number of factors that need to be optimized to fully optimize the media for a specific hMSC type. Another limitation of the OFAT approach is its inability to account for interactions among different components and conditions. Consequently, even if all OFAT experiments were performed, the combined results may not identify the optimal conditions for an ideal proliferation and hMSC potency.

Design of Experiments (DoE) is a successful and well-known tool for the development and optimization of cell culture media for bioprocessing [160]. Sequential strategies are effective in bioprocessing, whereby multiple components are first screened, followed by the characterization of the relationships between the variables, and ultimately their optimization [207–210]. However, when optimizing culture media for primary cells, the DoE approaches could not always provide reliable predictions due to a high donor variability, and have to be confirmed with each donor, or must be slightly adapted [211].

Some studies have already used a DoE approach to successfully differentiate hematopoietic stem cells and promote their expansion [198, 212–214], but specific data for a DoE approach to promote the proliferation of hMSCs have not yet been published. A good sequential approach to optimizing a serum-free medium for the growth of mouse embryonic stem cells has been reported [215]. The authors investigated the effects of 11 growth media factors using different design models. They started with a fractional factorial screening design (resolution IV) to identify the main positive or negative effects of the 11 factors. Subsequently, a full factorial design was used to identify possible two-way interactions. Finally, a Response Surface Model (RSM) was used to gain full knowledge of the concentration range. The study found that insulin and leukemia inhibitor factor (LIF) had a positive effect on cell growth, whereas zinc and cysteine had only a small effect [215].

For screening approaches with a large number of factors (as it is the case for medium optimization), the Plackett-Burman design is particularly useful for identifying the main effects in a small number of experiments, although it does not allow for the investigation of interactions between factors. It is often used as an initial screening step to identify the most important factors that should be further investigated, using more complex experimental designs, such as response surface methodology or factorial designs.

Another approach could be a fractional factorial screening design (resolution IV), which is also a type of DoE that is used to efficiently screen a large number of factors or variables, while minimizing the number of experiments required. In a fractional factorial design, only a subset of all possible experimental conditions is tested, while still providing enough information to estimate the effect of each factor on the outcome of interest. A resolution IV fractional factorial design involves testing each factor at two levels and using a design matrix that is constructed so that each factor is tested at least twice at each level. This allows for the estimation of main effects and two-way interactions among factors, while still keeping the number of experimental conditions manageable. For example, a resolution IV fractional factorial design with five factors would require only 16 experimental conditions, compared to the 32 conditions that would be required for a full factorial design. After the identification of the main factors with one of the screening designs, the final mathematical model can be created using a response surface model (RSM). This model can optimize complex relationships between the factors and response. Within this model, different designs could be chosen to define the design space, such as a central composite design (CCD) or a face-centered design (FCD), to predict the optimal conditions that result in the desired response. The RSM involves fitting a regression equation to the experimental data collected at different levels of the input variables, which can be represented as a surface in a 3D space. This surface can be visualized as a contour or a 3D plot, with the response as the z-axis and the input variables as the x- and y-axes. The RSM can be used to identify the optimal combination of input variables that will result in the maximum or minimum value of the response variable. It can also help to understand the relationships between the input variables and the response, including the presence of any interactions or non-linear effects. Summarizing the DoE approach is a powerful tool for media or process optimization in general, especially when the many parameters are involved that show complex interactions.

Continuous Bioreactor Processes Including Automation and Control Strategies to Ensure the Appropriate Amount of hMSCs for Cell Therapy

Since a high amount of hMSCs is needed for each dose, an expansion process must be highly efficient. To do so, the only viable option is a sophisticated dynamic process using bioreactors. As the hMSC is a strictly adherent cell, growth surface must be provided in bioreactor set-ups. How the growth surface is provided, is different depending on the bioreactor system. A diverse range of dynamic bioreactors are employed to expand hMSCs, including hydraulically and mechanically driven systems (Fig. 13.2).

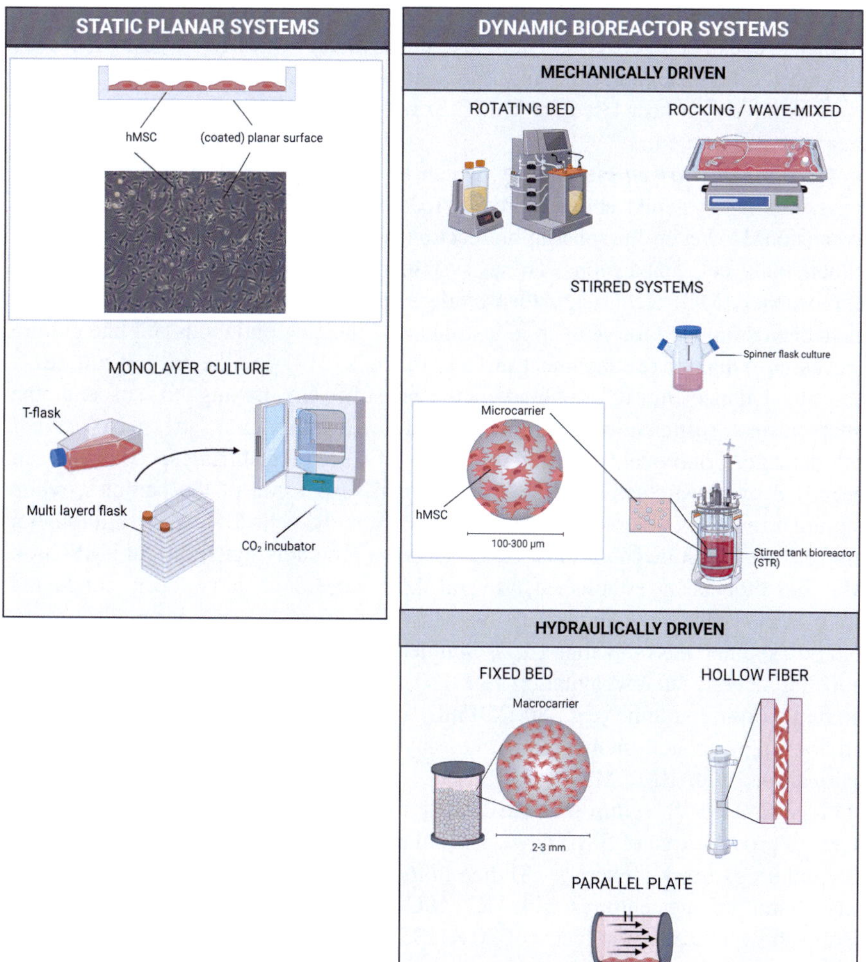

Fig. 13.2 Different cultivation systems and bioreactor types used for hMSC expansion

Various Bioreactor Types Are Suitable for hMSC Expansion

Hydraulically driven bioreactors utilize pumps to generate power, and hMSCs can grow on perfused plastic surfaces (parallel plate bioreactors), hollow fibers, or packed particles (fixed bed bioreactors.). Hollow fiber bioreactors are known to be high cell density continuous perfusion culture systems, where small tubes are placed in a cartridge shell and the cell culture medium is pumped inside. The hMSCs grow in the space outside or around the fibers, which provide a large surface area in a small volume. However, a major challenge is an effective cell detachment. Hollow fiber bioreactors and fixed bed bioreactors have also been explored for hMSC expansion (Table 13.5). For all of these bioreactors, a distinction can be made between reusable and disposable systems. In fixed-bed bioreactors, cells are immobilized or encapsulated in a matrix of particles packed in a cylindrical column, which may consist of macroporous beads, porous ceramic or glass beads, glass fibers, polyester disks, alginate beads, or hydrogels. The cell culture medium gently flows to the cells through the bed-packing, applying only a low shear towards the cells. Fixed bed systems have also been used for hMSC expansion [216], but the available growth area was still limited.

Mechanically driven bioreactors, which include stirred tank bioreactors (STR), wave mixing systems, and rotating bioreactors, are frequently used for hMSC expansion. However, the rotating bioreactors are mainly used for tissue engineering rather than cell expansion. These systems have often been combined with microcarrier (MC) technology to provide a growth surface for the hMSCs. MC have been employed for years in industrial and adherent continuous cell line culture processes to maximize attachment surfaces for cells [217]. In adherent cell cultures, a delicate balance must be found between sufficient mixing to suspend the microcarriers, sufficient oxygen transfer, and low hydrodynamic stresses to prevent cell damage. Consequently, different choices of stirrer and stirring speed have been proposed by researchers to achieve an optimal suspension of the particles, while minimizing hydrodynamic stresses experienced by the cells [218–221]. Although a uniform set of culture conditions (concentrations, protocols, agitation) for hMSCs on MC has not been established, several MC screenings have been conducted [117, 222]. These screening were often carried out in small-scale dynamic systems, such as spinner flasks (Table 13.5). Spinner flasks with a stirrer or magnetic rod impeller operate at low agitation (30–100 rpm). Because spinner flasks have a limited working volume (e.g., 80–250 mL) and lack of monitoring in basic configurations, they are not suitable for a large-scale hMSC production. Numerous spinner cultivations with BM-hMSCs, AT-hMSCs or hMSC-TERT in serum-containing [117, 222–229] or serum-free [230–234] media have been performed with an average process time of 5–10 days. Thus, the average expansion factor in a serum-free culture is much higher (4–18) than in media containing FCS (2.5–8.4). On the other hand, spinner cultures with UC-MSCs and FCS substitutes such as human serum show an expansion factor of 20.8 [235], with human platelet lysate (HPL) of 15 [15], and in serum-free culture between 5 and 13 [185, 236].

Table 13.5 Cell culture conditions for hMSC expansion in spinner flasks, discontinuous (*batch*) and continuous (*fed-batch* /*perfusion*) bioreactors

Adherence/growth surface ($k = cm^2 \cdot mL^{-1}$)		Working volume	MSC source	Medium supplementation	Duration (days)	Agitation (rpm)	*EF*	References
Spinner flasks								
Non-porous plastic	$k = 7.2$	80 mL	AT-hMSC	StemPro MSC-SFM	9	40	14	[230]
Cultisphere S	/	N/A	BM-hMSC	FBS	10	N/A	8	[223]
Cytodex-3	$k = 8.1$	50 mL	BM-hMSC	FBS	10	50	4	[224]
pNIPAM MC	$k = 13.5$	50 mL	BM-hMSC	FBS	5	40	2.5	[225]
Cytodex-3	$k = 13.5$	100 mL	BM-hMSC	FBS	5	40	2.5	[225]
Cytodex-1	$k = 20$	100 mL	BM-hMSC	FBS	9	30	4	[222]
pNIPAM MC	$k = 4.2$	100 mL	BM-hMSC	FBS	7	85	5	[226]
Cytodex-3	$k = 4.2$	35 mL	BM-hMSC	FBS	7	85	5	[226]
Synthemax II	$k = 5–12$	80 mL	BM-hMSC	Mesencult – ACF plus	40	30	6	[233]
Enhanced attachment	$k = 4.5$	80 mL	BM-hMSC	StemPro MSC-SFM	8	40	7	[231]
Plastic	$k = 7.2$	80 mL	BM-hMSC	StemPro MSC-SFM	8	40	7	[231]
Synthemax II	$k = 4.5$	80 mL	BM-hMSC	StemPro MSC-SFM	8	40	7	[231]
Plastic P-102 L	$k = 9$	100 mL	BM-hMSC	HPL	6	30	4	[227]

(continued)

Table 13.5 (continued)

Adherence/growth surface ($k = cm^2 \cdot mL^{-1}$)		Working volume	MSC source	Medium supplementation	Duration (days)	Agitation (rpm)	*EF*	References
Plastic P-102 L	$k = 5$	100 mL	BM-hMSC	Prime-XV SFM	9	30	5	[232]
Synthemax II	$k = 5.8$	120 mL	BM-hMSC	Mesencult – ACF Plus	7	45	10	[234]
Non-porous plastic	$k = 5.4$	80 mL	BM-hMSC	StemPro MSC-SFM	9	40	18	[230]
Enhanced attachment	$k = 5.4$	100 mL	TERT-hMSC	FBS	6	100	4	[117, 228]
Glass coated	$k = 5.4$	100 mL	TERT-hMSC	FBS	6	100	4	[117]
ProNectin	$k = 5.4$	100 mL	TERT-hMSC	FBS	6	100	3	[117]
Synthemax	$k = 5.4$	100 mL	TERT-hMSC	FBS	6	100	3	[117]
Biosilon/P102 L/Rapidcell	/	250 mL	TERT-hMSC	FBS	9	30	11	[229]
Plastic Plus	$k = 9$	100 mL	UC-hMSC	HPL	8	40	15	[15]
Plastic	$k = 4.3$	80 mL	UC-hMSC	UltraGro	6	40	13	[236]
Cytodex-1	$k = 1.8$	100 mL	UC-hMSC	SFM	7	38	5	[185]
Plastic P-102 L	$k = 11$	100 mL	UC-hMSC	Human serum	8	30	21	[235]
Discontinuous (batch)								
Plastic MC	/	STR (3.75 L) Batch	AT-hMSC	FBS	20	35	7	[264]

Collagen coated MC	/	STR (3.75 L) Batch	AT-hMSC	FBS	20	35	14	[264]
ProNectin-F coated MC	$k = 1.8–13.3$	STR (35 L) Batch	AT-hMSC	FBS	8	14	35	[265]
Plastic P-102 L	$k = 4–10$	STR (2.5 L) Batch	BM-hMSC	FBS	12	75	6	[266]
Plastic P-102 L	$k = 5$	STR (0.1 L) Batch	BM-hMSC	Prime-XV	6	115	4	[147]
Plastic P-102 L	$k = 5$	Ambr systems Batch	BM-hMSC	Prime-XV	9	400	10	[232]
Synthemax II	$k = 5.8–17$	PBS-VW (2.2 L) Batch	BM-hMSC	Mesencult – ACF Plus	14	45	12	[267]
Cytodex-1	$k = 5.6$	STR (1.5 L) Batch	UC-hMSC	FBS	6	80	7	[238]
Cultispher S	/	Wave bioreactor (0.6 L) Batch	UC-hMSC	FBS	10	24	12–26	[268]
Cultispher S	$k = 40$	STR (0.8 L)batch	UC-hMSC	StemPro MSC-SFM	6	50	6	[239]
Plastic P-102 L	$k = 3.6$	STR (0.8 L)batch	UC-hMSC	Human serum	7	30	9	[235]
Continuous (fed-batch / perfusion)								
Non-porous SoloHill® plastic MC coated with CELLstart™ substrate	$k = 4.1–14.4$	STR (2 L)fed-batch/ perfusion	BM-hMSC	StemPro MSC-SFM	9–13	40–60	6–11	[269]
Corning Synthemax II-coated dissolvable microcarriers	$k = 4$	STR (2 L)fed-batch/ perfusion	UC-hMSC	N/A	11–12	70	43–60	[241]
Non-porous plastic MC	$k = 7.2$	STR (0.8 L) Fed-batch	AT-hMSC	StemPro MSC-SFM	7	60	9	[245]
Non-porous plastic MC	$k = 7.2$	STR (0.8 L) Fed-batch	BM-hMSC	StemPro MSC-SFM	7	60	22	[245]

(continued)

Table 13.5 (continued)

Adherence/growth surface ($k = cm^2 \cdot mL^{-1}$)		Working volume	MSC source	Medium supplementation	Duration (days)	Agitation (rpm)	*EF*	References
Synthemax II	$k = 5.8$	STR (2 L) Perfusion	BM-hMSC	Mesencult – ACF Plus	7	100	14	[234]
Collagen coated MC	$k = 5.4$	STR (50 L) Fed-batch	hMSC	HPL	11	64	43	[97]
Polycarbonate plates	$k = 17.65$	Rotating bed (0.34 L) Perfusion	UC-hMSC	FBS	9	2	39	[270]
Poly-ε-caprolactone (PCL)	/	STR (0.3 L) Fed-batch	UC-hMSC	FBS	7	80	9	[141]
Cytodex-3	$k = 10.8$	STR (0.3 L) Fed-batch	UC-hMSC	FBS	7	80	8	[141]
Hollow fiber bioreactors								
/	/	Hollow fiber (quantum, Terumo BCT)	AT-hMSC	FBS	17	/	5	[271]
/	/	Hollow fiber (quantum, Terumo BCT)	BM-hMSC	HPL	20	/	20	[272]
/	/	Hollow fiber (quantum, Terumo BCT)	BM-hMSC	HPL	13	/	10–20	[273]
/	/	Hollow fiber (quantum, Terumo BCT)	BM-hMSC	FBS	7	/	6–14	[274]
Fixed-bed								
Fibra-Cell® disks	10 g	Fixed bed (0.5 L) (New Brunswich)	UC-hMSC	FBS	7	/	7	[216, 275]
Fibra-cell® disks	25 g	Fixed bed (2.5 L) (New Brunswich, Celligen 310)	BM-hMSC	FBS	9	/	9	[276]

Non-porous borosilicate glass spheres (2 mm)	$A_{FB/}$ $V_{FB:}$18.3 cm^2/ cm^3 60–300 cm^3	Fixed bed bioreactor	TERT-hMSC	FBS	7	/	17–26	[146, 277]
Rocking/wave bag bioreactor								
Cytodex-3; Cultispher S	/	Rocking/wave bag (2 L) CultiBag (Sartorius-Stedim) or Cellbag (wave, GE life healthcare)	Placenta hMSC	FBS	7	/	10–15	[278]
Cytodex-3	/	Rocking/wave bag (0.5–1.5; L-50 L) BIOSTAT CultiBag RM (Sartorius-Stedim)	AT-hMSC BM-hMSC	FBS	9	/	30–52	[279, 280]

Higher expansion factors are achieved in an STR rather than in spinner flasks (Table 13.5). The STR uses a top or bottom-mounted rotating agitator system with different impellers depending on the mixing parameters selected and is typically equipped with sensors to monitor and control various parameters, such as temperature, pH, dissolved oxygen (DO), carbon dioxide concentrations, gas and liquid flow rates, and impeller or rocking speed. The temperature can be precisely monitored and controlled, e.g., at 37 °C, the pH values can be maintained between 7.2 and 7.4, and hypoxic conditions are preferred for hMSC cultures, with the DO value set at 20% of air saturation (4% of pure O_2). Furthermore, the expansion factor differs depending on the process mode (Table 13.5). STRs allow for advanced process modes and control, including bead-to-bead transfer, the ability to detach from the microcarrier during culture upon division and attach to another microcarrier, for the development of MC-based hMSC expansion processes [237].

Advanced Process Modes for hMSC Expansion Are Rapidly Evolving

Traditionally, hMSC cultures have been performed in *batch* mode using bioreactors ranging in volume from 1 to 10 L, without the addition of a culture medium or MC. However, UC-hMSC cultures in batch mode last only a few days due to a rapid nutrient depletion and a lack of available surface growth [238, 239]. The limited productivity of a *batch* culture can be improved by using an advanced process mode, such as *fed-batch* or *perfusion*.

The *fed-batch* operation mode involves a gradual increase in the working volume of the bioreactor as the substrates are consumed by the cells and can be performed using various feeding strategies (e.g., *repeated batch* or *pulsed feed, fixed, linear,* or *exponentially increasing feed rate*). In the context of hMSC cultures in bioreactors, the *fed-batch* mode is commonly used to improve cell adhesion within a small volume of cell culture medium, followed by a gradual increase in the volume of the culture medium. This allows for a continuous supply of nutrients and helps to dilute the concentration of toxic metabolites (e.g., lactate with a critical concentration of 34 mM or ammonia – NH_3 with 2.4 mM) produced by the cells [240], thereby enhancing cell viability and productivity.

During *Perfusion*, cells are typically retained within the STR via spin-filter [241] or TFF [242] approaches, while the used medium is continuously removed. One of the primary advantages of perfused STRs is the ability to precisely control the feed and removal rates of the cell culture medium, up to several volumes of reactor per day. In contrast, continuous modes rely on cell growth kinetics to determine these rates [243]. Serra et al. [244] described the use of a perfused stirred tank bioreactor for the expansion of undifferentiated human embryonic stem cells (hESC) in 2010. Their experiments used a probe adapted to the bioreactor cap and an automated gravimetric media perfusion control system. This approach allowed for the renewal of nutrients and growth factors, as well as the removal of metabolic waste byproducts. Dos Santos et al. (2014) [245] tested a perfusion mode in STR with BM-hMSC, applying a perfusion rate of 0.25 day^{-1} and maintaining a working volume of 400 mL. A cell device retention system based on a modified alternating tangential flow (ATF) system (Repligen Corporation, Massachusetts, USA) was used to maintain hMSCs and MC within the STR, resulting in an expansion factor of 18.5. Finally, Cunha et al. (2015) [242] tested a Tangential Flow Filtration (TFF) system to maintain cells and MC in the STR, which allowed a higher expanded cell ratio for a longer time. It should be noted that all of these reported studies focused primarily on substrate limitations, rather than growth surface limitations.

However, the expansion of hMSCs in chemically defined media is accompanied by some CPPs that may prevent maximum cell density. These include cell-cell agglomeration of hMSCs, unoccupied MC, or loss of viability. Some publications have reported supplemental feeding of MC to allow bead-to-bead transfer, partly coupled with a 25% medium feeding strategy [233, 245]. This allowed a long stable expansion and may control aggregate formation. Aggregation was observed only at high cell densities. Intermittent stirring can also result in a homogeneous distribution of cells on the MC. Reducing the surface/volume ratio at the beginning of the cultivation is also often used to promote adhesion [231, 232]. hMSC expansion using a continuous process mode in a chemically defined medium is not established yet and has still room for improvements.

Real-Time Monitoring Is Essential for Continuous Production Control

The use of an STR with an advanced process mode is a promising approach for the expansion of therapeutic hMSCs. A key advantage of STRs is the ability to monitor important parameters such as temperature, pH, and oxygen levels in real-time, which can ensure a certain process stability. However, while online monitoring provides valuable data, many CPPs for hMSC expansion still rely on offline measurements to ensure quality control. As the concept of Process Analytical Technology (PAT), proposed by the FDA in 2004, has been endorsed by other major regulatory agencies such as the European Medicines Agency (EMEA) and the Ministry of Health, Labour & Welfare (MHLW) of Japan, there is a need for advanced monitoring and control tools. The primary objectives of PAT are to reduce production costs, improve biological yields, ensure real-time product release, and provide routine process audits. PAT involves the use of various analytical chemistry tools to monitor and control CPPs during the production process. Process knowledge is one part of PAT, but process control using *on-line* analytical sensors based on near-infrared, FTIR (MIR) [246], Raman [85, 247–250], or dielectric spectroscopy [251, 252] is a key element even for advanced process modes.

The application of spectroscopy-based techniques such as UV-vis spectroscopy, fluorescence spectroscopy, infrared spectroscopy, Raman spectroscopy, and dielectric spectroscopy for an *on-line* monitoring of cell viability has been demonstrated. Some of the technologies have already been applied for hMSC grown on MC [253]. For fed-batch and continuous processes, precise regulation is required to ensure an optimal process control. Most spectroscopic sensors are effective for continuous process monitoring and provide detailed information on the molecular, physical, and metabolic states of the process. Because they do not require reagents or sampling, they are non-invasive in situ techniques. Soft sensors, usually in combination with bioinformatic data analysis, can estimate multiple biological variables besides viability from the same set of measured data. However, most of the presented sensors are still in the research stage, and only a few are commercially available. High accuracy, reproducibility, selectivity, sensitivity, robustness, and stability are challenges for soft sensors to provide a reliable replacement for off-line methods. Therefore, validation of soft sensors is important. Soft sensors have advantages over special sensors in terms of easier maintenance and lower acquisition costs [253]. Increasingly, computational and artificial intelligence-based approaches are focused on designing automated bioreactors that can produce functional cells and tissues more efficiently [254]. Some recent studies have shown promising results in this area [255–259]. Computational modeling is also proving to be effective in improving the predictability of the clinical-grade cell product [260–263]. These advances in bioprocessing technology are expected to lead to improved predictability and clinical outcomes for hMSC-based therapies.

Conclusion

Multiple challenges must be addressed to create reliable processes for expanding hMSCs in bioreactors. These challenges include the need for a better understanding of the CPPs that affect hMSC production, as well as improved real-time monitoring tools to control these parameters and measure their influence on CQAs such as cell viability, potency, and secretory profiles. To meet the requirements of PAT for GMP manufacturing, it is crucial to develop a standardized process that mimics the natural hMSC niche but can also be scaled up for clinical trials without sacrificing CQAs such as cell functionality and potency.

One of the remaining challenges in the development of hMSC therapies is the production of cell mass. While bioreactor-based production systems can improve the scalability of hMSC manufacturing, there is a need to further increase cell densities and productivity to meet the demands of clinical trials. Achieving high cell densities can be challenging due to the sensitivity of hMSCs to shear stress and their tendency to aggregate in bioreactor culture. Moreover, mass production is limited by the finite proliferative capacity of hMSC *in vitro* and their tendency to fail the CQAs with an increasing number of divisions. The development of affordable and chemically defined media that can support the growth of hMSCs without compromising their quality will be essential for the widespread adoption of hMSC therapies. The use of chemically defined media will also help to address the regulatory concerns associated with the use of animal-derived supplements.

The heterogeneity of hMSCs obtained from various donors and tissues, resulting in variable phenotypes and functions, is also a challenge for manufacturing, making it very complicated to achieve a standardized and reproducible platform process. On the one hand, the development of well-controlled *in vitro* expansion processes in bioreactors can help reduce batch-to-batch variation in hMSCs, ensuring that cell products remain more homogeneous. To produce hMSCs with standardized properties, it is essential to develop validated potency assays that facilitate a comparison across different laboratories [63]. On the other hand, genetic modification techniques can generate more homogeneous hMSCs or "priming" hMSCs with small molecules that can enhance their therapeutic function exogenously [281]. The next generation of hMSC therapeutics is expected to consist of complex ATMPs that may include a combination of cell therapy products, genetic engineering products (viral and/or non-viral vectors), tissue engineering products (biomaterials/scaffolds), tissue architecture techniques (3D bioprinting and decellularized organs), and/or medical devices. Although this is an exciting prospect for the future, these approaches will be challenging to implement, due to their complexity, both in terms of GMP, process scalability, and regulatory aspects [156].

Successful clinical translation of hMSC therapy relies on robust processes for hMSC expansion in bioreactors with an improved understanding of the CPPs affecting hMSC CQAs and the accurate real-time monitoring tools to control these CPPs to create well-controlled conditions that reduce batch-to-batch variation.

Acknowledgements The authors thanks Catherine Meckel-Oschmann for proof-reading the manuscript. We further thank for funding within the EU Horizon EuroStar program (Project acronym: Carrier4TSC).

Conflict of Interests The authors declare no conflict of interest.

References

1. Caplan AI (1991) Mesenchymal stem cells. J Orthop Res 9(5):641–650
2. Kern S, Eichler H, Stoeve J, Klüter H, Bieback K (2006) Comparative analysis of mesenchymal stem cells from bone marrow, umbilical cord blood, or adipose tissue. Stem Cells 24(5): 1294–1301
3. Weiss ML et al (2006) Human umbilical cord matrix stem cells: preliminary characterization and effect of transplantation in a rodent model of Parkinson's disease. Stem Cells 24(3): 781–792
4. Herrera MB et al (2006) Isolation and characterization of a stem cell population from adult human liver. Stem Cells 24(12):2840–2850
5. Perry BC et al (2008) Collection, cryopreservation, and characterization of human dental pulp-derived mesenchymal stem cells for banking and clinical use. Tissue Eng C Methods 14(2): 149–156

6. Maerz JK et al (2016) Bone marrow-derived mesenchymal stromal cells differ in their attachment to fibronectin-derived peptides from term placenta-derived mesenchymal stromal cells. Stem Cell Res Ther 7(1):29
7. Liu Y et al (2018) Biological characteristics of human menstrual blood-derived endometrial stem cells. J Cell Mol Med 22(3):1627–1639
8. Dulak J, Szade K, Szade A, Nowak W, Józkowicz A (2015) Adult stem cells: hopes and hypes of regenerative medicine. Acta Biochim Pol 62(3):329–337
9. Petrenko Y et al (2020) A comparative analysis of multipotent mesenchymal stromal cells derived from different sources, with a focus on Neuroregenerative potential. Sci Rep 10(1): 4290
10. Donders R et al (2018) Human Wharton's jelly-derived stem cells display a distinct immunomodulatory and proregenerative transcriptional signature compared to bone marrow-derived stem cells. Stem Cells Dev 27(2):65–84
11. Maijenburg MW et al (2010) Cell cycle and tissue of origin contribute to the migratory behaviour of human fetal and adult mesenchymal stromal cells. Br J Haematol 148(3): 428–440
12. Ketterl N et al (2015) A robust potency assay highlights significant donor variation of human mesenchymal stem/progenitor cell immune modulatory capacity and extended radio-resistance. Stem Cell Res Ther 6(1):236
13. Calle A et al (2018) Iberian pig mesenchymal stem/stromal cells from dermal skin, abdominal and subcutaneous adipose tissues, and peripheral blood: in vitro characterization and migratory properties in inflammation. Stem Cell Res Ther 9(1):178
14. Menard C et al (2013) Clinical-grade mesenchymal stromal cells produced under various good manufacturing practice processes differ in their immunomodulatory properties: standardization of immune quality controls. Stem Cells Dev 22(12):1789–1801
15. Petry F, Smith JR, Leber J, Salzig D, Czermak P, Weiss ML (2016) Manufacturing of human umbilical cord mesenchymal stromal cells on microcarriers in a dynamic system for clinical use. Stem Cells Int 2016:4834616
16. Olsen TR, Ng KS, Lock LT, Ahsan T, Rowley JA (2018) Peak MSC-Are we there yet? Front Med 5(June):178
17. Elseberg C, Leber J, Weidner T, Czermak P (2017) The challenge of human mesenchymal stromal cell expansion: current and prospective answers. In: New insights into cell culture technology. IntechOpen, Rejika
18. García-Bernal D et al (2021) The current status of mesenchymal stromal cells: controversies, unresolved issues and some promising solutions to improve their therapeutic efficacy. Front Cell Dev Biol 9:650664
19. Arutyunyan I, Elchaninov A, Makarov A, Fatkhudinov T (2016) Umbilical cord as prospective source for mesenchymal stem cell-based therapy. Stem Cells Int 2016:6901286
20. Troyer DL, Weiss ML (2008) Concise review: Wharton's jelly-derived cells are a primitive stromal cell population. Stem Cells 26(3):591–599
21. Emnett RJ et al (2016) Evaluation of tissue homogenization to support the generation of GMP-compliant mesenchymal stromal cells from the umbilical cord. Stem Cells Int 2016: 3274054
22. Stenderup K, Justesen J, Clausen C, Kassem M (2003) Aging is associated with decreased maximal life span and accelerated senescence of bone marrow stromal cells. Bone 33(6): 919–926
23. Mueller SM, Glowacki J (2001) Age-related decline in the osteogenic potential of human bone marrow cells cultured in three-dimensional collagen sponges. J Cell Biochem 82(4):583–590
24. Knudtzon S (1974) In vitro growth of granulocytic colonies from circulating cells in human cord blood. Blood 43(3):357–361
25. McElreavey KD, Irvine AI, Ennis KT, McLean WHI (1991) Isolation, culture and characterisation of fibroblast-like cells derived from the Wharton's jelly portion of human umbilical cord. Biochem Soc Trans 19(1):29S

26. Wang H et al (2004) Mesenchymal stem cells in the Wharton's jelly of the human umbilical cord. Stem Cells 22(7):1330–1337
27. Dominici M et al (2006) Minimal criteria for defining multipotent mesenchymal stromal cells. The International Society for Cellular Therapy position statement. Cytotherapy 8(4):315–317
28. Kim DW, Staples M, Shinozuka K, Pantcheva P, Kang SD, Borlongan CV (2013) Wharton's jelly-derived mesenchymal stem cells: phenotypic characterization and optimizing their therapeutic potential for clinical applications. Int J Mol Sci 14(6):11692–11712
29. Seshareddy K, Troyer D, Weiss ML (2008) Method to isolate mesenchymal-like cells from Wharton's jelly of umbilical cord. Methods Cell Biol 86:101–119
30. Mitchell KE et al (2003) Matrix cells from Wharton's jelly form neurons and glia. Stem Cells 21(1):50–60
31. Tong CK et al (2011) Generation of mesenchymal stem cell from human umbilical cord tissue using a combination enzymatic and mechanical disassociation method. Cell Biol Int 35(3): 221–226
32. Wang L, Seshareddy K, Weiss ML, Detamore MS (2009) Effect of initial seeding density on human umbilical cord mesenchymal stromal cells for fibrocartilage tissue engineering. Tissue Eng A 15(5):1009–1017
33. Wang L, Zhao L, Detamore MS (2011) Human umbilical cord mesenchymal stromal cells in a sandwich approach for osteochondral tissue engineering. J Tissue Eng Regen Med 5(9): 712–721
34. Nirmal RS, Nair PD (2013) Significance of soluble growth factors in the chondrogenic response of human umbilical cord matrix stem cells in a porous three dimensional scaffold. Eur Cells Mater 26:234–251
35. Chen X et al (2013) Chondrogenic differentiation of umbilical cord-derived mesenchymal stem cells in type I collagen-hydrogel for cartilage engineering. Injury 44(4):540–549
36. Conconi MT et al (2006) CD105(+) cells from Wharton's jelly show in vitro and in vivo myogenic differentiative potential. Int J Mol Med 18(6):1089–1096
37. Friedman R, Betancur M, Boissel L, Tuncer H, Cetrulo C, Klingemann H (2007) Umbilical cord mesenchymal stem cells: adjuvants for human cell transplantation. Biol Blood Marrow Transplant 13(12):1477–1486
38. Peng J et al (2011) Human umbilical cord Wharton's jelly-derived mesenchymal stem cells differentiate into a Schwann-cell phenotype and promote neurite outgrowth in vitro. Brain Res Bull 84(3):235–243
39. Hsieh JY et al (2013) Mesenchymal stem cells from human umbilical cord express preferentially secreted factors related to neuroprotection, neurogenesis, and angiogenesis. PLoS One 8(8):e72604
40. Chao KC, Chao KF, Fu YS, Liu SH (2008) Islet-like clusters derived from mesenchymal stem cells in Wharton's jelly of the human umbilical cord for transplantation to control type 1 diabetes. PLoS One 3(1):e1451
41. Wang HS et al (2011) Transplantation of insulin-producing cells derived from umbilical cord stromal mesenchymal stem cells to treat NOD mice. Cell Transplant 20(3):455–466
42. Lu LL et al (2006) Isolation and characterization of human umbilical cord mesenchymal stem cells with hematopoiesis-supportive function and other potentials. Haematologica 91(8): 1017–1028
43. Jyothi Prasanna S, Sowmya Jahnavi V (2011) Wharton's jelly mesenchymal stem cells as off-the-shelf cellular therapeutics: A closer look into their regenerative and immunomodulatory properties. Open Tissue Eng Regen Med J 4(SPEC. ISSUE 1):28–38
44. Weiss ML et al (2008) Immune properties of human umbilical cord Wharton's jelly-derived cells. Stem Cells 26(11):2865–2874
45. Prasanna SJ, Gopalakrishnan D, Shankar SR, Vasandan AB (2010) Pro-inflammatory cytokines, IFNγ and TNFα, influence immune properties of human bone marrow and Wharton jelly mesenchymal stem cells differentially. PLoS One 5(2):e9016

46. Ménard C, Tarte K (2011) Immunosuppression and mesenchymal stem cells: back to the future. Med Sci 27(3):269–274
47. He X et al (2020) Effect of intramyocardial grafting collagen scaffold with mesenchymal stromal cells in patients with chronic ischemic heart disease: a randomized clinical trial. JAMA Netw Open 3(9):e2016236
48. Gao LR et al (2015) Intracoronary infusion of Wharton's jelly-derived mesenchymal stem cells in acute myocardial infarction: double-blind, randomized controlled trial. BMC Med 13(1):162
49. Baker M (2009) How to fix a broken heart? Nature 460(7251):18–19
50. Psaltis PJ, Zannettino ACW, Worthley SG, Gronthos S (2008) Concise review: mesenchymal stromal cells: potential for cardiovascular repair. Stem Cells 26(9):2201–2210
51. Nekanti U, Rao VB, Bahirvani AG, Jan M, Totey S, Ta M (2010) Long-term expansion and pluripotent marker array analysis of Wharton's jelly-derived mesenchymal stem cells. Stem Cells Dev 19(1):117–130
52. Chen MY, Lie PC, Li ZL, Wei X (2009) Endothelial differentiation of Wharton's jelly-derived mesenchymal stem cells in comparison with bone marrow-derived mesenchymal stem cells. Exp Hematol 37(5):629–640
53. Wu KH et al (2009) Cardiac potential of stem cells from whole human umbilical cord tissue. J Cell Biochem 107(5):926–932
54. Bartolucci J et al (2017) Safety and efficacy of the intravenous infusion of umbilical cord mesenchymal stem cells in patients with heart failure: a phase 1/2 randomized controlled trial (RIMECARD trial [randomized clinical trial of intravenous infusion umbilical cord mesenchymal stem cells on cardiopathy]). Circ Res 121(10):1192–1204
55. Hare JM et al (2009) A randomized, double-blind, placebo-controlled, dose-escalation study of intravenous adult human mesenchymal stem cells (prochymal) after acute myocardial infarction. J Am Coll Cardiol 54(24):2277–2286
56. Ulus AT et al (2020) Intramyocardial transplantation of umbilical cord mesenchymal stromal cells in chronic ischemic cardiomyopathy: a controlled, randomized clinical trial (HUC-HEART trial). Int J Stem Cells 13(3):364–376
57. Can A et al (2015) Human umbilical cord mesenchymal stromal cell transplantation in myocardial ischemia (HUC-HEART trial). A study protocol of a phase 1/2, controlled and randomized trial in combination with coronary artery bypass grafting. Stem Cell Rev Rep 11(5):752–760
58. Lipsitz YY, Timmins NE, Zandstra PW (2016) Quality cell therapy manufacturing by design. Nat Biotechnol 34(4):393–400
59. Rathore AS, Winkle H (2009) Quality by design for biopharmaceuticals. Nat Biotechnol 27(1):26–34
60. Rathore AS (2009) Roadmap for implementation of quality by design (QbD) for biotechnology products. Trends Biotechnol 27(9):546–553
61. Maillot C, Sion C, De Isla N, Toye D, Olmos E (2021 Sept) Quality by design to define critical process parameters for mesenchymal stem cell expansion. Biotechnol Adv 50:107765
62. Barekzai J, Petry F, Czermak P, Salzig D (2021) Process design for human mesenchymal stem cell products in stirred-tank bioreactors. In: Pörtner R (ed) Cell culture engineering and technology, Cell engineering, vol 10. Springer, Cham, pp 307–333
63. Barekzai J, Petry F, Zitzmann J, Czermak P, Salzig D (2020) Bioprocess development for human mesenchymal stem cell therapy products. In: New advances on fermentation processes. IntechOpen, Rejika
64. Nery AA, Nascimento IC, Glaser T, Bassaneze V, Krieger JE, Ulrich H (2013) Human mesenchymal stem cells: From immunophenotyping by flow cytometry to clinical applications. Cytom A 83 A(1):48–61
65. Al-Nbaheen M et al (2013) Human stromal (mesenchymal) stem cells from bone marrow, adipose tissue and skin exhibit differences in molecular phenotype and differentiation potential. Stem Cell Rev Rep 9(1):32–43

66. Kornicka K, Marycz K, Tomaszewski KA, Marędziak M, Smieszek A (2015) The effect of age on osteogenic and adipogenic differentiation potential of human adipose derived stromal stem cells (hASCs) and the impact of stress factors in the course of the differentiation process. Oxidative Med Cell Longev 2015:309169
67. Torre ML et al (2015) Ex vivo expanded mesenchymal stromal cell minimal quality requirements for clinical application. Stem Cells Dev 24(6):677–685
68. Galipeau J et al (2015) International Society for Cellular Therapy perspective on immune functional assays for mesenchymal stromal cells as potency release criterion for advanced phase clinical trials. Cytotherapy 18(2):151–159
69. Ferragut F, Vachetta VS, Troncoso MF, Rabinovich GA, Elola MT (2021) ALCAM/CD166: a pleiotropic mediator of cell adhesion, stemness and cancer progression. Cytokine Growth Factor Rev 61:27–37
70. Brinkhof B, Zhang B, Cui Z, Ye H, Wang H (2020) ALCAM (CD166) as a gene expression marker for human mesenchymal stromal cell characterisation. Gene 763:100031
71. Mildmay-White A, Khan W (2017) Cell surface markers on adipose-derived stem cells: a systematic review. Curr Stem Cell Res Ther 12(6):484–492
72. Halfon S, Abramov N, Grinblat B, Ginis I (2011) Markers distinguishing mesenchymal stem cells from fibroblasts are downregulated with passaging. Stem Cells Dev 20(1):53–66
73. Wuchter P et al (2015) Standardization of good manufacturing practice-compliant production of bone marrow-derived human mesenchymal stromal cells for immunotherapeutic applications. Cytotherapy 17(2):128–139
74. Masterson C et al (2018) Syndecan-2-positive, bone marrow-derived human mesenchymal stromal cells attenuate bacterial-induced acute lung injury and enhance resolution of ventilator-induced lung injury in rats. Anesthesiology 129(3):502–516
75. Psaltis PJ et al (2010) Enrichment for STRO-1 expression enhances the cardiovascular paracrine activity of human bone marrow-derived mesenchymal cell populations. J Cell Physiol 223(2):530–540
76. Pochampally R (2008) Colony forming unit assays for MSCs. Methods Mol Biol 449:83–91
77. Yang HJ et al (2014) The stem cell potential and multipotency of human adipose tissue-derived stem cells vary by cell donor and are different from those of other types of stem cells. Cells Tissues Organs 199(5–6):373–383
78. Stroncek DF et al (2020) Human mesenchymal stromal cell (MSC) characteristics vary among laboratories when manufactured from the same source material: a report by the cellular therapy team of the biomedical excellence for safer transfusion (BEST) collaborative. Front Cell Dev Biol 8:458
79. Smith JR, Pfeifer K, Petry F, Powell N, Delzeit J, Weiss ML (2016) Standardizing umbilical cord mesenchymal stromal cells for translation to clinical use: selection of GMP-compliant medium and a simplified isolation method. Stem Cells Int 2016:6810980
80. Kuo YC et al (2015) Oscillatory shear stress mediates directional reorganization of actin cytoskeleton and alters differentiation propensity of mesenchymal stem cells. Stem Cells 33(2):429–442
81. Zhao LG et al (2014) The MEK5/ERK5 pathway mediates fluid shear stress promoted osteoblast differentiation. Connect Tissue Res 55(2):96–102
82. Yeatts AB, Choquette DT, Fisher JP (2013) Bioreactors to influence stem cell fate: augmentation of mesenchymal stem cell signaling pathways via dynamic culture systems. Biochim Biophys Acta 1830(2):2470–2480
83. Bravery CA et al (2013) Potency assay development for cellular therapy products: an ISCT* review of the requirements and experiences in the industry. Cytotherapy 15(1):9–19.e9
84. Salzig D, Schmiermund A, Grace PP, Elseberg C, Weber C, Czermak P (2014) Enzymatic detachment of therapeutic mesenchymal stromal cells grown on glass carriers in a bioreactor. Open Biomed Eng J 7(1):147–158

85. Lazarević JJ, Kukolj T, Bugarski D, Lazarević N, Bugarski B, Popović ZV (2019) Probing primary mesenchymal stem cells differentiation status by micro-Raman spectroscopy. Spectrochim Acta A Mol Biomol Spectrosc 213:384–390
86. Kupcova Skalnikova H (2013) Proteomic techniques for characterisation of mesenchymal stem cell secretome. Biochimie 95(12):2196–2211
87. Heidari F, Razmkhah M, Razban V, Erfani N (2021) Effects of indoleamine 2, 3-dioxygenase (IDO) silencing on immunomodulatory function and cancer-promoting characteristic of adipose-derived mesenchymal stem cells (ASCs). Cell Biol Int 45(12):2544–2556
88. Lee S, Lai W, Yu S, Chen H, Shen P, Lin S (2019) Developing a flow cytometry-based quantitative ido assay to measure immune potency of mesenchymal stromal cells product for phase i clinical trial. Cytotherapy 21(5):S37–S38
89. Bulur P, Dietz A (2018) Secretion of indoleamine 2-3 deoxygenase by adipose derived mesenchymal stromal cells as a biomarker for immune suppressive capacity. Cytotherapy 20(5):S35
90. Zheng G et al (2017) Human adipose-derived mesenchymal stem cells alleviate obliterative bronchiolitis in a murine model via IDO. Respir Res 18(1):119
91. Ling W et al (2014) Mesenchymal stem cells use IDO to regulate immunity in tumor microenvironment. Cancer Res 74(5):1576–1587
92. Fu X et al (2019) Mesenchymal stem cell migration and tissue repair. Cell 8(8):784
93. Liang C-C, Park AY, Guan J-L (2007) In vitro scratch assay: a convenient and inexpensive method for analysis of cell migration in vitro. Nat Protoc 2(2):329–333
94. Hamilton J, Vaes B (2016) The effect of scale-up on cell phenotype: comparability testing to optimize bioreactor usage and manufacturing strategies. In: Bioprocessing for cell based therapies. Wiley Blackwell, Chichester, pp 65–89
95. Yang SH et al (2009) Soluble mediators from mesenchymal stem cells suppress T cell proliferation by inducing IL-10. Exp Mol Med 41(5):315–324
96. Demircan PC, Sariboyaci AE, Unal ZS, Gacar G, Subasi C, Karaoz E (2011) Immunoregulatory effects of human dental pulp-derived stem cells on T cells: comparison of transwell co-culture and mixed lymphocyte reaction systems. Cytotherapy 13(10):1205–1220
97. Lawson T et al (2017) Process development for expansion of human mesenchymal stromal cells in a 50 L single-use stirred tank bioreactor. Biochem Eng J 120:49–62
98. Nicotra T et al (2020) Mesenchymal stem/stromal cell quality control: validation of mixed lymphocyte reaction assay using flow cytometry according to ICH Q2(R1). Stem Cell Res Ther 11(1):426
99. Akiyama K et al (2012) Mesenchymal-stem-cell-induced immunoregulation involves FAS-ligand-/FAS-mediated T cell apoptosis. Cell Stem Cell 10(5):544–555
100. Chabannes D et al (2007) A role for heme oxygenase-1 in the immunosuppressive effect of adult rat and human mesenchymal stem cells. Blood 110(10):3691–3694
101. Maughon TS et al (2022) Metabolomics and cytokine profiling of mesenchymal stromal cells identify markers predictive of T-cell suppression. Cytotherapy 24(2):137–148
102. Xue Q et al (2010) The negative co-signaling molecule b7-h4 is expressed by human bone marrow-derived mesenchymal stem cells and mediates its T-cell modulatory activity. Stem Cells Dev 19(1):27–38
103. Jin P et al (2016) Interferon-γ and tumor necrosis factor-α polarize bone marrow stromal cells uniformly to a Th1 phenotype. Sci Rep 6:26345
104. Uccelli A, Moretta L, Pistoia V (2008) Mesenchymal stem cells in health and disease. Nat Rev Immunol 8(9):726–736
105. Oliver-Vila I, Ramírez-Moncayo C, Grau-Vorster M, Marín-Gallén S, Caminal M, Vives J (2018) Optimisation of a potency assay for the assessment of immunomodulative potential of clinical grade multipotent mesenchymal stromal cells. Cytotechnology 70(1):31–44
106. Chinnadurai R et al (2018) Potency analysis of mesenchymal stromal cells using a combinatorial assay matrix approach. Cell Rep 22(9):2504–2517

107. Carmen J, Burger SR, McCaman M, Rowley JA (2012) Developing assays to address identity, potency, purity and safety: cell characterization in cell therapy process development. Regen Med 7(1):85–100
108. Barkholt L et al (2013) Risk of tumorigenicity in mesenchymal stromal cell-based therapies – bridging scientific observations and regulatory viewpoints. Cytotherapy 15(7):753–759
109. Prockop DJ et al (2010) Defining the risks of mesenchymal stromal cell therapy. Cytotherapy 12(5):576–578
110. Capelli C et al (2014) Frequent occurrence of non-malignant genetic alterations in clinical grade mesenchymal stromal cells expanded for cell therapy protocols. Haematologica 99(6): e94–e97
111. Block TJ et al (2017) Restoring the quantity and quality of elderly human mesenchymal stem cells for autologous cell-based therapies. Stem Cell Res Ther 8(1):239
112. Ng TT, Mak KHM, Popp C, Ng RK (2020) Murine mesenchymal stromal cells retain biased differentiation plasticity towards their tissue of origin. Cell 9(3):756
113. Ko E, Lee KY, Hwang DS (2012) Human umbilical cord blood-derived mesenchymal stem cells undergo cellular senescence in response to oxidative stress. Stem Cells Dev 21(11): 1877–1886
114. Stab BR et al (2016) Mitochondrial functional changes characterization in young and senescent human adipose derived MSCs. Front Aging Neurosci 8(December):299
115. Sotiropoulou PA, Perez SA, Salagianni M, Baxevanis CN, Papamichail M (2006) Characterization of the optimal culture conditions for clinical scale production of human mesenchymal stem cells. Stem Cells 24(2):462–471
116. Mareschi K et al (2012) Multipotent mesenchymal stromal stem cell expansion by plating whole bone marrow at a low cellular density: a more advantageous method for clinical use. Stem Cells Int 2012:920581
117. Leber J, Barekzai J, Blumenstock M, Pospisil B, Salzig D, Czermak P (2017) Microcarrier choice and bead-to-bead transfer for human mesenchymal stem cells in serum-containing and chemically defined media. Process Biochem 59(August):255–265
118. Pachón-Peña G et al (2011) Stromal stem cells from adipose tissue and bone marrow of age-matched female donors display distinct immunophenotypic profiles. J Cell Physiol 226(3): 843–851
119. Ikeda Y, Keese SM, Tanaka K (1985) Molecular heterogeneity of variant isovaleryl-CoA dehydrogenase from cultured isovaleric acidemia fibroblasts. Proc Natl Acad Sci U S A 82(20):7081–7085
120. Li YM et al (2007) Effects of high glucose on mesenchymal stem cell proliferation and differentiation. Biochem Biophys Res Commun 363(1):209–215
121. Wuertz K, Godburn K, Neidlinger-Wilke C, Urban J, Iatridis JC (2008) Behavior of mesenchymal stem cells in the chemical microenvironment of the intervertebral disc. Spine (Phila. Pa. 1976) 33(17):1843–1849
122. Weil BR, Abarbanell AM, Herrmann JL, Wang Y, Meldrum DR (2009) High glucose concentration in cell culture medium does not acutely affect human mesenchymal stem cell growth factor production or proliferation. Am J Physiol Regul Integr Comp Physiol 296(6): R1735–R1743
123. Nauman EA, Deorosan B (2011) The role of glucose, serum, and three-dimensional cell culture on the metabolism of bone marrow-derived mesenchymal stem cells. Stem Cells Int 2011:429187
124. Liang C et al (2012) Responses of human adipose-derived mesenchymal stem cells to chemical microenvironment of the intervertebral disc. J Transl Med 10(1):49
125. Tan L, Liu X, Dou H, Hou Y (2022) Characteristics and regulation of mesenchymal stem cell plasticity by the microenvironment—specific factors involved in the regulation of MSC plasticity. Genes Dis 9(2):296–309
126. Kim N, Cho SG (2016) Overcoming immunoregulatory plasticity of mesenchymal stem cells for accelerated clinical applications. Int J Hematol 103(2):129–137

127. Tsuji K, Kitamura S, Wada J (2018) Secretomes from mesenchymal stem cells against acute kidney injury: possible heterogeneity. Stem Cells Int 2018:1–14
128. García-Sánchez D, Fernández D, Rodríguez-Rey JC, Pérez-Campo FM (2019) Enhancing survival, engraftment, and osteogenic potential of mesenchymal stem cells. World J Stem Cells 11(10):748–763
129. Grayson WL, Zhao F, Izadpanah R, Bunnell B, Ma T (2006) Effects of hypoxia on human mesenchymal stem cell expansion and plasticity in 3D constructs. J Cell Physiol 207(2): 331–339
130. Potier E et al (2007) Hypoxia affects mesenchymal stromal cell osteogenic differentiation and angiogenic factor expression. Bone 40(4):1078–1087
131. Ciria M et al (2017) Mesenchymal stem cell migration and proliferation are mediated by hypoxia-inducible factor-1α upstream of notch and SUMO pathways. Stem Cells Dev 26(13): 973–985
132. Lv B et al (2017) Hypoxia inducible factor 1α promotes survival of mesenchymal stem cells under hypoxia. Am J Transl Res 9(3):1521–1529
133. Antebi B et al (2018) Short-term physiological hypoxia potentiates the therapeutic function of mesenchymal stem cells. Stem Cell Res Ther 9(1):265
134. Felka T, Schäfer R, Schewe B, Benz K, Aicher WK (2009) Hypoxia reduces the inhibitory effect of IL-1β on chondrogenic differentiation of FCS-free expanded MSC. Osteoarthr Cartil 17(10):1368–1376
135. Tsai CC et al (2011) Hypoxia inhibits senescence and maintains mesenchymal stem cell properties through down-regulation of E2A-p21 by HIF-TWIST. Blood 117(2):459–469
136. Boregowda SV et al (2012) Atmospheric oxygen inhibits growth and differentiation of marrow-derived mouse mesenchymal stem cells via a p53-dependent mechanism: implications for long-term culture expansion. Stem Cells 30(5):975–987
137. Estrada JC et al (2012) Culture of human mesenchymal stem cells at low oxygen tension improves growth and genetic stability by activating glycolysis. Cell Death Differ 19(5): 743–755
138. Vertelov G, Kharazi L, Muralidhar MG, Sanati G, Tankovich T, Kharazi A (2013) High targeted migration of human mesenchymal stem cells grown in hypoxia is associated with enhanced activation of RhoA. Stem Cell Res Ther 4(1):5
139. Dalloul A (2013) Hypoxia and visualization of the stem cell niche. Methods Mol Biol 1035: 199–205
140. Bader AM et al (2015) Hypoxic preconditioning increases survival and pro-Angiogenic capacity of human cord blood mesenchymal stromal cells in vitro. PLoS One 10(9):e0138477
141. Lam ATL et al (2017) Biodegradable poly-ε-caprolactone microcarriers for efficient production of human mesenchymal stromal cells and secreted cytokines in batch and fed-batch bioreactors. Cytotherapy 19(3):419–432
142. Wuertz K, Godburn K, Iatridis JC (2009) MSC response to pH levels found in degenerating intervertebral discs. Biochem Biophys Res Commun 379(4):824–829
143. Buxboim A, Rajagopal K, Brown AEX, Discher DE (2010) How deeply cells feel: methods for thin gels. J Phys Condens Matter 22(19):194116
144. Tsai AC, Jeske R, Chen X, Yuan X, Li Y (2020) Influence of microenvironment on mesenchymal stem cell therapeutic potency: from planar culture to microcarriers. Front Bioeng Biotechnol 8:640
145. Kraume M (2002) Mischen und Rühren: Grundlagen und moderne Verfahren. Wiley-VCH, Weinheim
146. Weber C et al (2010) Expansion of human mesenchymal stem cells in a fixed-bed bioreactor system based on non-porous glass carrier – Part B: modeling and scale-up of the system. Int J Artif Organs 33(11):782–795
147. Heathman TRJ, Nienow AW, Rafiq QA, Coopman K, Kara B, Hewitt CJ (2018) Agitation and aeration of stirred-bioreactors for the microcarrier culture of human mesenchymal stem cells and potential implications for large-scale bioprocess development. Biochem Eng J 136:9–17

148. Czermak P, Pörtner R, Brix A (2009 Jan) Special engineering aspects. In: Cell and tissue reaction, A Contrib. by Martin Fussenegger Wilfried Weber. Springer, Berlin, pp 83–172
149. Kanda Y, Hinata T, Kang SW, Watanabe Y (2011) Reactive oxygen species mediate adipocyte differentiation in mesenchymal stem cells. Life Sci 89(7–8):250–258
150. Pattappa G, Heywood HK, de Bruijn JD, Lee DA (2011) The metabolism of human mesenchymal stem cells during proliferation and differentiation. J Cell Physiol 226(10):2562–2570
151. Cierpka K, Elseberg CL, Niss K, Kassem M, Salzig D, Czermak P (2013) HMSC production in disposable bioreactors with regards to GMP and PAT. Chemie-Ingenieur-Technik 85(1–2): 67–75
152. Cherry RS, Papoutsakis ET (1986) Hydrodynamic effects on cells in agitated tissue culture reactors. Bioprocess Eng 1(1):29–41
153. Zwietering TN (1958) Suspending of solid particles in liquid by agitators. Chem Eng Sci 8(3–4):244–253
154. Nienow AW, Rafiq QA, Coopman K, Hewitt CJ (2014) A potentially scalable method for the harvesting of hMSCs from microcarriers. Biochem Eng J 85:79–88
155. Liepe F, Sperling R, Jembere S (1998) Rührwerke theoretische Grundlagen, Auslegung und Bewertung, 1. Aufl. Fachhochsch, Köthen SE
156. Sanz-Nogués C, O'Brien T (2021 June) Current good manufacturing practice considerations for mesenchymal stromal cells as therapeutic agents. Biomater Biosyst 2:100018
157. Giri J, Galipeau J (2020) Mesenchymal stromal cell therapeutic potency is dependent upon viability, route of delivery, and immune match. Blood Adv 4(9):1987–1997
158. Galleu A et al (2017) Apoptosis in mesenchymal stromal cells induces in vivo recipient-mediated immunomodulation. Sci Transl Med 9(416). https://doi.org/10.1126/scitranslmed.aam7828
159. Arora M (2013) Cell culture media: A Review. Mater Methods 3(175). https://doi.org/10.13070/mm.en.3.175
160. Yao T, Asayama Y (2017) Animal-cell culture media: history, characteristics, and current issues. Reprod Med Biol 16(2):99–117
161. Fekete N, Rojewski MT, Lotfi R, Schrezenmeier H (2014) Essential components for ex vivo proliferation of mesenchymal stromal cells. Tissue Eng C Methods 20(2):129–139
162. Gottipamula S, Ashwin KM, Muttigi MS, Kannan S, Kolkundkar U, Seetharam RN (2014) Isolation, expansion and characterization of bone marrow-derived mesenchymal stromal cells in serum-free conditions. Cell Tissue Res 356(1):123–135
163. Jung S, Panchalingam KM, Wuerth RD, Rosenberg L, Behie LA (2012) Large-scale production of human mesenchymal stem cells for clinical applications. Biotechnol Appl Biochem 59(2):106–120
164. Crapnell K, Blaesius R, Hastings A, Lennon DP, Caplan AI, Bruder SP (2013) Growth, differentiation capacity, and function of mesenchymal stem cells expanded in serum-free medium developed via combinatorial screening. Exp Cell Res 319(10):1409–1418
165. Wu X, Ma Z, Wu D (2020) Derivation of clinical-grade mesenchymal stromal cells from umbilical cord under chemically defined culture condition – platform for future clinical application. Cytotherapy 22(7):377–387
166. Salzig D, Leber J, Merkewitz K, Lange MC, Köster N, Czermak P (2016) Attachment, growth, and detachment of human mesenchymal stem cells in a chemically defined medium. Stem Cells Int 2016:5246584
167. Wu X, Jiang J, Gu Z, Zhang J, Chen Y, Liu X (2020) Mesenchymal stromal cell therapies: immunomodulatory properties and clinical progress. Stem Cell Res Ther 11(1):345
168. Wu X, Kang H, Liu X, Gao J, Zhao K, Ma Z (2016) Serum and xeno-free, chemically defined, no-plate-coating-based culture system for mesenchymal stromal cells from the umbilical cord. Cell Prolif 49(5):579–588
169. Blázquez-Prunera A, Almeida CR, Barbosa MA (2019) Fibroblast growth factor improves the motility of human mesenchymal stem cells expanded in a human plasma-derived xeno-free medium through αVβ3 integrin. J Tissue Eng Regen Med 13(1):36–45

170. Jung S, Sen A, Rosenberg L, Behie LA (2010) Identification of growth and attachment factors for the serum-free isolation and expansion of human mesenchymal stromal cells. Cytotherapy 12(5):637–657
171. Cimino M et al (2018) Optimization of the use of a pharmaceutical grade xeno-free medium for in vitro expansion of human mesenchymal stem/stromal cells. J Tissue Eng Regen Med 12(3):e1785–e1795
172. Ng F et al (2008 July) PDGF, tgf-2. And FGF signaling is important for differentiation and growth of mesenchymal stem cells (mscs): transcriptional profiling can identify markers and signaling pathways important in differentiation of MSCs into adipogenic, chondrogenic, and oste. Blood 112(2):295–307
173. Ahn HJ, Lee WJ, Kwack KB, Do Kwon Y (2009) FGF2 stimulates the proliferation of human mesenchymal stem cells through the transient activation of JNK signaling. FEBS Lett 583(17): 2922–2926
174. Kang YJ et al (2005) Role of c-Jun N-terminal kinase in the PDGF-induced proliferation and migration of human adipose tissue-derived mesenchymal stem cells. J Cell Biochem 95(6): 1135–1145
175. Johnson GL, Nakamura K (2007) The c-Jun kinase/stress-activated pathway: regulation, function and role in human disease. Biochim Biophys Acta, Mol Cell Res 1773(8):1341–1348
176. Jian H, Shen X, Liu I, Semenov M, He X, Wang XF (2006) Smad3-dependent nuclear translocation of β-catenin is required for TGF-β1- induced proliferation of bone marrow-derived adult human mesenchymal stem cells. Genes Dev 20(6):666–674
177. Sabri A, Ziaee AA, Ostad SN, Alimoghadam K, Ghahremani MH (2011) Crosstalk of EGF-directed MAPK signalling pathways and its potential role on EGF-induced cell proliferation and COX-2 expression in human mesenchymal stem cells. Cell Biochem Funct 29(1): 64–70
178. Hemmings BA, Restuccia DF (2012) PI3K-PKB/Akt pathway. Cold Spring Harb Perspect Biol 4(9):a011189
179. Doorn J et al (2013) Insulin-like growth factor – I enhances proliferation and differentiation of human mesenchymal stromal cells in vitro. Tissue Eng A 19(15–16):1817–1828
180. Zhang W, Lee JC, Kumar S, Gowen M (1999) ERK pathway mediates the activation of Cdk2 in IGF-1-induced proliferation of human osteosarcoma MG-63 cells. J Bone Miner Res 14(4): 528–535
181. Ramos JW (2008) The regulation of extracellular signal-regulated kinase (ERK) in mammalian cells. Int J Biochem Cell Biol 40(12):2707–2719
182. Kong X et al (2010) VEGF promotes the proliferation of bone marrow derived mesenchymal stem cells through ERK1/2 signal pathway. Zhongguo Shi Yan Xue Ye Xue Za Zhi 18(5): 1292–1296
183. Blahnova VH, Filova E, Dankova J, Rampichova M (2020) Combinations of growth factors for human mesenchymal stem cell proliferation and osteogenic differentiation. Bone Jt Res 9(7):412–420
184. Collins JM, Goldspink PH, Russell B (2010) Migration and proliferation of human mesenchymal stem cells is stimulated by different regions of the mechano-growth factor prohormone. J Mol Cell Cardiol 49(6):1042–1045
185. Tan KY, Teo KL, Lim JFY, Chen AKL, Reuveny S, Oh SKW (2015) Serum-free media formulations are cell line-specific and require optimization for microcarrier culture. Cytotherapy 17(8):1152–1165
186. Forte G et al (2006) Hepatocyte growth factor effects on mesenchymal stem cells: proliferation, migration, and differentiation. Stem Cells 24(1):23–33
187. Wen Q, Zhou L, Zhou C, Zhou M, Luo W, Ma L (2012) Change in hepatocyte growth factor concentration promote mesenchymal stem cell-mediated osteogenic regeneration. J Cell Mol Med 16(6):1260–1273

188. Wang H, Liu J, Zhou Y, Tan W (2022) Effect of hepatocyte growth factor on proliferation of human mesenchymal stem cells in vitro. Huadong Ligong Daxue Xuebao/J East China Univ Sci Technol 48(6):790–796
189. Sartori T et al (2020) Branched chain amino acids improve mesenchymal stem cell proliferation, reducing nuclear factor kappa B expression and modulating some inflammatory properties. Nutrition 78:110935
190. Liu CH, Wu ML, Hwang SM (2007) Optimization of serum free medium for cord blood mesenchymal stem cells. Biochem Eng J 33(1):1–9
191. Lee JH, Um S, Jang JH, Seo BM (2012) Effects of VEGF and FGF-2 on proliferation and differentiation of human periodontal ligament stem cells. Cell Tissue Res 348(3):475–484
192. Fujimoto Y, Yokozeki T, Yokoyama A, Tabata Y (2020) Basic fibroblast growth factor enhances proliferation and hepatocyte growth factor expression of feline mesenchymal stem cells. Regen Ther 15:10–17
193. Martin C et al (2017) Revisiting MSC expansion from critical quality attributes to critical culture process parameters. Process Biochem 59:231–243
194. Xu X et al (2022) Therapeutic effects and repair mechanism of HGF gene-transfected mesenchymal stem cells on injured endometrium. Stem Cells Int 2022:5744538
195. Cao Z, Xie Y, Yu L, Li Y (2020) Hepatocyte growth factor (HGF) and stem cell factor (SCF) maintained the stemness of human bone marrow mesenchymal stem cells (hBMSCs) during long-term expansion by preserving mitochondrial function via the PI3K/AKT, ERK1/2, and STAT3 signaling pathways. Stem Cell Res Ther 11(1):329
196. Fang Meng H, Jin J, Wang H, Sheng Wang L, Tse Wu C (2022) Recent advances in the therapeutic efficacy of hepatocyte growth factor gene-modified mesenchymal stem cells in multiple disease settings. J Cell Mol Med 26(18):4745–4755
197. Feng X et al (2014) Insulin-like growth factor 1 can promote proliferation and osteogenic differentiation of human dental pulp stem cells via mTOR pathway. Develop Growth Differ 56(9):615–624
198. Lim M, Shunjie C, Panoskaltsis N, Mantalaris A (2012) Systematic experimental design for bioprocess characterization: elucidating transient effects of multi-cytokine contributions on erythroid differentiation. Biotechnol Bioprocess Eng 17(1):218–226
199. Barhanpurkar-Naik A, Mhaske ST, Pote ST, Singh K, Wani MR (2017) Interleukin-3 enhances the migration of human mesenchymal stem cells by regulating expression of CXCR4. Stem Cell Res Ther 8(1):168
200. Dorronsoro A et al (2020) Intracellular role of IL-6 in mesenchymal stromal cell immunosuppression and proliferation. Sci Rep 10(1):21853
201. Tamm C, Galitó SP, Annerén C (2013) A comparative study of protocols for mouse embryonic stem cell culturing. PLoS One 8(12):e81156
202. Chase LG, Lakshmipathy U, Solchaga LA, Rao MS, Vemuri MC (2010 Apr) A novel serum-free medium for the expansion of human mesenchymal stem cells. Stem Cell Res Ther 1(1): 1–11
203. Xu J, Lian W, Chen J, Li W, Li L, Huang Z (2020) Chemical-defined medium supporting the expansion of human mesenchymal stem cells. Stem Cell Res Ther 11(1):125
204. Mimura S et al (2011) Growth factor-defined culture medium for human mesenchymal stem cells. Int J Dev Biol 55(2):181–187
205. Jung S, Panchalingam KM, Rosenberg L, Behie LA (2012) Ex vivo expansion of human mesenchymal stem cells in defined serum-free media. Stem Cells Int 2012:123030
206. McKeehan R, McKeehan W, Ham K (1981) The relationship between defined low molecular weight substances and undefined serum-derived factors in the multiplication of untransformed fibroblasts. In: Waymouth PCC, Ham R (eds) The growth requirements of vertebrate cells in vitro. Cambridge University Press, New York, pp 223–243
207. Castro PML, Hayter PM, Ison AP, Bull AT (1992) Application of a statistical design to the optimization of culture medium for recombinant interferon-gamma production by Chinese hamster ovary cells. Appl Microbiol Biotechnol 38(1):84–90

208. Kim EJ, Kim NS, Lee GM (1998) Development of a serum-free medium for the production of humanized antibody from Chinese hamster ovary cells using a statistical design. Vitr Cell Dev Biol Anim 34(10):757–761
209. Liu CH, Chu IM, Hwang SM (2001) Factorial designs combined with the steepest ascent method to optimize serum-free media for CHO cells. Enzym Microb Technol 28(4–5): 314–321
210. Hammett K, Kuchibhatla J, Hunt C, Holdread S, Brooks JW (2007) Developing chemically defined media through DOE: complete optimization with increased protein production in less than 8 months. In: Cell technology for cell products. Springer, Dordrecht, pp 683–691
211. Grzesik P, Warth SC (2021) One-time optimization of advanced T cell culture media using a machine learning pipeline. Front Bioeng Biotechnol 9:614324
212. Panuganti S, Schlinker AC, Lindholm PF, Papoutsakis ET, Miller WM (2013) Three-stage ex vivo expansion of high-ploidy megakaryocytic cells: toward large-scale platelet production. Tissue Eng A 19(7–8):998–1014
213. Lim M, Panoskaltsis N, Ye H, Mantalaris A (2011) Optimization of in vitro erythropoiesis from CD34+ cord blood cells using design of experiments (DOE). Biochem Eng J 55(3): 154–161
214. Decaris ML, Leach JK (2011) Design of experiments approach to engineer cell-secreted matrices for directing osteogenic differentiation. Ann Biomed Eng 39(4):1174–1185
215. Knöspel F, Schindler RK, Lübberstedt M, Petzolt S, Gerlach JC, Zeilinger K (2010) Optimization of a serum-free culture medium for mouse embryonic stem cells using design of experiments (DoE) methodology. Cytotechnology 62(6):557–571
216. Mizukami A, Orellana MD, Caruso SR, de Lima Prata K, Covas DT, Swiech K (2013) Efficient expansion of mesenchymal stromal cells in a disposable fixed bed culture system. Biotechnol Prog 29(2):568–572
217. Van Wezel AL (1967) Growth of cell-strains and primary cells on micro-carriers in homogeneous culture. Nature 216(5110):64–65
218. Petry F, Salzig D (2021) Impact of bioreactor geometry on mesenchymal stem cell production in stirred-tank bioreactors. Chemie-Ingenieur-Technik 93(10):1537–1554
219. Collignon ML, Delafosse A, Crine M, Toye D (2010) Axial impeller selection for anchorage dependent animal cell culture in stirred bioreactors: methodology based on the impeller comparison at just-suspended speed of rotation. Chem Eng Sci 65(22):5929–5941
220. Delafosse A, Loubière C, Calvo S, Toye D, Olmos E (2018) Solid-liquid suspension of microcarriers in stirred tank bioreactor – experimental and numerical analysis. Chem Eng Sci 180:52–63
221. Loubière C, Delafosse A, Guedon E, Toye D, Chevalot I, Olmos E (2019) Optimization of the impeller design for mesenchymal stem cell culture on microcarriers in bioreactors. Chem Eng Technol 42(8):1702–1708
222. Schop D et al (2010) Expansion of human mesenchymal stromal cells on microcarriers: growth and metabolism. J Tissue Eng Regen Med 4(2):131–140
223. Eibes G et al (2010) Maximizing the ex vivo expansion of human mesenchymal stem cells using a microcarrier-based stirred culture system. J Biotechnol 146(4):194–197
224. Caruso SR et al (2014) Growth and functional harvesting of human mesenchymal stromal cells cultured on a microcarrier-based system. Biotechnol Prog 30(4):889–895
225. Yang HS, Jeon O, Bhang SH, Lee SH, Kim BS (2010) Suspension culture of mammalian cells using thermosensitive microcarrier that allows cell detachment without proteolytic enzyme treatment. Cell Transplant 19(9):1123–1132
226. Yuan X, Tsai AC, Farrance I, Rowley JA, Ma T (2018) Aggregation of culture expanded human mesenchymal stem cells in microcarrier-based bioreactor. Biochem Eng J 131:39–46
227. Heathman TRJ et al (2016) Scalability and process transfer of mesenchymal stromal cell production from monolayer to microcarrier culture using human platelet lysate. Cytotherapy 18(4):523–535

228. Leber J, Barekzai J, Weidner T, Salzig D, Czermak P (2016) Bead-to-bead transfer as scale-up strategy for human mesenchymal stem cell expansion in serum-containing and chemically defined media. New Biotechnol 33:S12
229. Weber C et al (2007) Expansion and harvesting of hMSC-TERT. Open Biomed Eng J 1(1): 38–46
230. Dos Santos F et al (2011) Toward a clinical-grade expansion of mesenchymal stem cells from human sources: a microcarrier-based culture system under xeno-free conditions. Tissue Eng C Methods 17(12):1201–1210
231. Carmelo JG, Fernandes-Platzgummer A, Diogo MM, da Silva CL, Cabral JMS (2015) A xeno-free microcarrier-based stirred culture system for the scalable expansion of human mesenchymal stem/stromal cells isolated from bone marrow and adipose tissue. Biotechnol J 10(8): 1235–1247
232. Rafiq QA et al (2017) Process development of human multipotent stromal cell microcarrier culture using an automated high-throughput microbioreactor. Biotechnol Bioeng 114(10): 2253–2266
233. Hervy M et al (2014) Long term expansion of bone marrow-derived hMSCs on novel synthetic microcarriers in xeno-free, defined conditions. PLoS One 9(3):e92120
234. Cunha B et al (2017) Bioprocess integration for human mesenchymal stem cells: from up to downstream processing scale-up to cell proteome characterization. J Biotechnol 248:87–98
235. Tozetti PA et al (2017) Expansion strategies for human mesenchymal stromal cells culture under xeno-free conditions. Biotechnol Prog 33(5):1358–1367
236. de Soure AM et al (2017) Integrated culture platform based on a human platelet lysate supplement for the isolation and scalable manufacturing of umbilical cord matrix-derived mesenchymal stem/stromal cells. J Tissue Eng Regen Med 11(5):1630–1640
237. Schop D, Janssen FW, Borgart E, De Bruijn JD, Van Dijkhuizen-Radersma R (2008) Expansion of mesenchymal stem cells using a microcarrier-based cultivation system: growth and metabolism. J Tissue Eng Regen Med 2(2–3):126–135
238. Hupfeld J et al (2014) Modulation of mesenchymal stromal cell characteristics by microcarrier culture in bioreactors. Biotechnol Bioeng 111(11):2290–2302
239. Mizukami A et al (2016) Stirred tank bioreactor culture combined with serum−/xenogeneic-free culture medium enables an efficient expansion of umbilical cord-derived mesenchymal stem/stromal cells. Biotechnol J 11(8):1048–1059
240. Schop D et al (2009) Growth, metabolism, and growth inhibitors of mesenchymal stem cells. Tissue Eng A 15(8):1877–1886
241. Sion C et al (2021 Nov) A new perfusion mode of culture for WJ-MSCs expansion in a stirred and online monitored bioreactor. Biotechnol Bioeng 118(11):4453–4464
242. Cunha B et al (2015 Nov) Exploring continuous and integrated strategies for the up- and downstream processing of human mesenchymal stem cells. J Biotechnol 213:97–108
243. Wagner A, Marc A, Engasser JM, Villermaux S, Einsele A (1990) Continuous production of prourokinase in feed harvest and perfusion cultures. Biotechnol Bioeng 36(6):623–629
244. Serra M et al (2010) Improving expansion of pluripotent human embryonic stem cells in perfused bioreactors through oxygen control. J Biotechnol 148(4):208–215
245. Dos Santos F et al (2014 June) A xenogeneic-free bioreactor system for the clinical-scale expansion of human mesenchymal stem/stromal cells. Biotechnol Bioeng 111(6):1116–1127
246. Rosa F et al (2016) Monitoring the ex-vivo expansion of human mesenchymal stem/stromal cells in xeno-free microcarrier-based reactor systems by MIR spectroscopy. Biotechnol Prog 32(2):447–455
247. Whelan J, Craven S, Glennon B (2012) In situ Raman spectroscopy for simultaneous monitoring of multiple process parameters in mammalian cell culture bioreactors. Biotechnol Prog 28(5):1355–1362
248. Abu-Absi NR et al (2011) Real time monitoring of multiple parameters in mammalian cell culture bioreactors using an in-line Raman spectroscopy probe. Biotechnol Bioeng 108(5): 1215–1221

249. Calvet A, Ryder AG (2014) Monitoring cell culture media degradation using surface enhanced Raman scattering (SERS) spectroscopy. Anal Chim Acta 840:58–67
250. Rangan S et al (2018) Types of cell death and apoptotic stages in Chinese hamster ovary cells distinguished by Raman spectroscopy. Biotechnol Bioeng 115(2):401–412
251. Petiot E, El-Wajgali A, Esteban G, Gény C, Pinton H, Marc A (2012) Real-time monitoring of adherent Vero cell density and apoptosis in bioreactor processes. Cytotechnology 64(4): 429–441
252. Justice C, Leber J, Freimark D, Pino Grace P, Kraume M, Czermak P (2011) Online- and offline-monitoring of stem cell expansion on microcarrier. Cytotechnology 63(4):325–335
253. Rösner LS, Walter F, Ude C, John GT, Beutel S (2022) Sensors and techniques for on-line determination of cell viability in bioprocess monitoring. Bioengineering 9(12):762
254. Khan FA (2022) Stem cell bioreactors: design, structure, and operation of stem cell bioreactors. In: Stem cell production. Springer, Singapore, pp 113–123
255. Luciani N et al (2016) Successful chondrogenesis within scaffolds, using magnetic stem cell confinement and bioreactor maturation. Acta Biomater 37:101–110
256. Dikina AD, Lai BP, Cao M, Zborowski M, Alsberg E (2017) Magnetic field application or mechanical stimulation: via magnetic microparticles does not enhance chondrogenesis in mesenchymal stem cell sheets. Biomater Sci 5(7):1241–1245
257. Stoppel WL, Kaplan DL, Black LD (2016) Electrical and mechanical stimulation of cardiac cells and tissue constructs. Adv Drug Deliv Rev 96:135–155
258. Mellor LF, Steward AJ, Nordberg RC, Taylor MA, Loboa EG (2017) Comparison of simulated microgravity and hydrostatic pressure for chondrogenesis of hASC. Aerosp Med Hum Perform 88(4):377–384
259. Guo T et al (2016) Effect of dynamic culture and periodic compression on human mesenchymal stem cell proliferation and chondrogenesis. Ann Biomed Eng 44(7):2103–2113
260. Shakeel M, Matthews PC, Graham RS, Waters SL (2013) A continuum model of cell proliferation and nutrient transport in a perfusion bioreactor. Math Med Biol 30(1):21–44
261. Guyot Y, Papantoniou I, Luyten FP, Geris L (2016) Coupling curvature-dependent and shear stress-stimulated neotissue growth in dynamic bioreactor cultures: a 3D computational model of a complete scaffold. Biomech Model Mechanobiol 15(1):169–180
262. Nguyen BNB, Ko H, Fisher JP (2016) Tunable osteogenic differentiation of hMPCs in tubular perfusion system bioreactor. Biotechnol Bioeng 113(8):1805–1813
263. Nguyen BNB, Ko H, Moriarty RA, Etheridge JM, Fisher JP (2016) Dynamic bioreactor culture of high volume engineered bone tissue. Tissue Eng A 22(3–4):263–271
264. Hudson KD, Bonassar LJ (2017) Hypoxic expansion of human mesenchymal stem cells enhances three-dimensional maturation of tissue-engineered intervertebral discs. Tissue Eng A 23(7–8):293–300
265. Schirmaier C et al (2014) Scale-up of adipose tissue-derived mesenchymal stem cell production in stirred single-use bioreactors under low-serum conditions. Eng Life Sci 14(3):292–303
266. Rafiq QA, Brosnan KM, Coopman K, Nienow AW, Hewitt CJ (2013) Culture of human mesenchymal stem cells on microcarriers in a 5 l stirred-tank bioreactor. Biotechnol Lett 35(8): 1233–1245
267. Sousa MFQ et al (2015) Production of oncolytic adenovirus and human mesenchymal stem cells in a single-use, vertical-wheel bioreactor system: impact of bioreactor design on performance of microcarrier-based cell culture processes. Biotechnol Prog 31(6):1600–1612
268. de Sá da Silva J et al (2019) Improving wave-induced motion bioreactor performance for human mesenchymal stromal cell expansion. Process Biochem 84:143–152
269. Fernandes-Platzgummer A, Cunha R, Morini S, Carvalho M, Moreno-Cid J, García C, da Silva CL (2023) Optimized operation of a controlled stirred tank reactor system for the production of mesenchymal stromal cells and their extracellular vesicles. Biotechnol Bioeng 120(9): 2742–2755. https://doi.org/10.1002/bit.28449
270. Reichardt A, Polchow B, Shakibaei M, Henrich W, Hetzer R, Lueders C (2013) Large scale expansion of human umbilical cord cells in a rotating bed system bioreactor for cardiovascular tissue engineering applications. Open Biomed Eng J 7(1):50–61

271. Haack-Sørensen M et al (2016) Culture expansion of adipose derived stromal cells. A closed automated quantum cell expansion system compared with manual flask-based culture. J Transl Med 14(1):319
272. Hanley PJ et al (2014) Efficient manufacturing of therapeutic mesenchymal stromal cells with the use of the quantum cell expansion system. Cytotherapy 16(8):1048–1058
273. Nold P, Brendel C, Neubauer A, Bein G, Hackstein H (2013) Good manufacturing practice-compliant animal-free expansion of human bone marrow derived mesenchymal stroma cells in a closed hollow-fiber-based bioreactor. Biochem Biophys Res Commun 430(1):325–330
274. Lechanteur C (2014) Large-scale clinical expansion of mesenchymal stem cells in the GMP-compliant, closed automated quantum® cell expansion system: comparison with expansion in traditional T-flasks. J Stem Cell Res Ther 4(8). https://doi.org/10.4172/2157-7633.1000222
275. Mizukami A, Swiech K (2018) Mesenchymal stromal cells: from discovery to manufacturing and commercialization. Stem Cells Int 2018:4083921
276. Tsai AC, Liu Y, Ma T (2016) Expansion of human mesenchymal stem cells in fibrous bed bioreactor. Biochem Eng J 108:51–57
277. Weber C et al (2010) Expansion of human mesenchymal stem cells in a fixed-bed bioreactor system based on non-porous glass carrier – Part A: inoculation, cultivation, and cell harvest procedures. Int J Artif Organs 33(8):512–525
278. Timmins NE et al (2012) Closed system isolation and scalable expansion of human placental mesenchymal stem cells. Biotechnol Bioeng 109(7):1817–1826
279. Jossen V, Pörtner R, Kaiser SC, Kraume M, Eibl D, Eibl R (2014) Mass production of mesenchymal stem cells—impact of bioreactor design and flow conditions on proliferation and differentiation. In: Cells and biomaterials in regenerative medicine. IntechOpen, Rejika
280. Jossen V et al (2016) Theoretical and practical issues that are relevant when scaling up hMSC microcarrier production processes. Stem Cells Int 2016:4760414
281. Zhou T et al (2021 Feb) Challenges and advances in clinical applications of mesenchymal stromal cells. J Hematol Oncol 14(1):1–24

Chapter 14
Biomanufacturing Aspects of Gene Therapy

James N. Warnock

Abstract The notion that genetic material could be transferred to a host patient for therapeutic benefit has been around for over half a century. However, it wasn't until 2017 that the concept became a clinical reality, following the approval of the first CAR-T cell therapy in the US by the FDA and subsequent approval by the E.U. Since the approval of this first product, many more gene therapies have come to market, and the number of products in late-stage clinical trials indicate that gene therapy could become one of the fastest growing sectors in the biopharmaceutical industry. A key aspect for the commercial success of any biopharmaceutical is the ability to economically manufacture the therapeutic product at a large enough scale to meet market demand. To date, manufacturing processes have been able to produce gene therapy viral vectors at the necessary scale to satisfy the demands of clinical studies. However, current biomanufacturing processes will need to be scaled-up and optimized to meet commercial demand, especially for therapies that treat diseases with large patient populations.

Keywords Gene therapy · Viral vectors · Vector production · Bioreactors · Producer cells · Packaging cells

Abbreviations

AAV	Adeno-associated Virus
ADA	Adenosine Deaminase
AdV	Adenovirus
ALL	Acute Lymphoblastic Leukemia
BCG	Bacillus Calmette-Guérin
CAR	Chimeric Antigen Receptor

J. N. Warnock (✉)
School of Chemical, Materials and Biomedical Engineering, University of Georgia, Athens, GA, USA
e-mail: james.warnock@uga.edu

R. Pörtner (ed.), *Biopharmaceutical Manufacturing*, Cell Engineering 11,
https://doi.org/10.1007/978-3-031-45669-5_14

CIS	Carcinoma *in situ*
DLBCL	Diffuse Large B-cell Lymphoma
FDA	Food and Drug Administration
HEK	Human Embryonic Kidney cells
HIV-1	Human Immunodeficiency Virus Type 1
HSV-1	Herpes Simplex Virus
LBCL	Large B-Cell Lymphoma
LPLD	Lipoprotein Lipase Deficiency
LTR	Long Terminal Repeat
LV	Lentivirus
MCL	Mantle Cell Lymphoma
MLV	Murine Leukemia Viruses
NMIBC	Non-muscle Invasive Bladder Cancer
RV	Retrovirus
SCID	Severe Combined Immunodeficiency
SIN	Self-inactivating
SV40	Simian Vacuolating Virus 40 Large T Antigen
TFDF	Tangential Flow Depth Filtration
TU	Transducing Units
TV	Transfer Vector
VSV-G	Vesicular Stomatitis Virus Glycoprotein

Introduction

The idea that a gene could be transferred into host cells to replace missing or defective genes and that expression of the transferred gene could provide a therapeutic benefit was first introduced by Theodore Friedman [1] with the first clinical experiment being performed in 1970 [2]. Gene therapy was initially envisioned as a treatment for inherited diseases. However, alternative strategies for treating other diseases that have a genetic component have been conceived. Current models for gene therapy include: *(i)* gene replacement, the delivery of a functional gene to replace a non-working gene; *(ii)* gene silencing, inactivation of a mutated gene that has become toxic to cells; *(iii)* gene addition, over expression of a "foreign" or exogenous gene to impact cellular function; and *(iv)* gene editing, a permanent manipulation of a gene in a patient's genome [3].

During the past 50 years, gene therapy has taken a roller coaster ride, with many advances being followed by setbacks. In 1980, following the advent of recombinant DNA, scientists demonstrated it was possible to transform mouse bone marrow cells *in vitro* with rDNA and inject the transformed cells into irradiated mice, giving rise to a dominant marrow population [4]. The same technique was also used to transform bone marrow cells from two patients with the inherited disorder, β-thalassemia, with a recombinant human β-globin gene [5]. The first trial involving human patients used retroviral-mediated transfer of the adenosine deaminase (ADA) gene into T-cells of two children with severe combined immunodeficiency (SCID) [6]. Overall, the results were positive and provided proof of concept, that gene therapy was a safe and effective treatment for certain patients. However, the scientific complexities associated with gene transfer were made apparent in 1999 when Jesse Gelsinger, an 18 year old with a relatively mild form of the nitrogen metabolism disorder ornithine transcarbamylase (OT) deficiency, died from adenoviral vector associated toxicity while participating in a gene therapy trial [5, 7, 8]. Further concerns arose after five patients with severe combined immunodeficiency-X1 (SCID-X1) developed leukemia, one of whom died, after their $CD34^+$ cells were transfected *ex vivo* with a murine retroviral vector encoding the γ chain cytokine receptor subunit [9]. Twenty patients were included in this study and gene therapy did provide full correction of disease phenotype in two patients.

China's regulatory agency approved the world's first commercially available gene therapy, *Gendicine*, in 2003 to treat head and neck squamous cell carcinoma, a form of skin cancer. *Gendicine* is a virus engineered to carry a gene that has instructions for making a tumor-fighting protein [10]. This product did not receive approval outside China and it wasn't until 2012 that Europe approved the first gene therapy, *Glybera* [11]. *Glybera* utilizes a recombinant adeno-associated virus (AAV) that can be used to treat familial lipoprotein lipase deficiency (LPLD) [12]. However, as LPLD is a rare disorder and occurs in so few patients, the drug was unprofitable. Consequently, the manufacturer declined to renew the marketing authorization and *Glybera* was withdrawn from the market in 2017 [10]. On August 30, 2017, the US Food and Drug Administration (FDA) approved tisagenlecleucel (Kymriah), the first gene therapy to receive FDA approval in the United States [13]. Kymriah is a CD19-directed chimeric antigen receptor (CAR) T-cell immunotherapy for the treatment of patients aged $\leq$25 years with B-cell precursor Acute Lymphoblastic Leukemia (ALL) that is refractory, in second relapse, or later relapse [14]. Kymriah was subsequently approved by the E.U. in 2018 [10]. Table 14.1 provides a list of all gene therapies that have been approved by the FDA as of December 2022.

There are three general protocols for transfecting host cells with therapeutic genes (as shown in Fig. 14.1): *ex vivo*, *in vivo* and *in situ*. In *ex vivo* gene transfer, a patient's cells are isolated and transfected with the therapeutic gene. The transfected cells are expanded before being reinfused to the patient. This methodology has been effectively used for the treatment of childhood leukemia using CAR-T cells [16]. While *ex vivo* gene therapy is effective for treating blood-borne diseases, *in vivo* gene therapy is more effective for treating diseases of the brain, spinal canal or liver [1]. *In vivo* gene therapy administers a gene vector directly into the host, via the circulating blood or the cerebrospinal fluid, where it transfects target cells within the whole organism [17]. Two different types of viral vectors are used predominantly: adenovirus (AdV) vectors and adeno-associated virus (AAV) vectors. Like *in vivo* gene therapy, *in situ* gene therapy introduces the vector directly to the host patient; however, the primary difference is vectors are administered to a specific organ or area of the patient as opposed to intravenous administration.

Table 14.1 FDA approved gene therapy products as of December 2022 [15]

Drug name	Manufacturer	Disease condition(s)	Viral vector[a]	Dose	Transfection protocol
ABECMA (idecabtagene vicleucel)	Celgene Corporation	Multiple myeloma	LV	300 to 460 × 10^6 CAR-positive T cells	*Ex vivo*
ADSTILADRIN	Ferring Pharma-ceuticals A/S	Bacillus Calmette-Guérin (BCG)-unresponsive non-muscle invasive bladder cancer (NMIBC) with carcinoma *in situ* (CIS)	AdV	2.25 × 10^{13} viral particles	*In situ*
BREYANZI (lisocabtagene maraleucel)	Juno therapeutics	Large B-cell lymphoma (LBCL)	LV	90 to 110 × 10^6 CAR-positive viable T cells	*Ex vivo*
CARVYKTI	Janssen Biotech, Inc.	Multiple myeloma	LV	0.5–1.0 × 10^6 CAR-positive via-ble T cells per kg of body weight	*Ex vivo*
HEMGENIX	CSL Behring LLC	Hemophilia B (congenital factor IX deficiency	AAV	2 × 10^{13} genome copies (gc) per kilogram (kg) of 35 body weight	*In vitro*
IMLYGIC (talimogene laherparepvec)	BioVex, Inc.	Unresectable cutaneous, subcutaneous, and nodal lesions	HSV-1	4 × 10^6 PFU (initial dose)4 × 10^8 PFU (subsequent doses)	*In situ*
KYMRIAH (tisagenlecleucel)	Novartis Pharma-ceuticals Corporation	B-cell acute lymphoblastic Leukemia (ALL) diffuse large B-cell lymphoma (DLBCL)	LV	0.1 to 5.0 × 10^6 CAR-positive viable T cells per kg body weight	*Ex vivo*
LUXTURNA (voretigene neparvovec-rzyl)	Spark therapeutics, Inc.	Biallelic *RPE65* mutation-associated retinal dystrophy	AAV	1.5 × 10^{11} vector genomes (vg),	*In situ*
TECARTUS (brexucabtagene autoleucel)	Kite pharma, Inc.	Mantle cell lymphoma (MCL)	RV	2 × 10^6 CAR-positive viable T cells per kg of body weight	*Ex vivo*

YESCARTA (axicabtagene ciloleucel)	Kite pharma, Inc.	Large B-cell lymphoma (LBCL)	RV	2×10^6 CAR-positive viable T cells per kg body weight	*Ex vivo*
ZYNTEGLO (betibeglogene autotemcel)	Bluebird bio, Inc.	ß-thalassemia	LV	5.0×10^6 CD34+ cells per kg of body weight	*Ex vivo*
ZOLGENSMA (onasemnogene abeparvovec)	Novartis gene therapies, Inc.	Spinal muscular atrophy (type I)	AAV	1.1×10^{14} vector genomes per kilogram (vg/kg) of body weight	*In vitro*

[a]AAV Adeno-associated Virus, AdV Adenovirus, HSV-1 Herpes Simplex Virus, LV Lentivirus, RV Retrovirus

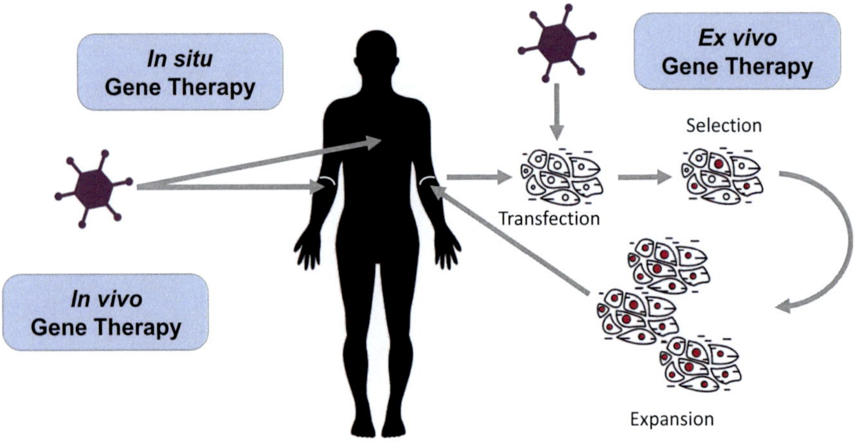

Fig. 14.1 Methods for the delivery of therapeutic genes and transfection of host cells

This has several advantages over *in vivo* gene therapy, such as reduction in potential side effects, the need for a lower vector dose, and, consequently, reduced costs [18].

The key to success for any gene therapy strategy is having a vector able to serve as a safe and efficient gene delivery vehicle [19]. Therapeutic genes can be delivered to cells via either viral or non-viral vectors. The most researched non-viral vectors are polymers, lipids, inorganic particles, or combinations of different types. Compared with a viral vectors, non-viral vectors are low in their cytotoxicity, immunogenicity, and mutagenesis. However, non-viral vectors do not carry ideal characteristics and have faced critical challenges, such as gene transfer efficiency, specificity, gene expression duration, and safety [20]. Non-viral vectors are not currently used in any of the FDA approved gene therapies. In contrast, viruses have complex and precise structural features, adjusted through natural evolution for efficient transfection of specific host cells or tissue [21]. Several virus types are currently used as gene delivery vectors in FDA approved gene therapies. These include retroviruses (γ-retroviral vectors derived from murine leukemia viruses (MLV) and lentiviral vectors (LV) mainly derived from HIV-1), adenoviruses, adeno-associated viruses and Herpes Simplex Virus Type-1 [17, 22–24]. The typical characteristics of these viral vectors are summarized in Table 14.2. The main limitation of viral vectors is the relatively low harvest from non-optimized processes. For viral gene therapy to be a clinical and commercial success, there must be biomanufacturing systems capable of efficient and economical virus production that can meet market demand.

Table 14.2 Typical characteristics of viral vectors used in gene therapy applications

Vector	Genome	Insert size	Infectivity	Expression
Adenovirus	ds DNA	5 kb (single deletion) 10 kb (double deletion) 30 kb (gutless)	Dividing and quiescent cells	Transient
Retrovirus	ss RNA	7–10 kb (oncogenic retrovirus)	Dividing cells	Permanent
		7–12 kb (lentivirus)	Dividing and quiescent cells	
Adeno-associated virus	ss DNA	4.7 kb	Quiescent cells	Long-term
Poxvirus	ds DNA	178 kb (MVA) 192 kb (NYVAC)	Dividing and quiescent cells	Permanent
Herpes simplex virus	ds DNA	20–30 kb	Dividing and quiescent cells	Episomal, but latent infection permanent

ss single stranded, ds double stranded

Lentivirus

Lentiviruses are a sub-family of the retroviridae family [25]. Retroviruses are unique compared to other viruses as they contain single stranded RNA, which is reverse transcribed during host cell infection. Unlike the oncovirinae sub-family, which was extensively investigated as a potential gene therapy vector in early clinical trials [17], lentivirinae can infect both quiescent and postmitotic cells and are categorized as complex retroviruses [26].

The retrovirus capsid is an enveloped protein shell that contains the viral genome and is 80–100 nm in diameter [27]. The envelope structure surrounding the capsid is a lipid bilayer that contains both virus-encoded surface glycoproteins and transmembrane glycoproteins that originate from the host cell [28]. The genome of the lentivirinae is linear, non-segmented, single-stranded RNA, 7–12 kb in length [26]. The genome of the simple retroviruses, such as the gammaretroviruses, contain three major coding segments and one small coding domain. The major segments contain three genes — *gag*, *pol*, and *env* — which code for proteins important in viral integration, replication, and encapsulation [27]. The small coding domain contains the *pro* gene, which encodes for viral protease [28]. The genome of the complex retroviruses, such as the lentiviruses, contain the same genes as the simple retroviruses (*gag*, *pol*, and *env*), but also include six additional genes - two regulatory genes and four accessory genes - that code for important viral replication, binding, infection, and release proteins. Table 14.3 outlines each of these six genes and the functions of their expressed proteins.

Retroviruses infect cells through interactions between glycoproteins anchored on the outer envelope and a specific cell receptor. Once the virus binds to the cell surface receptors, a series of events occurs that results in the incorporation of the

Table 14.3 Genes expressed in HIV-1 lentivirus in addition to the simple retrovirus genes, *gag*, *pol* and *env*. The lentiviral genes are non-essential and absent in gene therapy vectors. The *rev* gene, along with the simple genes, is expressed on plasmids that are present in packaging cells

Protein	Function
rev	An RNA-binding protein that acts to induce the transition from the early to the late phase of HIV gene expression
tat	An RNA-binding protein that enhances transcription 1000-fold
nef	Disturbs T-cell activation and stimulates HIV infectivity
vpr	Mediates HIV to infect non-dividing cells
vpu	Enhances the release of HIV-1 from the cell surface to the cytoplasm
Vif	A polypeptide necessary for the replication of HIV-1

virus particle into the cell lipid bilayer and subsequent unloading of the viral genetic material into the cell cytoplasm [29, 30]. However, some unique events take place in the life cycle of the lentivirus, which explains their ability to infect non-dividing cells [31]. First, gene expression occurs in two separate phases, known as the early and late phases, which are separated by the binding of the rev protein [32]. Second, the lentivirus infects non-dividing cells via proteins expressed from the *vpr* gene [33]. Finally, the *tat* gene, found only in complex retroviruses, is essential for the replication of HIV-1 [34].

Lentivirus Vectors in Gene Therapy

The most studied Lentivirus (LV) for use as a gene therapy vector is the human immunodeficiency virus type 1 (HIV-1) [35, 36]. Safety is of paramount concern and due to the pathogenicity of HIV-1 in humans, LV vector systems have been designed to be replication deficient [23]. Essentially all clinical and research efforts, including lentiviral vectors for production of CAR T-cells, have been based on the third-generation packaging system with the self-inactivating (SIN) configuration, and the design of state-of-the-art lentiviral vectors has not changed dramatically since the development of the third-generation lentiviral vectors [37]. Third-generation vectors are based on a four-plasmid system that consists of one gene transfer vector (TV plasmid) and three helper plasmids, as shown in Fig. 14.2. The TV plasmid is the only genetic material transferred to the target cells and consists of the LV backbone containing the transgene expression cassette flanked by *cis*-acting elements required for encapsidation, reverse transcription, and integration [23]. All accessory genes, *nef*, *vpr*, *vpu* and *Vif*, are removed, as they are unnecessary [38] and the regulatory *tat* gene is eliminated. The *rev* gene, which is required for replication, is supplied *in trans*, to create a conditional packaging platform [39]. Vectors are given an envelope surrounding the capsid structure that is composed of very specific glycoproteins, namely the vesicular stomatitis virus glycoprotein (VSV-G), which allows the vector a high tropism and the ability to infect a wide variety of cell types

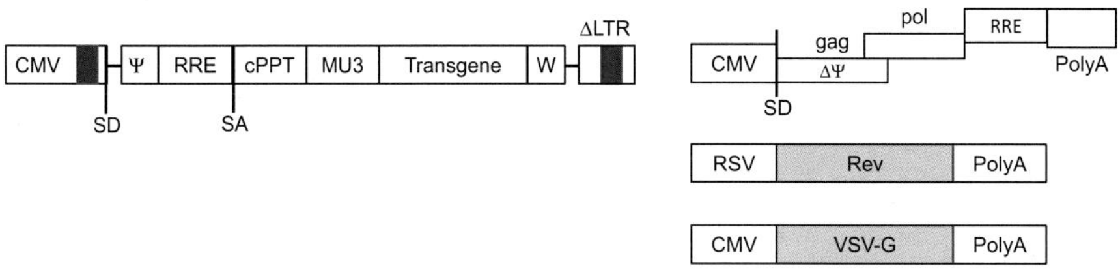

Fig. 14.2 Schematic diagram of the third generation Tat-independent Lentivirus vector system. The shown SIN transfer vector contains the cPPT for efficient nuclear import and uses the MSCV LTR promoter (MU3) as an internal promoter for driving expression of the transgene, as well as the WPRE (W) element for high-level transgene expression. The three packaging constructs encode for the HIV-1 gag-pol and rev proteins as well as for the VSV-g envelope glycoprotein [150]

[40]. Safety is further enhanced by the deletion of part of the 3′-LTR, which contains the TATA box and transcription factor-binding sites. Consequently, the vector packaging system becomes self-inactivating (SIN) [38, 41].

Producer Cell Lines

Most current production systems for LV vectors involve transient transfection of HEK293 or HEK293T cells. These cells are transfected with four plasmids containing *gag-pol*, *env*, *rev* and the TV plasmid [42–44]. The HEK293T cell line confers higher viral titers when compared to the HEK293 cell line and is therefore the preferred production platform [45]. The only difference between the HEK293 line and the HEK293T line is the latter carries the oncogenic simian vacuolating virus 40 (SV40) large T antigen. The mechanisms that instigate increased cell growth and transfection efficiency in cells expressing the SV40 T-antigen are unknown but this phenomenon has also been demonstrated between CV-1 (SV40 T-antigen negative) and COS-1 (SV40 T-antigen positive) cells [46]. However, the SV40 T-antigen is oncogenic and subsequently poses some safety concerns. Batch-to-batch variability using the four-plasmid transfection method in viral titers is also a significant challenge [3].

One of the first successful stable producer cell lines was the STAR packaging cell line. This cell line was generated by transfecting HEK293T cells with gammaretroviral vector, encoding a codon-optimized *gag-pol* gene followed by introduction of the genes for the HIV-1 regulatory proteins *Tat* and *Rev*, an envelope protein, and an HIV-1 LV vector [47]. Optimizing the structure of the transfected vector genome expression cassette lead to the development a producer cell line for a SIN LV vector [48]. Stable producer cells have several advantages over transient transfection including reproducibility between batches, scalability, and cost. The development of stable producer cell lines for LV vectors has proven to be challenging. Packaging genes and the gene silencing that occurs with HIV-1 based LV vectors causes cell toxicity during long culture periods [49]. Two different strategies have been employed to overcome this limitation. The first strategy uses a tetracycline-inducible system to control expression of VSV-G and *gal-pol* genes [50, 51]. The alternative strategy to inducible stable cell lines is the creation of constitutive packaging cell lines. These have been developed for a number of envelope proteins that are less toxic than VSV-G, such as cocal [52] and RD114-TR [53].

Lentivirus Vector Production

Lentivirus vectors are used in several FDA approved gene therapies. As shown in Table 14.1, these treatments all utilize *ex vivo* gene transfer and have dosage requirements as high as 10^9 CAR-T positive cells. Under FDA guidelines, the CAR-T positive cell is considered the final therapeutic product (also known as the medicinal product in the EU) and the LV vector is considered a key raw material [54]. Although the final cell product is autologous and must be produced individually for each patient, the LV vectors can be manufactured in large quantities and frozen at -80 °C for several years [55].

Commercial-scale production of LV vectors can be achieved either by the scale-out (i.e., the addition of supplementary production systems) of small-scale systems that support anchorage-dependent cells or the scale-up of systems that support expansion of cells adapted to grow in suspension. Currently, most large-scale productions rely on anchorage-dependent cells. As HEK293 cells are weekly anchorage-dependent, roller bottle cultures are not widely used [23]. Production is performed in parallel cultures of multi-tray systems – either Cell Factories or Cell Stacks. The preferred system is the 10-stack system [42, 44, 54, 56, 57]. In theory, a 40-stack system could be used but these systems present handling issues associated with their size and weight. Additionally, there can be uneven mass transport between the stacks, and this leads to variations in dissolved oxygen levels. The number of LV vector harvests can range from one final harvest to up to three harvests, producing between 20 L and 52 L from 12–24 10-stack cell factory devices [23]. There are some disadvantages to using cell factories to produce LV vectors, namely a limited harvest window, labor intensity, maintaining sterility in an open-system and the requirement for significant incubator space. To this end, investigators have looked at fixed-bed systems [58–61], hollow-fiber bioreactors [62] and the Wave bioreactor with cells immobilized on Fibra-cel discs [63]. Table 14.4 presents a comparison between the different production systems used in LV vector manufacturing.

Due to the inherent limitations in the scalability of anchorage-dependent cells, HEK-293 T cultures can be adapted to serum-free, suspension cultures for LV vector production and used in conventional stirred tank bioreactors [64]. The primary disadvantages of serum-free, suspension cultures are the lower cell densities and the reduction in virus titer, when compared to anchorage-dependent cell culture systems. Initial studies using suspension cultures showed viral titers being in the range of 10^6 TU/mL and below [65]. The cumulative yield of LV vectors can be

Table 14.4 Comparison of vector production yields in different anchorage-dependent culture systems

Culture system	Cell line	Production volume (L)	Virus titer (before downstream processing)
CF-10 [42, 56]	HEK293T	24–50	1–5 × 10^7 IG/mL
Hollow Fiber [62]	HEK293T	4	1.9 × 10^8 vg/mL
Fixed-Bed [61]	HEK295T	3.2	2 × 10^7 TU/ml
Wave Bioreactor [63]	HEK293T	~138	0.5–1 × 10^7 TU/mL

increased by up to 15-fold by operating the bioreactor in a perfusion mode versus batch mode, resulting in titers of 3.2 × 10^7 TU/mL, which is comparable to anchorage-dependent cell culture systems [66]. More recent studies have reported LV vector titers of 10^8 TU/mL via gradual adaptation in suspension media [67]. Another consideration when growing cells in suspension is separation of cells from virus in the supernatant. Perfusion cultures have made use of acoustic cell filter technology and Tangential Flow Depth Filtration (TFDF), which are able to retain cells within the bioreactor. Cumulative functional yields of 3.3 × 10^{11} and 3.9 × 10^{11} transducing units were achieved; the former over 6 days of LV production with 16.3 L of perfused media and the latter over 4 days with 16 L [68]. Batch cultures rely on the use of centrifugation and filtration post-harvest to remove cells [67].

Adenovirus

The adenovirus (Ad) was first identified in human adipose tissue in 1953 [69]. There are six species (A–F) that infect humans, which are further subdivided into over fifty infective serotypes [70]. The globular knob domain of the Ad capsid has a high affinity for the coxsackie/adenovirus receptor (CAR), which can be found on a variety of cells throughout the human body [71, 72]. Consequently, human Ads are thought to be responsible for 5–10% of all acute respiratory infections. The infectivity of Ads, along with their ability to exploit the cellular machinery to synthesize viral mRNA and proteins, has led to Ad vectors being used in approximately 50% of current clinical trials [3].

The Ad virus in a non-enveloped virus. The genome is linear, double-stranded DNA (dsDNA) ranging from 26 to 40 kb in length [73]. This linear form is organized into a compact, nucleosome-like structure within the viral capsid and has hairpin-like inverted terminal repeat (ITR) sequences on each end of the strand [74]. The ITRs vary in length and can be between 30 and 371 bp at its termini [75, 76]. As shown in Fig. 14.3, the viral genome comprises two major transcription regions, termed the early region and the late region [77, 78]. The early region of the genome contains four important transcription units (E1, E2, E3, and E4) that become activated upon

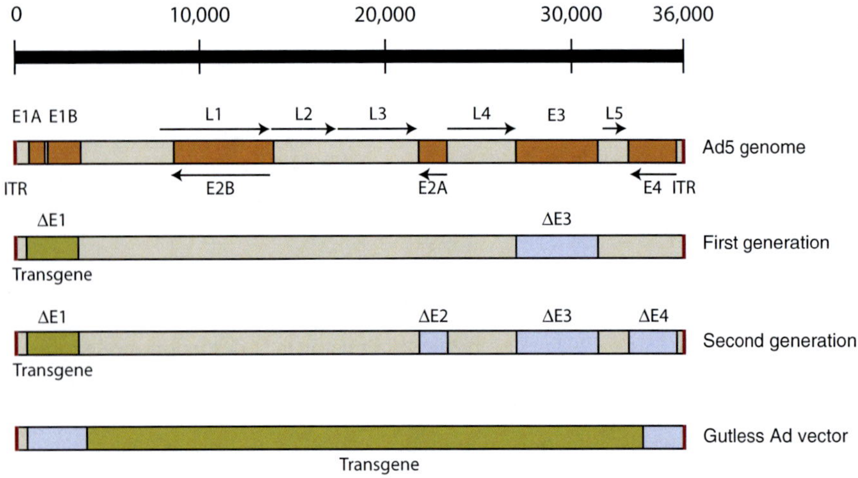

Fig. 14.3 Map of the Ad5 genome and different generations of Ad vectors

cell infection. Additionally, the genome contains two delayed early transcriptional units (IX and IVa2) and one major late (ML) transcription unit, which is processed to generate the late region mRNAs through post-translational processing (from L1 to L5) [79].

Adenovirus Vectors in Gene Therapy

First generation Ad vectors had the E1A/E1B region of the genome replaced with transgene cassettes of up to 4.5 kb of DNA [80, 81]. The E1A proteins are required for viral replication. Deletion of these genes from the viral genome consequently created a need to develop complimentary producer cell lines that provided these genes *in trans*, such as 293, 911, N52.E6 or PER.C6 [82].The E3 region is unnecessary for viral replication and can also be deleted to create additional space for transgene cassettes. These eliminations allow for the insertion of up to 8.3 kb of DNA into the vector [83]. First generation Ad vectors were beset with two limitations: (i) *de novo* expression of Ad proteins can activate the host immune response leading to clearance of virus-infected cells [84] and (ii) replication-competent Ads can be generated during gene amplification if spontaneous recombination occurs between the Ad vector and the engineered E1 region of the producer cell.

In addition to deletion of the E1 and E3 genes, second generation Ad vectors have the E2 and E4 genes deleted. Deletion of these additional genes allows for the insertion of transgene cassettes of up to 14 kb of DNA [85]. However, the E2 and E4 regions contain genes that transcribe proteins responsible for increased expression of virus-encoded genes. Subsequently, therapeutic gene expression is lower in second generation Ad vectors compared to first generation Ad vectors. As with the first generation vectors, cytotoxic T-cells attack and eliminate cells transduced with second generation Ad vectors *in vivo*, limiting the duration that the therapeutic gene is expressed [85, 86].

Third generation Ad vectors, also known as gutted or gutless vectors have all Ad DNA removed except for the ITR and packaging regions. This allows for the insertion of 36 kb of DNA into the vector. Replication requires packaging cells to be co-infected with a helper vector (HV), capable of producing all the necessary proteins, *in trans* [87]. The gutted Ad vector is produced in HEK293 producer cells that express the HV and Cre recombinase. The HV proteins permit replication and packaging of the gutted vector. The engineered Cre ensures the HV genome-packaging signal is excised by Cre-mediated recombination at the loxP sites and only the gutted Ad genome can be packaged [88].

Adenovirus Vector Production

Several upstream process parameters influence the quality and quantity of Ad vector production. These include the producer cell line, viral infection parameters, cell growth conditions and the mode of operation. Each of these parameters can each impact one another, requiring a holistic, systems-level approach to optimizing biomanufacturing processes [17].

Production of first-generation Ad vectors is commonly accomplished with HEK293 and PER.C6 cells [89]. The HEK293 cell line was generated by transfection of human embryonic kidney (HEK) cells with sheared human Ad5 DNA [90, 91], with nucleotides 1–4344, comprising the E1 region, being integrated on chromosome 19 [92]. These cells are highly versatile and can be cultivated in suspension or as adherent cells, and with serum free or serum containing culture media [89, 93]. It has been projected that suspension cultures of HEK293 cells can be produced at scales up to 10,000 L with virus yields in the range of 10^9–10^{10} VP/mL [94].

A serious disadvantage of the HEK293 cell line is the frequent production of replication competent adenovirus (RCA), caused by homologous recombination events during production [95]. The PER.C6 cell line was developed to alleviate the risk of RCA emergence during production. The PER.C6 cell line contains the Ad5 E1A and E1B encoding sequences (nucleotides 459–3510) under the control of the human phosphoglycerate kinase (PGK) promoter. RCA generation is overcome by combining cells with Ad vectors that do not have sequence overlap [96]. PER.C6 cells can be grown in suspension with and without serum-free medium [97, 98] and have become an industry standard.

Second generation Ad vectors, which lack E1, E2 and E4 regions can be produced with trans-complimenting cell lines derived from HeLa and Vero cells. Cell lines have been developed to produce E2 defective Ad vectors [99, 100] and E4 defective Ad vectors [101], respectively. Cell lines based on HEK293, 911 and E1 modified A549 cells have also been generated to allow propagation of second generation Ad vectors [102].

The third-generation or gutless Ad vectors, have all their genes removed except for the cis-acting sequences necessary for viral DNA replication and packaging. Vector production is complex due to the need for an appropriate producer cell line that can provide all the necessary viral elements *in trans*. The cytotoxicity of some Ad proteins is most likely the reason that work to create a cell line to compliment these vectors has failed [103]. It is therefore necessary to provide a helper Ad vector (HV), which provides the necessary packaging proteins of the viral particle [102]. Producer cell lines are derived from 293 cells and expresses Cre recombinase, while the HV is modified to contain *loxP* sites flanking the packaging site. In 293Cre cells, the packaging signal is efficiently excised, preventing the helper virus genome from being packaged [104]. However, excision efficiency is not perfect and contamination with the HV is a serious concern, with levels reported in the range of 0.01–1% [103].

Ad vector titers are affected by various factors, including multiplicity of infection (MOI), time of harvest (TOH) and cell concentration at infection (CCI). Once the producer cells are inoculated with the Ad vector, infection can be rapid, taking approximately 48 hours for a single infection cycle. Consequently, manufacturing processes are configured as single round infection cycles with MOI >1 to ensure all cells are infected. Optimal MOI should be determined for each system; however, typical values for Ad5 cultivation processes range from 10 to 200 virus particles per cell, resulting in an amplification ratio > 200 [89, 105, 106].

Cultivation of Per.C6 and 293 cells adapted to suspension culture can be achieved in stirred tank bioreactors. When cells are grown in batch culture, the optimal cell density for virus production is approximately 5×10^5 cells/mL, an order of magnitude lower than the theoretical maximum cell density possible [107]. This is referred to as the "*cell density effect*". Cell metabolism is hindered by nutrient depletion and metabolite accumulation and renders the cells unable to support viral production at high densities [89]. Exchanging media at the time of infection can reduce the cell density effect, as this replenishes substrates and removes metabolites. When utilizing a fed-batch mode, densities exceed 5×10^6 cells/mL at scales from 3 to 20 L [94, 108, 109]. By replacing at least 90% of the medium, optimal virus production can be achieved with $1 \text{ x}10^6$ cells/mL [110, 111]. Virus titer can be significantly improved through optimization of the physicochemical conditions in the bioreactor. Temperature, pH, dissolved oxygen (DO) and partial CO_2 pressure (pCO_2) have all been shown to be important parameters in cell growth, metabolism and Ad production [17, 112].

Adeno-associated Virus

The Adeno-associated virus (AAV) is a small, naturally replication defective virus, first discovered in 1965 as a co-infecting agent of the adenovirus. In order to replicate within the host cell nucleus, the AAV requires the assistance of either a helper virus, such as adenovirus or herpes virus, or some form of genotoxic stress

[113, 114]. There are 11 different AAV serotypes and over 100 AAV variants, although serotype AAV6 is almost identical to AAV1 and AAV10 and AAV11 are not well characterized [115]. Each serotype has its own characteristic capsid with a special affinity for certain host cell receptors, allowing it to be used to target a variety of tissue types [116–118].

The AAV genome is comprised of single-stranded DNA, approximately 5 kb in length. The genome lacks the essential genes needed for viral replication and expression of its own genome. An Ad provides these functions from the E1, E2a, E4 and VA RNA genes [119, 120]. The AAV genome houses four known open reading frames (ORFs), The four replication genes (*rep*) are contained on the first ORF, the *cap* gene that encodes three viral capsid proteins is found on the second ORF and the sub-genomic mRNAs found on the third and fourth ORFs are transcribed to produce the assembly-activating protein (AAP) [121–124]. Both ends of the genome are flanked by inverted terminal repeat (ITR) sequences that serve as self-priming structures for replication and provide the signals for Rep-mediated packaging [125].

Adeno-associated Virus Vectors in Gene Therapy

The process for creating vectors from AAVs begins with the deletion of genes coding for the Rep and Cap proteins. This deletion provides approximately 5 kb of packing space for foreign DNA. The new DNA is inserted into the "gutted" virus that contains only the ITRs. The ITRs contain all *cis*-acting elements necessary for replication and packaging in the presence of a helper virus. The Rep and Cap proteins and all necessary adenoviral helper genes are expressed on either one or two plasmids. The expression of Ad genes from a plasmid eliminates the need for co-infection with wild-type adenovirus. [114, 119, 126, 127].

Adeno-associated Virus Vector Production

Three different production systems are used to produce AAV vectors. The traditional method utilizes transient transfection of mammalian cells, most commonly HEK293. Cells are co-transfected with three plasmids in an equimolar ratio that contain the rAAV vector (i.e., the transgene flanked by the ITRs), the *rep* and *cap* genes, and the Adenovirus helper genes, E1, E2a, E4 and VA RNA, respectively [127, 128]. Variations of this protocol exist that utilize a two-plasmid system, combining the packaging and helper functions onto a single plasmid [129, 130] or a four-plasmid system that splits the *rep* and *cap* genes onto separate plasmids [131]. Production of rAAV can be enhanced through the addition of peptones (yeastolate, Trypton N1 or both), which increases production yield by 2.8- to 3.4-fold [132].

Serum-free, suspension cultures have been successfully used to produce rAAV vectors by transient transfection. Studies have shown rAAV vectors can be produced successfully using 20 L WAVE bioreactors with virus yields of 10^{14} VG/L [133]. This system could potentially be scaled up to 100 L, which is sufficient for programs involving relative low doses and for orphan disease indications such as LCA2. However, for successful clinical programs that are towards later phase clinical trials, and for disease applications with a greater number of patients and/or where higher per patient vector doses are needed, an estimated one to two order of magnitude increase in manufacturing capacity is predicted to be needed to meet the increasing number of disease indications that may be targeted by gene therapy [35, 134]. Increases in vector titer could be realized through increasing the proportion of productive cells. Despite obtaining high transfection efficiencies and nominal vector yields, studies suggest only 5%–10% of cells appear to produce measurable levels of assembled AAV capsids [135].

An alternative to the complex transient transfection methods, and one that could reduce costs, is the development of stable producer cell lines. These cells incorporate the viral and cargo genes into the cell, alleviating the need for transient transfection and eliminating batch-to-batch variations. Producer cell lines have been developed from A549 cells and HeLa cells [136, 137]. The packaging cells contain plasmids that include genes for the *cap*, *rep* and the transgene cassette flanked by the ITR sequences and a selectable marker gene. AAV production is initiated by infection with a replication competent Ad that provides the helper virus proteins [35]. Producer cell lines have been adapted to serum-free, suspension culture and can be grown in bioreactors with a yield of 10^4 virus particles/cell at the 15 L scale with the potential to scale-up to 500 L [137]. Other studies have produced rAAV from producer cells in 250 L stirred tank reactors with no foreseeable obstacles to scaling up to commercial scale 2000 L bioreactors or larger [138].

The Baculovirus Expression Vector System (BEVS) is another platform that has been successfully used to produce rAAV vectors. The main advantages to using BEVS over the mammalian producer cell lines is that no Ad helper virus is required. The first generation BEVS required co-infection of *S. frugiperda* (Sf9) cells with three baculovirus expression vectors (BEVs), each delivering one of the three essential genes (*rep*: Bac-Rep, *cap*: Bac-Cap, and ITR-flanked gene of interest: Bac-ITR-GOI) for AAV generation [139]. The first-generation BEVS were beset by genetic instability of BEVs and consequent loss of AAV yield and the inability to produce functional AAV vectors of other serotypes. Consequently, they did not attain widespread use [140, 141]. Overcoming these limitations has been achieved through the development of systems that rely on one or two BEVs, namely, One-Bac/Mono-Bac and Two-Bac.

The Mono-Bac systems requires Sf9 cells to undergo infection with a single BEV harboring all three essential AAV genes [142]. This is a different approach to the One-Bac1.0 protocol that utilizes a baculovirus (BV) inducible stable Sf9 packaging cell containing the *rep2* (AAV2 *rep*) and *capX* (X = serotype of interest) genes [143]. Cells are infected with a single baculoviral vector carrying an ITR-flanked *transgene* cassette. The resulting expression was induced from a combination of the amplification of integrated resident genes and the enhancement of the expression mediated by the immediate-early *trans*-regulator 1 (IE-1) encoded by BEV. The One-Bac 1.0 systems produce up to ten-fold higher cell-specific yield of AAV2 vector particles compared to the Three-Bac [139]. The Two-Bac system uses a single BEV containing the *rep* and *cap* expression cassettes (Bac-Rep-Cap or Bac-*In*-Rep-Cap), therefore now requiring co-infection of only two baculoviruses [144]. This method can yield 7×10^4 purified rAAV particles per cell from suspension cultures of recombinant baculovirus–infected insect cells [145].

Future Directions and Challenges

In recent years, the number of approved gene therapies has increased exponentially, and the indicators suggest that we will continue to see more products successfully complete clinical trials and come to market. Lentivirus and adeno-associated virus are the most viable vector options for future gene therapy research, due to their limited adverse reactions and favorable immunogenic profiles [146]. Despite the recent successes of gene therapy, several biomanufacturing challenges still exist.

Much of the research to date has focused on optimizing the design of the viral vectors and efficient producer and packaging cell systems. Future research will need to focus on Process Analytical Technologies (PAT) that can effectively monitor and control critical process parameters (CPP) that affect critical quality attributes (CQA), leading to improved consistency in the final product and optimization of the bioprocess production methods [147]. The 2016 National Roadmap (and the subsequent 2017 and 2019 updates) from the National Cell Manufacturing Consortium (NCMC) – a public-private consortium of industry, government, clinical, and academic leaders - identified several core challenges across Process Monitoring and Control, Cell Processing and Automation, Supply Chain and Transport Logistics, and Workforce Development, that must be addressed at a National level in order to realize the future of cell and gene manufacturing [148]. Among the core challenge identified were: (i) insufficient bioprocessing models to predict the impact of manufacturing conditions and materials on cell behavior; (ii) inadequate surrogate markers for understanding cell critical quality attributes (CQAs) and (iii) inadequate Process Analytic Technologies (PATs) and assays to assess cell characteristics. These challenges – which are the most critical research and development areas identified by industry, academic, and government stakeholders – require significant skills in Artificial Intelligence (AI) (which includes Machine Learning Theory and Automation Robotics and Controls), data analytics, and predictive modeling. Process control needs to move beyond the traditional physiochemical process control parameters (pH, dissolved O_2, and temperature) and incorporate -omics, real-time analysis of cell metabolites, and non-destructive measurements of endogenous factors, surface markers, and gene expression [149]. New PATs will elucidate which manufacturing or process variables can yield a consistent and reproducible product with appropriate CQAs. This can, in turn, lead to Quality-by-Design (QbD) manufacturing processes, and consequently improve large-scale, reproducible, low-cost production of high-quality and safe gene therapies.

Acknowledgements The author would like to acknowledge partial support for this work from the National Science Foundation Engineering Research Center for Cell Manufacturing Technologies (NSF EEC 1648035).

References

1. Papanikolaou E, Bosio A (2021) The promise and the hope of gene therapy. Front Genome Edit 3:618346
2. Friedmann T (1992) A brief history of gene therapy. Nat Genet 2(2):93–98
3. Bulcha JT et al (2021) Viral vector platforms within the gene therapy landscape. Signal Transduct Targeted Ther 6(1):53
4. Mercola KE et al (1980) Insertion of a new gene of viral origin into bone marrow cells of mice. Science 208(4447):1033–1035
5. Sheridan C (2011) Gene therapy finds its niche. Nat Biotechnol 29(2):121–128
6. Blaese RM et al (1995) T lymphocyte-directed gene therapy for ADA- SCID: initial trial results after 4 years. Science 270(5235):475–480
7. Baker AH, Herzog RW (2020) Did dendritic cell activation, induced by adenovirus-antibody complexes, play a role in the death of jesse gelsinger? Mol Ther 28(3):704–706
8. Gelsinger J (2016) Calculated risks. Nature 534(590):10.1038
9. Cavazzana-Calvo M et al (2000) Gene therapy of human severe combined immunodeficiency (SCID)-X1 disease. Science 288(5466):669–672
10. Daley J (2019) Gene therapy arrives. Nature 576(7785):S12–S13
11. Moran N (2012) First gene therapy approved. Nat Biotechnol 30(12):1153
12. Salmon F, Grosios K, Petry H (2014) Safety profile of recombinant adeno-associated viral vectors: focus on alipogene tiparvovec (Glybera®). Expert Rev Clin Pharmacol 7(1):53–65
13. Florko N (2017) FDA approves first CAR-T gene therapy treatment. InsideHealthPolicy Com's FDA Week 23(35):1–8
14. Fala L (2018) Kymriah (tisagenlecleucel) for young patients with acute lymphoblastic leukemia: first FDA-approved gene therapy
15. Approved Cellular and Gene Therapy Products (2022) Available from: https://www.fda.gov/vaccines-blood-biologics/cellular-gene-therapy-products/approved-cellular-and-gene-therapy-products
16. Levine B (2015) Performance-enhancing drugs: design and production of redirected chimeric antigen receptor (CAR) T cells. Cancer Gene Ther 22(2):79–84
17. Warnock JN, Daigre C, Al-Rubeai M (2011) Introduction to viral vectors. Methods Mol Biol 737:1–25
18. Santos HS et al (2021) In situ gene therapy. Curr Gene Ther 21(5):406–430
19. Friedmann T (1997) Overcoming the obstacles to gene therapy. Sci Am 276(6):96–101
20. Zu H, Gao D (2021) Non-viral vectors in gene therapy: recent development, challenges, and prospects. AAPS J 23(4):78
21. Navarro J et al (1998) Gene delivery systems: bridging the gap between recombinant viruses and artificial vectors. Adv Drug Deliv Rev 30(1–3):5–11
22. Gopinath C, Nathar TJ, Nelson EJR (2018) Retroviral vectors in gene therapy. In: Jayandharan GR (ed) Gene and cell therapy: biology and applications. Springer, Singapore, pp 3–28
23. Merten O-W, Hebben M, Bovolenta C (2016) Production of lentiviral vectors. Mol Ther Meth Clin Develop 3:16017
24. Syyam A et al (2022) Adenovirus vector system: construction, history and therapeutic applications. BioTechniques 73(6):297–305
25. Kim SH, Kim S, Robbins PD (2000) Retroviral vectors. Adv Virus Res 55:545–563

26. Coffin JM, Hughes SH, Varmus HE (1997) Retroviruses. Cold Spring Harbor Laboratory Press
27. Osten P, Grinevich V, Cetin A (2007) Viral vectors: a wide range of choices and high levels of service. Handb Exp Pharmacol 178:177–202
28. Zhang X, Godbey WT (2006) Viral vectors for gene delivery in tissue engineering. Adv Drug Deliv Rev 58(4):515–534
29. Stein BS et al (1987) pH-independent HIV entry into CD4-positive T cells via virus envelope fusion to the plasma membrane. Cell 49(5):659–668
30. Sinangil F, Loyter A, Volsky D (1988) Quantitative measurement of fusion between human immunodeficiency virus and cultured cells using membrane fluorescence dequenching. FEBS Lett 239(1):88–92
31. Buchschacher GL (2004) Safety considerations associated with development and clinical application of lentiviral vector systems for gene transfer. Curr Genom 5:19–35
32. Malim MH et al (1989) The HIV-1 rev trans-activator acts through a structured target sequence to activate nuclear export of unspliced viral mRNA. Nature 338(6212):254–257
33. Heinzinger NK et al (1994) The Vpr protein of human immunodeficiency virus type 1 influences nuclear localization of viral nucleic acids in nondividing host cells. Proc Natl Acad Sci U S A 91(15):7311–7315
34. Feng S, Holland EC (1988) HIV-1 tat trans-activation requires the loop sequence within tar. Nature 334(6178):165–167
35. Sharon D, Kamen A (2018) Advancements in the design and scalable production of viral gene transfer vectors. Biotechnol Bioeng 115(1):25–40
36. Ferreira MV, Cabral ET, Coroadinha AS (2021) Progress and perspectives in the development of lentiviral vector producer cells. Biotechnol J 16(1):2000017
37. Wolff JH, Mikkelsen JG (2022) Delivering genes with human immunodeficiency virus-derived vehicles: still state-of-the-art after 25 years. J Biomed Sci 29(1):79
38. Zufferey R et al (1998) Self-inactivating lentivirus vector for safe and efficient in vivo gene delivery. J Virol 72(12):9873–9880
39. Dull T et al (1998) A third-generation lentivirus vector with a conditional packaging system. J Virol 72(11):8463–8471
40. Wiznerowicz M, Trono D (2005) Harnessing HIV for therapy, basic research and biotechnology. Trends Biotechnol 23(1):42–47
41. Miyoshi H et al (1998) Development of a self-inactivating lentivirus vector. J Virol 72(10): 8150–8157
42. Merten OW et al (2011) Large-scale manufacture and characterization of a lentiviral vector produced for clinical ex vivo gene therapy application. Hum Gene Ther 22(3):343–356
43. Gándara C, Affleck V, Stoll EA (2018) Manufacture of third-generation lentivirus for preclinical use, with process development considerations for translation to good manufacturing practice. Hum Gene Ther Meth 29(1):1–15
44. Ausubel LJ et al (2012) Production of CGMP-grade lentiviral vectors. BioProcess Internat 10(2):32
45. Gama-Norton L et al (2011) Lentivirus production is influenced by SV40 large T-antigen and chromosomal integration of the vector in HEK293 cells. Hum Gene Ther 22(10):1269–1279
46. Smith SL, Shioda T (2009) Advantages of COS-1 monkey kidney epithelial cells as packaging host for small-volume production of high-quality recombinant lentiviruses. J Virol Methods 157(1):47–54
47. Ikeda Y et al (2003) Continuous high-titer HIV-1 vector production. Nat Biotechnol 21(5): 569–572
48. Throm RE et al (2009) Efficient construction of producer cell lines for a SIN lentiviral vector for SCID-X1 gene therapy by concatemeric array transfection. Blood 113(21):5104–5110
49. Ni Y et al (2005) Generation of a packaging cell line for prolonged large-scale production of high-titer HIV-1-based lentiviral vector. J Gene Med 7(6):818–834

50. Wielgosz MM et al (2015) Generation of a lentiviral vector producer cell clone for human Wiskott-Aldrich syndrome gene therapy. Mol Ther Meth Clin Develop 2:14063
51. Stewart HJ et al (2011) A stable producer cell line for the manufacture of a lentiviral vector for gene therapy of Parkinson's disease. Hum Gene Ther 22(3):357–369
52. Humbert O et al (2016) Development of third-generation cocal envelope producer cell lines for robust lentiviral gene transfer into hematopoietic stem cells and T-cells. Mol Ther 24(7): 1237–1246
53. Stornaiuolo A et al (2013) RD2-MolPack-Chim3, a packaging cell line for stable production of lentiviral vectors for anti-HIV gene therapy. Hum Gene Ther Meth 24(4):228–240
54. Levine BL et al (2017) Global manufacturing of CAR T cell therapy. Mol Ther Meth Clin Develop 4:92–101
55. Lamers C et al (2008) Retroviral vectors for clinical immunogene therapy are stable for up to 9 years. Cancer Gene Ther 15(4):268–274
56. Bellintani F et al (2008) Large scale process for the production and purification of lentiviral vectors for clinical applications. In: Human gene therapy, New Rochelle
57. Rout-Pitt N et al (2018) Large-scale production of lentiviral vectors using multilayer cell factories. J Biol Methods 5(2):e90
58. Wang X et al (2015) Large-scale clinical-grade retroviral vector production in a fixed-bed bioreactor. J Immunother 38(3):127–135
59. McCarron A et al (2019) Transient lentiviral vector production using a packed-bed bioreactor system. Hum Gene Ther Methods 30(3):93–101
60. Lesch HP, Valonen P, Karhinen M (2021) Evaluation of the single-use fixed-bed bioreactors in scalable virus production. Biotechnol J 16(1):2000020
61. Powers AD et al (2020) Lentiviral vector production from a stable packaging cell line using a packed bed bioreactor. Mol Ther Meth Clin Develop 19:1–13
62. Sheu J et al (2015) Large-scale production of lentiviral vector in a closed system hollow fiber bioreactor. Mol Ther Meth Clin Develop 2:15020
63. Greene MR et al (2012) Transduction of human CD34+ repopulating cells with a self-inactivating lentiviral vector for SCID-X1 produced at clinical scale by a stable cell line. Hum Gene Ther Part B Meth 23(5):297–308
64. Perry C, Rayat AC (2021) Lentiviral vector bioprocessing viruses. 13(2):268
65. Segura MM et al (2007) Production of lentiviral vectors by large-scale transient transfection of suspension cultures and affinity chromatography purification. Biotechnol Bioeng 98(4): 789–799
66. Manceur AP et al (2017) Scalable lentiviral vector production using stable HEK293SF producer cell lines. Hum Gene Ther Meth 28(6):330–339
67. Bauler M et al (2020) Production of lentiviral vectors using suspension cells grown in serum-free media. Mol Ther Methods Clin Dev 17:58–68
68. Tran MY, Kamen AA (2022) Production of lentiviral vectors using a HEK-293 producer cell line and advanced perfusion processing. Front Bioeng Biotechnol 10:887716
69. Rowe WP et al (1953) Isolation of a cytopathogenic agent from human adenoids undergoing spontaneous degeneration in tissue culture. Proc Soc Exp Biol Med 84(3):570–573
70. Shenk T (1996) In: Fields BN (ed) Adenoviridae: the viruses and their replication in fields virology. Lippincott-Raven Publishers, Philadelphia, PA, pp 2111–2148
71. Bergelson JM et al (1997) Isolation of a common receptor for coxsackie B viruses and adenoviruses 2 and 5. Science 275(5304):1320–1323
72. Tomko RP, Xu R, Philipson L (1997) HCAR and MCAR: the human and mouse cellular receptors for subgroup C adenoviruses and group B coxsackieviruses. Proc Natl Acad Sci U S A 94(7):3352–3356
73. Campos SK, Barry MA (2007) Current advances and future challenges in Adenoviral vector biology and targeting. Curr Gene Ther 7(3):189–204
74. Douglas JT (2007) Adenoviral vectors for gene therapy. Mol Biotechnol 36(1):71–80

75. Arrand JR, Roberts RJ (1979) The nucleotide sequences at the termini of adenovirus-2 DNA. J Mol Biol 128(4):577–594
76. Shinagawa M, Padmanabhan R, Padmanabhan R (1980) The nucleotide sequence of the right-hand terminal smaI-K fragment of adenovirus type 2 DNA. Gene 9(1–2):99–114
77. Boyer J, Ketner G (1999) Adenovirus late gene expression. In: Seth P (ed) Aednovirues: basic biology to gene therapy. R. G. Landes Bioscience, Austin, TX, pp 69–77
78. Rux JJ, Burnett R (1999) Adenoviruses: basic biology to gene therapy. In: Seth P (ed) Medical intelligence 15. R. G. Landes Bioscience, Austin, TX, pp 5–15
79. Giacca M, Zacchigna S (2012) Virus-mediated gene delivery for human gene therapy. J Control Release 161(2):377–388
80. Berkner KL (1988) Development of adenovirus vectors for the expression of heterologous genes. BioTechniques 6(7):616–629
81. Kovesdi I et al (1997) Adenoviral vectors for gene transfer. Curr Opin Biotechnol 8(5): 583–589
82. Danthinne X, Imperiale MJ (2000) Production of first generation adenovirus vectors: a review. Gene Ther 7(20):1707–1714
83. Russell WC (2009) Adenoviruses: update on structure and function. J Gen Virol 90(Pt 1):1–20
84. Yang Y et al (1994) Cellular immunity to viral antigens limits E1-deleted adenoviruses for gene therapy. Proc Natl Acad Sci 91(10):4407–4411
85. Bessis N, GarciaCozar FJ, Boissier MC (2004) Immune responses to gene therapy vectors: influence on vector function and effector mechanisms. Gene Ther 11(Suppl 1):S10–S17
86. Liu Q, Muruve DA (2003) Molecular basis of the inflammatory response to adenovirus vectors. Gene Ther 10(11):935–940
87. Kreppel F (2014) Production of high-capacity adenovirus vectors. Methods Mol Biol 1089: 211–229
88. Hartigan-O'Connor D, Amalfitano A, Chamberlain JS (1999) Improved production of gutted adenovirus in cells expressing adenovirus preterminal protein and DNA polymerase. J Virol 73(9):7835–7841
89. Silva AC et al (2010) Adenovirus vector production and purification. Curr Gene Ther 10(6): 437–455
90. Graham FL et al (1977) Characteristics of a human cell line transformed by DNA from human adenovirus type 5. J Gen Virol 36(1):59–74
91. Lin Y-C et al (2014) Genome dynamics of the human embryonic kidney 293 lineage in response to cell biology manipulations. Nat Commun 5(1):4767
92. Louis N, Evelegh C, Graham FL (1997) Cloning and sequencing of the cellular-viral junctions from the human adenovirus type 5 transformed 293 cell line. Virology 233(2):423–429
93. Kallel H, Kamen AA (2015) Large-scale adenovirus and poxvirus-vectored vaccine manufacturing to enable clinical trials. Biotechnol J 10(5):741–747
94. Kamen A, Henry O (2004) Development and optimization of an adenovirus production process. J Gene Med 6(Suppl 1):S184–S192
95. Lochmuller H et al (1994) Emergence of early region 1-containing replication-competent adenovirus in stocks of replication-defective adenovirus recombinants (delta E1 + delta E3) during multiple passages in 293 cells. Hum Gene Ther 5(12):1485–1491
96. Fallaux FJ et al (1998) New helper cells and matched early region 1-deleted adenovirus vectors prevent generation of replication-competent adenoviruses. Hum Gene Ther 9(13):1909–1917
97. Xie L et al (2003) Large-scale propagation of a replication-defective adenovirus vector in stirred-tank bioreactor PER.C6 cell culture under sparging conditions. Biotechnol Bioeng 83(1):45–52
98. Xie L et al (2002) Serum-free suspension cultivation of PER.C6(R) cells and recombinant adenovirus production under different pH conditions. Biotechnol Bioeng 80(5):569–579
99. Klessig DF, Grodzicker T, Cleghon V (1984) Construction of human cell lines which contain and express the adenovirus DNA binding protein gene by cotransformation with the HSV-1 tk gene. Virus Res 1(2):169–188

100. Brough DE, Cleghon V, Klessig DF (1992) Construction, characterization, and utilization of cell lines which inducibly express the adenovirus DNA-binding protein. Virology 190(2): 624–634
101. Weinberg DH, Ketner G (1983) A cell line that supports the growth of a defective early region 4 deletion mutant of human adenovirus type 2. Proc Natl Acad Sci U S A 80(17):5383–5386
102. Kovesdi I, Hedley SJ (2010) Adenoviral producer cells. Viruses 2(8):1681–1703
103. Alba R, Bosch A, Chillon M (2005) Gutless adenovirus: last-generation adenovirus for gene therapy. Gene Ther 12(1):S18–S27
104. Parks RJ et al (1996) A helper-dependent adenovirus vector system: removal of helper virus by Cre-mediated excision of the viral packaging signal. Proc Natl Acad Sci U S A 93(24): 13565–13570
105. Dormond E, Kamen AA (2011) Manufacturing of adenovirus vectors: production and purification of helper dependent adenovirus. In: Merten O-W, Al-Rubeai M (eds) Viral vectors for gene therapy: methods and protocols. Humana Press, Totowa, NJ, pp 139–156
106. Chen K-D et al (2018) Process optimization for the rapid production of adenoviral vectors for clinical trials in a disposable bioreactor system. Appl Microbiol Biotechnol 102(15): 6469–6477
107. Garnier A et al (1994) Scale-up of the adenovirus expression system for the production of recombinant protein in human 293S cells. Cytotechnology 15(1–3):145–155
108. Cote J et al (1998) Serum-free production of recombinant proteins and adenoviral vectors by 293SF-3F6 cells. Biotechnol Bioeng 59(5):567–575
109. Nadeau I et al (1996) Improvement of recombinant protein production with the human adenovirus/293S expression system using fed-batch strategies. Biotechnol Bioeng 51(6): 613–623
110. Zhang S et al (2001) Method for the production and purification of adenoviral vectors. Introgen Therapeutics, Inc.
111. Liu L, Shoupeng L (2001) Method of producing adenoviral vector stocks. Genvec, Inc., US
112. Warnock JN, Merten OW, Al-Rubeai M (2006) Cell culture processes for the production of viral vectors for gene therapy purposes. Cytotechnology 50(1–3):141–162
113. Buning H, Braun-Falco M, Hallek M (2004) Progress in the use of adeno-associated viral vectors for gene therapy. Cells Tissues Organs 177(3):139–150
114. Goncalves MA (2005) Adeno-associated virus: from defective virus to effective vector. Virol J 2:43
115. Wu Z, Asokan A, Samulski RJ (2006) Adeno-associated virus serotypes: vector toolkit for human gene therapy. Mol Ther 14(3):316–327
116. Ding W et al (2005) Intracellular trafficking of adeno-associated viral vectors. Gene Ther 12(11):873–880
117. Kotin RM et al (1991) Mapping and direct visualization of a region-specific viral DNA integration site on chromosome 19q13-qter. Genomics 10(3):831–834
118. McCarty DM et al (2003) Adeno-associated virus terminal repeat (TR) mutant generates self-complementary vectors to overcome the rate-limiting step to transduction in vivo. Gene Ther 10(26):2112–2118
119. Matsushita T et al (1998) Adeno-associated virus vectors can be efficiently produced without helper virus. Gene Ther 5(7):938–945
120. Xiao X, Li J, Samulski RJ (1998) Production of high-titer recombinant adeno-associated virus vectors in the absence of helper adenovirus. J Virol 72(3):2224–2232
121. Smith RH (2008) Adeno-associated virus integration: virus versus vector. Gene Ther 15(11): 817–822
122. Naumer M et al (2012) Properties of the adeno-associated virus assembly-activating protein. J Virol 86(23):13038–13048
123. Sonntag F, Schmidt K, Kleinschmidt JA (2010) A viral assembly factor promotes AAV2 capsid formation in the nucleolus. Proc Natl Acad Sci 107(22):10220–10225

124. Linden RM, Berns K (2000) Molecular biology of adeno-associated viruses. Contrib Microbiol 4:68–84
125. Lusby E, Fife K, Berns K (1980) Nucleotide sequence of the inverted terminal repetition in adeno-associated virus DNA. J Virol 34(2):402–409
126. Russell DW, Kay MA (1999) Adeno-associated virus vectors and hematology. Blood 94(3): 864–874
127. Merten O-W et al (2014) Manufacturing of viral vectors for gene therapy: part I. Upstream Process Pharm Bioprocess 2(2):183–203
128. Rashnonejad A et al (2016) Large-scale production of adeno-associated viral vector serotype-9 carrying the human survival motor neuron gene. Mol Biotechnol 58(1):30–36
129. Grimm D, Kay MA, Kleinschmidt JA (2003) Helper virus-free, optically controllable, and two-plasmid-based production of adeno-associated virus vectors of serotypes 1 to 6. Mol Ther 7(6):839–850
130. Tang Q et al (2020) Two-plasmid packaging system for recombinant adeno-associated virus. BioResearch Open Access 9(1):219–228
131. Emmerling VV et al (2016) Rational plasmid design and bioprocess optimization to enhance recombinant adeno-associated virus (AAV) productivity in mammalian cells. Biotechnol J 11(2):290–297
132. Zhao H et al (2023) Process development of recombinant adeno-associated virus production platform results in high production yield and purity. Hum Gene Ther 34(1–2):56–67
133. Grieger JC, Soltys SM, Samulski RJ (2016) Production of recombinant adeno-associated virus vectors using suspension HEK293 cells and continuous harvest of vector from the culture media for GMP FIX and FLT1 clinical vector. Mol Ther 24(2):287–297
134. van der Loo JCM, Wright JF (2015) Progress and challenges in viral vector manufacturing. Hum Mol Genet 25(R1):R42–R52
135. Dash S et al (2022) Only a small fraction of cells produce assembled capsids during transfection-based manufacturing of adeno-associated virus vectors. Biotechnol Bioeng 119(6):1685–1690
136. Martin J et al (2013) Generation and characterization of adeno-associated virus producer cell lines for research and preclinical vector production. Hum Gene Ther Meth 24(4):253–269
137. Farson D et al (2004) Development and characterization of a cell line for large-scale, serum-free production of recombinant adeno-associated viral vectors. J Gene Med 6(12):1369–1381
138. Thorne BA, Takeya RK, Peluso RW (2009) Manufacturing recombinant adeno-associated viral vectors from producer cell clones. Hum Gene Ther 20(7):707–714
139. Joshi PRH et al (2021) Advancements in molecular design and bioprocessing of recombinant adeno-associated virus gene delivery vectors using the insect-cell baculovirus expression platform. Biotechnol J 16(4):2000021
140. Kohlbrenner E et al (2005) Successful production of pseudotyped rAAV vectors using a modified baculovirus expression system. Mol Ther 12(6):1217–1225
141. Urabe M et al (2006) Scalable generation of high-titer recombinant adeno-associated virus type 5 in insect cells. J Virol 80(4):1874–1885
142. Galibert L, Merten O-W, Jacob A (2018) Baculovirus system for the expression of a gene therapy vector. Google Patents
143. Aslanidi G, Lamb K, Zolotukhin S (2009) An inducible system for highly efficient production of recombinant adeno-associated virus (rAAV) vectors in insect Sf9 cells. Proc Natl Acad Sci 106(13):5059–5064
144. Chen H (2008) Intron splicing–mediated expression of AAV rep and cap genes and production of AAV vectors in insect cells. Mol Ther 16(5):924–930
145. Smith RH, Levy JR, Kotin RM (2009) A simplified baculovirus-AAV expression vector system coupled with one-step affinity purification yields high-titer rAAV stocks from insect cells. Mol Ther 17(11):1888–1896
146. Alexandra LGM et al (2022) In: Molecular Cloning D, Sadık TH, Mercimek A, Melis Sumengen O (eds) Viral vectors in gene therapy and clinical applications. Rijeka, IntechOpen

147. Williams T et al (2023) Machine learning and metabolic modelling assisted implementation of a novel process analytical technology in cell and gene therapy manufacturing. Sci Rep 13(1): 834
148. Consortium NCM (2030) Cell manufacturing roadmap:2019
149. Wang B et al (2022) Process analytical technologies in cell therapy manufacturing: state-of-the-art and future directions. J Advan Manufact Process 4(1):e10106
150. Ramezani A, Hawley RG (2002) Overview of the HIV-1 lentiviral vector system. Curr Prot Mol Biol 60(1):16.21. 1–16.21. 15

Chapter 15
Manufacturing and Quality of Advanced Therapeutics: Cellular Therapies, CAR-T and Gene Therapies

Christian van den Bos, Robert Keefe, Mary Keville, and Michael McCaman

Abstract Advanced therapeutics include cellular and gene therapies, fields originally distinct but now sometimes converging as exemplified by CAR-T approaches.

Common to all facets of advanced therapies is that the active pharmaceutical ingredients are vastly more complex than those of common biologics such as antibodies and hence pose elevated challenges regarding their manufacture and quality control.

We discuss prominent examples of advanced therapies and their technology platforms, delineate their historic and technological origin as well as possible paths forward so as to understand the evolution of manufacturing processes seen today.

Importantly, using this approach we draw lessons from past experiences with the intent to steer the current and future manufacturing and quality control professional away from potentially problematic and towards likely successful approaches.

Given the complexity of advanced therapeutics, regulatory guidelines evolved accordingly and we discuss the current state; it is with a degree of pride to see that many of the experiences we and others in the field shared with authorities eventually contributed to regulatory guidelines such as the EU ATMP guide.

Finally, we hope to share the profound satisfaction we experience by working in a field at the intersection of cutting-edge science and production technology, thereby providing crucial medicinal products, most recently in the context of the Covid19 pandemic, at a pace previously unimaginable.

C. van den Bos (✉)
Mares Ltd., Greven, Germany
e-mail: cvandenbos@maresltd.eu

R. Keefe
Bluespherebio, Pittsburgh, PA, USA
e-mail: keefe@bluespherebio.com

M. Keville
Mary Keville Consulting, LLC, Plymouth, PA, USA
e-mail: mary@ourhybridconcepts.com

M. McCaman
Phoenix BioService Consulting, Fanwood, NJ, USA
e-mail: michael.mccaman@orionthx.com

R. Pörtner (ed.), *Biopharmaceutical Manufacturing*, Cell Engineering 11,
https://doi.org/10.1007/978-3-031-45669-5_15

Keywords Manufacture · Quality control · Cell therapy · Gene therapy · Advanced therapy

Abbreviations

AAV	Adeno-Associated Virus
APH	Apheresis
ATMP	Advanced Therapy Medicinal Product
BSC	Biological Safety Cabinet
BSE	Spongiform Encephalitis
CAPA	Corrective Action, Preventive Action
CAR-T	Chimeric antigen receptor T-cells
CFR	Code of Federal Regulations
cGMP	Current Good Manufacturing Practice (US)
CJD	Creutzfeld-Jacob Disease
CMC	Chemistry, Manufacture, Control
COC	Chain of Custody
COGS	Cost of Goods
Crispr/Cas9	Clustered regularly interspaced short palindromic repeats/CRISPR associated protein 9
DMSO	Dimethylsulfoxide
DNA	Deoxynucleic Acid
EBR	Electronic Batch Record
eTCR	Endogenous T-Cell Receptor
EVs	Extracellular Vesicles
FDA	Food and Drug Administration
GMP	Good Manufacturing Practice
GvHD	Graft vs. Host Disease
HDR	Homology-Dependent-Recombination
HLA	Human Leucocyte Antigens
IDO	Indoamine-2,3-dioxydase
IFN-γ	Interferon Gamma
IL-10	Interleukin 10
IPCs	In-Process-Controls
IPSC	Induced Pluripotent Stem Cells
ITR	Inverted Terminal Repeats
Kb	Kilobases
kDa	Kilo Dalton
LN2	Liquid Nitrogen
MES	Manufacturing Execution System
MOI	Multiplicity of Infection
mRNA	Messenger Ribonucleic Acid
MSCs	Mesenchymal Stromal Cell/Mesenchymal stem cells

OKT3	Octamer-binding transcription factor 4
PAT	Process Analytical Technology
PBMCs	Peripheral Blood Mononuclear Cells
PDL	Population Doubling Level
POC	Point of care
QbD	Quality by Design
QC	Quality Control
QP	Qualified Person
rAAV	Recombinant Adeno-Associated Virus
RBA	Risk-Based-Approach
RCL	Replication-Competent Lentivirus
SUB	Single-Use-Bioreactor
TCR	T-cell receptor
TFF	Tangential Flow Filtration
TGFβ	Transforming Growth Factor b
Trail	TNF-related apoptosis-inducing ligand
VEGF	Vascular Endothelial Growth Factor
VP1,2,3	Viral Protein 1,2,3
VSV	Vesicular Stomatitis Virus

Historical

Development of Early Production Elements

The first examples of cellular therapies were perhaps blood transfusions anecdotally known since the 1600s and were based on a scientific foundation due to Landsteiner's discovery of blood groups published in 1901 [1, 2]; this rendered transfusions reliable and made them widely available. At a minimal level, the production process is limited to drawing blood from one individual -the donor- and injecting it into another -the recipient-, the need to do so on a large scale and in controlled ways added blood typing, microbial control/testing and, importantly, logistics, i.e., storing, transport and standardized handling to the production process of blood transfusions; this, and the vast need for blood transfusions compounded by belligerent events triggered an explosion in developing tools and processes, amongst them standardized sterile materials such as syringes, bags and related, eventually in single-use variants which are in common use today and have informed tools and processes in many areas of pharmaceutical and biotechnological industries.

Bone Marrow Transplants

Since the 1940s reports of bone marrow transplants have been published, the intention being to provide therapy to patients suffering from disorders of the blood system such as leukemia. Similar to the development of blood transfusions insights, clinical success was achieved only once the theoretical basis was established; here, this included the understanding that donors and recipients have be matched according to the Human Leucocyte Antigen (HLA) system [3]. Following this, successful bone marrow transplantations have been reported since the late 1960s and certain cancers caused by cells belonging to the blood system became treatable. The process involves, first, obtaining biopsies of donor bone marrow, e.g., by drawing it from the bone marrow cavity of the iliac crest, enriching samples for hematopoietic stem cells, cryopreserving the transplant, conditioning the patient, transporting the transplant to the location of the patient under cryogenic conditions, thawing the transplant under controlled conditions and infusing the transplant [4]. At a minimal level, the process included that described for blood transfusions plus steps required for cold-chain/cryogenic preservation of transplants and their transport under also cold/cryogenic conditions as well as controlled thawing prior to infusion. Subsequently, hematopoietic stem cells were identified by surface antigens [5, 6] so that it became possible to characterize, quantify and further purify the cell fraction in bone marrow transplants responsible for regenerating the blood system, I.e., the actual active ingredient; these cells became widely known as CD34 positive cells acc. to the Cluster of Differentiation terminology employed in the hematologic lineage. Similar to the above situation described for blood transfusions, the ability to transplant bone marrow reliably triggered the development of equipment supporting these new needs, particularly improved harvesting devices, enrichment procedures and devices, cryogenic formulations, controlled rate freezing devices, improved cryogenic storage units and HLA-based testing equipment.

Bioproduction Contributions

In the 1970s it became possible to recombinantly produce proteins, insulin being an early and prominent such product [7]. Much of the kit developed for bioproduction has a direct bearing for industrial production of cell and gene therapy products and so will be discussed in some detail.

While the availability of recombinant proteins produced in microbes was a great success, it quickly became apparent that microbial expression systems are subject to severe limitations and so researchers turned to mammalian expression systems [8]. These cells were originally adherent cells, hence required surfaces to adhere to in order to survive, proliferate and synthesize the protein of interest [9]. Initial cultures likely have occurred in typical lab scale cell culture devices such as T-Flasks. Given their limited surface area, often 225 cm^2 and less, these quickly became limiting and larger formats with simplified handling were developed. Well known examples of these are cell factories/cell stacks comprising 1–40 layers of cell growth area incorporated into 1 device with central feeding and venting so that cell suspensions may be seeded into such a stack by e.g. pumping the required volume/ titer into the stack through one opening. The opening is closed, and the suspension then distributed to the individual layers by turning and tilting the stack. Closing the stack using a suitable sterile filter arrangement of 0.22 μm pore size permits ventilation and allows the stack to use regular CO_2-controlled cell culture incubators.

Such stacks are available from various vendors and sizes; for practical reasons, 10-stacks and smaller are often used for manual operations; 40 stacks tend to require mechanical aids, and these are also available from stack providers. It should be noted that output from such stacks does not necessarily scale linearly.

The principle of stacking multiple growth surfaces together and into one device has sprouted various permutations with altered geometric configurations, enhanced throughput and volume and with specific applications [10–12].

Bioproduction has long since moved beyond stacked systems designed for adherend cell culture by employing single-use suspension-type bioreactors at current at a scale of currently up to 2 m^3 in conjunction with fully synthetic media [13]. Thus, therapeutic approaches based on hematological needs provided key aspects for the toolbox required for industrialized therapeutics based on living cells, albeit on a small scale. Production scale, an important parameter for the availability of any medicine including cell-based therapies, and adaptations from bioproduction do provide the tools for such scale-up.

Stem Cells and Lessons Learned

Various cell types have been discussed for therapeutic purposes, amongst them embryonic stem cells. The allure of such embryonic stem cells was in part that these cells can give rise to all tissues of the body and so it was postulated that they might also be able to regenerate all diseased tissues of the body. This school of thought may have informed considerations of many other cell types considered for therapeutic purposes resulting in a sometimes extravagant combinations of claims [14, 15]: their stemness would allow such cells to home to injured tissues and integrate there, thus repairing defective tissues and be spared from recognition and disposal by the recipient's immune system – essentially being a self-guided spare part of broad therapeutic potential. A consequence of these concepts was a focus on "stemness" as it was considered key to the therapeutic effect of cells. Consequently, and sometimes extending to this day, stemness tests remain release criteria for cells possessing a certain plasticity. While there are cell-biological tests for stem cell features, testing for such features is very much different from a well-defined mechanism of action and pertinent quantitative tests and we consider its relationship to therapeutic efficacy to be tenuous at best. The exception to this view are induced pluripotent stem cells, which are, however, not a medicinal product in themselves but rather a precursor for desirable cell types [16].

Many cell types were discussed in the context of therapeutic cells and applied to a vast range of indications; it would be impractical to consider them individually and so below we discuss examples of blood and non-blood derived therapeutic cell types and begin with the latter; mesenchymal stem cells may serve as a prototype for therapeutically valuable cells and may serve to illustrate manufacturing steps, pitfalls, and solutions.

Mesenchymal Stem Cells

Mesenchymal stem cells, sometimes also referred to as mesenchymal stromal cells (MSCs), were originally described by Friedenstein and co-workers [17, 18] and later proposed as the basis for novel cellular therapeutics [19]. They are fibroblastic cells derived from tissues including bone marrow, fat, or umbilical cord. Crude cell preparations from such tissues can be generated and seeded onto static culture surfaces, e.g., T-flasks. Such cell preparations contain a mixture of proliferating and quiescent, as well as adherent and non-adherent cells. Under standard culture conditions, MSCs adhere and proliferate so that e.g. three media changes suffice to remove non-adherent cells; of the remaining fraction and in the presence of suitable growth factors, MSCs tend to be the only proliferating cells so that such cultures after perhaps 2 weeks comprises mainly MSCs [20, 21]. Such initial MSC cultures can then be expanded by common cell culture expansion protocols – harvesting, splitting, re-seeding, and repeating the cycle.

A Case in Point: MSC-Based Investigational Medicinal Products

We consider it likely that the above stemcell-related misconceptions have led to significant consequences in the development of MSC-based therapies and base this not only on personal experience but also on the relationship between publications, clinical trials and approved products. One would fully expect a chronological order starting with publications, followed by clinical trials followed by approved products; likewise, one would expect clinical trials to be of sometimes substantial duration so that the chronological order referred to may well be spread out. Nonetheless, taking into account data presented in Fig. 15.1, namely nearly 1500 clinical trials, and considering the low number of approved products, namely 9 at the time of this writing (*https://alliancerm.org/available-products/*), we suggest that these numbers indicate a unfortunate ratio between effort and success and consider it likely that one root cause is likely to be grounded in the above consideration: adhering to

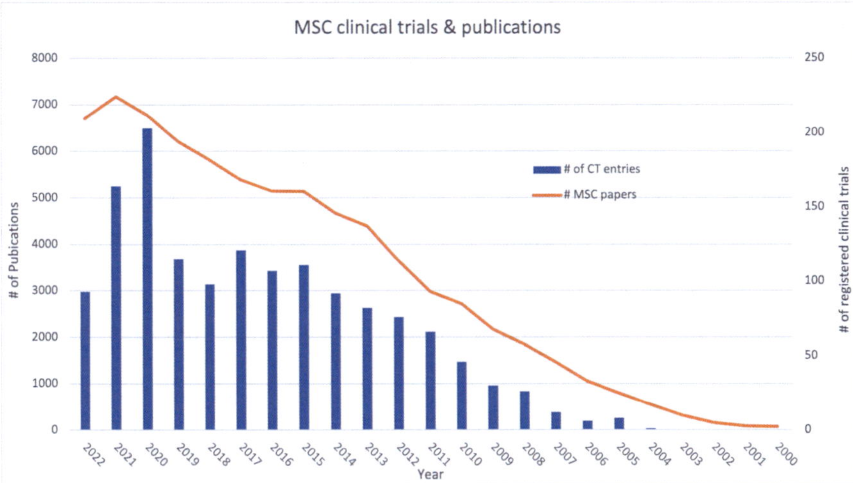

Fig. 15.1 Number of publications and clinical trials published for MSCs per year; data derived from queries of www.clinicaltrials.gov as well as from medline.com conducted in January 2023. Total numbers were 1457 and 75,342 entries for clinical trials and publications, respectively. Data obtained from websites clinicaltrials.gov and pubmed.com in January 2023. We hypothesize that the number of products is low when compared to the vast body of clinical trials

extravagant claims regarding cell capabilities rather than to pertinent and quantitative functional specifications likely resulted in investigational medicinal products which were overburdened and hence did not live up to expectations; eventually this likely resulted in a substantial number of inconclusive clinical trials. We emphasize this point because we consider this to be a crucial lesson for current and future work with cellular/tissue engineering/gene therapy investigational medicinal products: To maximize the probability of success, we consider it paramount to adhere to rational and quantitative considerations and specifications of the product.

First Generation MSC Products and Manufacturing Processes: Static/Planar Culture Systems

Production often occurred in static/planar culture systems, i.e., initially in T-flasks and with increasing cell quantities, moving into the above multi-stack systems, as these offer direct scale-up from T-flasks w/o the need for specialized and expensive equipment beyond what is typically found in laboratories performing cell culture – often the origin of new therapeutic developments.

Using such culture systems both analytical as well as initial production needs can be served – for analytical requirements, where few cells per sample often suffice and many multiples are desirable, 96-well plates with about 0.3 cm^2 growth area per well and up to e.g. 2000 cells are often found useful and compatible with analytical equipment such as spectrophotometers or fluorometers so that many data points may be acquired quickly; for specialized applications higher throughput can be achieve using denser formats such as 384 plates. For manufacturing purposes, often a high cell number is required. A 10-stack provides approximately 6600 cm^2 of growth area and this may yield 150–200 × 10^6 MSCs; this corresponds to the number of MSCs often used for a single adult dose so one might calculate 1 10-stack per dose.

It is the nature of cell expansion in such static/planar systems that initially few cells are seeded and left to proliferate until they have reached a desired cell density; during this period, it may be beneficial to perform a media exchange. Once the desired density is reached, cells can be harvested by removing the medium, washing cells with a buffered saline solution and then liberating cells from their growth surface by e.g. enzymatically cleaving their connections to the growth surface, washing the resulting suspension to remove the enzyme, re-suspending them in an appropriate volume of culture medium and re-seeding them at a split ratio of e.g. 1:5, i.e., distributing the number of cells harvested from 1 culture vessel to 5 such vessels. This process can be repeated until a desired number of cells has been obtained; at that point, cells can be prepared either for long term storage, i.e. cryopreservation, see below, or direct application.

Process Considerations

Complex Animal Ingredients

Since the earliest cell culture successes in Gey's laboratory [22], cell culture media contained various complex ingredients of animal origin and over time, the addition of fetal bovine serum became the standard approach. Apart from ethical considerations regarding the sourcing of the material, considerations of safety became paramount due to the advent of Bovine Spongiform Encephalopathy (BSE) and its human counterpart Creutzfeldt–Jakob disease (CJD) [23]. Thus, this complex animal derived reagent posed safety risks and, in turn, regulatory hurdles, became scarce due to restrictions on sourcing and, consequently, expensive. Unlike bioproduction cell lines, at current we are not aware of a fully synthetic formation for a culture medium serving adherent cells. As of now, bovine serum is often replaced by using human platelet lysate, which provides similar benefits, is not (non-human) animal sourced and available in pharmaceutical quality [24]. Media formulations containing it are often referred to as "Xeno-Free" media [25] and likely present the current formulation of choice.

Cell Age/Capability

Hayflick proposed in the 1960s that non-cancerous cells can undergo cell divisions only up to an upper limit [26] and, unsurprisingly, this is also observed in attempts to generate large number of therapeutic MSCs. Each period of cells seeding, culture expansion and harvest has commonly been referred to as a “passage”, essentially a crude measure of the number of cell divisions a particular cell culture has been subjected to. When observing MSCs during extended culture expansion processes, one readily observes that MSCs, expanding into higher passage numbers, perhaps 6–10, frequently exhibit changes including altered morphologies and reduced proliferation rates [27].

It seems likely that, in addition to any biological limits to the number of divisions MSCs in culture may undergo, these phenomena are influenced also by process specific parameters and events, some of them inherent in planar/static cell culture.

Cell Density

It is well established that cell density has a substantial impact on cell proliferation and survival, and it has been suggested that paracrine effects are responsible [28]. Similarly, cell-cell contact is also known as a key regulator of key cellular process, and it has been suggested that cell-cell contact is part of the regulatory network controlling cells within a tissue [29].

In planar/static culture, cells are subjected to repeated oscillations in cell density – they are seeded at low density and harvested at high density at each passage and so are repeatedly exposed to conditions similar to those of a coordinated tissue where proliferation is low. It seems likely that this repeated exposure results in reduced or even absent proliferation.

Cell Culture Medium pH: The Gas Phase

When starting a culture, particularly in multi-stack systems, it takes time for the passive gas exchange to establish the incubators’ conditions within the stack. This is important since the gas mixture, specifically its CO_2 concentration, dictates the pH value of the cell culture medium. Cells typically require a near-neutral pH value of their culture medium. Using a bicarbonate buffer system in 5% CO_2 this is accomplished; air contains only approximately 0.04% of CO_2 and so a bicarbonate buffer-system based culture medium will drift towards to alkaline in air. Many cells do not tolerate this well and so, to rapidly establish the desired conditions within such cell culture systems, these can be initially flushed with an appropriate gas mix, e.g. air containing 5% CO_2, thereby establishing the desired atmosphere and pH value of the media rapidly.

Metabolism

The mammalian body tightly controls levels of oxygen, glucose, lipids and other fuels and nutrients as well as that of metabolic by- or, indeed, waste products such as ammonia. The dramatic impact of disturbed control in diseases such as diabetes is well known. In planar/static culture, however, no such control occurs, media are often batch-fed and so cells experience a recurring exposure to drastic changes in fuel and waste concentration; ammonia may be of particular importance as it is known to inhibit oxidative phosphorylation forcing cells to generate energy via the much less efficient pathway of glycolysis.

Thus, cells generated via planar/static culture have been subjected to a recurring exposure to a multitude of non-physiological and destructive influences.

Taking this into account and, further, considering the lack of stringent means of gauging the therapeutic quality of cells produced discussed above, it is likely that secondary parameters such as quantity became the primary measure of such processes. It is tempting to speculate that this situation led to product which was meeting its cell number specifications but may have suffered from a lack of control regarding its therapeutic efficacy.

Second Generation MSC Products and Manufacturing Processes: Bioreactors

We referred to bioproduction as a source of tools, processes and equipment upon which cell therapy manufacturing processes can built. A typical bioproduction process starts with a small number of cells, perhaps a few millions, which are thawed, culture-expanded first in small devices such as shake flasks, then transferred to small bioreactors of perhaps up to 50 l, from thereon to a bioreactor of final size – at current, of up to 2 m^3. The principal configuration of single-use bioreactors available are either stirred tank types or rocker-based ones. Reports for successful manufacture of MSCs have been published for both types – in our hands, stirred type bioreactors worked well and did not require any modification [30].

A benefit of bioreactors is that they offer a much higher degree of control than static/planar systems. Not only is the gas mixture well controlled, but also that of nutrient levels as bioreactor control devices can add needed ingredients to the culture.

Another benefit of bioreactors is that they are closed systems. While there may be open steps at the beginning of the process, this minimizes the number of open aseptic manipulations drastically compared to static/planar culture, hence reduces the probability of errors and of microbial contamination.

Adherent Cells in Bioreactors

Bioproduction systems are developed with suspension cells in mind. As we discussed above, MSCs require adhesion surfaces to survive and function and this gap is closed by using microcarriers; the latter are also part of the bioproduction toolbox, albeit from a time when production cell lines were still adherend. The need for microcarriers for cellular therapies triggered a wave in novel microcarrier development so that there are now vast varieties to choose from. In choosing a microcarrier, these considerations appear useful:

- MSCs are comparatively susceptible to shear stress and so this needs to be minimized. A key cause of shear stress is agitation provided by e.g. impellers. Agitation is required to create a continually mixing environment so that cells have access to fresh nutrients and waste products are dispersed and microcarriers are suspended. To accomplish this with minimal agitation, microcarriers of a density similar to that of the medium may be chosen.
- MSCs require a suitable adhesion surface, hence, microcarriers which support MSCs adhesion may be chosen.
- MSCs need to be harvested at the end of the bioreactor run. Therefore, microcarriers should be chosen which support release of MSCs under conditions compatibly with high MSC viability. We have had excellent results with protein-based microcarriers which were digestible using appropriate enzymes so that cells adhering to them were released.
- MSCs should be expanded in a medium devoid of animal components. Since such components provided adhesion proteins, lipids, and growth factors, microcarriers likely can provide a degree of substitution, certainly for adhesion proteins, but perhaps also as slow-release sources of lipids and growth factors.
- With these topics in mind, microcarriers are part of a pharmaceutical manufacturing process and hence need to be of GMP quality.

Cell Age/Capabilities

Using a bioreactor system with microcarriers to manufacture MSCs voids the concept of a passage. At the same time, the need to describe the number of cell divisions a cell population has undergone remains and so an improved terminology is used: The Population Doubling Level (PDL). This term describes the average number of cell divisions a population of cells has undergone during an expansion step. Not only is it much more accurate than the term passage, but it also is not restricted to planar/static culture systems, so it is now widely used, also in planar culture. It can be calculated as follows:

Equation 15.1: calculating the population doubling level of a culture of expanded cells. N = number of cells at the end of a expansion; N0 = number of cells at the beginning of an expansion. We put forward this calculus to emphasize the importance of true cell age in commercial production; the principle may be challenged anew by IPSCs and the associated difficulty in using single-cell suspensions in culture. IPSCs may accumulate genetic changes during extended periods of cell expansion.

$$PDL = log\,10\left(\frac{N}{N0}\right) \times 3.33 \qquad (15.1)$$

Using PDL, a universally comparable description of cell age is at hand, and it is fully compatible with 3D culture expansion; of note, it also permits a comparison of a cell preparation's number of divisions between planar/static and 3D culture.

Mode of Action and Pertinent Controls

As discussed above, MSCs and other plastic cells were initially considered to perform therapeutic feats through their "stemness". In our view, this concept has not proven itself and so should be disregarded. At the same time, it is well established that MSCs secrete copious amounts of bioactive factors, several of which are known to be key agents in regenerative effects. Indeed, not only secrete MSCs numerous bioactive factors, they also shed vesicles containing bioactive factors, nucleic acids and other materials – exosomes [31]. Therefore, it seems likely that MSCs' mode of action is through the secretion of such factors and vesicles.

The release of such bioactive factors, either soluble or in vesicles, can be quantified and hence presents excellent opportunities for the quantitative characterization of cell manufactured. For example, vascular Endothelial Growth Factor (VEGF) is a well-known key trigger for the de-novo formation of blood vessels [32]. MSCs are known to produce and secrete VEGF in copious quantities and so it has been suggested that MSCs' mode of action for perfusion-related indications is to release VEGF and thereby to trigger the formation of new blood vessels so that perfusion is regenerated. Since required functional parameters of cells produced have to be defined *a priori* -specified-, a minimum rate of VEGF secretion by cells manufactured is likely to be a meaningful and quantitative specification for cells intended for such an indication.

In another example, MSCs may be used to treat the indication Graft vs Host disease (GvHD), a severe autoimmune disease. In this indication, the likely mechanism of action is to modulate activities of the immune system mediated by lymphocytes. The latter require the amino acid tryptophan for their activity and proliferation and MSCs are known to express an enzyme catabolizing tryptophan, namely indolamine 2,3-dioxydase (IDO) [33]. This enzyme has been shown to be required for nixing rejection reactions and thus presents a convincing candidate for characterizing MSCs intended for such an indication. Therefore, a minimum IDO activity is likely to be a meaningful and quantitative specification for cells intended for such an indication.

We believe that these are two quantitative and rational examples of MSC-product specifications and opine that following this logic in identifying suitable specifications for other indications is likely to be helpful in developing successful pertinent products. The quest for such rational and quantifiable tests and specifications is particularly pertinent for advanced therapies: their complexity vastly exceeds that of e.g. antibodies, not to mention small molecules [34, 35].

In addition to functional markers as discussed above, during culture expansion, in-process-controls (IPCs) should be taken to facilitate monitoring a culture's progress and viability – particularly during process development. In-process-control monitoring is also used to show process consistency from batch-to-batch and will be needed for marketing applications. Such IPCs may include:

- Cell counts – monitoring cell titers and hence the expansion of cells. During development, this provides information as to the culture system's capabilities. Similar to a microbial system, cells should be harvested before they enter a stationary phase, or, indeed, drop in their viability. Note that cell counts are a late indication of the cellular state. Therefore we recommend to also determine earlier markers:
- Markers indicative of cell lysis. Cell counts are a late indication of untoward events in a cell culture. Well before cell counts drop off, individual cells exhibit signs of degradation, typically release of intracellular material. Often, the cytosolic enzyme lactate dehydrogenase's activity in the medium has been used to gauge cell degradation. Even simpler tests determine the presence of double stranded DNA in the culture, in our experience a very useful tool.
- Metabolic markers. Often, concentrations of glucose, lactate and ammonia are measured. Glucose is a major fuel, lactate the end product of glycolysis and changes in the ratio of glucose consumption and lactate production may be indicative of insufficient oxidative phosphorylation, the main energy source of most mammalian cells. Ammonia is known to inhibit oxidative phosphorylation and so its concentration needs to be tightly controlled; it is generated either by spontaneous decomposition of e.g. glutamine or by cellular metabolism.

Expansion-Free MSCs

Every expansion process involves cost, time and reagents subject to regulatory review. It is therefore interesting to consider approaches which make do without most of these aspects:

Ossium Health, a company developing MSC-based therapies has been able to source much larger collections of MSC rich tissue from deceased donors by using organ transplant guidance and infrastructure. This starting material apparently suffers much less from T-cell contamination than conventional ileal crest marrow aspirates. Thus, this MSC manufacturing process may start with many more cells, perhaps as much as 10 times more, in their pre-differentiated and natural state, thus greatly reducing the demand for *in vitro* expansion, e.g. from 3 passages to one, and may provide a population of MSCs much less affected by expansion effects and possibly with enhanced therapeutic properties.

Third Generation MSC Products and Manufacturing Processes: Exosomes and Engineered MSCs

There is a body of evidence indicating that MSCs' therapeutic efficacy is linked to their secretome, i.e. growth factors and cytokines they synthesize and directly release into their local environment. MSCs, like many other cell types, can also shed or release larger, enveloped bodies known as extracellular vesicles of various sizes and names (EVs, exosomes, or microbodies) that can also carry the growth factors and cytokines to other cells [36]. This observation explains why several companies are exploring the collection of exosomes from MSCs and delivery of these "packaged" cytokines in a manner that allows alternative uptake of these messenger molecules by the target cells. In this scenario MSCs are the behind-the-scenes biofactories producing a Drug Substance (exosomes) but are not themselves delivered to patients.

This approach may also have limitations if the character of the EV cargo changes as MSCs are cultured, and specific desired growth factors or other messenger molecules are diminished as cell population doublings increase. Currently, Clinical trial.gov website lists 296 registered studies utilizing exosomes, of which only 24 appear to study exosomes derived from MSCs.

Throughout this chapter, we advocate use of rational, plausible and quantitative measures to gauge the functionality of cells in relation to their therapeutic target(s). We do believe that following the above concept of rational and quantitative specifications is likely to result in increased success rates and should be pursued. However, there is an even more direct route to ensure that MSCs produce desirable factors – genetic manipulation.

Genetically Engineered MSCs

As discussed above, the safety profile of MSCs is excellent and so it is tempting to consider therapeutic indications beyond those "built-in". Various technologies exist for introducing genetic material into mammalian cells including electroporation, retro- or lentiviral transduction or, more recently, use of adeno-associated viruses and these technologies have been employed to enhance MSCs' therapeutic profile (Table 15.1).

Zhou et al. reviewed the clinical state of MSC engineering and discussed 5 examples of transduced MSCs, e.g. to secrete interferon gamma (IFN-γ) so to target ovarian cancer [37]. They further listed 7 examples of Crispr/Cas9 modified MSCs, e.g. interleukin 10 (IL-10) secretion targeting myocardial infarction; Crispr modified MSCs were still at the R&D level, presumably because of the recent advent of the technology.

Damasceno and co-workers reviewed MSC transduction approaches and listed 187 reports of genetically modified MSCs [38]; a Medline search for combinations of Crispr/Cas and mesenchymal stem cells yielded 65 entries [39]. Irrespective of the precise number for each approach, it exemplifies that genetic modification of MSCs

Table 15.1 Genetically engineered MSC products in clinical testing (2021), modified from Zhou et al. [37]

Delivery system	Indication	Sponsor
MSCs secreting IFN-β	Ovarian cancer	M.D. Anderson Cancer Center, Dallas TX
MV-NIS infected adipose tissue– DERIVed MSCs	Recurrent ovarian cancer	Mayo Clinic, Roches- ter MN
Bone marrow- derived autolo- gous MSCs infected with ICOVIR5, an onco- lytic ade- novirus (CELYVIR)	Metastatic and refractory solid tumours	Hospital Infantil Universitario Niño Jesús, Madrid, Spain
MSCs genetically modified to express TRAIL	Lung adenocarci- noma	University College, London
Autologous human MSCs geneti- cally mod- ified to express HSV-TK	Advanced gastroin- testinal cancer	Apceth GmbH & Co. KG, Germany

Genetic modifications of MSCs cover a wide variety of targets and may prove successful in the future

is a field of intense interest and one might speculate that a proportion of these reports will proceed to clinical trials.

To ensure that such products can come to clinical studies and to market, it is important to understand the additional requirements these approaches pose to manufacturing. In principle, one might transduce or otherwise alter the MSC batch early in the expansion process so that only a comparatively small number of cells needs to be engineered. Successfully engineered cells would then be expanded as usual, formulated, and stored. While this would be attractive from the genetic engineering point of view, not all approaches are compatible with cells proliferating following genetic engineering and thus the desired effect may be lost; in such cases one would be limited to genetically engineer cells either prior to or after the final harvest. For example, for the 50 l bioreactor example discussed above, the yield was approximately 37×10^9 cells, a quantity of cells which even at high cell titer might be several liters in volume. Thus, to transduce such a quantity of cells requires corresponding quantities of material – e.g. plasmids in appropriate quality as well as large scale transfection mechanisms. Often chemical transfection routes are chosen and more recently, electroporation devices working at scale have been developed. Also, most countries regulate the use of genetically altered organisms and both a bacterial culture used to create recombinant plasmids as well as the resulting engineered cell will be considered GMOs, hence such work will require corresponding approvals as well as suitable facilities, training, and decontamination procedures for spills and related emergencies; Germany, for example, considers on-site inactivation of GMO contaminated materials by physical means, i.e., autoclaving the rule. Therefore, such needs should be clarified up front – retrofitting suitable equipment may be costly and time consuming.

Chemically Engineered MSCs

As stated above, it is tempting to enhance MSCs beyond their natural state and another means of doing so might be referred to as chemical engineering – treating MSCs with chemical compounds to elevate their therapeutic efficacy. For example, MSCs have been treated with a small molecule compound exhibited dramatically improved therapeutic efficacy so that recovery approached near-normal levels [40]. MSCs were culture-expanded and, prior to harvest, co-cultured in the presence of the compound. Cells were then harvested and washed so that the compound's concentration was reduced prior to application of MSCs. It was suggested that the treatment enhanced MSC survival and minimized necrotic events (Cvdb, pers. com.). At the same time, the additional process of treatment was simple and, importantly, this procedure has been subjected to regulatory review by authorities and a straightforward regulatory path emerged so that clinical use of such technology appears to be a realistic option.

Thus, it appears that engineering MSCs, genetically or chemically, is already a reality and that a great many new approaches are being tested and a proportion of these will likely move into the clinic. Some of the additional manufacturing steps involved seem straight forward while others may involve substantial investments and advanced planning.

Cryopreservation

For MSCs to be cryopreserved, cells can be suspended in a suitable cryogenic medium comprising often complex carbohydrates as well as a cryoprotectant, often DMSO at e.g. 5%. The latter protects cells against freezing damage by controlling ice crystal formation; it is, however, also cytotoxic and so needs to be handled appropriately: parameters to be controlled include (a) cells being present as a single cell suspension as much as possible so that DMSO has access to all cells, (b) the maximum titer of cells so that sufficient DMSO is available, (c) limits to time in DMSO prior to freezing including time needed for filling cryovials and (d) cold storage while in DMSO and prior to freezing. Freezing should be a controlled process and, while research laboratories sometimes use procedures as simple as storing cell vials in an insulated container in a −80 °C deep freezer or a simple device containing a volume of cold isopropanol to slow down cooling, therapeutic settings most often use controlled-rate freezing devices; cells contained in closed vials are placed e.g. in a rack into the freezing chamber and the temperature is controlled by controlling the influx of liquid nitrogen. These devices permit tight control of the freezing process, for controlling the steepness of freezing curves as well as for compensations, e.g. to buffer heat released from samples when they cross the glass temperature of about −130 °C [41].

Cryopreservation for MSCs manufactured through bioreactor processes proceeds in the same fashion as described for planar/static culture, i.e. via cell harvest, cell washing and formulation in cryogenic medium. We discussed the toxicity of the common cryoprotectant DMSO resulting in a limited duration of exposure prior to freezing. Manufacturing MSCs using a bioreactor typically results in larger batches and so cryopreservation requires diligent planning and logistics as cells, once frozen, need to enter cryogenic storage w/o any intermediate elevation in temperature or, indeed, thawing.

For example, a 50-l bioreactor run expanding MSCs on microcarriers with a total of 35 l of medium yielded approximately 37×10^9 cells, that is, approx. 185 doses at 200×10^6 cells/dose [30]. Assuming that 30% of the product is consumed by various IPCs and QC tests, this leaves 129 doses to be frozen under controlled conditions and to be transferred into cryogenic storage, all while avoiding any substantial warming during the transfer. The challenge of doing so can be substantial and warrant appropriate organization and equipment.

Blood-Cell Derived Therapies

Previously, cell therapy referred to applying unmodified cells to particular indications and gene therapy to correcting faulty genetic dispositions of patients. In recent years, the distinction between these two fields has blurred as therapeutic cells are successfully engineered genetically ex-vivo and later returned to the donor/recipient, an approach also discussed earlier in this chapter for MSCs.

Advances in the Field

Cell processing has progressed rapidly and impressively in the 20 years or so since the inception of Chimeric antigen receptor T cell (CAR-T) concepts with 5 FDA approved products at current (January 2023) and a similar number approved by EMA [42]. Progress has been made in hematological/oncological indications using CAR, although the manufacturing processes continue to be cumbersome, expensive, and inconsistent. Solid tumors remain elusive targets, resistant to CAR-T alone. Advances in understanding of the immunosuppressive tumor microenvironment are informing more advanced therapeutics, with armoring strategies, for example, as well as Neo-antigen TCR discovery for true driver mutation targeting. Below we discuss current, advanced and expected future developments and refer to these as stage 1,2 and 3.

STAGE 1: Kymriah to Tecarta

T-cell culture initially was performed using open manipulations, with labor-intensive unit operations requiring highly skilled operators. T-cells had traditionally been maintained in 2D plastic vessels, e.g. T-flasks, primarily designed for adherent cultures in a humidified CO_2 incubator and manipulated in a biological safety cabinet (BSC). GE and Sartorius Wave Bioreactors (now the Xuri Cell Expansion System (Cytiva)) allowed large-scale closed expansion of T-cells with sampling and feeding ports. Culture media (generally RPMI or equivalent) with up to 10% FBS were used, along with high doses of IL-2. Cells were derived from manual Ficoll fractionation of peripheral blood mononuclear cells (PBMCs). The onset of serum-free defined media such as X-VIVO (Lonza), and TexMACS (Miltenyi), enabled expansion without serum, although Human AB serum is still often employed, without much obvious benefit, reviewed by [43]. Using the Elutra (Terumo) device, i.e. counter flow elutriation, a more defined population compared to PBMC could be used for starting cultures. The advent of the Miltenyi magnetic micro beads and the CliniMACS isolation/culture system allowed GMP-compliant isolation of highly defined cell types (CD34+, CD19+, CD3/4/8+). Expansion of cells was greatly advanced by the introduction of Dynabeads from Life Tech/Invitrogen (TF) stimulating CD3/CD28 receptors on target cells. This approach outperforms the then-standard – soluble CD3 (OKT3) [44].

Gene modification to produce CAR-T was pioneered by Carl June and others at UPenn, e.g. the gene modification was performed *ex vivo* with low-titer viral vectors, either lentiviral or retrovirus. Harvesting and formulating the cell-based drug product was generally accomplished through centrifugation and pelleting followed by manual filling. Occasionally, the COBE 2991 (Terumo) was used to concentrate and formulate the harvest. Release testing was slow and cumbersome, or sometimes non-existent. The definition of potency and other quality attributes developed in parallel with the transition from murine to clinical products. Sterile welding and sealing from Terumo's TSCD welder greatly advanced the speed and ease of manipulations, allowing several operations outside of the BSC. Controlled rate freezing using the CryoMed (TF) is an important final element. These technologies were implemented on the first stage of commercial cell therapies, allowing Novartis and Kite Pharma to obtain initial BLA approvals in 2017 and move the therapies from academic to commercial settings.

STAGE 2: Autolus to Allogene

While the technological landscape of T-cell therapy was established in the first stage, bottlenecks in scaling out, and critical quality attribute specifications demanded more advances in the technology. Several advances began emerging in the 2000s, and concepts of design space, Quality by Design (QbD), and Process Analytical Technology (PAT) began to find adaptations to the less predictable T-cells.

Gas-permeable bags from Saint Gobain and Baxter, and the GREX from Wilson Wolf, which introduced semi-closed gas permeable membrane from the bottom of the vessel, allowed static culture with high magnitude of T-cell number expansion. The Miltenyi Prodigy is an expansion of the CliniMACS PLUS that allows the same isolation procedure but combines culture chamber and washing and harvesting, programmed transduction and cytokine feeding, with customized user programs in a GMP-compliant fashion. Such an end-to-end closed processing scenario enables apheresis to cryopreserved drug product manufacturing in clinical, i.e. non-classified, settings. In most development laboratory and contract manufacturer settings there remain requirements for aseptic manipulations, such as the preparation of the growth factors/media feeds, and vector prep that necessitate the continued reliance on BSCs. While the Prodigy is potentially a complete system, many products also continue to use the Sepax or newer Sefia processing systems (Cytiva) to harvest cells, reduce liquid volumes, and formulate cells, both upfront on the apheresis starting material, as well as the final Drug Product. Working with a sole-source technology does streamline and consolidate the technical operations but does have limitations. Committing to a single unit platform makes process improvements with alternative technologies more challenging. While a company such as Miltenyi is often updating the platform, adding things like a bag filler or electroporator, these add-ons may not be as efficient as other similar competing technologies. Also, increasing the operations of the unit occupies the device longer and bottlenecks the manufacturing suite changeover to the next patient while the unit continues to be dedicated to a single patient. Other closed systems continue to improve success rates and their impact, such as the Cocoon (Lonza) and the Xuri [45, 46]. The Cytiva systems include the Chronicle and Unicorn software systems, hence allowing a full Manufacturing Execution System (MES) package during manufacturing.

Sepax and Sefia are technological and operational advances from Cytiva, based on an internal cylindrical processing device with pores of fixed size that allow the separation of RBC and platelets from therapeutic cell types, and these units can reduce volumes in a closed, programmed fashion [47, 48]. The choice of Sepax or Sefia is largely dependent on the process volume and cell number being processed, the Sefia being more suitable for larger cell numbers. Using these devices simplifies cryopreserving the apheresis starting material (APH); although not required, it is a more controlled addition of the cryoprotectant/DMSO formulation. Cryopreserving the APH comes with several implications: freeze-thawing generally reduces the number of granulocytes while reducing T-cells viability only minimally. The most important aspect of cryopreserved APH is to allow a predictable, planned manufacturing start day and time, which is key to process logistics and operational success. Other volume reduction technologies increasingly used include the LOVO (Fresenius) and the GatherX (Wilson-Wolf).

LOVO is a harvest and wash device that uses a rotating porous cylinder to retain cells and push out media. It is part of a closed system, with volume reduction capabilities down to about 150 mL.

Several devices are now available which can fill cryobags in an automated, closed, and precise manner. Finia (Terumo) is likely the most well-known and can fill standard or custom volumes in a total of 4 bags in a temperature-controlled manner. Signata™ CT-5 (Sexton Biotechnologies) is a similar device and includes bag manifolds which hold smaller pouches for QC testing. Both devices can de-bubble the bags for freezing. The Rotea and Sepax can potentially perform this, but in a custom application. Miltenyi Prodigy is releasing a module that can also fill the bags after washing, making it a complete end-end device.

Other isolation and washing technologies are new to the landscape and include the Rotea (TF), a scaled-down and simplified counterflow centrifugal system, that uses a fluidized bed to harvest T-cells, which is much gentler towards cells. Gibco CTS DynaCellect Magnetic Separation System, which includes an alternative to the Miltenyi microbeads, based on the Dynabeads CD3/28 beads, and including a de-beader. Aptamers are an affinity based nucleic acid-based system that has yet to be fully exploited.

Media. Serum-free media has become a staple of the cell therapy industry, and now extends to many high-quality product offerings, such as CTS, X-VIVO, TexMACS. These media are animal-component free and highly defined. Cryostor™ (BioLife Solutions) is an oft used buffered formulation containing DMSO that has been used successfully for cryopreservation of a multitude of cell types.

Gene Modification

Lentiviral and Retroviral vectors continue to be the most heavily utilized method of introducing genes into T-cells, especially CAR or TCR transgenes. AAV is also heavily used, but is more for straight-up gene therapy, such as Spark, UniCure or Regenxbio. Lentiviral vectors can encode a CAR or TCR payload of up to 9 kb, but for good yields, practically 5-6 kb. The lentiviral and retro vectors hold can also accommodate a bi-cistronic delivery, for either tags like the Autolus RQR8, which can be used to sort, stain, or also function as a safety switch; or a "backpack" or armoring strategy to overcome the immunosuppressive tumor microenvironment by, for example, expressing either cytokine receptors, e.g. TGFβ receptor dominant negative or releasing cytokines; such strategies are referred to as armored CARs. The production and testing of these drug substances has advanced considerably since Stage 1, with companies such as Oxford Biomedica and Yposkesi leading the forefront of advances. Suspension culture of 200 L Single Use Bioreactors (SUBs) is common, with high titer-harvests achieved using depth filters, TFF, and Ion Exchange chromatography and followed by sterile filtration. Dedicated producer cell lines can have a substantial and positive impact and contribute to consistency of manufacturing. For example, a single (HEK 293) line can be engineered to contain both the p93 gag/helper gene and the VSVg pseudotyping gene such that only the transgene plasmid needs to be transfected. VSVg can be somewhat toxic, so it is often used under an inducible promoter, leaving expression off until right before the transfection.

Replication Competent lentivirus (RCL) testing remains a major bottleneck and slows down rapid iterations of innovations with product lines. Turnaround time for the RCL test for vectors is around 8 weeks, due to the cumbersome passaging of cell lines. Actual volumes of supernatant/end of production cells required for these tests are often a matter of considerable regulatory debates. Continued dialog with regulatory bodies such as the FDA is necessary to maintain therapeutic advances, as there has never been a single documented case of recombination in current vector production scenarios; indeed, FDA may waive this requirement after demonstrating 10 or more acceptable batches. New fourth generation lentiviral systems will add pressure to reduce and simplify RCL testing. In addition, transformation events after lentiviral transduction have not been observed, calling into question the perhaps overly cautious limitations of vector copy number (recommended as <5 per Transgene+ cells) and Multiplicity of infection (MOI) used during the manufacturing step [49]. For these reasons, and to continue addressing future unmet needs as well as to be able to make cell therapies broadly available, new methods of gene modification that do not require RCl tests are needed [50].

Electroporation

Non-viral methods offer attractive alternatives to lentiviral transduction. CRISPR, Transposons, and mRNA are all viable payload options for genetic engineering and have been introduced into cells using electroporation. Electroporator platforms vary but seem to perform similarly. The MaxCyte Expert™ system, Lonza 4D™, Thermo-Fisher Xenon™, and newer methods such as Kytopen are all used extensively. Simple plasmid DNA can be used, but is generally toxic to T-cells, and losses in viability prohibit extensive use [51]. Transposons, however, can be highly efficient, and insert permanently, similar to a lentiviral vector. In addition, the payload limit of the transposon system, such as Sleeping Beauty™ (Lonza), Leap-IN™ (Atum), or TcBuster™ (BioTechne) is dramatically higher, with no known upper limit. The larger plasmid size still confers a reduction in viability, but cells can be cultured into recovery. Smaller mini-genes and mini-circle DNAs can attenuate the toxicity to some degree [52]; minicircle DNA is a circular DNA format sold by Plasmid Factory (Biefeld, Germany), licensed from Nature Technology Corporation (Aldevron). Another system, the "doggybone" (Touchlight) features a double-stranded linear DNA format, is generated in a cell-free system and amenable of up to 20 kb payload. Prominent companies pursuing transposons in the clinic include Catamaran, Precigen, and Poseida, among others.

CRISPR

CRISPR-based methods are highly effective tools to knock out genes through Homology Dependent Recombination (HDR), an effective technology for eliminating single-allele heterozygous disease manifestations. It's use in inserting genes has become more common, with ssDNA accompanying (via an AAV vector) the gRNA and Cas9 to insert a new gene at the site of breakage. Efficiencies are generally lower than with lentiviral vectors, but cells can also be selected at the terminus of the manufacturing process. This can be useful in inserting a CAR or T-Cell Receptor (TCR) into the natural TCR alpha locus, regulating expression physiologically, and eliminating the endogenous TCR (eTCR). Removing the e-TCR is necessary in an allogeneic setting, but of lesser concern for autologous or haplo-matched situations. While it is thought that knock out of eTCR allows for higher levels of Transgenic TCR expression by eliminating competition to the CD3 complex, it is not clear that this is altogether necessary. In addition, the cost of licensing and producing the CRISPR reagents is quite high and must be considered in the context of the expected or potential reimbursement [53].

mRNA

The use of mRNA for generating CAR is a relatively new concept that has taken on added interest with the powerful success of mRNA COVID19 vaccines. Using mRNA for gene modification has obvious advantages of speed and cost of production. Release testing is much more limited due to lessened safety concerns, not the least of which is the bypassing the need for RCL testing as demonstrated in an early example from the University of Pennsylvania [54]. This product used a one-day manufacturing process of PBMC, which were harvested from apheresis, electroporated, and then frozen. However, release testing still took another 2 weeks to complete. These therapies are by nature transient. Due to the limited intracellular stability of the exogenous mRNA, CAR expression *in vivo* did not surpass 2 weeks, which is in stark contrast to the lentiviral CAR therapies which can be permanent. However, if multiple doses are infused over time, it is possible that an immune cascade can re-ignite the dormant and anergic immune system to become self-sustaining. It would seem that cost and speed benefits of this approach likely counterbalance current uncertainties.

STAGE 3: Future Directions of Manufacturing Technologies

Unmet needs and ongoing operational and cost constraints continue to push further development of technology. While response rates of CAR-T products are impressive, relapse rates are also fairly high [55] and can be caused by loss of antigen or by cell burnout and exhaustion after differentiated effector T-cells experience chronic antigen exposure. One common approach to mitigating the differentiation and exhaustion is to use lower doses of IL-2 (10–20 IU/mL), and to include IL-7, -15, -or -21, all common gamma chain family cytokines that confer maintenance of stem cell memory phenotype. The addition of GMP growth factors as critical raw materials does increase the cost of goods and supply chain risk, however this may be offset if the therapy has a profound impact on clinical durability.

For the existing therapeutics, the current direction of the field is toward shorter manufacturing processes. The Novartis T-charge, Gracell Fast-CAR, and BMS Nex-T platforms all utilize very short manufacturing times, with limited cell expansion [56–58]. Quite simply, these T-cells are harvested, gene-modified, and cryopreserved. While this trend is somewhat contrary to the original Stage 2 manufacturing paradigm, the potency of these drug products appears to be dramatically higher (10–50X) than traditionally cultured T-cells. This is likely due to the cells staying in the Central Memory or Stem Cell Memory state and so are more likely to persist long-term in vivo while not terminally differentiating yet sustaining a durable response. The operational advantages of rapid manufacturing are also considerable. Shorter processes mean less clean room time and shift work per product manufactured. So, in a given manufacturing area, more patient products can be produced during a given time. Rapid processes greatly reduce COGS due to reduced need for media, cytokines, and labor per product lot. Finally, using closed devices allows for moving production into a class C clean-room environment so that reduced environmental monitoring further lowers COGS. So far, clinical responses have been encouraging for these rapid platforms (https://clinicaltrials.gov/ct2/show/NCT04231747).

Manufacturing products faster is required but not sufficient to get products to patients faster and rapid release is also required. To this end, the Bac-T system (BioMerieux) is an advance over compendial USP72, but still takes 9–10 days.

The Millipore MilliFlex system is membrane-based, and generally best suited for mAbs, but a gentle lysis solution can potentially adapt this 3-day method to cell therapy. Ultimately however, a 1-day PCR method directed at bacterial ribosomal DNA sequences will enable release at a speed corresponding to that of rapid manufacturing. Other rapid testing technologies include Accelix, which can give real-time flow cytometry results using pre-loaded cartridges (standard or custom) and can be used in the clean room. Besides rapid release, rapid QA review and disposition are needed. Use of electronic batch records (EBR) is likely key to reduce technical review of batch records, especially regarding GDP errors. Deviations, RCA, and CAPA still need an agile QA department to manage risk and allow patients to receive the product infusions.

Chain of Custody has historically been managed with paper-based documentation. Systems such as Trakcel, can streamline the COC and prepare for high throughput commercialization. Paper-based systems can still suffice for small Phase 1 clinical trials where the number of products is low. Progression into the commercial phase may be associated with much higher numbers of products, in which case a mix-up or mistake risk is elevated. COC software can help minimize these risks. Logistics and shipping are also critical components of the supply chain. EVO™ and CREDO™ controlled shipping units seem well suited for blood products. Cryoport has an arsenal of LN2 dry shippers for the final drug product. These now have real-time temperature and location tracking, a great advantage over previous situations where e.g. cryogenic vaults were discovered only upon product arrival.

Table 15.2 Summary of current cutting-edge technologies in cell/gene therapy

Device	Company	Use
Cell shuttle	Cellares	Automated, closed cell expansion and testing
ErbiBio	Millipore	Miniaturized perfusion reactor
Prodigy	Miltenyi	Closed, automated cell isolation and culture
Cocoon	Lonza	Closed, automated cell isolation and culture
Mobile clean room	GermFree	Mobile clean rooms
Alta	Invetech	Closed, automated cell isolation and culture
Solupure	Avectas	Closed, automated cell isolation and culture
Cell box	Orbi	Closed, automated cell isolation and culture
AdvaX	AdvaBio	Closed, automated cell isolation and culture
Quantum	Terumo	Closed, automated cell expansion reactor
Xuri	Cytiva	Closed, automated cell expansion reactor
CellCellect CT5	Gibco	
Cellisium	MicrofluidX	Closed, automated cell expansion reactor
Lipid nanoparticles	Lupagen	Bedside MFG
FloDesign Sonics	Millipore	Cell isolation using sonic waves
Accelix	Accelix	Pre-programmed cartridge rapid flow

The field is rapidly advancing and can be represented only as a snapshot

Automated closed processing technology continues to increase as the field expands and more and more innovative minds and funding push the boundaries and expand options. AdvaBio includes a built-in online analyzer for metabolites and nutrients. Miniaturization and novel technologies include FloDesign Sonics (Millipore), microfluidic systems (MicrofluidX), ErbiBio (Millipore). Cell encapsulation is an upcoming technology with companies such as Celvivo; Denmark, and TreeFrog Therapeutics, France, using these approaches. Cell expansion of primary T-cells using a non-beveled stir tank up to 10 L has been showing encouraging results using BioBlu (Eppendorf) or Ambr250 (Sartorius), which may align more closely with established PAT concepts [59].

A summary of cutting-edge technologies is listed in Table 15.2.

Induced Pluripotent Stem Cells

They are, similar to embryonic stem cells, intended as a pluripotent precursor to therapeutic cells. Thus, IPSCs are not a therapeutic themselves but rather an intermediary in the production of cellular therapeutics.

Once produced, the question raised above (Genetically engineered MSCs) is relevant here, too: in principal, IPSCs can be induced to assume their desired phenotype prior to culture expansion or afterwards. Constraints to this decision are quite similar, too, and often IPSCs have to be expanded prior to differentiation.

Consequently, IPSC manufacturing processes contain elements of those discussed for MSCs in regard to initial small-scale culture and later scaled-up processes. Distinguishing elements include, first, derivation of IPSCs from donor cells by means of by introducing selected transcription factors acc. to Yamanaka [60]. During this step, IPSC colonies tend to form over a period of days and, if using suspension donor cells, can be distinguished readily by forming adherend colonies. These can be expanded e.g. in small-scale T-flasks, characterized and cryopreserved as cell banks following the principles of cryopreservation discussed above. Samples of this initial cell bank can be retrieved when needed to expand them to larger numbers. Akin to the processes described above, these adherend cells may be expanded in planar/static culture using the tools discussed; one exception is caused by the biology of IPSCs: whereas other cells are released from their adhesion surface and rendered single-cell suspensions prior to re-seeding, IPSCs to not tolerate lack of cell-cell contact well and so tend to be split as aggregates rather than single cell suspensions. Newer protocols emulate the path previously taken by bioproduction and MSCs and culture expand IPSCs in bioreactors, e.g. as suspended aggregates or microencapsulated [61, 62].

Key to culture expanding IPSCs is using culture media which conform with pharmaceutical requirements, in particular, with the need to use fully characterized media, i.e. media devoid of complex animal derived ingredients. Here, too, great progress has been made with initial commercial offerings such as mTeSR media from Stemcell Technologies based on a formulation containing bovine albumin as well as human proteins [63] followed by a simplified E8 medium containing human proteins [64] and B8 [65] containing recombinant human proteins, hence, probably to be referred to as a chemically defined medium.

Following expansion, IPSCs might be manipulated/differentiated such that they assume the phenotype of the desired therapeutic cell type. Once IPSCs are differentiated into desired phenotypes, these cell preparations need to be devoid of undifferentiated IPSCs as these may cause uncontrolled growth. Therefore, the absence of stemness markers in the final product is an important specification for cellular medicinal products derived from IPSCs.

Gene therapy Product Manufacture

Introduction

Gene therapy aims to correct pathologies via genetic manipulation, e.g. by providing correct versions of faulty genes or by inducing the patient's body to produce desirable responses to pathological situations. This can occur either by treating a patient's cells extracorporeally as described for various therapies such as Yescarta (Kite Pharma), see also in section STAGE 1: Kymriah to Tecarta, or by introducing vectors into the human body to affect either many or select cells and tissues such as Zolgensma(Novartis) [66, 67].

Short- and Long-Term Therapies

Most recently, mRNA technologies have gained much deserved prominence, particularly due to their success in rapidly providing powerful covid 19 vaccines [68]. They operate by delivering mRNA coding for desired gene products/proteins to particular tissues, e.g. muscle, so that cells in this tissue express this gene product/tissue. In case of Covid 19, this led to display of fragments of the Covid 19 spike protein on the cell surface which, in turn, triggered desired immune responses. The expression of such protein fragments has been shown to be detectable for several days [69]; while this is often sufficient to trigger immune responses, it is likely less helpful if one wishes to provide for a long-term therapy, for example correcting genetic defects or providing therapeutics over a long period of time.

Thus, to induce the patient's body to produce certain proteins for long periods of time, viral tools have been used to deliver a genetic payload to cells of the human body and these include Herpes Simplex Virus, Lentivirus, Adenovirus, Adeno-Associated Virus as well as retroviruses [70]. Considering the pros and cons of these platforms eluded to in Section "Induced pluripotent stem cells", adeno-associated viruses have merged as a tool with a particularly favorable profile [71].

Adeno-Associated Viruses: Biology

Part of the attractive profile of AAVs is their apparent innocuousness: AAVs have been reported to be endemic at a level of approximately 70% and limited associated pathogenicity has been reported [72, 73]. Currently, 3 products have been FDA approved [74] while www.clinicaltrials.gov yields 285 entries; hence, one might anticipate a substantial number of new rAAV-based products to enter the market.

To provide context for production-related topics discussed below, we briefly describe Adeno-associated virus' (AAVs) features. AAVs are non-enveloped viruses of the Parvovirus/Dependovirus family/genus. Their capsid comprises three proteins, namely VP1, VP2 and VP3 ranging from 60 to 90 kDa molecular weight with a total of approximately 60 monomers and at a ratio of 1:1:10 [75], although recently stochastic elements have been discussed [76]. Depending on the nature of the capsid proteins, currently 8 serotypes have been reported [77]. The AAV genome comprises a single stranded DNA molecule of approximately 4.7 kB flanked by inverted terminal repeats (ITRs), the latter serving as start sites for viral genome replication and packaging signals. The AAV genome contains 2 genes coding for 9 proteins making use of alternative translation start sites, alternative splicing and alternative open reading frames [78]. This set of genes/proteins is insufficient for replication and, in order to replicate, AAVs require part of the adenovirus-replication machinery – hence, adeno-associated virus [79, 80]. The molecular weight of empty and full capsids has been reported to approximate 3670 and 4751 kDa, respectively; the diameter of intact AAVs to be 26 nm

[78]. For expression of the AAV genome, there is a choice of either episomal or integration into the host genome at a specific location on chromosome 19 [81, 82]; episomal DNA tends to be lost during mitosis, hence, the episomal path is more suitable for slowly diving cells if expression of the gene of interest is desired for longer periods of time.

Depending on the nature of the capsid proteins, different serotypes have been described. These seem to exhibit only a limited degree of tissue specificity so that localized application of AAVs appears to be the immediate route if tissue specificity is desired; however, engineered capsid proteins incorporating structures mediating tissue specificity may likely be an enhancement [83, 84].

Adeno-Associated Viruses: Gene therapy Vectors

To use AAVs as a vector for gene therapy purposes, both viral genes are summarily deleted to make room for the therapeutic payload and to ensure that thusly generated recombinant viruses cannot self-replicate thereby minimizing the risk of uncontrolled events. This results in a recombinant viral genome comprising the gene of interest flanked by the inverted terminal repeats; the size of the inverted terminal repeats has been reported to be 145 bases each, leaving most of the approximately 4.7 kb of its genome available for payload.

Since the recombinant AAV (rAAV) genome lacks both AAV genes as well as those otherwise required from an adenovirus for replication, these have to be supplied during the production process. Initially, this was accomplished using adenovirus as a helper virus. However, adding adenovirus to production systems increased the potential for a mixed virus harvest. As of 1997 methods were devised to supply the needed viral replication cycle genes by distributing them into 3 plasmids: (1) the AAV vector plasmid carrying the payload, (2) the AAV helper plasmid carrying both the AAV rep and cap genes and (3) the adenovirus helper plasmid carrying the adenovirus-derived genes required for AAV replication. To facilitate rAAV production, all three plasmids are then co-transfected into a production cell line such as the human cell line HEK293 [85].

Initially, production cell lines were cultivated in planar/2D culture systems such as cell factories or hyperflasks/stacks, a configuration in which e.g. calcium/DNA precipitation transfection works particularly well. As discussed in section first generation MSC products and manufacturing processes: static/planar culture systems, such platforms offer compatibility with equipment often found in cell culture laboratories such as CO_2- and temperature-controlled incubators and hence are suitable for early work. They are, however, limited, in their ability to support scale-up and so large-scale production of rAAVs as required for commercial purposes tends to follow the path of bio- and cell production discussed in section second generation MSC products and manufacturing processes: bioreactors, i.e. tends to move into single-use bioreactors. For rAAVs, and likely for many other gene therapy products, use of established single-use platforms represents a straight-forward path to scale-up and likely further alleviates concerns regarding possible cross-contamination with other, similar, products.

Above, we discussed the particular suitability of transfection by calcium precipitation of DNA for planar systems. Conversely, it may not be as suitable for suspension systems and hence alternative approaches to transfection have been explored. Mass electroporation is a possibility but current instrumentation is volume-limited and so, likely for reasons of practicality, chemical transfection agents are now frequently used [86, 87].

As discussed above, three plasmids are frequently transfected to provide for production of recombinant AAVs. Transfection with a single plasmid might occur at a reasonably high transfection rate. Attempting transfection with three plasmids is likely to lower the transfection efficiency notably. This may be countered to some degree by adjusting process parameters such as concentrations of cells, plasmids and transfection reagents or it may simple be accepted. One approach to circumvent this bottleneck is to generate production cell lines which are already expressing the machinery required for rAAV replication so that these cells need to be transfected only with one plasmid containing the gene of interest or, indeed, none at all in case of the Elevecta system marketed by Cevec/Cytiva [88, 89].

Down-Stream Processing

Once recombinant viral particles have been produced, e.g. in the above single-use bioreactor systems, they have to be purified, formulated, filled and packaged.

AAV has been reported not to be efficiently released from production cells and so the first step is often to release rAAVs from their production cells by means of detergents or mechanical manipulation [90]. The resulting crude lysate is often treated enzymatically to break down production-cell derived nucleic acids, for example, benzonase in particular is used widely [91]. Not only does this treatment increase the efficiency of subsequent filtration steps, by destroying production cell-line DNA, it minimizes risks of introducing undesirable nucleic acids/genes as contaminants into the product.

In a second step, the resulting suspension of rAAVs and remaining cellular debris are often subjected to depth filtration [92].

In a third step, rAAV preparations are further purified, e.g. by a sequence of chromatography steps such as ion-exchange, affinity-chromatography using matrices coupled to antibodies targeting the rAAV capsid proteins and additional ion-exchange or size-exclusion. Such purification processes tend to yield highly purified rAAVs [93].

Production processes have been reported to yield a mix of complete and incomplete capsids, i.e. capsids with and without DNA and the proportion of empty capsids has been reported to sometimes be as high as 80%. Empty capsids serve no function and so their presence in the product reduces its potency. Consequently, methods have been devised to remove empty capsids and these include density centrifugation, flow-through ultra-centrifugation and specialized ion-exchange chromatography; however, while these approaches have been reported to be successful at small scale, their utility for large scale operations is sometimes debated and likely makes for a trade-off between viral quality and quantity – a high level of full and intact capsids at the expense of product quantity or a significant presence of cargoless (inactive) capsids [94].

The final step is often tangential flow filtration used to concentrate product to the desired titer and to suspend it in the desired buffer solution. At this step, the resulting bulk drug substance can be stored frozen until further processing. With the above purification steps in mind, yields of 30–60% of capsids produced in culture vs present in the final drug substance have been reported [93].

Formulation, filling etc. ensue; these steps follow established processes. Often, the bulk drug substance is considered low-bioburden (rather than sterile) and, once formulation has been completed, the bulk drug product can be sterilized by sterile filtration – as discussed above, the rAAV is approximately 26 nm in diameter, i.e. approximately a tenth of the pore size of a sterile filter and so quite amenable to sterile filtration.

ATMP-Specific Quality Requirements

European Union

The European Commission published its long *awaited Eudralex Volume 4, Guidelines on Good Manufacturing Practice Specific to Advanced Therapy Medicinal Products* in November 2017 with an implementation date of May 2018. The guidelines cover both authorized (those with an approved marketing application) and investigational clinical products. Topics included are all those generally discussed within the standard Pharmaceutical GMPs, with a more risk-based approach (RBA) in some cases; the guideline also provides some specific GMP examples that apply to ATMPs.

The general principles do not differ from standard GMP in that ATMP manufacturers are responsible to assure the quality, safety and efficacy of the products they manufacture. There is an allowance for simpler pharmaceutical quality systems based on phase of clinical development, complexity of manufacturing and size of the company. The pharmaceutical quality system is discussed in general terms and includes the basic elements of GMP: Personnel requirements, facilities and equipment that are suitable for use, documentation system with clear specifications and instructions, manufacturing process that is consistent, a quality control unit and system independent of manufacturing, and systems for recordkeeping, change control, deviations, traceability of materials and periodic product quality review.

There is specific guidance included for a Risk Based Approach (RBA) which allows a degree of flexibility based on manufacturing technology, setting of manufacturing, and clinical phase. The RBA is intended to permit flexibility required by the technical/biological needs of ATMPs; however, it is not intended to weaken GMP requirements and, consequently, does not exempt the manufacturer from doing a thorough risk assessment where flexibility may be desired, and the need to include justification to ensure product safety and quality. RBA may be applied to facilities, equipment, process, materials, release and stability testing. Commonly, RBA is also used when product needs to be delivered to the patient/clinical subject "fresh", soon after manufacture and while full release testing may not be available. The guidance specific to sterility tests is consistent with US FDA guidance on this topic, including the possibility to use rapid tests, testing for sterility earlier in the process - during intermediate process steps, and establishing a protocol with clinical investigators to manage additional information on sterility test outcome, including contamination detected post-application/infusion.

Requirements for clean rooms, aseptic practice, personnel training and hygiene, gowning, and aseptic process simulations are in principle the same as described in Annex I. The specific section on aseptic manufacturing gives pertinent examples of room classification required for ATMP-specific equipment. However, requirements for open manipulations, fermenters/bioreactors, and closed systems are the same as would be required for any sterile product or biologic. The guidance and examples in the section on multi-product facilities is helpful in that it gives numerous ATMP-specific scenarios on separation of products batches based on segregation by time or space. The sections on documentation and starting/raw materials give specific and useful examples pertinent to ATMPs. The section on prevention of cross-contamination of products, starting materials, is particularly helpful, and include scenarios for autologous products. Sections on process validation, assay validation and cleaning validation describe specific requirements for both clinical (some flexibility) and commercial ATMP products.

United States

FDA has not published any specific GMPs for ATMPs. *FDA Guidance for Industry CGMP for Phase 1 Investigational Products* (July 2008) provides a framework for the minimum GMP requirements for first in human clinical trials. This Guidance includes annexes that describe expectations for multi-product facilities, adventitious agent control, cellular and gene therapy products and sterile/aseptically processed products. The Guidance refers specifically to *FDA Guidance for Industry: Sterile Products Produced by Aseptic Processing – Current Good Manufacturing Practice* (September 2004), therefore the expectations around aseptic processing are identical to requirements for commercial products. Beyond Phase 1, the expectation is for full cGMP (21CFRPart 211). Note that for all phases of clinical production, regulators will usually accept phase-appropriate levels of validation with justification. *FDA Guidance for Industry: Chemistry, Manufacturing and Control (CMC) Information for Human Gene Therapy Investigational New Drug Applications (INDs)* (January, 2020) is comprehensive and provides some GMP expectations including specific guidance on adventitious agents, raw materials assessments, autologous and allogeneic cells, cell and viral banking, process and assay validation, safety testing, sterility testing, and protocols for assessing sterility of fresh products and response to contamination detected post-infusion.

Release of clinical and commercial Products in EU and US follow the standard GMPs with the same expectations. However, notably, in the EU certification of each batch must be done by a Qualified Person (QP), and according to *Eudralex Volume 4 Annex 16, Certification by a Qualified Person and Batch Release* which defines the requirements for certification. In the US, the manufacturer appoints the independent Quality-responsible staff who release batches.

Conclusions

This chapter concerned itself with cell-, cell-based- as well as viral gene therapy, all of which accelerate since their inception and this is reflected by the number of therapeutics under development, the number of clinical trials and of approved products. So, too, does the sophistication of the manufacturing process. From the earliest days, using 2D plastic, current technologies have expanded to include more sophisticated, semi-automated, closed processing systems. It is of considerable interest to see that each new technology starts with similar stages of equipment and tooling and then successively advances to more sophisticated approaches which have been developed for preceding technologies. These advances have enabled faster, more robust, and consistent production of all therapeutic entities discussed. Development of more robust and sensitive analytic tools have provided the marketplace and regulators with a deeper biological characterization of such therapeutics. This provides more efficient methods of determining comparability and process performance qualification (PPQ). Despite these advances, further progress is required to democratize such therapeutics, sustain a marketplace and address unmet needs through ambitious innovations. Future directions likely include advances in therapeutic efficacy, product handling and storage, in-vivo gene modification using vectors, lipid nanoparticles as well as Point-Of-Care (POC) fresh manufacturing.

Statements in Regard to Funding, Conflict of Interest and Ethical Approval The authors did not receive funding for writing this manuscript, they declare no conflict of interest and no ethical approvals were required.

References

1. Landsteiner K (1901) Ueber Agglutinationserscheinungen normalen menschlichen Blutes. Wien Klin Wochenschr, pp 1132–1134

2. Landsteiner K (1961) On agglutination of normal human blood. Transfusion (Paris) 1:5–8. https://doi.org/10.1111/j.1537-2995.1961.tb00005.x
3. Brown CJ, Navarrete CV Clinical relevance of the HLA system in blood transfusion. https://doi.org/10.1111/j.1423-0410.2011.01474.x
4. Robert E, Hardy EVI (1989) Bone marrow transplantation: a review. Cell Transplant 81:518–523. https://doi.org/10.1177/096368979300200110
5. Civin CI, Strauss LC, Brovall C et al (1984) Antigenic analysis of hematopoiesis. III. A hematopoietic progenitor cell surface antigen defined by a monoclonal antibody raised against KG-1a cells. J Immunol 133:157–165
6. Tindle RW, Nichols RAB, Chan L et al (1985) A novel monoclonal antibody BI-3C5 recognises myeloblasts and non-B non-T lymphoblasts in acute leukaemias and CGL blast crises, and reacts with immature cells in normal bone marrow. Leuk Res 9:1–9. https://doi.org/10.1016/0145-2126(85)90016-5
7. Riggs AD (2021) Making, cloning, and the expression of human insulin genes in bacteria: the path to humulin. Endocr Rev 42:374–380. https://doi.org/10.1210/endrev/bnaa029
8. Hartley JL (2012) Why proteins in mammalian cells? Methods Mol Biol 801:1–12. https://doi.org/10.1007/978-1-61779-352-3_1
9. A Brief History of Adherent Cell Culture: Where We Come From and Where We Should Go – BioProcess International. https://bioprocessintl.com/upstream-processing/bioreactors/a-brief-history-of-adherent-cell-culture-where-we-come-from-and-where-we-should-go/. Accessed 26 Mar 2023
10. Corning® CellSTACK® Culture Chambers | Corning. https://ecatalog.corning.com/life-sciences/b2c/US/en/Browse-Products-by-Application/Cell-Therapy-and-Vaccines/CellSTACK%C2%AE/Corning%C2%AE-CellSTACK%C2%AE-Culture-Chambers/p/corningCellSTACKCultureChambers. Accessed 20 Mar 2023
11. 20036 | Corning® CellBIND® HYPERStack® – 36 Layer Cell Culture Vessel, 2 per Case | Corning. https://ecatalog.corning.com/life-sciences/b2c/US/en/Browse-Products-by-Application/Cell-Therapy-and-Vaccines/HYPERStack%C2%AE/HYPERFlask%C2%AE/Corning%C2%AE-HYPERStack%C2%AE-Cell-Culture-Vessels/p/20036?clear=true. Accessed 20 Mar 2023
12. Rossi Bilodeau A, Ferrie AM, Shyu J (2022) 36-layer Vessels using the corning automated manipulator system. Corning application notes 188
13. Single-Use Bioreactors | Sartorius. https://www.sartorius.com/en/products/fermentation-bioreactors/single-use-bioreactors. Accessed 20 Mar 2023
14. Kögler G, Sensken S, Airey JA, et al (2004) A new human somatic stem cell from placental cord blood with intrinsic pluripotent differentiation potential. J Exp Med 200:123–135. https://doi.org/10.1084/jem.20040440
15. Jiang Y, Vaessen B, Lenvik T et al (2002) Multipotent progenitor cells can be isolated from postnatal murine bone marrow, muscle, and brain. Exp Hematol 30:896–904
16. Shi Y, Inoue H, Wu JC, Yamanaka S Induced pluripotent stem cell technology: a decade of progress. https://doi.org/10.1038/nrd.2016.245
17. Friedenstein AY, Chailakhlan RK, Lalykina KS, Horn EC (1970) The development of fibroblast colonies of fibroblast colonies in monolayer cultures of Guinea-pig bone marrow and spleen cells. Cell Tissue Kinetics 3:393–403. https://doi.org/10.1111/j.1365-2184.1970.tb00347.x
18. Afanasyev BV, Elstner EE, Zander AR et al (2010) A. J. Friedenstein, founder of the mesenchymal stem cell concept. Transplantation 1:35–38. https://doi.org/10.3205/ctt-2009-en-000029.01
19. Caplan AI. Mesenchymal stem cells. J Orthop Res. 1991 Sep;9(5):641–650. https://doi.org/10.1002/jor.1100090504. PMID: 1870029.
20. Bruder SP, Jaiswal N, Haynesworth SE (1997) Growth kinetics, self-renewal, and the osteogenic potential of purified human mesenchymal stem cells during extensive subcultivation and following cryopreservation. J Cell Biochem 64:278–294. https://doi.org/10.1002/(sici)1097-4644(199702)64:2<278::aid-jcb11>3.0.co;2-f

21. van den Bos C, Mosca JD, Winkles J et al (1997) Human mesenchymal stem cells respond to fibroblast growth factors. Hum Cell 10:45–50
22. von Owens HO, Gey MK, Gey GO (1954) Growth of cells in agitated fluid medium. NY Acad Sci 58(7):1039–1055
23. Variant Creutzfeldt-Jakob Disease (vCJD) | Prion Diseases | CDC. https://www.cdc.gov/prions/vcjd/index.html. Accessed 26 Mar 2023
24. Fernandez-Rebollo E, Mentrup B, Ebert R et al Human platelet lysate versus fetal calf serum: these supplements do not select for different mesenchymal stromal cells OPEN. https://doi.org/10.1038/s41598-017-05207-1
25. Mallon BS, Park KY, Chen KG et al (2006) Toward xeno-free culture of human embryonic stem cells. Int J Biochem Cell Biol 38:1063–1075. https://doi.org/10.1016/j.biocel.2005.12.014
26. Hayflick L, Moorhead PS (1961) The serial cultivation of human diploid cell strains. Exp Cell Res 25:585–621
27. Wagner W, Horn P, Castoldi M et al (2008) Replicative senescence of mesenchymal stem cells: a continuous and organized process. PLoS One 3:2213. https://doi.org/10.1371/journal.pone.0002213
28. van den Bos C, Silverstetter S (1998) p21(cip1) rescues human mesenchymal stem cells from apoptosis induced by low-density culture. J Cell Physiol 293:463–470
29. Pavel M, Renna M, Park SJ et al Contact inhibition controls cell survival and proliferation via YAP/TAZ-autophagy axis. https://doi.org/10.1038/s41467-018-05388-x
30. Schirmaier C, Jossen V, Kaiser SC et al (2014) Scale-up of adipose tissue-derived mesenchymal stem cell production in stirred single-use bioreactors under low-serum conditions. Eng Life Sci 14:292–303. https://doi.org/10.1002/elsc.201300134
31. Lee Y, El Andaloussi S, Wood MJA (2012) Exosomes and microvesicles: extracellular vesicles for genetic information transfer and gene therapy. Hum Mol Genet 21:R125–R134. https://doi.org/10.1093/hmg/dds317
32. Couffinhal T, Silver M, Zheng LP et al (1998) Mouse model of angiogenesis. Am J Pathol 152:1667–1679
33. Munn DH, Zhou M, Attwood JT et al (1998) Prevention of allogeneic fetal rejection by tryptophan catabolism. Science 281:1191–1193
34. van den Bos C, Keefe R, Schirmaier C, McCaman M (2014) Pillars of regenerative medicine: therapeutic human cells and their manufacture. In: Industrial scale suspension culture of living cells. Wiley Blackwell, Weinheim, pp 464–502
35. van den Bos C, Keefe R, Schirmaier C, McCaman M (2013) Therapeutic human cells: manufacture for cell therapy/regenerative medicine. Adv Biochem Eng Biotechnol 138:61–97
36. Kim HO, Choi SM, Kim HS (2013) Mesenchymal stem cell-derived secretome and microvesicles as a cell-free therapeutics for neurodegenerative disorders. Tissue Eng Regen Med 10:93–101. https://doi.org/10.1007/s13770-013-0010-7
37. Zhou T, Yuan Z, Weng J et al (2021) Challenges and advances in clinical applications of mesenchymal stromal cells. J Hematol Oncol 14:1–24. https://doi.org/10.1186/s13045-021-01037-x
38. Damasceno PKF, de Santana TA, Santos GC et al (2020) Genetic engineering as a strategy to improve the therapeutic efficacy of mesenchymal stem/stromal cells in regenerative medicine. Front Cell Dev Biol 8:1–24. https://doi.org/10.3389/fcell.2020.00737
39. crispr/cas "mesenchymal stem cell*" – Search Results – PubMed. https://pubmed.ncbi.nlm.nih.gov/?term=crispr%2Fcas+%22mesenchymal+stem+cell*%22. Accessed 26 Mar 2023
40. Christian Van Den Bos, Barbara REINISCH, Judith Schenk, Claudia Rosenbaum (2016) Compositions and methods for enhancing the therapeutic potential of stem cells EPO EP2777703B1 https://worldwide.espacenet.com/patent/search/family/047900872/publication/EP2777703B1?search_type=patents&q=EP2777703B1&op=search. Accessed Jan 2023
41. Kilbride P, Meneghel J, Fonseca F, Morris J (2021) The transfer temperature from slow cooling to cryogenic storage is critical for optimal recovery of cryopreserved mammalian cells. PLoS One 16:1–12. https://doi.org/10.1371/journal.pone.0259571

42. Jommi C, Bramanti S, Pani M et al (2022) CAR T-cell therapies in Italy: patient access barriers and recommendations for health system solutions. Front Pharmacol 13:1–11. https://doi.org/10.3389/fphar.2022.915342
43. Karnieli O, Friedner OM, Allickson JG et al (2017) A consensus introduction to serum replacements and serum-free media for cellular therapies. Cytotherapy 19:155–169. https://doi.org/10.1016/j.jcyt.2016.11.011
44. Trickett A, Kwan YL (2003) T cell stimulation and expansion using anti-CD3/CD28 beads. J Immunol Methods 275:251–255. https://doi.org/10.1016/S0022-1759(03)00010-3
45. Cocoon® Platform | Cell Therapy Manufacturing | Lonza. https://www.lonza.com/cell-and-gene/cocoon-platform. Accessed 1 Apr 2023
46. Xuri Cell Expansion System W25 | Cytiva. https://www.cytivalifesciences.com/en/us/shop/cell-therapy/systems/xuri-cell-expansion-system-w25-p-06192. Accessed 1 Apr 2023
47. Sepax S-100 (Sepax 2) Cell Separation System | Cytiva. https://www.cytivalifesciences.com/en/us/shop/cell-therapy/systems/sepax-s-100-sepax-2-cell-separation-system-p-08417. Accessed 1 Apr 2023
48. Sefia Cell Processing System | Cytiva. https://www.cytivalifesciences.com/en/us/solutions/cell-therapy/products-and-technology/cell-processing/sefia-cell-processing-system. Accessed 1 Apr 2023
49. Li L, Gao Y, Srivastava R et al (2020) Lentiviral delivery of combinatorial CAR/CRISPRi circuit into human primary T cells is enhanced by TBK1/IKKϵ complex inhibitor BX795. J Transl Med 18:1–12. https://doi.org/10.1186/s12967-020-02526-2
50. Cornetta K, Duffy L, Turtle CJ et al (2018) Absence of replication-competent lentivirus in the clinic: analysis of infused T cell products. Mol Ther 26:280–288. https://doi.org/10.1016/j.ymthe.2017.09.008
51. Atsavapranee ES, Billingsley MM, Mitchell MJ (2021) Delivery technologies for T cell gene editing: applications in cancer immunotherapy. EBioMedicine 67:103354. https://doi.org/10.1016/j.ebiom.2021.103354
52. Alves CPA, Prazeres DMF, Monteiro GA (2021) Minicircle biopharmaceuticals–an overview of purification strategies. Front Chem Eng 2:1–15. https://doi.org/10.3389/fceng.2020.612594
53. Katti A, Diaz BJ, Caragine CM et al (2022) CRISPR in cancer biology and therapy. Nat Rev Cancer 22:259–279. https://doi.org/10.1038/s41568-022-00441-w
54. Beatty GL, Haas AR, Maus MV et al Mesothelin-specific chimeric antigen receptor mRNA-engineered T cells induce antitumor activity in solid malignancies. Cancer Immunol Res 2. https://doi.org/10.1158/2326-6066.CIR-13-0170
55. Wang Z, Wu Z, Liu Y, Han W New development in CAR-T cell therapy. https://doi.org/10.1186/s13045-017-0423-1
56. Harris E (2023) A breakdown of scaling cell therapies part 2. Cell & Gene. https://www.cellandgene.com/doc/a-breakdown-of-scaling-cell-therapies-part-0001. Accessed 1 Apr 2023
57. Plieth J (2022) Ash 2022 – fast production fails to cure Car-T's problem | Evaluate. Evaluate Vantage. https://www.evaluate.com/vantage/articles/events/conferences-trial-results/ash-2022-fast-production-fails-cure-car-ts. Accessed 1 Apr 2023
58. Stanton D (2020) CAR T-cell therapies: allogenic the way to go? – BioProcess International. https://bioprocessintl.com/manufacturing/emerging-therapeutics-manufacturing/car-t-cell-therapies-is-allogeneic-the-way-to-go/. Accessed 1 Apr 2023
59. Gatla H, Uth N, Levinson Y et al (2022) Enabling allogeneic T cell-based therapies: scalable stirred-tank bioreactor mediated manufacturing. Front Med Technol 4:1–15. https://doi.org/10.3389/fmedt.2022.850565
60. Okita K, Ichisaka T, Yamanaka S (2007) Generation of germline-competent induced pluripotent stem cells. Nature 448:313–317. https://doi.org/10.1038/nature05934
61. Huang S, Razvi A, Anderson-Jenkins Z et al (2020) Process development and scale-up of pluripotent stem cell manufacturing. Cell Gene Ther Insights 6:1277–1298. https://doi.org/10.18609/CGTI.2020.141

62. Wilson JL, McDevitt TC (2012) Stem cell microencapsulation for phenotypic control, bioprocessing, and transplantation. https://doi.org/10.1002/bit.24802
63. Ludwig T, Thomson JA (2007) Defined, feeder-independent medium for human embryonic stem cell culture. Curr Protoc Stem Cell Biol Chapter 1. https://doi.org/10.1002/9780470151808.SC01C02S2
64. Chen G, Gulbranson DR, Hou Z, Bolin JM, Ruotti V, Probasco MD, Smuga-Otto K, Howden SE, Diol NR, Propson NE, Wagner R, Lee GO, Antosiewicz-Bourget J, Teng JM, Thomson JA. Chemically defined conditions for human iPSC derivation and culture. Nat Methods. 2011 May;8(5):424–429. https://doi.org/10.1038/nmeth.1593. Epub 2011 Apr 10. PMID: 21478862; PMCID: PMC3084903.
65. Kuo HH, Gao X, DeKeyser JM et al (2020) Negligible-cost and weekend-free chemically defined human iPSC culture. Stem Cell Rep 14:256–270. https://doi.org/10.1016/j.stemcr.2019.12.007
66. Yescarta | European Medicines Agency. https://www.ema.europa.eu/en/medicines/human/EPAR/yescarta. Accessed 13 Mar 2023
67. Zolgensma | European Medicines Agency. https://www.ema.europa.eu/en/medicines/human/EPAR/zolgensma. Accessed 13 Mar 2023
68. Covid-19 vaccines saved an estimated 20m lives during their first year. The Economist. https://www.economist.com/graphic-detail/2022/07/07/covid-19-vaccines-saved-an-estimated-20m-lives-during-their-first-year. Accessed 12 Mar 2023
69. CHMP E (2021) Committee for medicinal products for human use (CHMP) assessment report comirnaty common name: COVID-19 mRNA vaccine (nucleoside-modified)
70. Bulcha JT, Wang Y, Ma H et al (2021) Viral vector platforms within the gene therapy landscape. Signal Transduct Target Ther 6:1–24. https://doi.org/10.1038/s41392-021-00487-6
71. Wang D, Tai PWL, Gao G (2019) Adeno-associated virus vector as a platform for gene therapy delivery. Nat Rev Drug Discov 18:358–378. https://doi.org/10.1038/s41573-019-0012-9
72. Flotte TR (2005) Adeno-associated virus-based gene therapy for inherited disorders. Pediatr Res 58:1143–1147. https://doi.org/10.1203/01.PDR.0000189226.03684.FE
73. Miyazawa D (2022) Potential mechanisms by which adeno-associated virus type 2 causes unexplained hepatitis in children. J Med Virol 94:5623–5624. https://doi.org/10.1002/jmv.28082
74. Approved Cell- and gene therapy products https://www.fda.gov/vaccines-blood-biologics/cellular-gene-therapy-products/approved-cellular-and-gene-therapy-products. Accessed Oct 2023
75. Becerra SP, Koczot F, Fabisch P, Rose JA (1988) Synthesis of adeno-associated virus structural proteins requires both alternative mRNA splicing and alternative initiations from a single transcript. J Virol 62:2745–2754. https://doi.org/10.1128/jvi.62.8.2745-2754.1988
76. Wörner TP, Bennett A, Habka S, et al (2020) Adeno-associated virus capsid assembly is divergent and stochastic. bioRxiv. https://doi.org/10.1101/2020.10.09.332619
77. Wu Z, Asokan A, Samulski RJ (2006) Adeno-associated virus serotypes: vector toolkit for human gene therapy. Mol Ther 14:316–327. https://doi.org/10.1016/j.ymthe.2006.05.009
78. Dhungel BP, Bailey CG, Rasko JEJ (2021) Journey to the center of the cell: tracing the path of AAV transduction. Trends Mol Med 27:172–184. https://doi.org/10.1016/j.molmed.2020.09.010
79. Atchison RW, Casto BC, Hammon WM (1965) Adenovirus-associated virus. Science 149:754–755
80. Hoggan MD, Blacklow NR, Rowe WP (1966) Studies of small DNA viruses found in various adenovirus preparations: physical, biological, and immunological characteristics. Proc Natl Acad Sci U S A 55:1467–1474. https://doi.org/10.1073/PNAS.55.6.1467/ASSET/FEF8FBA2-C260-4A75-986D-0A0FF7BAA276/ASSETS/PNAS.55.6.1467.FP.PNG
81. Surosky RT, Urabe M, Godwin SG et al (1997) Adeno-associated virus rep proteins target DNA sequences to a unique locus in the human genome. J Virol 71:7951–7959. https://doi.org/10.1128/jvi.71.10.7951-7959.1997

82. Penaud-Budloo M, Le Guiner C, Nowrouzi A et al (2008) Adeno-associated virus vector genomes persist as episomal chromatin in primate muscle. J Virol 82:7875–7885. https://doi.org/10.1128/jvi.00649-08
83. Zincarelli C, Soltys S, Rengo G, Rabinowitz JE (2008) Analysis of AAV serotypes 1-9 mediated gene expression and tropism in mice after systemic injection. Mol Ther 16:1073–1080. https://doi.org/10.1038/MT.2008.76
84. Feiner RC, Teschner J, Teschner KE et al rAAV Engineering for capsid-protein enzyme insertions and mosaicism reveals resilience to mutational, structural and thermal perturbations. Int J Mol Sci. https://doi.org/10.3390/ijms20225702
85. Ferrari FK, Xiao X, McCarty D, Samulski RJ (1997) New developments in the generation of ad-free, high-titer raav gene therapy vectors. Nat Med 3:1295–1297. https://doi.org/10.1038/nm1197-1295
86. Wright JF (2009) Transient transfection methods for clinical adeno-associated viral vector production. Hum Gene Ther 20:698–706. https://doi.org/10.1089/hum.2009.064
87. Guan J-S, Chen K, Si Y et al (2022) Process improvement of adeno-associated virus production. Front Chem Eng 4:1–10. https://doi.org/10.3389/fceng.2022.830421
88. Lee Z, Lu M, Irfanullah E et al (2022) Construction of an rAAV producer cell line through synthetic biology. ACS Synth Biol 11:3285–3295. https://doi.org/10.1021/ACSSYNBIO.2C00207
89. Coronel J, Patil A, Al-Dali A et al (2021) Efficient production of rAAV in a perfusion bioreactor using an ELEVECTA® stable producer cell line. 41:S23. https://home.liebertpub.com/gen. https://doi.org/10.1089/GEN.41.S2.07
90. Penaud-Budloo M, François A, Clément N, Ayuso E (2018) Pharmacology of recombinant adeno-associated virus production. Mol Ther Methods Clin Dev 8:166–180. https://doi.org/10.1016/J.OMTM.2018.01.002
91. Zincarelli C, Soltys S, Rengo G, Rabinowitz JE (2008) Analysis of AAV serotypes 1-9 mediated gene expression and tropism in mice after systemic injection. Mol Ther 16:1073–1080. https://doi.org/10.1038/mt.2008.76
92. Thorne BA, Takeya RK, Peluso RW (2009) Manufacturing recombinant adeno-associated viral vectors from producer cell clones. Hum Gene Ther 20:707–714. https://doi.org/10.1089/hum.2009.070
93. Hebben M (2018) Downstream bioprocessing of AAV vectors: industrial challenges & regulatory requirements. Cell Gene Ther Insights 4:131–146. https://doi.org/10.18609/cgti.2018.016
94. Srivastava A, Mallela KMG, Deorkar N, Brophy G (2021) Manufacturing challenges and rational formulation development for AAV viral vectors. J Pharm Sci 110:2609–2624. https://doi.org/10.1016/j.xphs.2021.03.024

Index

R. Pörtner (ed.), *Biopharmaceutical Manufacturing*, Cell Engineering 11, https://doi.org/10.1007/978-3-031-45669-5

D

The manufacturer's authorised representative in the EU is Springer Nature Customer Service Centre GmbH, Europaplatz 3, 69115 Heidelberg, Germany. If you have any concerns regarding our products, please contact ProductSafety@springernature.com

Printed and bound by CPI Group (UK) Ltd, Croydon, CR0 4YY
30/09/2026
02224059-0006